NANOFABRICATION IN THE ERA OF MACHINE LEARNING

Advances and Emerging Frontiers

Other Titles by the Editor

Nanofabrication: Fundamentals And Applications

Tip-Based Nanofabrication: Fundamentals and Applications

NANOFABRICATION IN THE ERA OF MACHINE LEARNING

Advances and Emerging Frontiers

Editor

Ampere A Tseng

Arizona State University, USA

World Scientific

NEW JERSEY • LONDON • SINGAPORE • BEIJING • SHANGHAI • HONG KONG • TAIPEI • CHENNAI • TOKYO

Published by

World Scientific Publishing Co. Pte. Ltd.
5 Toh Tuck Link, Singapore 596224
USA office: 27 Warren Street, Suite 401-402, Hackensack, NJ 07601
UK office: 57 Shelton Street, Covent Garden, London WC2H 9HE

Library of Congress Control Number: 2026009831

British Library Cataloguing-in-Publication Data
A catalogue record for this book is available from the British Library.

NANOFABRICATION IN THE ERA OF MACHINE LEARNING
Advances and Emerging Frontiers

ISBN 978-981-98-3075-6 (hardcover)
ISBN 978-981-98-3076-3 (ebook for institutions)
ISBN 978-981-98-3077-0 (ebook for individuals)

For any available supplementary material, please visit
https://www.worldscientific.com/worldscibooks/10.1142/14809#t=suppl

Desk Editor: Shaun Tan Yi Jie

Typeset by Stallion Press
Email: enquiries@stallionpress.com

https://doi.org/10.1142/9789819830763_fmatter

Preface

Originally, the publisher planned to release a revised edition of the widely used monograph *Nanofabrication: Fundamentals and Applications*, first published in 2008. During the preparation process, however, it became evident that the field had undergone a profound transformation. Many of the original contributors had retired or moved in new research directions, and more importantly, nanofabrication itself had entered a new technological era—one increasingly shaped by machine learning (ML) and data-driven methodologies.

In response to this shift, the present editor invited a new group of experts to contribute chapters that reflect both recent technical advances and the growing influence of ML across nanofabrication. This effort resulted in ten entirely new chapters. Among them, five focus directly on ML-enabled approaches, while the remaining chapters address major fabrication processes and applications that are increasingly influenced by intelligent modeling, advanced metrology, and adaptive process control. Given the scale of this transformation, the new title, *Nanofabrication in the Era of Machine Learning: Advances and Emerging Frontiers*, more accurately captures the spirit and direction of the field.

The ten chapters are organized into three thematic areas: (1) semiconductor manufacturing (Chapters 1–3), (2) emerging nanofabrication processes (Chapters 4–7), and (3) nanoparticle synthesis and on-line sensing and characterization (Chapters 8–10).

Each thematic section is anchored by an overview chapter that highlights the role of ML within that domain. Together, these ML-focused chapters demonstrate how data-driven techniques are redefining nanofabrication through predictive modeling, adaptive optimization, intelligent metrology, and autonomous control. By improving precision, efficiency, yield, and sustainability, ML is enabling the fabrication of high-quality nanoscale structures and devices with unprecedented reliability and reduced cost. The accompanying chapters provide in-depth discussions of specific fabrication technologies, offering a balanced integration of intelligent methodologies and process-level expertise.

In semiconductor manufacturing, Chapters 1–3 examine the convergence of advanced lithography, precision process control, and ML. Chapter 1 reviews ML integration across key fabrication steps, emphasizing gains in yield prediction, uniformity control, and sustainability. Chapter 2 presents an ML-based framework for thin film deposition control using SEM-derived metrology, enabling accurate thickness prediction even on complex and rough surfaces. Chapter 3 surveys ML applications in electron beam writing and manipulation, demonstrating adaptive control, enhanced metrology, and predictive optimization at atomic and nanoscale dimensions. Collectively, these chapters illustrate how semiconductor manufacturing is being reshaped by intelligent, data-driven strategies.

Chapters 4–7 explore emerging nanofabrication processes that extend beyond conventional lithography. Chapter 4 provides a comprehensive review of ML applications in nanoimprinting, tip-based nanofabrication, focused ion beam processing, laser nanofabrication, and nanocapillary methods, underscoring ML's role in predictive control and defect mitigation. Chapter 5 introduces superplastic nanomolding as a thermomechanical route for producing highly ordered metallic nanostructures, with applications in nanophotonics and chiral plasmonics. Chapter 6 reviews atomic force microscopy-based nanofabrication, tracing its evolution into a versatile multi-loading platform while addressing ongoing challenges in reproducibility, resolution, and throughput. Chapter 7 examines nanoscale additive manufacturing for architected materials and mechanical metamaterials, highlighting process–structure–property relationships and data-driven inverse design strategies that enable tailored mechanical performance.

Chapters 8–10 address nanoparticle synthesis and advanced characterization, both of which are essential for reliable and scalable nanofabrication. Chapter 8 synthesizes insights from multiple reviews on ML-assisted nanoparticle synthesis, demonstrating how ML accelerates materials discovery and enables precise control over nanoparticle size, shape, and functionality. Chapter 9 discusses on-line characterization techniques—including atomic force microscopy, scanning electron microscopy, ellipsometry, and Raman spectroscopy—that provide real-time, non-destructive process monitoring and feedback control. Chapter 10 explores microwave-assisted scanning probe microscopy, detailing near-field sensing and quantum-based approaches that extend atomic force microscopy capabilities for subsurface characterization and nanoscale fabrication. Together, these chapters highlight the integration of intelligent synthesis, advanced sensing, and adaptive diagnostics within modern nanofabrication systems.

As the editor of this monograph, I am deeply grateful to all contributors for their professionalism, patience, and collaboration throughout the preparation process. I also sincerely thank the reviewers for their careful evaluations and constructive suggestions, which significantly strengthened the quality of each chapter. I hope this volume faithfully reflects the contributors' efforts; any remaining shortcomings are entirely my responsibility.

Finally, as this year marks the 50th anniversary of my professional career, this monograph will most likely be my last major scholarly effort before fully entering retirement. I am profoundly grateful to my wife, Maggie, for her unwavering support and for making my professional and personal life smooth, joyful, and meaningful. I also thank my beloved children, Claire, Karen, and Miles, for the love, encouragement, and happiness they have brought to me throughout this journey.

Ampere A. Tseng
Arizona State University, Tempe, Arizona, USA
January 2026

https://doi.org/10.1142/9789819830763_fmatter

About the Editor

Dr. Ampere A. Tseng is Emeritus Professor in the School for Engineering of Matter, Transport and Energy at Arizona State University, USA. He is an elected Fellow of the American Society of Mechanical Engineers. He has published over 250 peer-reviewed journal papers, edited more than 10 books, delivered more than 200 invited seminars/keynote speeches, and holds nine patents. As of late 2025, his research impact is reflected in a Google h-index of 44.

Throughout his distinguished career, Professor Tseng has held a prestigious Chair Professorship from National Tsinghua University, Taiwan; National Taiwan University of Science and Technology, Taiwan; Brno University of Technology, Czech Republic; and Henan University, China. Notably, he also held visiting professorships at China's elite universities, including Tsinghua University (Beijing), Shanghai Jiao Tong University, and University of Science and Technology of China (Hefei) at a time when these institutions rarely granted chair positions to international faculty. For the past five years, he is frequently ranked in the top 1% globally within the subfield of Materials, according to the Stanford/Elsevier "World's Top 2%" scientists list.

https://doi.org/10.1142/9789819830763_fmatter

Contents

https://doi.org/10.1142/9789819830763_0001

CHAPTER 1

Transforming Semiconductor Manufacturing with Machine Learning: Insights and Applications

Ampere A. Tseng[1,*], Ying-Tung Chen[2]
[1]Manufacturing Institute, Arizona State University, Tempe, Arizona 85287-6106, USA
[2]Abrasive Total Solutions, Taoyuan, Taiwan 33042, ROC
*Corresponding author: ampere.tseng@asu.edu

Abstract

This chapter examines the transformative role of machine learning in semiconductor manufacturing, focusing on its integration across key nanofabrication processes such as lithography, chemical-mechanical planarization, ion implantation, and silicon crystal growth. It also reviews two related areas: photolithographic mask design and predictive process modeling. Drawing on 19 selected studies, the review demonstrates how machine learning techniques including neural networks, decision trees, support vector machines, k-nearest neighbors, multivariate linear regression, and various ensemble or hybrid models have advanced predictive metrology, defect detection, process optimization, and equipment control. These methods have significantly improved overlay prediction accuracy, process uniformity, and manufacturing throughput while reducing turnaround time and fabrication-induced variability. The review also notes current challenges, including dependence on simulated data, limited generalizability, and integration difficulties across the manufacturing pipeline. Recommendations for mitigating these challenges are provided. Indeed, this review illustrates that machine learning has become a critical enabler of precision, efficiency, and sustainability in integrated circuit fabrication, offering clear guidance for future research and industrial innovation.

Keywords: chemical-mechanical planarization, ion implantation, machine learning, photolithography, predictive modeling, photomask, semiconductor manufacturing

1.1 Introduction

Semiconductor manufacturing, or integrated circuit (IC) nanofabrication, is an exceptionally complex and highly automated process that produces chips and micro- to nanoelectronic devices. The continuous drive for smaller feature sizes has propelled major innovations across fabrication processes. For instance, ongoing reductions in lithographic wavelengths and increases in numerical aperture (NA) have greatly advanced patterning resolution. State-of-the-art techniques such as extreme ultraviolet (EUV) lithography now achieve feature sizes well below the limits of deep ultraviolet (DUV) lithography, down to less than 2 nm as of 2025. However, these advances introduce new technical challenges. EUV processes demand finer dose and overlay control, tighter sensor integration, and more frequent post-lithographic metrology.[1,2] Similarly, developments in photolithography increasingly rely on real-time sensor feedback, generating vast, multidimensional datasets at every step of the manufacturing process. Consequently, even incremental improvements in equipment performance or process development now face unprecedented complexity.

Traditional methods for process control and equipment optimization have become inadequate in managing such intricate data environments. The explosion of data in IC fabrication has accelerated the adoption of machine learning (ML), which has demonstrated remarkable potential in equipment development, process optimization, defect detection, and metrology. By extracting patterns from high-dimensional datasets, ML enables precision control across multiple fabrication stages, helping to overcome longstanding bottlenecks in micro- and nano-electronic device production.

This chapter systematically reviews 19 selected studies on ML applications in semiconductor manufacturing. Beyond summarizing individual findings, it provides a structured synthesis of ML's contributions across fabrication domains, including equipment design, process improvement, optimization methodologies, and problem-solving strategies. The reviewed studies are grouped into three sections: the first covers six works on photolithography; the second discusses three on chemical-mechanical planarization, two on ion implantation, and one on monocrystalline silicon growth; and the third examines four on mask design for photolithography and three on predictive modeling. Each review is presented chronologically in its respective subject. The final section, Summary and Conclusions, consolidates key insights, identifies challenges, and offers recommendations. For ease of reference, major data and findings are further summarized in three comparative tables. While electron beam patterning and thin film deposition are vital to IC nanofabrication, they are detailed in Chapters 3 and 2, respectively. Consequently, they fall outside the scope of this specific chapter.

The 19 studies included were identified through comprehensive searches of Google Scholar, Engineering Village, and IEEE Xplore. Initial searches yielded more than 100 publications on ML-assisted semiconductor manufacturing published over the past ten years. The final selection was narrowed to 19 studies that focus directly on manufacturing processes rather than design, maintenance, packaging, testing, or cost analysis. Scholarly judgment was also applied in choosing between journal and conference papers, emphasizing citation impact, methodological rigor, and recency of publication.

The value of this review lies in clarifying a rapidly expanding yet fragmented research landscape. Although individual studies demonstrate innovative ML applications, their findings are often isolated, making it difficult to discern overarching trends, methodological variations, or research gaps. By consolidating current knowledge, this chapter maps the domains in which ML has been most effectively applied, such as lithography, yield prediction, and process optimization, and highlights areas where potential remains underexplored.

The broader context underscores the significance of this work. According to the Semiconductor Industry Association, global semiconductor sales reached USD 627.6 billion in 2024 and are projected to grow to USD 697 billion in 2025, with the market expected to surpass USD 1 trillion by 2030.[3] Because semiconductors underpin virtually every sector of the global economy, advancing manufacturing technologies is not only a technical imperative but also an economic and strategic priority.

By providing a comprehensive overview of ML applications, identifying effective strategies, outlining unresolved challenges, and making solvable recommendations, this review serves as a practical reference for researchers, engineers, and industry professionals. The insights derived here aim to accelerate progress toward greater competitiveness, efficiency, and sustainability in semiconductor, microelectronic, and nanofabrication industries.

1.2 Machine Learning-Aided Improvements in Photolithography

This section explores six key advancements utilizing ML to enhance the photolithography process in semiconductor manufacturing. Photolithography is the foundational technology for creating intricate circuit patterns on a wafer, but it inherently faces significant challenges including overlay errors, limitations in patterning accuracy, and issues with process stability as feature sizes continue to shrink. This review focuses on how ML techniques are being leveraged to directly address these manufacturing bottlenecks.

We review six seminal articles that apply ML to improve lithography. For context, a photolithography process typically involves several key steps: a surface preparation and cleaning phase; the application of a photoresist via spin coating, followed by soft baking; alignment of the photomask and the critical exposure to UV light; the development of the photoresist to reveal the pattern; an etching step to transfer the pattern to the underlying material; and a final strip of the remaining photoresist. The ability of ML to model and control this complex, multi-step process is the central theme of this review.

The reviews of these six articles are presented chronologically in the following subsections. For each review, we first introduce its scope and objectives, followed by a discussion of the key ML models and algorithms applied. We specifically emphasize the models' roles and performance in addressing the challenges identified by the authors, such as improvements in patterning accuracy or overlay control. Finally, for readers' convenience, the key data and major findings of each study are consolidated into a single row of Table 1.1, which provides a chronological, comparative overview of the reviewed literature.

1.2.1 *Machine Learning-Enhanced Virtual Metrology for Overlay Prediction and Outlier Detection*

Wang et al. developed a ML-enhanced virtual metrology (VM) model for overlay (OVL) prediction and outlier detection in

Table 1.1. ML-aided improvements in lithography.*

Section. Author (Year)[Ref.]	Subject and Process	ML Models, Algorithms, and Platforms	Training and Testing Data	Key Equipment	Key Findings for Model Performance
1.2.1. Wang et al. (2022)[4]	Overlay prediction and outlier detection for lithography exposure control	Tree-based LightGBM, GBDT	Data from 636 wafers	Exposure and metrology tools	LightGBM achieved 100% in outlier detection and moderate accuracy (R^2 = 0.41–0.76) in overlay prediction
1.2.2. Wu et al. (2022)[6]	Process stability classification for photolithography	ANNs, autoencoder, gradient descent	Measured SEM images (number unspecified)	CD-SEM, lithography exposure tools	Autoencoder achieved 87% in classification accuracy and reduced 91.7% in turnaround time
1.2.3. Yoshida et al. (2022)[7]	Overlay error prediction for DUV exposure in fabricating 3D-NAND flash memory	VMOM	22,600 data sets from 51 wafers for training, 18 wafers for validation	Canon 6300ES6a lithography system	VMOM achieved high accuracy (MAE = 4.46–6.76 nm) in overlay error prediction (target accuracy is 10 nm)
1.2.4. Oak et al. (2023)[8]	Quantifying edge placement error in EUV double and high-NA patterning	ANNs	400 million simulation data sets for training	Unspecified	ANNs helped to achieve edge placement error targets below 7 nm

1.2.5. Raza et al. (2025)[9]	Predicting fab-induced degradation for e-beam and DUV lithography	DNNs, FAID, pre-fab, MGC	Unspecified	JEOL JBX-8100FS EBL, DUV lithography	FAID reduced insertion loss by 50–86.7% in fabricating SWGTS converters by considering fab-induced degradation
1.2.6. Lau et al. (2025)[10]	Process-aware prediction of CD-SEM image for DUV lithography and plasma etching	cGAN, H-Pix2Pix, Pix2Pix, U-Net, ResNet	11,184 experiment SEM images for training and 2,798 for testing	ASML DUV stepper, Hitachi CD-SEM, Oxford NGP100 Cobra etcher	H-Pix2Pix achieved IoU = 0.689–0.865, accuracy = 0.952–0.962, and precision = 0.8056–0.9324 in overlap prediction and CD errors less than 10 nm

*The representing words or phrases for the acronyms or abbreviations used in the above table are defined below in alphabetical order: 1) ANNs = Artificial neural networks, 2) CD = Critical dimension, 3) cGAN = Conditional generative adversarial network, 4) DNNs = Differentiable neural networks, 5) DUV = Deep ultraviolet, 6) EBL = Electron beam lithography, 7) EUV = Extreme ultraviolet, 8) FAID = Fabrication-aware inverse design, 9) GBDT = Gradient boosting decision tree, 10) H-Pix2Pix = HyperPix2Pix, 11) IoU = Intersection over union, 12) MAE = Mean absolute error, 13) MGC = Mentor Graphics Calibre, 14) R^2 = Coefficient of determination, 15) SEM = Scanning electron microscope, 16) SWGTS = Subwavelength grating-to-strip, and 17) VMOM = Virtual machine overlay metrology.

photolithography pattern transfer in semiconductor manufacturing.[4] Overlay, the misalignment between different wafer layers during lithography, is a key parameter affecting yield. However, traditional overlay metrology is costly and typically applied to only a few wafers per lot. Outlier detection, meanwhile, identifies wafers or lots that deviate significantly from expected behavior. The proposed ML-enhanced VM model addresses these limitations by predicting overlay for unmeasured wafers using historical data from exposure and process equipment, thereby improving cost efficiency and overlay control. It also enables early detection of abnormal wafers or lots.

The authors employed the Light Gradient Boosting Machine (LightGBM) algorithm to enhance the VM model for overlay prediction and outlier detection. LightGBM, based on gradient boosting decision trees, efficiently handles large datasets without data normalization, supports parallel computation, and offers high-speed training with low memory usage.[5]

The dataset used for training and testing was collected over eight months in a high-volume manufacturing environment, comprising 166 lots with 636 wafers (3.8 wafers per lot). It included exposure data, process corrections, overlay metrology results, and contextual information (e.g., tool and recipe data). The dataset was slightly skewed toward outliers ($\approx$5% of lots). Data were gathered from Exposure Tools, Archer OVL Metrology Tools, and Process Equipment (e.g., chemical-mechanical planarization and etching), with analysis and VM prediction performed using the KLA Software Suite.

The performance of the LightGBM-based VM model was evaluated using both regression and classification metrics. Based on regression results, the lot-level overlay predictions achieved

$R^2 = 0.64$ (OVL-X) and 0.41 (OVL-Y), while the wafer-level predictions improved to $R^2 = 0.76$ (OVL-X) and 0.65 (OVL-Y), a 10–25% gain attributed to the inclusion of wafer-level metrology data. Here, OVL-X and OVL-Y denote overlay misalignments along the X and Y directions, respectively, reflecting the alignment accuracy between two semiconductor layers.

Using classification metrics, the LightGBM model converted continuous outputs into binary labels to identify outlier wafers. It achieved a true positive rate of 100% for both OVL-X and OVL-Y, with corresponding false positive rates of 13% and 15%.

Although the model showed a perfect classification accuracy in true-positive outlier detection, the moderate R^2 values indicate that this tree-based ensemble algorithm (LightGBM) may not be optimal for this predictive task, leaving room for further improvement in overlay prediction accuracy. Nevertheless, the authors reported that the LightGBM model outperformed conventional random-lot sampling methods, though quantitative comparisons were not provided.

Overall, the study demonstrated that ML-enhanced VM approaches offer moderate predictive accuracy but significant advantages in overlay control, outlier detection, and cost efficiency compared with traditional metrology techniques.

1.2.2 *Machine Learning for Process Stability Diagnosis through Image Analysis in Lithography*

Wu et al. applied ML to automate image analysis and classification for process stability diagnosis (PSD) in the exposure process in photolithography.[6] Specifically, they implemented an autoencoder algorithm to enhance the existing PSD system for reconstructing scanning electron microscopy (SEM) images, improving classification accuracy by minimizing manual intervention. PSD monitors deviations in printed feature sizes from their target values, using metrology tools and SEM images to ensure that exposure conditions and other parameters remain within acceptable limits for producing high-quality patterns.

The autoencoder, a type of artificial neural network (ANN) used in unsupervised learning, enables efficient data representation through dimensionality reduction and feature extraction. In this study, the trained autoencoder processed SEM images by encoding them into compact feature vectors. These encoded vectors were then compared to assess image similarity and classify pattern profiles as either "good" or "bad," thereby automating what was traditionally a manual task.

Training and testing data for the autoencoder were SEM images and critical dimension (CD) values collected from a CD-SEM in the lithography process. SEM images and corresponding CD values were linked with their focus-exposure information during data preparation. The unlabeled SEM images were encoded into feature vectors, which were analyzed to classify pattern profiles based on similarity scores. Model optimization was performed using gradient descent, with hyperparameters — such as learning rate, batch size, and number of epochs — tuned for performance.

For validation, a user manually classified the same images for testing to create a reference label map, which was then compared against the model's classifications to determine accuracy. Although the authors did not specify the number of training or validation samples, they confirmed that both datasets were derived from the same lithography process and imaging conditions, ensuring process relevance and robust performance.

The autoencoder-assisted PSD system achieved notable results, having a classification accuracy of 87% compared with manual classification and a turnaround time reduction from one day to under two hours (a 91.7% decrease). Furthermore, adjusting the similarity threshold allowed fine-tuning of class maps to better align with reference labels.

Overall, the ML-enabled integration of the CD-SEM into an automated framework for lithography process stability proved highly effective in minimizing manual intervention and enhancing process consistency, operational efficiency, and cost-effectiveness in semiconductor fabrication.

1.2.3 *Machine Learning-Aided Overlay Prediction for Optimizing Lithography Exposure Processes*

Yoshida et al. investigated a ML-aided Virtual Machine Overlay Metrology (VMOM) system designed to predict overlay measurement results after exposure in lithography processes.[7] The primary objective of the ML-aided system was to optimize exposure processes in lithography by accurately predicting overlay errors and subsequently adjusting relevant process parameters. The system's performance was evaluated in fabricating 3D-NAND flash memory, with the specific goal of achieving a mean absolute error (MAE) below the target value of 10 nm between predicted and overlay measurements. MAE, a standard statistical measure, quantifies the average magnitude of prediction errors.

Although the authors did not disclose the specific ML algorithms or models employed, they noted that the VMOM system was jointly developed by Canon and Macronix to predict overlay measurements in semiconductor manufacturing. The Canon 6300ES6a Lithography System (a DUV system), equipped with an overlay measurement instrument, was used to generate the exposure, real-time log, and overlay metrology data necessary for ML model development. A total of 22,600 data points collected from 51 wafers in the same 3D-NAND process were used to train the ML-based VMOM model, enabling it to learn the complex relationships between exposure system variables and resulting overlay errors.

The trained model was then validated using 18 additional wafers from the same process. Predicted overlay errors were compared with actual measurement data obtained after applying optimized exposure parameters derived from the model. The ML-assisted VMOM system achieved MAE values of 4.46 nm (Overlay X) and 6.76 nm (Overlay Y), meeting the target accuracy threshold of less than 10 nm for overlay prediction. However, the authors noted that the model faced challenges in predicting differences between wafers with varying tendencies, indicating the need for further model refinement to address wafer-to-wafer variability.

Overall, despite these challenges, the ML-aided VMOM system demonstrated strong potential for improving production performance through ML-driven exposure optimization. The authors emphasized that continued model enhancement will be essential to fully account for process variations and further advance overlay control in semiconductor manufacturing.

1.2.4 Machine Learning-Aided Quantification of Process Sensitivity for Double and High Numerical Aperture Patterning in Extreme Ultraviolet Lithography

Oak et al. developed an ML-based framework combining Monte Carlo simulations and ANNs to quantify process sensitivities in EUV lithography, focusing on double patterning and high-NA EUV single patterning.[8] The objective was to optimize process parameters, predict scaling limitations for advanced nodes, and support tighter edge placement error (EPE) budgets, thereby improving patterning accuracy in future IC manufacturing. In this context, the EPE budget represents the allowable margin or tolerance for EPE in EUV patterning. As feature sizes shrink, variability increases, requiring tighter EPE budgets to ensure device reliability. Understanding EPE sensitivity thus helps identify process bottlenecks and prioritize parameter optimization.

The ANN model developed consisted of a three-layer neural network architecture with Exponential Linear Unit activation functions, and data normalization was implemented to improve learning efficiency. Monte Carlo simulations were used to generate data for training and testing the ANN model by simulating one high-NA EUV single patterning and three EUV double patterning techniques: litho-etch-litho-etch, self-aligned double patterning, and self-aligned litho-etch-litho-etch. Each double patterning technique involved multiple lithographic and etching steps to achieve finer feature sizes beyond the limits of single patterning.

A total of 400 million data points were generated by assuming Gaussian distributions for individual EPE components such as optical proximity correction (OPC), line edge roughness, overlay, critical dimension uniformity, and proximity bias average. For each simulation, 100,000 local errors were sampled for 4,000 global error values, yielding 400 million EPE samples. The trained ANNs were designed to predict the standard deviation of EPE as a function of key process parameters and led to map the EPE budget against the most sensitive parameters (e.g., OPC, line edge roughness, overlay, critical dimension uniformity) to visualize the relationship between these parameters and the EPE budget. Thus, the ANN model helped identify the parameter ranges required to achieve specific EPE targets (e.g., below 6 nm or 7 nm).

However, the study relied entirely on simulated data and did not involve actual device fabrication using EUV lithography. Moreover, *the authors did not report any performance metrics*, such as model accuracy, loss, or computational efficiency. Therefore, future work should validate the ANN-based framework using experimental EPE measurements from fabricated devices and compare its performance with other ML-based models to assess predictive accuracy and generalizability.

1.2.5 *Machine Learning-Aided Fabrication-Aware Inverse Design for Photonic Devices*

Raza et al. introduced an ML-aided approach termed fabrication-aware inverse design (FAID) to optimize the design of photonic IC devices by integrating two computational lithography models into the inverse design (ID) process.[9] This integration allows the design to account for fabrication-induced variations, thereby improving real-world device performance.

Traditional ID methods often fail to address fabrication-induced performance degradation caused by process variations such as corner rounding, proximity effects, and sidewall roughness. The ML-aided FAID framework overcomes these limitations by incorporating two ML-based lithography models, Pre-Fab and Mentor Graphics Calibre (MGC), into the optimization workflow to predict fabrication-induced variations during electron beam lithography (EBL) and DUV lithography, respectively.

Pre-Fab is a differentiable neural network-based model that enables FAID outputs to be mathematically differentiated with respect to input parameters. This differentiability allows gradient calculations essential for tuning design parameters and mitigating performance degradation during EBL. In contrast, MGC accounts for deterministic lithographic effects, including corner rounding, proximity effects, and line shortening, that degrade device performance. Since Pre-Fab and MGC were not developed by the authors, no details were provided regarding their training or model development procedures.

The FAID framework relies heavily on gradient-based optimization to iteratively refine design parameters and minimize lithography-induced degradation. It was fully integrated into the ID software, requiring no external scripting beyond invoking the lithography model and computing its gradients. However, the paper did not include details on FAID's model training or testing process.

To validate the ML-aided FAID, two types of photonic IC devices were fabricated: a Y-branch waveguide and a subwavelength grating-to-strip (SWGTS) converter, where a JEOL JBX-8100FS EBL system was used for EBL, while the facilities for DUV lithography were not mentioned. For comparison, each device type was fabricated in two variants: one based on standard lithography models integrated into the ID process, and another based on the ML-aided FAID-predicted geometry. Devices were also fabricated for operation with different wavelengths by both EBL and DUV lithography. Feature dimensions were carefully controlled to account for fabrication-induced rounding and under-etch effects predicted by the lithography model. Etching steps were optimized to define the Y-branch, subwavelength gratings, and strip waveguides, ensuring smooth transitions and minimal insertion loss, which was measured using the cutback method.

The study highlighted significant performance improvements with FAID compared to standard ID, particularly in mitigating fabrication-induced degradation. For simulated insertion loss at 1320 nm, FAID consistently outperformed standard ID. For the Y-branch fabricated via EBL, FAID achieved a predicted loss of 0.067 dB, matching the ideal ID value, whereas standard ID yielded 0.072 dB (a 6.9% reduction). Similarly, for DUV lithography, FAID reduced the predicted loss to 0.068 dB compared to 0.089 dB for standard ID (7.6% reduction).

Even greater improvements were observed for the SWGTS converter at 1320 nm. Under EBL, FAID reduced the predicted insertion loss to 0.036 dB, compared to 0.072 dB with standard ID (50% reduction). For DUV lithography, the predicted loss decreased dramatically from 0.836 dB to 0.111 dB (86.7% reduction). Measured insertion losses across 1305–1325 nm further validated FAID's effectiveness. For EBL-fabricated SWGTS converters, FAID achieved a measured loss of 0.033 dB versus 0.081 dB for standard ID (59.3% reduction). For DUV-emulated fabrication, FAID reduced the measured loss from 1.321 dB to 0.611 dB (53.7% reduction).

For smoother devices such as the Y-branch, FAID provided modest improvements (6.9–7.6% reduction in insertion loss), as these structures are less susceptible to lithography-induced variations. In contrast, for the SWGTS converter optimized for DUV lithography, FAID achieved an 86.7% improvement in simulated insertion loss at 1320 nm and a 53.7% improvement in measured insertion loss from 1305 to 1325 nm, as compared to standard ID.

Overall, the ML-aided FAID framework markedly outperformed conventional ID by accounting for fabrication-induced variations for reducing insertion loss, particularly in small-feature devices fabricated via DUV lithography. By integrating differentiable and deterministic lithography models, FAID improved the alignment between design predictions and experimental performance. However, the absence of detailed descriptions regarding the underlying ML algorithms, model architectures, and training datasets for the integrated lithography models limits the transparency and reproducibility of the study. Without this information, it is difficult to fully assess the robustness, generalizability, and potential transferability of FAID to other device types and fabrication conditions.

1.2.6 *Machine Learning-Driven Process-Aware Prediction for Deep Ultraviolet Photolithography and Plasma Etch*

Lau et al. developed a process-aware digital twin (PADT) model, HyperPix2Pix, to predict output CD-SEM images in DUV lithography and plasma etch processes by incorporating equipment and process parameters.[10] A PADT is a virtual replica of a real-world process that integrates multimodal learning to capture the influence of various process conditions, thereby enabling accurate simulation of outcomes such as CD-SEM images in DUV lithography and plasma etch.

HyperPix2Pix extends the conventional Pix2Pix model by embedding process parameters into the image-to-image translation task. While the original Pix2Pix framework employs a conditional generative adversarial network for general image translation, HyperPix2Pix employs multimodal learning for process awareness. In this study, U-Net, a convolutional neural network (CNN), served dual functions: as the generator in HyperPix2Pix for predicting CD-SEM images, and within the metrology pipeline to binarize post-lithography CD-SEM images and extract CDs such as line widths and gaps. A ResNet-based generator was also implemented as an alternative architecture for image translation tasks.

Training datasets were generated from two physical experiments conducted at Cornell University. Experiment 1 (Photolithography) employed a 100 mm silicon wafer coated with UV210-0.6 photoresist and patterned using an ASML PAS 5500/300C 248 nm DUV stepper, producing 10,061 paired images (8,032 for training, 2,029 for testing). Experiment 2 (Plasma Etching) used eight 100 mm silicon wafers processed with an Oxford PlasmaPro NGP100 Cobra etcher, yielding 3,921 images (3,152 for training, 769 for testing). All images were captured with a Hitachi CS4800 CD-SEM and curated to eliminate defects before training.

Model performance was evaluated using intersection over union (IoU), accuracy, precision, and recall. For photolithography, HyperPix2Pix achieved an IoU of 0.6890, accuracy of 0.9523, precision of 0.8056, and recall of 0.8762. For plasma etching, results improved to an IoU of 0.8646, accuracy of 0.9616, precision of 0.9324, and recall of 0.9176. CD errors were below 7.8 nm (~2 pixels) for photolithography and within 10 nm (~2.5 pixels) for plasma etching, while gap errors were approximately 10–11.5 nm (~2.5 pixels). Predicted feature distributions closely matched experimental measurements. Here, IoU is the overlap between the predicted segmentation and the ground truth segmentation in an SEM image.

Although IoU and precision decreased slightly for narrow sidewalls and minor image shifts, HyperPix2Pix effectively captured structural variations driven by process parameters, accurately reproducing linewidth and gap dimensions with minimal deviation. Overall, the study demonstrates a robust and flexible PADT framework capable of simulating and optimizing semiconductor fabrication processes, thereby reducing dependence on iterative physical experimentation and improving process predictability.

1.3 Machine Learning-Aided Enhancements in Chemical-Mechanical Planarization, Ion Implantation, and Silicon Growth

This section reviews six studies that explore the application of ML in enhancing three critical semiconductor manufacturing processes: chemical-mechanical planarization (CMP), ion implantation, and monocrystalline silicon growth.

CMP is a hybrid polishing process that combines mechanical abrasion with chemical etching to achieve a flat, smooth wafer surface. A slurry containing abrasive particles and reactive chemicals is applied to the wafer, which is then polished using a rotating pad. This step is essential for achieving interlayer planarity and preparing the wafer for subsequent fabrication stages.

In contrast, ion implantation introduces dopant atoms into the wafer by accelerating ions and embedding them into the substrate. This process modifies the wafer's electrical properties and requires precise control of parameters such as ion energy, dose, and beam shape to ensure accurate implantation profiles.

One study considers the crystal lifting and rotating in the growth of monocrystalline silicon, which uses high-purity molten silicon to form a large single crystal, typically via the Czochralski method. A seed crystal is dipped into the melt and slowly lifted while rotating, solidifying into a cylindrical ingot that is later sliced into wafers for semiconductor and solar cell manufacturing.

In each process, reviews are presented in chronological order in the following subsections. Key information and major findings from each study are then consolidated in Table 1.2, which organizes the data chronologically for ease of reference.

1.3.1 *Machine Learning for Automated Overlay Variation Diagnostics*

Giollo et al. investigated ML-assisted automated diagnostics for addressing overlay variation issues in IC manufacturing, particularly in CMP and wafer alignment during exposure.[11] Overlay variation refers to the misalignment between successive layers of a semiconductor wafer, which is an issue that can severely impact device functionality and performance. When this variation exceeds acceptable limits, overlay excursions occur.

To address this, the authors integrated ML into an automated workflow to monitor and analyze large volumes of process data,

Table 1.2. ML-aided enhancements in CMP and ion implantation.*

Section. Author (Year)[Ref.]	Subject and Process	ML Models, Algorithms, and Platforms	Training and Testing Data	Key Equipment	Key Findings for Model Performance
1.3.1. Giollo et al. (2017)[11]	Overlay variation detection for CMP and wafer alignment	LASSO	150 × 900 experimental data sets	CMP, optical metrology tools	LASSO exhibited poor performance ($r = 0.3$) with low accuracy in overlay prediction (RMSE = 2.37 nm)
1.3.2. Yu et al. (2021)[12]	Closed loop control and optimization for CMP	CNNs, DBNs, MLR, NNs, SVM, transfer learning	Surface profiles using 36 test chips (number unspecified)	Reflexion LK CPM, atomic force long-range profiler	NNs achieved $R^2 = 0.87$ in accuracy, reduced 84.9% in dishing residuals, and improved 46% accuracy in predicting thickness
1.3.3. Lin et al. (2022)[13]	Classifying recipes and detecting anomaly in ion implantation	AdaBoost, DT, GNB, LR, random forest	68 recipes (≈ 530,000 data sets) from foundry	Eaton NV6200 ion implanter	Random forest and AdaBoost respectively achieved 100% and 99.99% accuracy for detection and classification tasks
1.3.4. Takemura et al. (2022)[14]	Beam shape and uniformity control for ion implantation	Gradient-boosting regressor, Scikit-Learn	20,000 beam width measurements	Nissin BeyEX ion implanter, XY monitor	The gradient-boosting regressor achieved high accuracy ($R^2 \geq 0.90$) in beam width prediction and 88% success rate in beam shape control

(*Continued*)

Table 1.2. (*Continued*)

Section. Author (Year)[Ref.]	Subject and Process	ML Models, Algorithms, and Platforms	Training and Testing Data	Key Equipment	Key Findings for Model Performance
1.3.5. Timoney et al. (2023)[16]	Copper CMP automated process control	Supervised learning	Trench height and resistance measurement (number unspecified)	CMP with integrated metrology	Supervised learning improved trench height prediction accuracy by ~50% (R^2 ~0.58 vs. R^2 ~0.38)
1.3.6. Mu et al. (2025)[17]	Condition monitoring of silicon crystal growth	k-nearest neighbor	16,000 data sets for training and testing	Czochralski crystal growth furnace	k-nearest neighbor achieved >95% accuracy in bearing faults detection

*The representing words or phrases for the acronyms or abbreviations used in the above table are defined below in alphabetical order: 1) CMP = Chemical-mechanical planarization, 2) DBNs = Deep belief networks, 3) DT = Decision tree, 4) GNB = Gaussian naive Bayes, 5) LASSO = Least absolute shrinkage and selection operator, 6) LR = Linear regression, 7) MLR = Multiple linear regression, 8) NNs = Neural networks, 9) r = Mean Pearson correlation, 10) R^2 = Coefficient of determination, 11) RMSE = Root mean square error, and 12) SVM = Support vector machine.

ultimately building a predictive model for overlay variation. Their workflow employed interpretable linear models, specifically the Least Absolute Shrinkage and Selection Operator (LASSO), to rank features and identify key contributors to overlay variation.

LASSO, a regression technique known for its ability to perform feature selection and regularization, was used to isolate the most relevant variables influencing overlay variation while discarding irrelevant ones. To mitigate data imbalance, a sample-weighting function emphasized the importance of "off-spec" (out-of-specification) wafers. LASSO's regularization term, which penalizes large coefficients, also helped prevent overfitting.

The dataset included 900 wafers, each characterized by 150 process parameters, such as overlay variation, nitrate thickness, and shallow trench isolation depth. Most wafers showed overlay variations between 0–4 nm, while about 5% exhibited larger variations (4–12 nm). Missing data affected roughly 75% of the features, mainly due to throughput constraints. Here, nitrate thickness represents the thickness of the nitrate layer in the wafer stack, while shallow trench isolation depth corresponds to the silicon dioxide depth in the shallow trench isolation region. Both were found to be key contributors to overlay variation measured after CMP.

Data were split into 80% training, 10% validation, and 10% testing sets. The training set was used to fit the LASSO model; the validation set optimized the regularization parameter; and the testing set with unseen samples evaluated model generalizability. Although the authors did not specify the metrology tools or software used, optical metrology tools are likely, as they are standard for alignment accuracy measurements in semiconductor fabrication.

The LASSO model achieved a mean Pearson correlation (r) of 0.3 ± 0.1, indicating a weak positive linear relationship between predicted and actual overlay values (perfect correlation: $r = 1$; no correlation: $r = 0$). The mean root mean square error (RMSE) in overlay predictions was 2.37 ± 0.2 nm, suggesting relatively high prediction errors given that 95% of overlay variations fall between 0 and 4 nm. While this level of accuracy may be acceptable within certain process tolerances, it highlights considerable room for improvement.

Overall, this study indicated that ML-based diagnostics could only marginally predict overlay excursions. The limited performance of the LASSO model — being inherently linear — suggests that overlay variation is likely governed by complex, nonlinear relationships among process parameters. Consequently, employing nonlinear ML models could better capture these interactions and substantially improve predictive accuracy.

1.3.2 *Optimizing Chemical-Mechanical Planarization Processes with Machine Learning*

Yu et al. studied the application of ML to optimize CMP processes with closed-loop control (CLC) in IC manufacturing.[12] The study aimed to develop ML-assisted CMP CLC models to enhance prediction accuracy for dishing and erosion, improve thickness control and wafer uniformity, and detect high-risk design areas (hot spots). Achieving these goals enables real-time process adjustments that reduce yield loss and strengthen layer integrity. Early identification of risky areas during layout design further accelerates validation and yield ramp-up by minimizing systematic hot-spot defects. Here, dishing refers to concave metal-line deformation after CMP, while erosion indicates dielectric loss surrounding these lines. Since CMP combines mechanical abrasion and chemical etching to produce flat wafer surfaces, both defects degrade surface quality and yield.

Four ML algorithms were explored: support vector machines, neural networks (NNs), multiple linear regression (MLR), and transfer learning (TL). Support vector machines initially used a radial basis function kernel to optimize the CMP CLC process, but underperformed relative to MLR, while NNs effectively captured nonlinear behavior and became the core algorithm for predicting dishing and erosion trends across varying line/space ratios and pattern densities for optimization. The MLR model, used in conventional CMP CLC, served as a baseline. TL was applied to adapt NNs to be trained on test-vehicle data — experimental data from specially designed chips with controlled patterns and densities — to enhance predictive performance. Previous studies using CNNs for surface topography modeling and deep belief networks for material removal analysis demonstrated broader ML potential in CMP optimization.

Training data for the NN model was obtained through controlled experiments using test chips with different line widths (1–3 μm) and pattern densities (25–80%). These chips were polished under various process settings (slurry type, down force, over-polish time, rotation speed), and surface profiles were measured by atomic force profilers to quantify dishing and erosion. Data were organized by wafer regions (center, middle, edge) to capture within- and between-wafer variations, then preprocessed for model training and validated using real product wafers. TL further refined model adaptation to real-world conditions using test-vehicle data. Industrial-grade equipment supported this work, including an Applied Materials Reflexion LK CMP system with a Titan Evo head (seven polish zones) and inline metrology for real-time feedback. Note that the conventional methods were based on an MLR model and only trained by limited data provided by an *ex situ* metrology tool.

The ML-assisted or NN model showed marked improvement over conventional methods. For CMP CLC prediction, it achieved target thickness on the first wafer and maintained a 50–90 nm range by the third wafer, whereas the traditional model required six wafers to stabilize within ~130 nm (~46% improvement). For dishing and erosion prediction, R^2 values reached 0.87 and 0.73, respectively, with error residuals reduced by 18.3%. Applying TL further lowered erosion residuals from 33.6 nm to 7.2 nm (78.4% improvement) and dishing residuals from 16 nm to 2.41 nm (84.9% improvement). The model also successfully identified high-risk design areas, minimizing systematic defects and enhancing yield ramp-up.

Overall, the ML-assisted CMP CLC model significantly outperformed traditional methods in predictive accuracy and defect mitigation. This ML-driven approach enables superior process control, faster adjustments, and enhanced optimization — addressing key challenges in advanced IC manufacturing.

1.3.3 *Machine Learning for Recipe Classification and Anomaly Detection in Ion Implantation*

Lin, Zhao, and Horng developed a ML-enhanced framework for automatically monitoring ion implantation processes in semiconductor manufacturing, focusing on recipe classification and anomaly detection to minimize downtime and improve efficiency.[13] Ion implantation introduces impurities into semiconductor wafers, and any fault during the process can render a wafer unusable. In this context, a recipe refers to a parameter set guiding an ion implanter to perform a specific process. Recipe classification ensures that the predicted recipe matches the designated one for each wafer, thereby preventing anomalies, reducing false alarms, safeguarding wafer quality, and supporting real-time fault detection for optimal yield.

The authors evaluated seven ML algorithms and analyzed five in detail for recipe classification and anomaly detection: linear regression, decision tree, random forest, Gaussian Naive Bayes, and adaptive boosting (AdaBoost). Linear regression modeled linear relationships between features and the target variable but was unsuitable for anomaly detection due to its assumption of linear separability. Decision tree used decision rules to predict targets, offering interpretability and the ability to pinpoint specific conditions leading to anomalies. Random forest aggregated results from multiple decision trees via ensemble learning, effectively mitigating overfitting and improving accuracy on complex datasets. Gaussian Naive Bayes applied probabilistic modeling to classify data as normal or anomalous based on feature likelihoods; it was computationally simple but less effective on complex data. AdaBoost combined multiple weak learners (e.g., decision trees) into a strong classifier by iteratively adjusting sample weights to improve accuracy and anomaly detection. The authors preferred these models over neural network-based methods because they provide more interpretable and actionable insights for process improvement.

The study used data from a modern wafer foundry in Taiwan, collected from an Eaton NV6200 ion implanter via the SEMI

Equipment Communications Standard Part 2 (SECS-II) interface. Two datasets were analyzed: one with 26 recipes (≈200,000 records) and another with 42 recipes (≈330,000 records). 70% of the data was used for training and 30% for testing. Based on domain expertise, only 12 attributes from four subsystems (ion source, extractor, mass analysis, and accelerator) were selected. Although the implementation platform was not specified, the models were likely developed using standard libraries such as Scikit-Learn, TensorFlow, or PyTorch.

Experimental results showed that linear regression performed poorly for both recipe classification and anomaly detection due to its inability to handle nonlinear separability (94.43% accuracy for the 26-recipe dataset and 92.91% for the 42-recipe dataset). Decision tree achieved 99.99% accuracy for both datasets after hyperparameter optimization but was somewhat prone to overfitting with smaller sample size. Random forest reached 100% accuracy for both datasets after hyperparameter optimization and demonstrated excellent stability even with limited data. Gaussian Naive Bayes performed modestly, with 5–10% recipe classification errors. AdaBoost achieved 99.99% accuracy for both datasets after hyperparameter optimization, effectively reducing recipe classification errors and detecting anomalies, though it showed mild overfitting in small datasets.

Here, hyperparameters are settings or configurations manually set before training. Hyperparameter optimization is essential for achieving best model performance and balancing underfitting and overfitting, leading to better generalization. This optimization process involves defining a search space, selecting a search strategy, and evaluating combinations (often via cross-validation) to find the best configuration.

In conclusion, both random forest and AdaBoost demonstrated outstanding performance, achieving 100% and 99.99% accuracy, respectively. Random forest was the most stable across varying data volumes, whereas AdaBoost showed slight instability. Overall, these two ML models emerged as the top performers for recipe classification and anomaly detection in ion implantation — a critical process in IC manufacturing.

1.3.4 *Machine Learning-Aided Beam Shape Control for Ion Implantation*

In 2022, Takemura, Sakai, and Murayama developed ML-aided models for beam shape control in ion implantation for IC device fabrication.[14] The precise shape and size of the ion beam are critical to achieving uniform implantation and consistent sheet resistance reproducibility. Their system employed the Gradient Boosting Regressor (GBR) algorithm to optimize beam tuning parameters such as ion species, energy, and beam current, enhancing beam setup accuracy and maintaining beam shape within predefined tolerances. Since sheet resistance reproducibility depends on consistently achieving target values across multiple implantations, accurate beam-size control is also essential.

The authors implemented the GBR module from the Scikit-Learn library within the NISSIN BeyEX medium-current ion implanter. The GBR model iteratively minimized prediction errors via gradient descent. It was trained using approximately 20,000 measurement results from the XY monitor, with beam conditions and operational parameters as inputs and horizontal and vertical beam sizes as outputs. The trained GBR model achieved high predictive reliability, with R^2 values of 0.90 for horizontal beam width and 0.94 for vertical beam width in test data. Furthermore, the ML system achieved an 88% success rate in controlling beam size within ±10% of the target, whereas conventional non-ML control showed substantial deviations similar to out-of-control cases.

However, the 2022 GBR model presented above relied solely on 1D beam measurements, evaluating beam quality based on the width at 50% of the peak current. While effective for independent horizontal and vertical assessments, it could not capture 2D non-uniformities in beam shape. To overcome this limitation, Takemura et al. in 2025 extended their approach by introducing the Mura value, a comprehensive metric representing 2D beam uniformity.[15] The Mura value accounts for the number, density, width, and height of each peak in the beam profile, offering a more holistic measure of beam quality.

The 2025 model integrated image processing techniques to reconstruct 2D beam shapes from superimposed slit data using asymmetric Gaussian distributions to identify and characterize individual peaks.[15] This approach enabled accurate prediction of both beam size and shape quality. The enhanced system could estimate the Mura value prior to implantation and automatically halt the process if excessive non-uniformity was detected, thus preventing defects. These improvements directly addressed the shortcomings of the 2022 model, resulting in better beam uniformity, higher implantation quality, and reduced defect rates in mass production.

Overall, both the 2022 and 2025 ML-based beam control models by Takemura and colleagues demonstrated significant advances in uniformity, reproducibility, and operational efficiency in ion implantation, a critical process in semiconductor manufacturing.

1.3.5 *Machine Learning-Enabled Trench Height Prediction for Enhanced Process Control in Copper Chemical-Mechanical Planarization*

Timoney et al. explored the application of supervised ML to enhance process control in copper CMP, a critical step in IC fabrication.[16] Their work focused on leveraging ML-enabled measurements to improve trench height (Trench_H) monitoring and automated process control (APC). The study compared the correlation of ML-derived Trench_H measurements with conventional optical critical dimension (OCD) metrology against electrical test (e-test) resistance for two product categories across multiple metallization layers. In this context, OCD uses optical properties such as reflection, refraction, and scattering of light to indirectly determine wafer feature dimensions, including trench height. APC refers to control systems that dynamically adjust CMP process parameters in real or near-real time based on *in situ* measurement data.

As a brief three-page article, the authors provided limited details on the specific ML algorithms or hardware configurations used. However, it noted the application of supervised learning (SL) to generate Trench_H predictions and enhance the correlation between Trench_H prediction and e-test resistance measurements. Spectral data from integrated metrology modules and corresponding e-test data were used to train the SL models, while the predicted Trench_H was subsequently correlated with e-test resistance for validation. CMP process tools were used for trench height measurement and metallization layer processing, and 100% wafer-level sampling was conducted through integrated metrology to enable comprehensive data collection for advanced process control.

The authors reported that, when compared with e-test resistance measurements of Trench_H, the ML-enabled Trench_H model predictions (R^2 = 0.63–0.52) consistently outperformed conventional OCD measurements (R^2 = 0.45–0.30). Consequently, using ML-based model predictions as feedback signals for APC yielded better performance than using OCD-derived measurements. Moreover, APC employing the SL Trench_H model achieved improved centering for both Trench_H and e-test resistance relative to OCD-based APC. Here, "better centering" refers to more accurate targeting of process parameters toward their desired values during CMP operations. In contrast, conventional OCD-based APC introduced baseline deviations in e-test resistance, leading to inconsistent process centering across different product categories.

Overall, ML-based modeling moderately improved the correlation between predicted Trench_H and e-test resistance, enabling more precise process control and enhanced targeting across various product categories and metallization layers in Cu CMP. These improvements contribute to higher manufacturing efficiency and yield in advanced IC fabrication.

1.3.6 *Machine Learning-Assisted Condition Monitoring for Monocrystalline Silicon Growth*

Mu et al. developed a ML-assisted condition monitoring technique for the crystal lifting and rotating mechanism (CLRM) in the Czochralski crystal growth furnace used for monocrystalline silicon growth.[17] The study focused on signal expansion and feature extraction to enhance diagnostic accuracy.

Vibration signals collected by sensors mounted on the CLRM were analyzed to assess operational status. The ML algorithm k-nearest neighbor (kNN) was used to classify the CLRM's operating conditions based on features extracted from vibration signal data. Signals were expanded using the variational mode canonical polyadic decomposition (VMCPD) method, and features were extracted via permutation entropy before being fed into the kNN classifier. The kNN model effectively handled nonlinear data without distributional assumptions, making it well-suited for detecting subtle and inconspicuous signal variations. In the offline phase, historical data trained the model, while in the online phase, real-time signals were classified to diagnose the mechanism's operational state.

Three datasets were used for kNN model training and testing. The first, from Case Western Reserve University, included six classes of vibration signals (1,000 samples per class). The second contained six classes under different motor directions and speeds, also with 1,000 samples per class. The third dataset, collected from an actual Czochralski crystal growth furnace process, included four classes — three failure types and one normal condition — with 1,000 samples each. 75% of data were used for training and 25% for testing, totaling 16,000 samples.

The kNN classifier combined with VMCPD achieved outstanding condition monitoring performance for bearing faults across all three datasets. For example, on the Case Western Reserve University data, recognition rates exceeded 99.4%, with harmonic mean (F1) of precision and recall scores above 99.4% and specificity reaching 99.96% for certain feature combinations. These results demonstrated the method's robustness and high diagnostic accuracy, even with limited or low-amplitude vibration signals.

Overall, the kNN-based condition monitoring system achieved over 95% accuracy across all datasets, effectively minimizing misclassifications. The VMCPD-enhanced signal expansion and feature extraction enabled reliable fault diagnosis from small, subtle datasets, demonstrating excellent performance for real-time monitoring of the CLRM in silicon crystal growth.

1.4 Machine Learning in Mask Design for Photolithography and Predictive Modeling

This section reviews seven studies that demonstrate the application of ML in two key areas of semiconductor manufacturing: mask design for photolithography and predictive modeling for process control. The first four studies focus on ML-assisted mask design, while the remaining three explore predictive metrology and property modeling.

Mask design is a critical stage in photolithography, involving the creation of photomasks that accurately transfer intricate patterns onto semiconductor wafers. Effective design must account for pattern distortions caused by diffraction, optical interference, and resist process effects. ML techniques have been increasingly adopted to enhance mask optimization, reduce design cycle time, and improve pattern fidelity.

Predictive metrology leverages ML to forecast key wafer parameters or feature dimensions based on upstream process data, enabling proactive adjustments to minimize deviations and improve yield. Similarly, predictive property modeling applies ML to estimate material or device properties, such as electrical characteristics or wafer acceptance test outcomes, thereby supporting process and quality control.

For clarity and comparison, the key datasets and major findings of each study are summarized in Table 1.3.

1.4.1 *Machine Learning for Sub-Resolution Assist Feature Placement in Full-Chip Lithography*

Wang et al. explored the use of ML, specifically deep convolutional neural networks (DCNNs), for sub-resolution assist feature (SRAF) placement in full-chip lithography.[18] SRAFs are small patterns added to photomasks during photolithography to enhance image contrast and enlarge the process window for target features, though they are not intended to print on the wafer. Proper SRAF placement is crucial for lithographic performance, stability, and yield in high-volume IC manufacturing. Full-chip lithography refers to photomask generation for an entire chip design, ensuring accurate pattern transfer across the wafer.

The ML-SRAF model was built on a DCNN architecture and trained by simulation data via SL using TensorFlow. The simulation data was generated using Tachyon SMO/iOPC software by ASML. Training employed mini-batch stochastic gradient descent over 10,000 iterations, with input data consisting of design target images and the simulated continuous transmission mask (CTM) images from Tachyon SMO/iOPC as the output. These iterations are necessary because CTM employs an iterative process to refine the mask image to achieve optimal lithographic performance for SRAF placement. A $15 \times 15\ \mu m^2$ layout was used for training, and 200 iterations of CTM were run to generate the simulation data sets. Here, CTM uses a grayscale representation to model electric field transmission after the mask plane. Through iteration,

Table 1.3. ML in mask design for photolithography and predictive modeling.*

Section. Author (Year)[Ref.]	Subject and Process	ML Models, Algorithms and Platform	Training and Testing Data	Key Equipment or Software	Key Findings for Model Performance
1.4.1. Wang et al. (2017)[18]	SRAF placement for mask design in lithography	DCNNs, ML-SRAF	200 iterations of simulated CTM (number unspecified)	ASML Brion Tachyon platform	ML-SRAF reduced runtime by 85.7%, raised process window by 35% and improved consistency
1.4.2. Utzny (2017)[19]	Improving critical dimension uniformity in mask making	BS, CT, HRM, LR, PCA, RP, TSM	Experimental critical dimension data sets (number unspecified)	Electron beam writer, mask fabrication tools	HRM improved compensation residuals by 25% and reduced update time by 56%; TSM achieved 0.1 nm better than BS in compensation residuals
1.4.3. Lin et al. (2023)[20]	Optical proximity correction for optimizing mask designs	Multi-layer perceptron neural network	343 data sets from mask pattern experiment	Hitachi SU-8010 SEM machine	Multi-layer perceptron neural network reduced MSE 17–67% in predicting resist dimensions in mask design
1.4.4. Habib et al. (2024)[21]	Resolution enhancement by photomask modification in lithography	TPM-RET, CNNs	Image pixel data computational created (number unspecified)	Nvidia Quadro RTX 4000 (GPU)	TPM-RET achieved high accuracy in predicting optical proximity correction and higher than 97% parallel efficiency

1.4.5. Rana et al. (2015)[22]	Predictive metrology for better control of RIE, CMP, and lithography	MVLR, NNs, partial least squares	Critical dimension, DTC, and MLR measurements (number unspecified)	MBIR, scatterometry tool, CMP, RIE, Litho, AFM, SEM	NNs predicted DTC better than MVLR with low TMU (0.753 AU vs. 1.058 AU); partial least squares ($R^2 = 0.973$) and NNs ($R^2 = 0.923$) were strong in MLR prediction
1.4.6. Pratik et al. (2022)[23]	C-V curve prediction by considering oxidation and wafer cleaning	SSMTL, transfer learning, multilayer perceptron	124,320 simulated data sets and 4,896 experiment data sets	PVD, Canon line stepper, TCP etching machine	SSMTL and TL achieved high accuracy ($R^2 > 0.925$) in predicting C-V curves by adding simulation data in training
1.4.7. Wei et al. (2025)[24]	Predicting wafer acceptance test outcomes	AutoML, GPR, SMART-WAT, XGBoost LightGBM	9,720 data sets from experiment	Wafer acceptance test and chip probing tools, IC line	SMART-WAT's RMSE was 0.77% to 18.90% lower than AutoML's and 10.43%–42.79% lower than GPR's in predicting parameters in wafer acceptance test

*The representing words or phrases for the acronyms or abbreviations used in the above table are defined below in alphabetical order: 1) AFM = Atomic force microscopy, 2) AU = Arbitrary units, 3) BS = Balanced sampling, 4) CMP = Chemical-mechanical planarization, 5) CNNs = Convolutional neural networks, 6) CT = Clustering techniques, 7) CTM = Continuous transmission mask, 8) C-V= Capacitance-voltage, 9) DCNNs = Deep convolutional neural networks, 10) DTC = Deep trench capacitance, 11) GPR = Gaussian process regression, 12) HRM = High-resolution model, 13) LR = Linear regression, 14) MBIR = Model-based infrared reflectometry 15) ML = Machine learning, 16) MLR = Metal line resistance, 17) MSE = Mean square error, 18) MVLR = Multivariate linear regression, 19) NNs = Neural networks, 20) PCA = Principal component analysis, 21) PVD = Physical vapor deposition, 22) R^2 = Coefficient of determination, 23) RIE = Reactive ion etching, 24) RMSE = Root mean square error, 25) RP = Recursive partitioning, 26) SRAF = Sub-resolution assist feature, 27) SSMTL = Semi-supervised multitask learning, 28) TCP = Transformer coupled plasma, 29) TMU = Total measurement uncertainty, 30) TPM-RET= True pixel-based machine learning resolution enhancement techniques, and 31) TSM = Tool-specific model.

it optimizes lithographic metrics like image contrast and process window, enabling advanced mask designs.

For testing, the trained DCNN (ML-assisted SRAF) predicted SRAF guidance maps on unseen layouts. Data augmentation further improved model accuracy, and the framework integrated smoothly into existing OPC processes. The exact number of individual simulation data sets for training or testing was not explicitly reported.

The ML-assisted SRAF placement model was implemented on the ASML Brion Tachyon platform for computational lithography. Compared with traditional computational lithography techniques such as rule-based SRAF and model-based SRAF, which often struggle with complex layouts, ML-SRAF enabled faster runtimes, higher accuracy, better consistency, and seamless integration into ASML's OPC workflows. OPC is a computational lithography technique that adjusts mask layouts to counteract optical diffraction, resist effects, and etching distortions.

In performance benchmarking, ML-SRAF surpassed traditional approaches. Compared to CTM, it reduced runtime by 85.7% (achieving only 1/7 of CTM runtime: 1 h vs. 7 h), enabling viability for full-chip applications. In process window metrics, ML-SRAF achieved a 35% improvement in depth-of-focus margin at 5% exposure latitude (70 nm) compared to the SRAF Guidance Map (SGM) method (52 nm), while maintaining similar performance to CTM. Higher exposure latitude values signify greater tolerance to exposure variation and thus a more robust process. For worst process variation bands, ML-SRAF consistently outperformed SGM and matched CTM across most features, including 1D-line and unmatched patterns. Additionally, it improved consistency, with reduced EPE variance compared to CTM and SGM.

Overall, ML-SRAF demonstrated significant improvements in runtime, process robustness, and patterning consistency. Its ability to integrate into production-ready OPC workflows while reducing computational overhead highlights its promise as a practical and scalable solution for full-chip lithography and advanced mask synthesis.

1.4.2 *Machine Learning for Critical Dimension Uniformity in Photomask Design and Fabrication*

Utzny introduced ML-enhanced models to improve control of CD uniformity in mask design for photolithography.[19] Mask design for photolithography presents unique challenges for ML applications due to high product variability and relatively low production volumes. To address these issues, three ML-enhanced models were developed: (1) the high-resolution model (HRM), (2) the tool-specific model (TSM), and (3) the balanced sample model (BSM). Each model was designed to enhance CD uniformity in mask production under different emphases or operational conditions.

Compared with an earlier low-resolution model (LRM), which used fewer spatial bins and lacked boundary-specific handling, the HRM incorporated two key ML algorithms: (1) recursive partitioning (RP) and (2) balanced sampling (BS) to improve boundary compensation. The HRM improved spatial resolution by dividing the mask field into smaller bins and introduced boundary-specific covariates to distinguish inner, outer, and non-boundary regions. RP generated decision trees to predict CD deviations using relevant covariates, while BS ensured that the training data maintained representative statistical properties by blending tool and process information.

The TSM was developed to create separate compensation maps for each tool, thereby improving CD control by accounting for tool-induced variations in CD signatures. It employed principal component analysis (PCA) to identify and classify global CD patterns and variation modes, along with RP and SL algorithms used in the HRM. In contrast, the BSM generated a unified compensation map by combining data from multiple tools, providing production flexibility and enabling interchangeable tool use while maintaining tight CD control. The BSM integrated three ML algorithms — PCA, RP, and BS — previously applied in the HRM and TSM.

Training and testing datasets for all three models were derived from multiple stages of the mask fabrication process. These included actual CD distribution measurements from manufactured masks, data from electron beam writers and other production tools, and key process parameters such as tool geometries, clear-field data, resist age, and other factors influencing CD uniformity. CD measurements were collected across the entire mask field to capture both global and local variations, while BS ensured statistical representativeness by accounting for tool mismatches and process variations. Specifically, the HRM used spatially resolved CD measurements emphasizing boundary effects (e.g., CD shifts near active area edges). The TSM relied on CD data from individual tools, such as Tool 1 and Tool 2, incorporating tool-specific parameters. The BSM blended CD data from multiple tools to preserve the statistical properties of the full dataset. This systematic approach ensured robust, representative data for model training and validation.

Performance evaluations demonstrated that the HRM outperformed the LRM baseline. For boundary effects (CD shifts near mask edges), the HRM reduced compensation residuals (CRs) by 25% and shortened model update time from 78 hours to 34 hours (a 56.4% reduction). Here, CRs refer to remaining uncorrected CD deviations, while update time refers to the duration required to integrate new data and generate updated correction maps. Increasing spatial resolution by 50% further improved boundary compensation accuracy, while the use of dedicated covariates reduced model complexity.

The TSM and BSM performances were assessed using two key metrics: CRs and the area under the curve (AUC), which measures how well the model's predictions align with observed data (with higher AUC values indicating better accuracy). The TSM slightly outperformed the BSM in both metrics, achieving a better residual improvement of 0.1 nm, though the performance gap was minimal.

In summary, the ML-enhanced models, particularly the HRM, effectively addressed key challenges in mask fabrication, such as boundary effects and tool mismatches, while improving production flexibility and computational efficiency. The study also demonstrated that the BSM achieved nearly the same accuracy as the TSM, offering greater operational versatility. Collectively, these ML-driven approaches provide fine-grained, data-driven strategies for optimizing CD uniformity control in advanced mask fabrication.

1.4.3 *Hybrid Human–Machine Learning Approach for Optical Proximity Correction*

Lin et al. investigated OPC to optimize photolithographic mask designs, aiming to enhance pattern transfer accuracy onto wafers by mitigating distortions from diffraction, interference, and resist process effects.[20] The authors proposed a hybrid optimization approach combining human intuition and ML, which outperforms both pure human decision-making (HDM) and standalone ML methods. In OPC, HDM plays a vital role in the early stages when ML models lack sufficient data, allowing experts to identify trends, correct stalled optimization, and leverage experiential insight. The hybrid approach thus integrates human creativity with ML's computational efficiency to handle complex, nonlinear mappings and fine-tune mask dimensions effectively.

However, the proposed approach is debatable. Relying on highly skilled experts for iterative decision-making contradicts one of the key goals of ML — to automate complex and costly processes. In practice, the scarcity and expense of such experts limit the scalability and practicality of the hybrid model. Furthermore, the effectiveness of HDM is inherently difficult to quantify or standardize, which poses challenges for evaluating the robustness of this approach.

The study employed a multi-layer perceptron neural network (MLPNN) with two hidden layers of 15 neurons each, a three-neuron input layer, and a single output neuron. The MLPNN was trained to predict optimal mask dimensions by learning the nonlinear relationship between mask geometry and resulting resist patterns, aiding fine-tuning in later optimization stages where human intuition becomes less effective.

Training and testing data were generated through experimental mask designs created using KLayout software and the Python Gdspy module, based on geometric parameters provided by mask designers (HDM inputs). A total of 343 mask patterns with varying dimensions were fabricated on n-type silicon wafers. After DUV exposure and resist development, pattern dimensions were measured via SEM, forming a dataset of 343 paired mask and resist dimensions.

The MLPNN was trained with the Adam optimizer for 1,000 epochs at a learning rate of 0.001, using the Scaled Exponential Linear Unit activation function. Model performance was evaluated by MSE between predicted and measured resist dimensions.

For two-bar patterns, MSE decreased from 6,360 (pure HDM) to 2,101 (ML-assisted HDM), a 67% improvement; For tri-line attacker patterns, from 7,132 to 5,931, a 17% improvement. At early stages, pure ML exhibited significantly higher MSE (26,019 for two-bar and 79,979 for tri-line patterns) than ML-assisted HDM (6,023 and 7,738, respectively).

The notably lower MSE values for ML-assisted HDM (e.g., 2,101 vs. 6,360) demonstrate that integrating ML with human expertise substantially improves OPC performance. Overall, the MLPNN served as the computational backbone of the hybrid framework, enabling accurate prediction and optimization of mask dimensions. Despite its reliance on expert input, the study effectively illustrates the complementary strengths of human intuition and ML in achieving superior OPC outcomes.

1.4.4 *Machine Learning-Based Pixel-Level Resolution Enhancement for Photolithography Optimization*

Habib et al. introduced true pixel-based machine learning resolution enhancement techniques (TPM-RET), an end-to-end ML workflow developed to improve the production readiness of RETs in photolithography for IC manufacturing.[21] RETs are a class of computational lithography methods that modify the initial IC layout on the photomask to compensate for optical, mechanical, and chemical distortions occurring during the lithographic process. Their goal is to ensure that the wafer patterns faithfully reproduce the intended circuit design.

As feature sizes shrink and chip areas expand, RETs become increasingly computationally intensive. To address this, TPM-RET provides a more efficient and accurate alternative by applying CNNs for pixel-by-pixel correction. This pixel-based strategy eliminates the boundary conflicts and inconsistencies often seen in conventional RET approaches, enabling higher accuracy and scalability with reduced computational cost. The framework utilizes MobileNetV3, a lightweight open-source CNN model, whose efficient architecture supports large-scale, full-chip corrections without the heavy resource demands of traditional models.

For training and testing the CNN model within the TPM-RET flow, the input IC design layout was first converted into a high-resolution image and divided into smaller tiles representing pixels and their neighborhoods. Non-uniform image compression was applied to reduce data size while retaining essential features. An ideal photomask was then generated using Calibre pxOPC, and an inverse intensity profile (IIP) map was produced by convolving the binary photomask with a predefined kernel. Each tile was paired with its corresponding IIP value, discretized into labeled classes. The resulting dataset was split into training, validation, and testing sets. Here, Calibre pxOPC refers to a computational lithography tool for OPC, one of the primary RETs, while the IIP function introduced in TPM-RET helps recover process-related information typically lost in binary masks. Using

TensorFlow and Keras, the CNN was trained and evaluated with standard metrics, including a confusion matrix.

Model performance was assessed by comparing TPM-RET's OPC predictions with the reference results from Calibre pxOPC. The confusion matrix (Fig. 5 in the article) demonstrated strong correlation between the TPM-RET outputs (OPC) and reference pxOPC values, indicating high model accuracy and stability. Further validation was performed on isolated line and line-space patterns for a 32 nm metal test pattern, where TPM-RET achieved close agreement with ideal pxOPC photomasks even for unseen pattern dimensions. In terms of effectiveness in utilizing multiple processors, TPM-RET achieved over 97% parallel efficiency when distributed across 128 CPUs, providing a 125× speedup compared to a single CPU.

The authors emphasized that TPM-RET effectively overcame major production challenges associated with previous ML-RET methods. While many ML-RET frameworks relied on generative adversarial networks (GANs) and image-based corrections that could introduce stitching errors and boundary artifacts, TPM-RET's pixel-based approach ensured consistent, seamless corrections across the full chip. Its incorporation of IIP data restored critical process context, reducing overfitting and enhancing model robustness. Moreover, TPM-RET's CPU scalability offered a cost-efficient alternative to GPU-dependent ML-RET systems.

Overall, TPM-RET demonstrated accurate, consistent, and scalable performance, making it a production-friendly, computationally efficient, and highly reliable solution for advanced lithography modeling and correction in IC manufacturing.

1.4.5 *Machine Learning-Driven Predictive Metrology in Integrated Circuit Nanofabrication*

Rana et al. introduced ML-driven predictive metrology to enhance process control and optimization in IC fabrication, focusing on reactive ion etching (RIE), photolithography, and CMP.[22] Their approach employed ML models to predict critical electrical performance metrics: deep trench capacitance (DTC) and metal line resistance (MLR), early in the manufacturing pipeline. Such early prediction is crucial for minimizing losses by proactively identifying process deviations, enabling timely corrective actions such as wafer rework or scrap.

Predictive metrology complements traditional hybrid metrology by leveraging multivariate data and ML to capture complex correlations that are difficult to model mathematically or measure directly. In this context, hybrid metrology refers to combining measurements from multiple toolsets to improve determination of key process parameters.

The authors developed three ML models: NNs, partial least squares (PLS), and multivariate linear regression (MVLR), to predict DTC and MLR. Data was collected from different metrology techniques across the IC fabrication flow, including post-Litho, post-RIE, and post-CMP processes; model-based infrared reflectometry measured trench geometry, including CDs at the top and bottom and trench depth. These attributes were used as input for predicting DTC. Additional CD measurements were obtained using atomic force microscopy and SEM.

For MLR prediction, raw reflectance spectra from scatterometry tools were analyzed to capture fine details of metal line geometry, dimensions, and material properties. Because these spectra are highly sensitive to structural variations, they provided robust data for ML training. The dataset was divided into training (80%) and testing (20%) subsets, with inline measurements and electrical performance data (capacitance and resistance) serving as validation. This approach enabled the prediction of electrical properties during fabrication, eliminating the need to wait until the end of the process. Notably, the authors did not specify the hardware platforms or software packages used for ML model development.

By comparing with measurement data post-RIE, NNs predicted DTC more accurately than MVLR, achieving lower total measurement uncertainty (0.753 AU vs. 1.058 AU, where AU stands for arbitrary units and lower values are better). The performances of NNs were assessed for predicting MLR at post-Litho, post-RIE, and post-CMP. By comparing with the measured spectra data, NNs' predictions showed $R^2 = 0.802$ for post-Litho, $R^2 = 0.923$ for post-Etch, and $R^2 = 0.916$ for post-CMP. Clearly, NNs' post-RIE predictions correlate better with the measurement data as compared with those of post-Litho and post-CMP. Moreover, the performance of NNs for predicting MLR was compared with that of PLS at post-Etch with $R^2 = 0.973$ for PLS and $R^2 = 0.923$ for NNs. Both PLS and NNs showed strong predictive accuracy, while PLS slightly outperformed both NNs.

In general, NNs demonstrated superior performance in terms of accuracy, generalizability, and robustness to outliers, particularly for DTC predictions. PLS showed slightly better correlation for MLR, while MVLR functioned primarily as a baseline model.

By enabling early and accurate predictions of both electrical performance and geometrical parameters, ML-based predictive metrology reduces costs, improves yield, and strengthens process control. These findings underscore its value as a critical enabler for advanced-node IC manufacturing.

1.4.6 *Machine Learning for Predicting Capacitance-Voltage Characteristics of Metal-Oxide-Semiconductor Capacitors*

Pratik et al. developed a physics-assisted ML model to establish relationships between semiconductor processing parameters (e.g., oxidation and wafer cleaning procedures) and device characteristics, specifically capacitance-voltage (C-V) curve, in fabricating metal-oxide-semiconductor capacitors (MOSCAPs).[23] The model applied semi-supervised multitask learning (SSMTL) and TL and leveraged theoretical physics data to improve prediction accuracy and mitigate experimental data sparsity in IC nanofabrication.

MOSCAPs store and release charge to control carrier movement in ICs, enabling data storage, photonic sensing, and voltage reference applications. Here, MOSCAPs were used to examine how process parameters affect C-V characteristics, which are critical for assessing device performance and stability. Typical fabrication steps include wafer cleaning, oxide growth, metal deposition, lithography, and etching. Variations in temperature, chemical concentration, or deposition methods during these steps lead to differences in MOSCAP C-V characteristics.

Theoretical physics data refer to values derived from physics theories, such as Poisson's equation and C-V relationships, used to generate training data rather than directly enhancing ML models. MATLAB software was applied to simulate the behavior of charges, capacitance, and voltage under various conditions of MOSCAPs. Specifically, MATLAB simulation provided capacitance data across 1 to 100 kHz. On the experimental side, MOSCAPs were fabricated in an IC production line. The fabricated MOSCAPs were then characterized with an Agilent 4284A LCR meter under illumination, measuring C-V at 1 and 100 kHz. For implementation, TensorFlow 2.0 with Keras, Python 3.7.7, and libraries such as NumPy, SciPy, Pandas, Scikit-Learn, and Matplotlib supported model training, data analysis, and visualization.

The SSMTL model was trained with both theoretical and measured data. While experimental data (totaling 4,896 data points) existed only at 1 and 100 kHz, theoretical data (totaling 144,288 data points) spanned additional frequencies (1, 3, 5, 10, 50, 100 kHz). This enabled the model to share learned physics between data types and accurately predict intermediate frequencies (3, 5, 10, 50 kHz) where measurements were absent. The TL model followed a two-step approach: pretraining on theoretical data across all frequencies and then fine-tuning with measured data (1 and 100 kHz). Unlike SSMTL, TL separated the two datasets, transferring pretrained weights and biases to improve accuracy with limited experimental input. Specifically, the size of the training data was 124,320 for the SSMTL model, 120,240 for the TL model, and 4,080 for the baseline MLP model, while the corresponding testing samples were 9,792 for all three models.

Both models outperformed the MLP model, which was trained only on measured data. SSMTL and TL achieved R^2 scores of 0.9442 and 0.9253, respectively, compared to 0.6108 for MLP. The authors attributed this to theoretical data integration, though their suggestion that simulated data may surpass measurements remains debatable. A fairer test might train MLP with both measured and simulated datasets to assess if it could also exceed $R^2 > 0.9$.

Overall, the study shows that ML approaches such as SSMTL and TL effectively map fabrication processes to device characteristics, reduce reliance on extensive experimental data, and enhance prediction accuracy — offering promising solutions for semiconductor manufacturing.

1.4.7 *Machine Learning-Driven Prediction of Wafer Acceptance Test Outcomes*

Wei et al. developed an ML-based SMART-WAT model to predict wafer acceptance test (WAT) results in semiconductor manufacturing, particularly targeting accurate predictions for non-tested shots.[24] The SMART-WAT framework integrates Gaussian process regression (GPR) and automated machine learning (AutoML) to address challenges in sparse sampling and in aligning die-level chip probing data with shot-level WAT measurements. WAT involves selectively testing wafer samples to detect defects or variations that may affect overall product quality.

In the SMART-WAT model, GPR captures complex, nonlinear relationships through Bayesian inference, effectively modeling both intra-wafer and inter-wafer variations. It employs kernel functions to evaluate correlations between data points. AutoML, on the other hand, combines multiple algorithms including NNs, LightGBM, and XGBoost to preprocess chip probing data by eliminating missing values and irrelevant features, thereby improving prediction robustness and generalization performance.

The datasets used for training and testing SMART-WAT, GPR, and AutoML were derived from two real-world semiconductor manufacturing studies. The first study included 155 lots, each with six tested wafers and ten fixed shots, resulting in 9,300 WAT data points. Chip probing data contained 24 dies per shot and 1,270 parameters across various test conditions. For model training, the inner four shots from the first and last wafers were used, while the remaining shots supported validation and testing. The data were split into 124 lots for training, 15 for validation, and 16 for testing. The second study used seven fully tested wafers from three lots (420 data points) to evaluate the model's performance under practical conditions.

Model performance was assessed using RMSE and MAE across four WAT parameters: V1, LV1, LV2, and R1, representing three voltage parameters and one electrical resistance parameter. For RMSE, the metrics are: V1 (SMART-WAT: 0.2822, GPR: 0.4083, AutoML: 0.2844), LV1 (SMART-WAT: 0.0103, GPR: 0.0115, AutoML: 0.0127), LV2 (SMART-WAT: 0.00482, GPR: 0.00761, AutoML: 0.00492), and R1 (SMART-WAT: 3.6234, GPR: 6.3393, AutoML: 3.7034). For MAE, the metrics are: V1 (SMART-WAT: 0.2268, GPR: 0.3377, AutoML: 0.2284), LV1 (SMART-WAT: 0.00821, GPR: 0.00922, AutoML: 0.00934), LV2 (SMART-WAT: 0.00391, GPR: 0.00616, AutoML: 0.00399), and R1 (SMART-WAT: 2.8980, GPR: 5.3383, AutoML: 2.9616).

SMART-WAT demonstrated consistent performance gains. Compared to AutoML, its RMSEs were 0.77%, 18.90%, 2.03%, and 2.16% lower for V1, LV1, LV2, and R1, respectively. Relative to GPR, SMART-WAT achieved even greater reductions: 30.91%, 10.43%, 36.66%, and 42.79%, respectively.

Overall, SMART-WAT outperformed both GPR and AutoML by combining their complementary strengths into an ensemble model. The hybrid framework reduced prediction errors by up to 43%, achieving robust and accurate WAT predictions across all parameters. This work highlights a practical and scalable ML solution for improving quality control and yield prediction in semiconductor manufacturing.

1.5 Summary and Conclusions

This chapter reviewed ML applications across the semiconductor manufacturing spectrum: from monocrystalline silicon growth and wafer cleaning to CMP, ion implantation, oxidation, etching, mask fabrication, and photolithography, targeting improvements in detectability, precision, efficiency, yield, and reliability. More than 30 ML models were analyzed for process control, optimization, predictive metrology, and root cause analysis. Collectively, these studies demonstrated consistent gains in prediction accuracy, defect prevention, yield improvement, and process stability, although the extent of improvement varied by application, data quality, and algorithm design.

Despite these achievements, several persistent challenges limit large-scale industrial deployment. Many studies rely on simulated or narrowly scoped experimental datasets, restricting generalizability to high-volume manufacturing conditions. Integration across the fabrication pipeline remains fragmented, as most models address single process steps rather than end-to-end workflows. Moreover, large-scale training requires vast, high-quality data and substantial computational resources, while cross-facility adaptation introduces calibration and standardization challenges.

Linear models such as LASSO exhibit limited predictive capability for nonlinear, high-dimensional process behaviors, whereas nonlinear and deep learning models struggle with wafer-to-wafer variability, process drift, and complex interdependencies. Overfitting, underfitting, data imbalance, and high computational demands remain common concerns. In addition, the limited interpretability of many deep models constrains their acceptance in mission-critical manufacturing operations.

A substantial portion of the reviewed studies originated from industrial research and development groups or were co-authored with semiconductor companies. While these contributions provide valuable real-world relevance, several studies omit key methodological details such as model architectures, hyperparameter settings, and data generation procedures, because much of this information is proprietary. This lack of disclosure, though understandable from a competitive standpoint, limits reproducibility and the broader educational value of these studies.

Addressing these challenges requires coordinated efforts among semiconductor engineers, data scientists, and regulatory experts to develop robust, interpretable, and scalable ML frameworks that ensure both technical reliability and operational efficiency. Future research should emphasize large-scale validation, standardized data integration, and improved model transparency, while fostering mechanisms for secure data sharing between industry and academia. Key recommendations include:

1) *Enhancing Data Quality*: Semiconductor fabs generate vast but often noisy, incomplete, or biased data. Systematic collection, cleansing, and fusion — combined with imputation strategies for missing data — are critical for improving model reliability. Integrating multimodal datasets (sensor, image, and process data) can yield a more comprehensive representation of process dynamics.

2) *Improving Model Transparency*: Interpretability methods such as attention visualization, feature attribution, and explainable artificial intelligence frameworks can increase user confidence and accelerate adoption. Encouraging partial disclosure of model design principles by industry or collaborative sharing under confidentiality agreements can further enhance reproducibility.
3) *Developing Adaptive and Integrated ML Models*: Ensemble and hybrid architectures such as graph neural networks and physics-informed neural networks can balance accuracy, interpretability, and efficiency. Physics-informed neural networks are particularly valuable when experimental data are scarce or costly, as they embed physical process constraints directly into training.
4) *Promoting Real-Time and Distributed Learning*: Scalable architectures that incorporate edge and federated learning can enable real-time decision-making and continuous monitoring across distributed fabrication lines, improving stability and yield.
5) *Industrial and Academic Cooperation*: Because many methodological details and datasets in industrial studies are proprietary, cooperation between semiconductor companies and academic researchers is essential. Controlled data-sharing frameworks and joint development programs would enable benchmarking, reproducibility, and innovation while protecting commercial interests.
6) *Advancing Toward Closed-Loop Manufacturing*: The ultimate goal of ML integration lies in creating intelligent, closed-loop manufacturing ecosystems that autonomously optimize multiple interdependent processes. Realizing this vision requires

sustained interdisciplinary collaboration and stronger partnerships between industrial and academic research teams to promote transparency and trust.

Finally, this chapter serves as a scoping review that maps ML applications across individual semiconductor processes rather than synthesizing results for the entire production line.[25] Its findings establish a foundation for future meta-analyses and collaborative studies aimed at achieving full-factory ML integration through greater openness and joint industry–academia innovation.

Acknowledgements

The first author gratefully acknowledges financial support from Pacific Technology, LLC (Phoenix, USA), which enabled the author to spend time to prepare substantial portions of this manuscript. The authors also employed Gemini (Google, Mountain View, CA) to address technical questions arising from the review of related literature and to help refine the language of the manuscript. All artificial intelligence-assisted content was critically evaluated, verified, and edited by the authors, who take full responsibility for the accuracy, originality, and integrity of the final manuscript.

References

1. Chen, Ying-Lin, Sara Sacchi, Bappaditya Dey, Victor Blanco, Sandip Halder, Philippe Leray, and Stefan De Gendt. "Exploring machine learning for semiconductor process optimization: a systematic review." *IEEE Transactions on Artificial Intelligence* 5, no. 12 (2024): 5969–5989.
2. Kazazis, Dimitrios, Jara Garcia Santaclara, Jan van Schoot, Iacopo Mochi, and Yasin Ekinci. "Extreme ultraviolet lithography." *Nature Reviews Methods Primers* 4, no. 1 (2024): 84.

3. SIA. "Global Semiconductor Sales Increase 19.1% in 2024; Double-Digit Growth Projected in 2025." Semiconductor Industry Association, Washington DC, 2025. https://www.semiconductors.org/global-semiconductor-sales-increase-19-1-in-2024-double-digit-growth-projected-in-2025/. (Accessed June 20, 2025.)
4. Wang, W. H., Irina Brinster, Mohsen Maniat, Fatima Anis, Yen-Hui Lee, Sven Boese, C. F. Tseng et al. "Machine learning modeling using process context and exposure data for overlay prediction." In *Metrology, Inspection, and Process Control XXXVI*, vol. 12053, pp. 549–560. SPIE, 2022.
5. Ke, Guolin, Qi Meng, Thomas Finley, Taifeng Wang, Wei Chen, Weidong Ma, Qiwei Ye, and Tie-Yan Liu. "Lightgbm: A highly efficient gradient boosting decision tree." In *Advances in Neural Information Processing Systems 30*, ed. by I. Guyon et al., NeurIPS, 2017. https://dl.acm.org/doi/10.5555/3294996.3295074
6. Wu, Peng, Qi Zhu, Jihong Yang, Changjie Sun, Yiming Zhu, Abhishek Vikram, Ye Chen et al. "A machine learning-based solution for modulated image analysis in lithography process stability diagnosis." In *DTCO and Computational Patterning*, vol. 12052, pp. 271–278. SPIE, 2022. https://doi.org/10.1117/12.2613403.
7. Yoshida, Masahiro, W. H. Wang, C. H. Huang, Elvis Yang, T. H. Yang, K. C. Chen, Yosuke Takarada et al. "Exposure process optimization using machine learning overlay prediction." In *Metrology, Inspection, and Process Control XXXVI*, vol. 12053, pp. 516–520. SPIE, 2022.
8. Oak, Apoorva, Ryan Ryoung-Han Kim, and Soobin Hwang. "Quantifying process sensitivities for EUV and high-NA using machine learning based analytics." In *DTCO and Computational Patterning II*, vol. 12495, pp. 293–306. SPIE, 2023.
9. Raza, Shaheer, Mustafa Hammood, Nicolas AF Jaeger, and Lukas Chrostowski. "Fabrication-aware inverse design with shape optimization for photonic integrated circuits." *Optics Letters* 50, no. 1 (2025): 117–120.
10. Lau, Christina L., Shuhan Ding, Yutong Xie, Edwin R. Law, Bahar Kor, Benyamin Davaji, Amit Lal, and Peter C. Doerschuk. "Process-Aware Digital Twins by Deep Learning for DUV Photolithography and Plasma Etch." *IEEE Transactions on Semiconductor Manufacturing* 38, no. 3 (2025): 634–641.
11. Giollo, Manuel, Auguste Lam, Dimitra Gkorou, Xing Lan Liu, and Richard van Haren. "Machine learning for fab automated diagnostics." In *33rd European Mask and Lithography Conference*, vol. 10446, pp. 146–153. SPIE, 2017.
12. Yu, Hsiang-Meng, Chih-Chen Lin, Min-Hsuan Hsu, Yen-Ting Chen, Kuang-Wei Chen, Tuung Luoh, Ling-Wuu Yang, Tahone Yang, and Kuang-Chao

Chen. "CMP process optimization engineering by machine learning." *IEEE Transactions on Semiconductor Manufacturing* 34, no. 3 (2021): 280–285.

13. Lin, Yu-Ling, Qiangfu Zhao, and Shih-Cheng Horng. "Machine learning-based approach for automatic ion implanter monitoring." In *2022 International Automatic Control Conference (CACS)*, pp. 1–6. IEEE, 2022.
14. Takemura, Shinya, Shigeki Sakai, and Eiichi Murayama. "Beam shape control system by machine learning on the NISSIN BeyEX medium current ion implanter." *MRS Advances* 7, no. 36 (2022): 1463–1467.
15. Takemura, S., M. Imura, and S. Sakai. Machine learning-based beam shape controlling system on NISSIN medium current ion implanter. *MRS Advances* 10 (2025): 481–485. https://doi.org/10.1557/s43580-025-01179-x
16. Timoney, Padraig, Joseph Luke, Mark Rovereto, Jordan Wyble, Cheng-Ting Lien, Zahir Alamgir, and Eswar Ramanathan. "Enhanced Cu CMP process control by machine learning enabled measurement on e-test macro." In *2023 34th Annual SEMI Advanced Semiconductor Manufacturing Conference (ASMC)*, pp. 1–3. IEEE, 2023.
17. Mu, Lingxia, Ding Liu, Shihai Wu, Yuyu Liu, Peiyuan Gao, Han Liu, and Youmin Zhang. "A Condition Monitoring Method via a New Signal Expansion Strategy for the Crystal Lifting and Rotating Mechanism." *IEEE Transactions on Semiconductor Manufacturing,* Early access article (2025). https://doi.org/10.1109/tsm.2025.3619539
18. Wang, Shibing, Jing Su, Quan Zhang, Weichun Fong, Dezheng Sun, Stanislas Baron, Cuiping Zhang et al. "Machine learning assisted SRAF placement for full chip." In *Photomask Technology 2017*, vol. 10451, pp. 95–101. SPIE, 2017.
19. Utzny, Clemens. "Splendidly blended: a machine learning set up for CDU control." In *33rd European Mask and Lithography Conference*, vol. 10446, pp. 137–145. SPIE, 2017.
20. Lin, Albert, Tejender Rawat, Chung-Yuan Chang, Han-Chun Tung, Hsueh-Li Liu, and Peichen Yu. "Optical proximity correction using machine learning assisted human decision." *IEEE Photonics Journal* 15, no. 1 (2023): 8500209. https://doi.org/10.1109/JPHOT.2022.3231426
21. Habib, Mohamed SE, Hossam AH Fahmy, and Mohamed F. Abu-ElYazeed. "Novel End-to-End Production-Ready Machine Learning Flow for Nanolithography Modeling and Correction." arXiv preprint arXiv:2401.02536 (2024). https://doi.org/10.48550/arXiv.2401.0253
22. Rana, Narender, Yunlin Zhang, Donald Wall, and Bachir Dirahoui. "Predictive data analytics and machine learning enabling metrology and process

control for advanced node IC fabrication." In *2015 26th Annual SEMI Advanced Semiconductor Manufacturing Conference (ASMC)*, pp. 313–319. IEEE, 2015.

23. Pratik, Sparsh, Po-Ning Liu, Jun Ota, Yen-Liang Tu, Guan-Wen Lai, Ya-Wen Ho, Zheng-Kai Yang, Tejender Singh Rawat, and Albert S. Lin. "Mapping oxidation and wafer cleaning to device characteristics using physics-assisted machine learning." *ACS Omega* 7, no. 1 (2022): 933–946.
24. Wei, Ming-Chun, Chun-Wei Shen, and Hsun-Ping Hsieh. "Spatial Modeling with Automated Machine Learning and Gaussian Process Regression Techniques for Imputing Wafer Acceptance Test Data." In *2025 Design, Automation & Test in Europe Conference (DATE)*, pp. 1–7. IEEE, 2025.
25. Munn, Zachary, Micah DJ Peters, Cindy Stern, Catalin Tufanaru, Alexa McArthur, and Edoardo Aromataris. "Systematic review or scoping review? Guidance for authors when choosing between a systematic or scoping review approach." *BMC Medical Research Methodology* 18, no. 1 (2018): 143.

https://doi.org/10.1142/9789819830763_0002

CHAPTER 2

Control of Thin Film Deposition Process Using Machine Learning-Based Analysis of Scanning Electron Microscopy Measurements

Alex Kondi[1,2,*], Efi-Maria Papia[1,2], Vassilios Constantoudis[1]
[1]Institute of Nanoscience and Nanotechnology, N.C.S.R. Demokritos, Agia Paraskevi, 15310, Greece
[2]Department of Physics, University of Athens, 15784 Athens, Greece
*Corresponding author: a.kondi@inn.demokritos.gr

Abstract

This study presents a novel approach utilizing machine learning methods to control the thin film deposition process through accurate measurements from top-down grayscale scanning electron microscopy images. Traditional measurement techniques often struggle with the inherent complexity of rough substrates, particularly after etching processes in nanomanufacturing. In response, we propose a machine learning framework trained on synthetic grayscale scanning electron microscopy-like images, realistically simulating pre- and post-deposition surface conditions. The methodology encompasses comprehensive feature extraction and analysis, enabling accurate prediction of film thickness irrespective of surface irregularities. Extensive testing and validation on diverse datasets confirm the applicability of the proposed approach, achieving high predictive accuracy. This machine learning-based technique improves the precision, reliability, and efficiency of thin film deposition process control, offering promising opportunities for quality assurance and optimization in semiconductor manufacturing.

Keywords: thin film deposition, machine learning, convolutional neural networks, scanning electron microscopy, synthetic scanning electron microscopy images, rough substrates, conformal coatings, surface roughness, self-affine surfaces, Fourier transform filtering, power-spectral density, thickness prediction, nanometrology, semiconductor manufacturing, in-line process control, virtual metrology, domain adaptation

2.1 Introduction

Thin films are ubiquitous in semiconductor nanomanufacturing, forming critical layers in devices from transistors to solar cells. Ensuring these films have precise thickness across a wafer is vital for device performance and yield.[1] Even nanometer-scale deviations can alter electrical properties or cause failures, so metrology techniques that accurately measure thin film thickness are a cornerstone of process control. Traditionally, optical methods like spectroscopic ellipsometry and reflectometry are widely used for thin film thickness measurement. Complementing these optical techniques, scanning electron microscopy (SEM) coupled with computational analysis has matured into a quantitative metrology platform capable of nanometer-level dimensional control in high-volume manufacturing.[2,3] Model-based SEM analyses already underpin linewidth and critical dimension monitoring in production fabs.[4,5]

However, they often assume smooth, uniform surfaces and can struggle when films are deposited on rough substrates. In advanced fabrication, surfaces may be intentionally or inherently rough to improve functionality. This introduces new complexity in measuring film thickness. Accurate thickness metrology on rough surfaces is not just a scientific challenge but a practical necessity: it enables tighter process tolerances, which in turn improves device consistency and manufacturing yield. Recent work underscores the importance of advancing metrology for such scenarios. For instance, a novel Raman spectroscopy technique was demonstrated to profile thin film thickness on rough silicon, achieving sub-nanometer precision.[6] Their approach highlights both the significance of the problem and the opportunity for innovation. Decades of digital image processing and stochastic geometry studies show that rich statistical information is encoded in SEM contrast,[7-11] making the images ideal input for data-driven algorithms. To create physically faithful synthetic SEM micrographs, Monte Carlo electron scattering simulators can be used.[12-15] Moreover, deep learning and machine learning (ML) methods have already automated nanoparticle sizing, defect inspection, and coating thickness estimation in SEM images with excellent speed and accuracy,[16-27] proving the viability of artificial intelligence-assisted SEM metrology.

Accurate prediction of coating thickness has also been studied in other particulate manufacturing settings, especially in pharmaceutics. One study introduced a real-time machine vision and deep learning approach capable of simultaneously measuring film coating thickness and classifying tablet defects with high accuracy.[28] Another study developed a non-invasive imaging system based on Mask R-CNN, enabling reliable, real-time monitoring of film thickness in complex environments.[29]

At the same time, ML has emerged as a powerful tool in semiconductor manufacturing for data analysis and control, motivating its application to metrology problems that are difficult for conventional physics-based methods. Recent work has shown that deep neural networks can infer deposition thickness directly from top-down binarized SEM images, achieving nanometer accuracy.[30] Complementary convolutional pipelines simultaneously denoise and locate edges in low-dose or synthetically generated SEM micrographs, providing clean training targets for thickness prediction models.[31]

Self-affine roughness alters local film growth and therefore must be explicitly encoded in synthetic data sets used for ML training.[32-34] Power spectral density-based roughness synthesis techniques commonly employed in line edge roughness studies[34] can supply these realistic edge statistics, while emerging ML-enabled nanometrology pipelines couple image-based inference with physics-informed constraints to deliver real-time feedback for process control.[35,36] A recent review further frames these challenges, highlighting how resolution enhancement strategies and stochastic surface analysis are converging into advanced computational nanometrology workflows.[37]

In this chapter, we explore a novel ML-based approach that addresses the challenge of predicting thin film thickness on rough substrates using models trained on synthetic SEM images. We first discuss why rough substrates complicate deposition and measurement, then describe how synthetic SEM data can be generated and used to train ML models. We detail the ML framework and evaluate its performance against traditional techniques. Finally, we consider the broader implications for semiconductor manufacturing and future research directions in this emerging intersection of nanometrology and artificial intelligence.

As summarized in Figure 2.1, the vertical arrow traces Sections 2.2 through 2.7. Section 2.2 analyzes deposition and metrology challenges introduced by rough substrates and motivates a data-driven approach. Section 2.3 describes generation of synthetic SEM images that supply labeled training data. Section 2.4 presents the ML framework that ingests top-down SEM micrographs and infers coating thickness. Section 2.5 reports validation and performance evaluation against conventional techniques demonstrating nanometer-level accuracy. Sections 2.6 and 2.7 discuss implications for semiconductor manufacturing together with conclusions and future research directions.

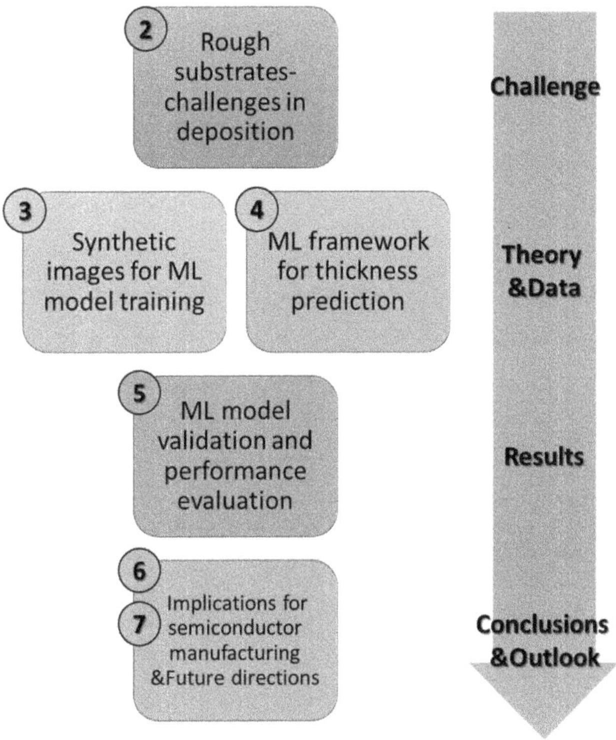

Figure 2.1. Schematic overview of the chapter and proposed workflow for ML prediction of thin film thickness on rough substrates.

2.2 Challenges in Thin Film Deposition on Rough Substrates

Deposition on a rough substrate surface inherently leads to non-uniform film growth. The substrate's topography can cause local variations in film thickness, as high points (peaks) and low points (valleys) receive different amounts of material during deposition. In physical vapor deposition processes, for example, surface asperities can shadow the largely line-of-sight incoming flux, resulting in thinner coverage in valleys.[38] Conversely, techniques like spin coating or atomic layer deposition might accumulate slightly thicker material in concave regions. The net effect is that a rough substrate often "dominates" the initial film morphology — the film tends to replicate the underlying surface roughness, at least for the first tens of nanometers.[39] Recent studies confirm that substrate roughness can markedly alter film microstructure. For instance, experiments with boron carbide coatings showed that using a very rough, bead-blasted substrate led to nodular defects in the film that would not appear on a polished substrate.[39,40] Similarly, TiO_2 films deposited on rough surfaces exhibit numerous surface nodules, with the density of these nodules increasing as substrate roughness increases.[41] These observations underscore that roughness at the substrate–film interface induces thickness variability and defectivity that are absent on smooth surfaces.

Surface roughness not only impacts how a film grows but also makes measuring its thickness far more challenging. Standard metrology tools often assume a well-defined, planar interface between film and substrate; roughness violates this assumption.[42] In optical metrology, rough surfaces scatter light, invalidating the specular reflection models that techniques like ellipsometry rely on. Specialized theoretical models (e.g., scalar diffraction theory or Rayleigh-Rice perturbation theory) are required to extract film parameters when roughness is on the order of the measurement wavelength. Ignoring roughness leads to systematic errors: the optical thickness extracted may be incorrect because the model conflates surface texture with film properties.

In practical terms, an ellipsometer might report a thinner or thicker film than actual if the substrate is rough, unless a roughness correction is applied. Beyond optical methods, other metrologies face difficulties too. X-ray reflectometry loses contrast in the interference fringes when interfaces are not smooth. Mechanical profilometry can record the combined surface profile but cannot easily distinguish how much of that profile is due to substrate vs. film. Likewise, cross-sectional SEM or transmission electron microscopy can directly image the film thickness on a rough substrate, but such techniques are destructive and low-throughput. A recent approach used spatially resolved Raman spectroscopy to map thin film thickness on rough substrates by measuring attenuation of the substrate's Raman signal. While remarkably precise, this method requires the substrate to be Raman-active and involves careful calibration.[6]

Clearly, there is a demand for metrology solutions that can cope with roughness in a more routine, non-destructive manner. The challenges include: (1) Thickness non-uniformity — the film thickness can vary significantly across small lateral distances, so any measurement is sensitive to where and how it is taken; (2) Surface scattering — techniques that rely on specular signals are less effective, as roughness causes diffuse scattering and signal loss;[42] and (3) Mixed material contrast — in electron microscopy, for example, a very thin film on a rough substrate might produce SEM contrast that is a mix of film and substrate, complicating direct interpretation. These issues motivate the exploration of ML-based approaches that can learn the complex relationship between observed signals (like SEM image intensities or textures) and actual film thickness on rough surfaces. By leveraging data-driven models, we might bypass the need for explicit analytical corrections and directly predict the film thickness even under conditions that foil conventional metrology. These realities motivate a shift to top-down SEM, where changes in the brightness and fractional coverage of surface "peaks" correlate monotonically with conformal film thickness. By training an ML model to learn this relationship, we can generate thickness maps without cross-sectioning, overcoming both throughput and statistical limitations while remaining fully non-destructive.

2.3 Synthetic Scanning Electron Microscopy Image Generation for Machine Learning Modeling

Gathering a large, labeled dataset of real SEM images for training a ML model is difficult in this domain. In semiconductor manufacturing, SEM images of thin films on rough substrates are available in limited numbers, and more importantly, it is challenging to label them with ground-truth film thickness. One would need to prepare many samples, each with known thickness (perhaps measured by a reference method like transmission electron microscopy or Raman mapping), and image them in the SEM under consistent conditions. This is time-consuming and costly. To overcome this data scarcity, our approach relies on synthetic grayscale SEM images. Synthetic data generation has proven effective in related nanoscience applications, where real data is scarce.[43]

To make these simulations realistic, we first reproduce the statistical signature of real substrate roughness and then emulate the conformal coating process digitally. This two-step workflow produces paired micrographs — one depicting the bare surface and the other the coated surface — whose geometric parameters, including the exact film thickness, are known by construction.

We render the bare substrates as self-affine Gaussian height maps using the Fourier transform filtering method.[44,45] The synthesis begins by specifying the root-mean-square roughness (rms), the isotropic correlation length ξ and the Hurst exponent α. These three numbers define the autocorrelation function:

$$ACF(r) = rms^2 e^{-\left(\frac{r}{\xi}\right)^{2a}} \quad (1)$$

with $r = \sqrt{(x^2 + y^2)}$, which fixes the required power spectrum. Zero-mean, unit-variance white noise is generated on an $N \times N$ grid, filtered with the square root spectrum in the frequency domain and transformed back to real space with an inverse fast Fourier transform. After linear rescaling to σ and quantisation to the 8-bit range 0–255, the resulting grayscale image faithfully reproduces the spatial texture of a real SEM view. Large ξ values yield broad, slowly varying mounds, whereas small ξ compresses correlations to only a few pixels and creates grainy textures. Lower α enhances short-wavelength modulations and makes the image appear more erratic, while $\alpha \approx 1$ suppresses local roughness and smooths the scene. Because all images are finally clipped to the same grayscale span, σ merely fixes the overall intensity range and plays a secondary visual role compared with ξ and α. Examples of SEM-like synthetic images of size 512×512 pixels generated with the aforementioned method are presented in Figure 2.2.

Conformal film growth is then modeled by grayscale morphological dilation. In this operation each pixel is replaced by the maximum value within a disk-shaped structuring element whose radius equals the nominal coating thickness d, expressed

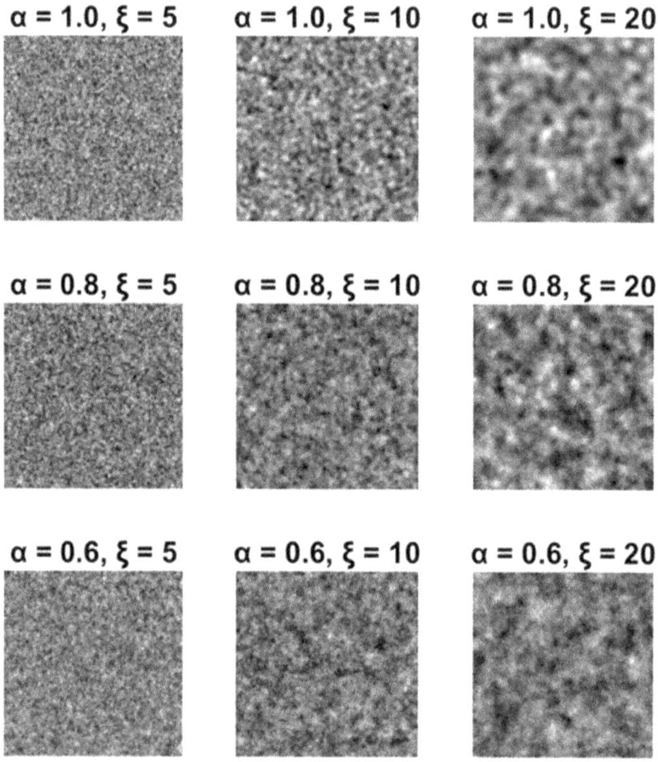

Figure 2.2. Typical examples of synthetic images of size 512 × 512 pixels generated from the Fourier filtering technique using various values of correlation length ξ and roughness exponent α. From left to right we show surfaces with ascending correlation length (=5, 10, 20 pixels) and three different roughness exponents (α = 1 top row, α = 0.8 middle row and α = 0.6 bottom row). As ξ grows, the surfaces evolve from noise-like to progressively smoother textures; lowering α amplifies high-frequency fluctuations and sharpens small-scale features.

in pixels. Valleys narrower than the disk are filled, ridges are uniformly raised, and the dilated map represents the new top surface of the coating. The difference between the dilated and original height fields provides a pixel-resolved thickness distribution, while the mean offset equals the prescribed d.

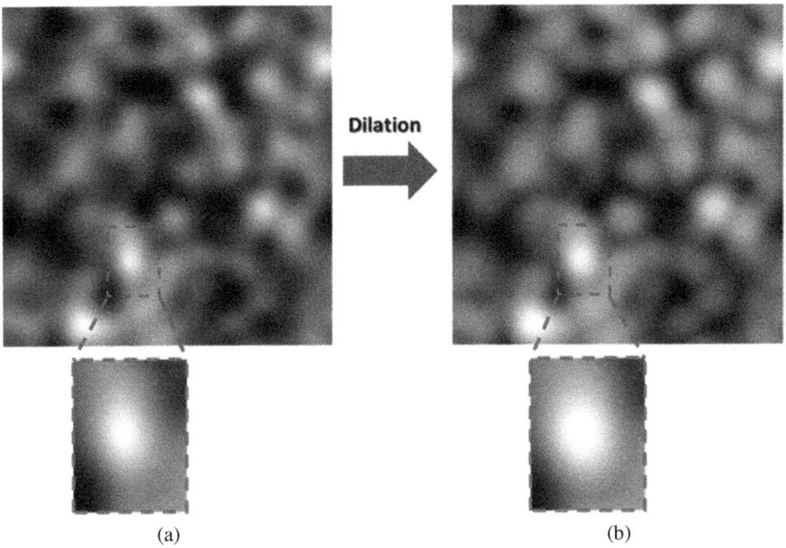

Figure 2.3. Schematic representation of the effect that the dilation operator has on a typical self-affine synthetic image generated with the Fourier filtering technique and of how deposition is simulated through use of this operator. (a) The full original self-affine synthetic image generated by the Fourier filtering technique along with the zoomed-in portion of the image before the dilation operator is applied; (b) the full self-affine synthetic image generated by the Fourier filtering technique along with the zoomed-in portion of the image after the dilation operator is applied.

Figure 2.3 illustrates the effect of this operation. Image (a) shows the full original self-affine synthetic image generated by Fourier filtering along with the zoomed-in portion of the image before the dilation operator is applied. Image (b) shows the full self-affine synthetic image along with the zoomed-in portion of the image after the dilation operator is applied. Together these views clarify how morphological dilation simulates conformal deposition and produces a pixelwise thickness map with mean equal to d.

2.4 Machine Learning Framework for Thickness Prediction

2.4.1 *Machine Learning in Scientific Research*

ML[46-48] has become a versatile tool across scientific disciplines, enabling automated analysis and prediction in ways previously impossible. In general, ML algorithms can interpret complex data patterns and make predictions without being explicitly programmed with physical laws or rules.[49,50] This makes them widely applicable in tasks such as image analysis, where ML models can classify or interpret images (e.g., identifying cells in a microscope image[51] or galaxies in telescope data[52]), signal processing for denoising or pattern recognition in time series data (such as seismic or sensor signals[53,54]), and even structural prediction problems.[55,56] A high-profile example is protein folding: deep learning models like AlphaFold[55] can predict 3D protein structures from amino acid sequences, a task that was historically very challenging. More broadly, ML's ability to approximate high-dimensional functions with impressive efficiency has opened new frontiers in scientific computing.[57] In other words, ML models can learn extremely complex, nonlinear relationships directly from data, which is invaluable when the underlying physics are not fully understood or are too complex to model directly. Unlike traditional equation-based approaches, ML can uncover patterns without requiring detailed physical understanding, focusing purely on the data correlations.[58]

2.4.2 *Impact of Machine Learning in Nanotechnology and Materials Science*

In the fields of nanotechnology and materials science, ML is driving a paradigm shift in how materials are characterized, designed, and discovered.[59] Thin film analysis is a prime example: supervised ML models trained on optical spectroscopy or imaging data can now infer thickness or composition in real time, eliminating the slow, iterative fitting that classical ellipsometry demands. Recent work shows that ML-enhanced spectroscopic ellipsometry[60] can achieve sub-nanometer accuracy for atomic layer deposition-grown TiO_2 films, while deep learning frameworks such as EllipsoNet[61] recover the complex refractive indices of multilayer coatings directly from reflectance spectra, even when the substrate is unknown. By learning the nonlinear mapping between spectral response and structure, these models deliver fast, non-destructive, and spatially resolved evaluations that go well beyond traditional point-based measurements.

ML has moved past thin film analysis to become a property-first engine for materials discovery. Graph-based and descriptor-based models now learn the mapping from atomic structure or chemical formula to target properties — whether an electronic bandgap, an elastic modulus, or a catalytic adsorption energy — without ever solving Schrödinger's equation.[62] Because inference is orders of magnitude faster than density functional theory, researchers can screen millions of hypothetical compounds *in silico*: random forest and gradient-boosting classifiers have already singled out promising high-T_c superconductors,[63] while interpretable ensemble models flag lithium-ion conductors with battery-grade conductivity before a single cell is built.[64] The same approach accurately predicts mechanical strength, such as Young's modulus in compositionally complex alloys[65] and catalytic turnover frequencies in heterogeneous catalysts.[66]

Another transformative impact of ML in materials science is the acceleration of simulations and modeling. Complex physics simulations (e.g., finite element analysis, molecular dynamics, or quantum calculations) can be very slow for large systems. ML offers a way to build surrogate models that approximate the results of these simulations much faster.

ML is rapidly accelerating physics-based simulations in materials science through surrogate modeling. Physics-informed neural networks can learn to solve the governing partial differential equations directly, yielding field solutions far faster than classical finite element solvers.[35,67] At the atomistic scale, deep potential molecular dynamics replaces expensive *ab initio* force evaluations, delivering quantum-accurate molecular dynamics trajectories for thousands of atoms at a fraction of the cost.[68] Similar ideas convert large finite element datasets into real-time predictors of stress or displacement, providing engineering "digital twins" that run instantly once trained.[69] In nanophotonics, a deep neural network trained on simulated scattering data maps metamaterial geometries directly to their optical spectra, effectively standing in for a full electromagnetic solver and collapsing compute time from hours to milliseconds.[70]

2.4.3 *Supervised Learning: Training Models via Loss Minimization*

Most of the successful ML applications in materials science use supervised learning, where the algorithm learns from examples of input data labeled with the correct output. Training proceeds by computing the gradient of a differentiable loss function with respect to every model parameter via back-propagation[71] and then updating those parameters (typically with stochastic gradient descent) to reduce the loss. In practical material workflows the data are partitioned into training and validation folds, and safeguards such as k-fold cross-validation, L_2/weight-decay regularization, and early stopping are applied so that the network captures underlying trends rather than memorizing the training set.[72-74]

2.4.4 *Convolutional Neural Networks for Image-Like Data*

Deep convolutional neural networks (CNNs) sit at the heart of modern image analytics. They extend the classic multi-layer perceptron with convolutional layers whose small, learnable kernels slide ("convolve") across an input grid, reusing the same weights at every spatial location. This weight sharing slashes the parameter count and imbues the network with translation invariance (the ability to recognize a feature no matter where it appears in the image).[75] After each convolution comes a pointwise nonlinearity (e.g., ReLU) and often a pooling layer that down-samples the feature map, further reducing computation while making the representation tolerant to small deformations. Stacking many such blocks yields a hierarchy of receptive fields: early layers detect edges and corners, mid-layers assemble textures or motifs, and deep layers capture whole objects or phase patterns. These stages were visualized directly, showing how CNNs build up progressively more abstract features.[76] The power of this architecture was first made unmistakable when "AlexNet" cut ImageNet error rates in half in 2012, igniting today's deep learning boom.[77]

CNNs are now routinely applied in several scientific disciplines. In materials science, they quantify phase fractions in welded steel micrographs with expert-level accuracy and predict stress or modulus fields of fiber-reinforced polymers directly from tomography images, reducing reliance on time-consuming finite element simulations.[78,79] In nanotechnology, a CNN trained on a large library of SEM images classifies nanoparticles, nanowires, thin films and 3D architectures with high accuracy, while also providing basic morphology metrics.[18] In biotechnology, CNNs match board-certified dermatologists in distinguishing malignant from benign skin lesions and deliver reliable cell masks for high-throughput fluorescence microscopy workflows, even when trained with weak annotations.[80,81] Together, these applications show that the core convolutional operations developed for general computer vision now underpin practical analysis pipelines across materials science, nanotechnology and biotechnology.

Building on these precedents, we apply a tailored CNN to a custom dataset of 5,000 pairs of simulated SEM-like images captured before and after the deposition step. The network's layer layout and hyper-parameters, detailed in Table 2.1, were selected to accommodate the image resolution and the paired-image training scheme used in this study.

As detailed in Table 2.1, the network ingests two stacked 512×512 SEM images (bare and coated) and processes them through four successive convolutional blocks. Each block applies a 3×3 convolution with ReLU activation and spatial downsampling (2×2 max pooling in the first three blocks, adaptive average pooling in the fourth), while expanding the feature map depth from 32 to 256 channels. A final 256-dimensional feature vector is then passed through a fully connected layer to produce 10 logits corresponding to discrete coating thickness classes.

The network resolves intensity variations indicative of nanometer-scale changes in coating thickness. Training was carried

Table 2.1. CNN architecture for classifying coating thickness into ten discrete classes from two-channel SEM inputs. Each block applies a 3×3 convolution followed by ReLU activation and spatial down-sampling (max pooling or adaptive average pooling), progressively increasing feature map depth ($32 \rightarrow 256$ channels) while reducing spatial dimensions ($512 \times 512 \rightarrow 1 \times 1$), before a final fully connected layer ($256 \rightarrow 10$ logits).

Stage	Operation	Output (Channels × Pixels)
Input	Two stacked SEM images (bare and coated)	$2 \times 512 \times 512$
Block 1	Conv $3 \times 3 \rightarrow 32$ + ReLU + MaxPool 2×2	$32 \times 256 \times 256$
Block 2	Conv $3 \times 3 \rightarrow 64$ + ReLU + MaxPool 2×2	$64 \times 128 \times 128$
Block 3	Conv $3 \times 3 \rightarrow 128$ + ReLU + MaxPool 2×2	$128 \times 64 \times 64$
Block 4	Conv $3 \times 3 \rightarrow 256$ + ReLU + Adaptive AvgPool 1×1	$256 \times 1 \times 1$
Head	Fully connected $256 \rightarrow 10$ (thickness classes)	10 logits

out for 200 epochs with the AdamW optimizer (initial learning rate 1×10^{-4}) under a cosine-annealing schedule and early stopping. Under these settings the model reaches more than 95% validation accuracy, with a mean absolute error at the nanometer scale on the synthetic test set. For deployment on real SEM data, a conservative approach may use lightweight domain adaptation and physics-informed regularization, for example session-wise intensity histogram matching and a monotonicity regularizer aimed at reducing intensity-thickness order violations.

2.5 Validation and Performance Evaluation

Assessing the reliability of any data-driven metrology tool requires a sober comparison with a physics-based reference. Here we use a reference that exploits the geometrical link between the bright footprints visible in top-down SEM images and the lateral growth of surface features after a conformal coating step.[82] For a solitary cylindrical pillar the thickness equals the difference between the pre- and post-deposition radii. For arbitrary shapes we appeal to the Frenet–Serret theorem: the extra area ΔA enclosed between two closed curves separated everywhere by the same offset d equals the mean of their perimeters times d. Solving for d gives a first-order estimator based solely on measurable pixel areas and perimeters. Real substrates, however, present many adjacent summits whose expanding footprints overlap as material accumulates; the lost arc length and area must therefore be restored analytically. By counting overlap events and inserting the corresponding corrections one arrives at a closed-form expression that remains valid even in crowded, highly corrugated landscapes. The equation is as follows:

$$Thickness_{corrected} = \frac{\Delta A + N_{overlaps} \times A_{overlap}}{(P_1 + P_2 + 2 \times N_{overlaps} \times ArcL_{overlap})/2} \tag{2}$$

where $N_{overlaps}$ is the number of overlaps, $A_{overlap}$ accounts for the mean area lost due to overlaps, and $ArcL_{overlap}$ accounts for the mean change in perimeter due to the overlapping regions. This formula ensures that the film thickness can be determined even when overlaps affect the area and perimeter measurements.

To apply the geometric formula, each SEM micrograph taken before and after deposition must be reduced to a binary mask. So, given a pair of SEM images one needs to find the optimal threshold to make the grayscale image binary and then calculate areas and perimeters of the objects in the image.

Determining the threshold that transforms a grayscale SEM image into a reliable binary mask is, however, far from trivial. The widely used global method of Otsu[83] assumes that the gray-level histogram is clearly bimodal and attempts to maximize the inter-class variance between foreground and background. In many SEM data sets this ideal condition is violated: topographic contrast, charging artefacts or detector inhomogeneity may broaden one peak or merge the two, such that a single global threshold either fragments the deposit into multiple islands or swallows parts of the substrate.

Even when the histogram exhibits two peaks, the gray-level populations can overlap substantially because of shot noise and edge blurring. Thin deposits, porous films and regions shadowed by relief features share intensities with the underlying substrate, making misclassification unavoidable if a single cut-off is enforced. The problem is aggravated in high-magnification images, where a moderate change in beam current or working distance can shift the entire intensity distribution.

ML methods sidestep the thresholding pitfalls altogether. Instead of imposing a hard foreground–background split, a CNN can be trained on paired grayscale SEM frames, one taken before deposition and one after, to learn the direct mapping from gray-level changes to deposition thickness. Once training is complete, the CNN is evaluated on an independent set of synthetically generated SEM-like images whose correlation lengths ξ and Hurst exponents α span a broad parameter grid that represents a wide spectrum of rough surface morphologies. The model's predictions are then benchmarked against the values obtained from the analytical equation; the resulting performance comparison is presented in Figures 2.4 and 2.5.

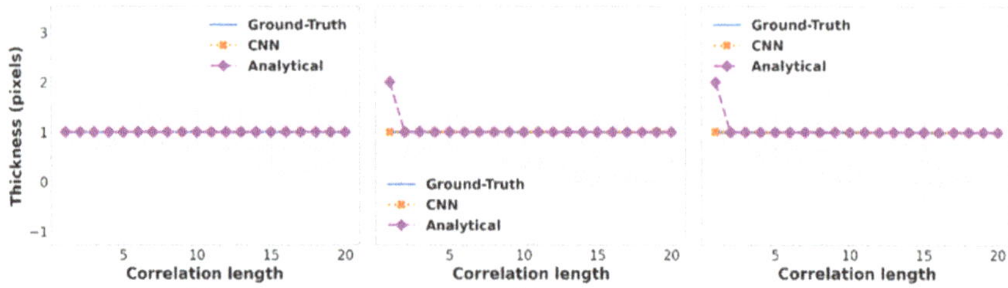

Figure 2.4. Predicted film thickness as a function of surface correlation length ξ for three roughness (Hurst) exponents, (a) $\alpha = 0.4$, (b) $\alpha = 0.6$ and (c) $\alpha = 0.8$, at a nominal deposition level of 1 pixel. The blue line marks the ground-truth thickness used to generate the synthetic SEM images. Orange star-dotted curves show the mean output of the CNN over 100 independently simulated surfaces; the associated standard deviations are smaller than the symbol size. Purple diamond curves represent the analytical model, with the shaded envelope indicating $\pm 1\ \sigma$ across the same ensemble. Across the entire ξ-α grid the CNN keeps its mean absolute error below 5% of the true thickness, whereas the analytical expression shows larger scatter and underperforms only at $\xi = 1$ pixel.

Figures 2.4 and 2.5 compare CNN predictions with the analytical model across the grid of correlation length ξ and Hurst exponent α at nominal deposition levels of 1 pixel and 7 pixels respectively. It should be noted that for a fair comparison, the predictions of the analytical model were rounded to the nearest integer. In both cases the orange CNN curves track the constant ground-truth thickness closely with mean absolute error below 5% for every combination of ξ and α. The analytical model (purple) exhibits broader scatter with performance deterioration concentrated at the smallest correlation lengths: limited to ξ = 1 pixel at the low deposition level and extending over a wider interval of small ξ when the nominal deposition is increased to

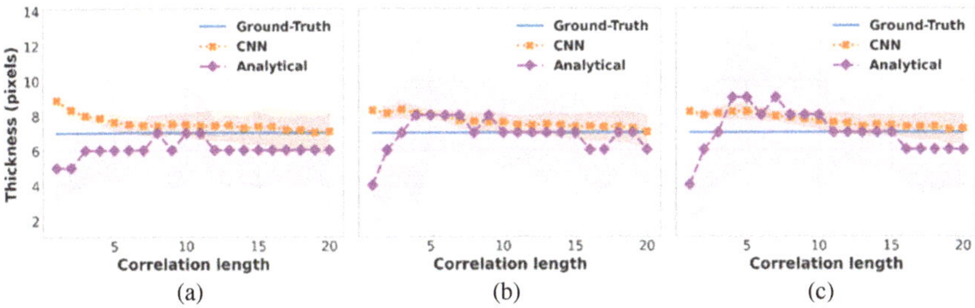

Figure 2.5. Predicted film thickness as a function of surface correlation length ξ for three roughness (Hurst) exponents, (a) $\alpha = 0.4$, (b) $\alpha = 0.6$ and (c) $\alpha = 0.8$, at a nominal deposition level of 7 pixels. The blue line marks the ground-truth thickness used to generate the synthetic SEM images. Orange star-dotted curves show the mean output of the CNN over 100 independently simulated surfaces; the associated standard deviations are smaller than the symbol size. Purple diamond curves represent the analytical model, with the shaded envelope indicating $\pm 1\ \sigma$ across the same ensemble. Across the entire ξ-α grid the CNN keeps its mean absolute error below 5% of the true thickness, whereas the analytical expression shows larger scatter and underperforms at smaller correlation lengths.

7 pixels. These trends indicate that the learned mapping generalizes robustly to diverse rough surface morphologies and maintains accuracy even as thickness increases, whereas the analytical expression becomes increasingly sensitive to short-range roughness.

In summary, the analytical expression yields good thickness estimates across much of the parameter space, especially at moderate and large correlation lengths, but its accuracy declines for short-range roughness and thicker deposits. The CNN offers similar or better accuracy where the analytical method performs well and shows improvement at small correlation lengths and higher thickness, making it a practical tool for subsequent analyses.

2.6 Implications for Semiconductor Manufacturing

The development of an accurate, image-based thickness prediction model has significant implications for semiconductor manufacturing, where control over thin film deposition is directly tied to device performance and yield. In the context of semiconductor processes, this requires training a new ML model using images of patterned wafers captured both before and after deposition. Semiconductor fabs typically have several inspection steps using CD-SEM (critical dimension SEM) or optical tools to monitor patterns and film properties. By deploying the ML model on an inspection SEM tool, manufacturers could measure film thickness on product wafers or test structures without extra processing. For example, after a deposition step on a structured wafer, a few sites could be imaged in the SEM and the thickness quickly inferred by the model. This could augment slower or less accurate techniques such as cross-section SEM (which requires cleaving or focused ion beam milling the sample) or ellipsometry (which might not be reliable on a rough surface). Because the model's inference is essentially instantaneous once the image is acquired, this approach aligns with the push toward high-throughput metrology in modern fabs and may provide a pathway for the SEM to complement its qualitative inspection role with quantitative thin film measurement capabilities.[84,85]

Integrating this capability into process control opens new pathways for feedback and feed-forward regulation. In semiconductor manufacturing, maintaining uniform film thickness within specifications is critical; if deviations are detected early, corrective actions can improve overall yields. With ML-based thickness metrology, one can envision a feedback loop in which, for instance, if the model predicts that a film on a monitor wafer is 5% thinner than the target, the deposition tool's parameters (time, precursor flow, etc.) are adjusted for subsequent wafers. Decision-based virtual metrology frameworks already demonstrate that coupling ML predictions to advanced process control can reduce excursion rates and cycle times[86] while machine learning virtual metrology studies on chemical vapor deposition thickness forecasting show practical accuracy on production lines.[87]

A recent review of generative ML process control strategies further underscores that data-driven sensors like ours are key enablers for the next generation of adaptive, run-to-run control in manufacturing.[88]

Another implication is multimodal process monitoring. Although the present model is trained solely for usage on SEM images, it can be fused with other sensors for more robust monitoring, for example, combining an optical spectrum or spectroscopic ellipsometry signal with the SEM-based ML prediction.

Using ML for thickness measurement on patterned substrates also contributes to yield improvement and defect reduction. Non-uniform or incorrect thickness can cause reliability issues (e.g., a too-thin diffusion barrier film on a rough trench). Because the model analyses images, it could be extended to flag other anomalies; deep learning ensembles have already achieved fast, high-accuracy defect classification directly on SEM images.[89] Thus, the same inference step that provides thickness data could simultaneously highlight delamination, particle defects, or voids.

From an operational perspective, adopting ML-based metrology requires robust model maintenance. Just as traditional metrology tools are periodically calibrated, the ML model's predictions must be validated against ground truth. Periodic cross-section SEM or ellipsometry on calibration wafers will detect model drift. Image statistics-based approaches to estimating section thickness, such as Gaussian process regression on electron microscopy data,[90] show how statistical checks can be embedded in an online monitoring loop, enabling auto-retraining when the process or materials change. Over time, continual learning on new wafers will expand the model's coverage and robustness.

In conclusion, ML-driven thickness prediction for rough substrates promises faster metrology, tighter process control, and higher yield. By embedding intelligence into existing SEM infrastructure and coupling it with complementary sensors, fabs move closer to autonomous, adaptive manufacturing lines capable of real-time quality assurance even on complex, textured surfaces.

2.7 Conclusion and Future Outlooks

The present study demonstrates that ML enhances existing efforts to transform qualitative top-down SEM into a quantitative technique capable of detecting nanometer-scale variations in film thickness even when the underlying substrate exhibits high roughness. By training a deliberately compact CNN on a large dataset of synthetic SEM micrographs, we were able to reproduce the statistical variability of real rough surfaces while retaining full knowledge of the ground-truth thickness used to generate each image pair. The resulting model learned to map subtle gray-level changes directly to absolute thickness values with a mean absolute error at the nanometer scale, a level of precision that rivals or surpasses many established optical methods yet requires no destructive cross-sectioning, no time-consuming sample preparation and no restrictive assumptions about interface planarity.

In practical terms, any SEM equipped with sufficient computational horsepower can be converted into a metrology station, allowing process engineers to measure conformal film thickness at dedicated test structures. Such a capability promises earlier excursion detection, tighter run-to-run control and, ultimately, higher yield in advanced semiconductor manufacturing where even slight deviations in layer uniformity propagate through multi-step process flows.

It is nevertheless important to recognize the limits of the present framework. The network is only as reliable as the domain it has experienced, and the synthetic data, although carefully parameterized, cannot capture every subtlety of real fabs: charging artefacts, drift in detector gain, changes in secondary electron yield caused by evolving chemistries, or multi-material interfaces that alter contrast in ways the simulator did not anticipate. Periodic calibration against traceable standards therefore remains essential. Cross-section measurements on a small number of witness wafers can act as anchor points, alerting engineers when the model begins to drift. In addition, subjecting the network to constant, real-world images invites the danger of concept drift, calling for a robust strategy that monitors prediction confidence, identifies out-of-distribution inputs and triggers retraining or fine-tuning only when warranted.

Several directions emerge naturally from these observations. One promising avenue is domain adaptation, where the synthetic training phase is followed by targeted fine-tuning on a handful of labeled real images gathered under fab-specific conditions. Techniques such as adversarial feature alignment or self-distillation can further close the gap between simulated and experimental domains, ensuring that the feature representations learned *in silico* genuinely reflect the physics of on-tool imaging. A complementary strand of research lies in self-supervised representation learning. Semiconductor fabs already generate vast archives of SEM data that remain mostly unlabeled; by pretraining encoders on contrastive or masked prediction tasks that require no thickness labels, one can distil features that are robust to process drift and detector variability, after which only a modest number of labeled examples are needed to specialize the model for quantitative metrology.

In parallel, the fusion of multiple sensors offers a path toward still greater resilience and accuracy. SEM imagery encodes high-frequency topographic information, whereas optical reflectance or ellipsometry spectra provide depth-integrated responses that depend sensitively on both thickness and refractive index. Combining these orthogonal signals within a unified learning architecture, be it through late-stage feature concatenation or attention-based cross-modal fusion, can allow one modality to compensate when the other is compromised. Incorporating physics directly into the loss function further enhances trustworthiness: imposing differentiable constraints derived from deposition kinetics or electron scattering theory encourages the network to respect conservation laws and suppresses chemically implausible predictions, a capability that grows increasingly vital as device geometries and material stacks become more exotic. Model reliability can be maintained by flagging unfamiliar inputs, showing how uncertain each prediction is, checking accuracy monthly at different wafer sites using cross-section or ellipsometry, and updating the model only if results cross set limits.

Over the longer horizon, the maturation of edge-accelerated inference will permit real-time deployment of these models directly within the SEM column controller, eliminating the need to stream high-resolution imagery across the network and thereby safeguarding intellectual property while slashing latency. Once the lag between image acquisition and quantitative assessment collapses to milliseconds, closed-loop control strategies that couple neural inference to recipe parameters become feasible. As predictive analytics propagate up the control hierarchy, fabs move closer to the vision of fully autonomous manufacturing lines in which every lithographic, etch and deposition step is continuously tuned by data-native sensors.

Finally, the synthetic training paradigm demonstrated here is not limited to conformal film thickness. The same methodology can be adapted to predict step coverage ratios in high aspect ratio features, quantify lateral etch roughness after plasma exposure or even estimate interfacial diffusion depth in multilayer stacks. More broadly, combining generative surface models with simulation of other imaging modalities, such as atomic force microscopy, X-ray photoelectron spectroscopy, or focused ion beam tomography, opens a rich design space in which artificial intelligence augments, rather than replaces, physics to create a new generation of high-fidelity, high-throughput metrology tools. In that future landscape, every pixel captured on the factory floor might serve as quantitative data for real-time decision engines, shortening the feedback cycle of innovation and ensuring that the relentless drive toward smaller, faster and more energy-efficient devices remains economically viable.

References

1. Venkatesan, Shashank, Michael A. Cullinan, and Michael Baldea. "Recent Advances in Continuous Nanomanufacturing: Focus on Machine Learning

Driven Process Control." *Reviews in Chemical Engineering* 41, no. 4 (2025): 311–331.

2. Zhou, Weilie, Robert Apkarian, Zhong Lin Wang, and David Joy. "Fundamentals of Scanning Electron Microscopy (SEM)." In *Scanning Microscopy for Nanotechnology: Techniques and Applications*, edited by Weilie Zhou and Zhong Lin Wang, pp. 1–40. Springer, 2007.
3. Krishnan, Kannan M. *Principles of Materials Characterization and Metrology*. Oxford University Press, 2021.
4. Davidson, Mark P., and Andras E. Vladar. "Inverse Scattering Approach to SEM Linewidth Measurements." In *Metrology, Inspection, and Process Control for Microlithography XIII*, vol. 3677, pp. 640–649. SPIE, 1999.
5. Frase, Carl Georg, Dominic Gnieser, and Harald Bosse. "Model Based SEM for Dimensional Metrology Tasks in Semiconductor and Mask Industry." *Journal of Physics D: Applied Physics* 42, no. 18 (2009): 183001.
6. Ledinský, Martin, Bertrand Paviet Salomon, Aliaksei Vetushka, Jonas Geissbühler, Andrea Tomasi, Matthieu Despeisse, Stefaan De Wolf, Christophe Ballif, and Antonín Fejfar. "Profilometry of Thin Films on Rough Substrates by Raman Spectroscopy." *Scientific Reports* 6 (2016): 37859.
7. Gonzalez, Rafael C., Richard E. Woods, and Steven L. Eddins. *Digital Image Processing Using MATLAB*. 3rd ed. Gatesmark Publishing, 2020.
8. Illian, Janine, Antti Penttinen, Helga Stoyan, and Dietrich Stoyan. *Statistical Analysis and Modelling of Spatial Point Patterns*. John Wiley & Sons, 2008.
9. Papia, Efi Maria, Vassilios Constantoudis, Dimosthenis Ioannou, Angelos Zeniou, Youmin Hou, Prexa Shah, Michael Kappl, and Evangelos Gogolides. "Quantifying Pore Spatial Uniformity: Application on Membranes before and after Plasma Etching." *Micro and Nano Engineering* 24 (2024): 100278.
10. Mavrogonatos, Argyris, Efi Maria Papia, Panagiotis Dimitrakellis, and Vassilios Constantoudis. "Measuring the Randomness of Micro and Nanostructure Spatial Distributions: Effects of Scanning Electron Microscope Image Processing and Analysis." *Journal of Microscopy* 289, no. 1 (2023): 48–57.
11. Mohsenzadeh, Rasool, Behzad Hashemi Soudmand, Amirhossein Najafi, and Fawwaz Hazzazi. "Morphology Driven Nanofiller Size Measurement Integrated with Micromechanical Finite Element Analysis for Quantifying Interphase in Polymer Nanocomposites." *ACS Applied Materials & Interfaces* 16, no. 30 (2024): 39927–39941.
12. Joy, David C. *Monte Carlo Modeling for Electron Microscopy and Microanalysis*. Oxford University Press, 1995.

13. Li, H. M., and Z. J. Ding. "Monte Carlo Simulation of Secondary Electron and Backscattered Electron Images in Scanning Electron Microscopy for Specimens with Complex Geometric Structure." *Scanning* 27, no. 5 (2005): 254–267.

14. Li, Y. G., S. F. Mao, H. M. Li, S. M. Xiao, and Z. J. Ding. "Monte Carlo Simulation Study of Scanning Electron Microscopy Images of Rough Surfaces." *Journal of Applied Physics* 104, no. 6 (2008): 064901.

15. Zhang, P., H. Y. Wang, Y. G. Li, S. F. Mao, and Z. J. Ding. "Monte Carlo Simulation of Secondary Electron Images for Real Sample Structures in Scanning Electron Microscopy." *Scanning* 34, no. 3 (2012): 145–150.

16. Ge, Mengshu, F. Su, Z. Zhao, and Dong Su. "Deep Learning Analysis on Microscopic Imaging in Materials Science." *Materials Today Nano* 11 (2020): 100087.

17. López de la Rosa, Francisco, Rubén Sánchez Reolid, José L. Gómez Sirvent, Rafael Morales, and Antonio Fernández Caballero. "A Review on Machine and Deep Learning for Semiconductor Defect Classification in Scanning Electron Microscope Images." *Applied Sciences* 11, no. 20 (2021): 9508.

18. Modarres, Mohammad H., Rossella Aversa, Stefano Cozzini, Roberto Ciancio, Angelo Leto, and Giuseppe P. Brandino. "Neural Network for Nanoscience Scanning Electron Microscope Image Recognition." *Scientific Reports* 7, no. 1 (2017): 13282.

19. Azimi, Seyed Majid, Dominik Britz, Michael Engstler, Mario Fritz, and Frank Mücklich. "Advanced Steel Microstructural Classification by Deep Learning Methods." *Scientific Reports* 8, no. 1 (2018): 2128.

20. Bals, Jonas, and Matthias Epple. "Deep Learning for Automated Size and Shape Analysis of Nanoparticles in Scanning Electron Microscopy." *RSC Advances* 13, no. 5 (2023): 2795–2802.

21. Dahy, Ghada, Mona M. Soliman, Heba Alshater, Adam Slowik, and Aboul Ella Hassanien. "Optimized Deep Networks for the Classification of Nanoparticles in Scanning Electron Microscopy Imaging." *Computational Materials Science* 223 (2023): 112135.

22. Anagun, Yildiray, Sahin Isik, Murat Olgun, Okan Sezer, Zekiye Budak Basciftci, and Nazife Gozde Ayter Arpacioglu. "The Classification of Wheat Species Based on Deep Convolutional Neural Networks Using Scanning Electron Microscope (SEM) Imaging." *European Food Research and Technology* 249, no. 4 (2023): 1023–1034.

23. Kim, Hyojin, Jinkyu Han, and T. Yong Jin Han. "Machine Vision Driven Automatic Recognition of Particle Size and Morphology in SEM Images." *Nanoscale* 12, no. 37 (2020): 19461–19469.

24. Baderot, J., M. Grould, D. Misra, N. Clément, A. Hallal, S. Martinez, and J. Foucher. "Application of Deep Learning Based Techniques for Automatic Metrology on Scanning and Transmission Electron Microscopy Images." *Journal of Vacuum Science & Technology B* 40, no. 5 (2022): 054003.

25. Gong, Yunli, Binghua Cao, Hong Zhang, Fengshan Sun, and Mengbao Fan. "Terahertz Based Thickness Measurement of Thermal Barrier Coatings Using Hybrid Machine Learning." In *Nondestructive Testing and Evaluation*, pp. 1–17. 2023.

26. Papia, Efi Maria, Alex Kondi, and Vassilios Constantoudis. "Inverse Design of Hexagonal Moiré Materials: Machine Learning for Tunable Pore Properties." In *Proceedings of the 13th Hellenic Conference on Artificial Intelligence*, pp. 1–6. 2024.

27. Papia, Efi Maria, Alex Kondi, and Vassilios Constantoudis. "Use Case of Symbolic Regression in Moiré Patterns Using Simulated Data." In *Proceedings of the 13th Hellenic Conference on Artificial Intelligence*, pp. 1–3. 2024.

28. Ficzere, Máté, Lilla Alexandra Mészáros, Nikolett Kállai-Szabó, Andrea Kovács, István Antal, Zsombor Kristóf Nagy and Dorián László Galata. "Real Time Coating Thickness Measurement and Defect Recognition of Film Coated Tablets with Machine Vision and Deep Learning." *International Journal of Pharmaceutics* 623 (2022): 121957.

29. Zhong, Liang, Hengqiang Cheng, Lele Gao, Lian Li, Wenping Yin, Hui Wang, Qiyi Miao, Yunshi Zhang, Lei Nie and Hengchang Zang. "Real Time Film Thickness Monitoring in Complex Environments Using Deep Learning Based Visual Imaging." *Powder Technology* 456 (2025): 120795.

30. Kondi, Alex, Efi Maria Papia, and Vassilios Constantoudis. "Machine Learning Applications in Nanotechnology Manufacturing: From Etching Accuracy to Deposition Prediction." In *Proceedings of the 13th Hellenic Conference on Artificial Intelligence*, pp. 1–6. 2024.

31. Chaudhary, Narendra, and S. A. Savari. "Simultaneous Denoising and Edge Estimation from SEM Images Using Deep Convolutional Neural Networks." In *2019 30th Annual SEMI Advanced Semiconductor Manufacturing Conference*, pp. 1–6. IEEE, 2019.

32. Mandelbrot, Benoit B. "Fractal Geometry: What Is It, and What Does It Do?" *Proceedings of the Royal Society of London A* 423, no. 1864 (1989): 3–16.

33. Barabási, Albert László, and H. Eugene Stanley. *Fractal Concepts in Surface Growth*. Cambridge University Press, 1995.

34. Cutler, Charlotte, James W. Thackeray, Peter Trefonas, Dan B. Millward, Choong Bong Lee, and Chris A. Mack. "Pattern Roughness Analysis Using

Power Spectral Density: Application and Impact in Photoresist Formulation." *Journal of Micro/Nanopatterning, Materials, and Metrology* 20, no. 1 (2021): 010901.

35. Raissi, Maziar, Paris Perdikaris, and George Em Karniadakis. "Physics Informed Neural Networks: A Deep Learning Framework for Solving Forward and Inverse Problems Involving Nonlinear Partial Differential Equations." *Journal of Computational Physics* 378 (2019): 686–707.
36. Guo, Guangxin, Hailong You, Cong Li, Zhengguang Tang, and Ouwen Li. "A Physics Informed Automatic Neural Network Generation Framework for Emerging Device Modeling." *Micromachines* 14, no. 6 (2023): 1150.
37. Kondi, Alex, Efi Maria Papia, Eleni Stai, and Vassilios Constantoudis. "Computational Methods in Nanometrology: The Challenges of Resolution and Stochasticity." *Frontiers in Nanotechnology* 7 (2025): 1559523.
38. Mattox, D M. *Handbook of Physical Vapor Deposition (PVD) Processing*. Oxford, UK: William Andrew, 2010.
39. Gambardella, A., M. Berni, A. Russo, and M. Bianchi. "A Comparative Study of the Growth Dynamics of Zirconia Thin Films Deposited by Ionized Jet Deposition onto Different Substrates." *Surface and Coatings Technology* 337 (2018): 306–312.
40. Merlo, J. B., J. B. Forien, K. Kawasaki, S. Gonzalez, S. J. Shin, L. R. Sohngen, G. V. Taylor, Z. R. Rodriguez, and S. O. Kucheyev. "High Rate Sputter Deposition of Ultrathick Boron Carbide Coatings on Rolling Spherical Substrates." *Fusion Science and Technology* (2025): 1–8.
41. Harada, Toshinori, Hiroshi Murotani, Shigeharu Matumoto, and Hiromitu Honda. "Influence of Substrate Surface Roughness on Light Scattering of TiO_2 Optical Thin Films." *Chinese Optics Letters* 11, no. s1 (2013): S10303.
42. Ohlídal, Ivan, Jiří Vohánka, Jan Dvořák, Vilma Buršíková, and Petr Klapetek. "Determination of Optical and Structural Parameters of Thin Films with Differently Rough Boundaries." *Coatings* 14, no. 11 (2024): 1439.
43. Kharin, A. Yu. "Deep Learning for Scanning Electron Microscopy: Synthetic Data for the Nanoparticles Detection." *Ultramicroscopy* 219 (2020): 113125.
44. Yang, Guoqing, Baotong Li, Yang Wang, and Jun Hong. "Numerical Simulation of 3D Rough Surfaces and Analysis of Interfacial Contact Characteristics." *Computer Modeling in Engineering & Sciences* 103, no. 4 (2014): 251–279.
45. Wang, Yuechang, Abdullah Azam, Mark C. T. Wilson, Anne Neville, and Ardian Morina. "Generating Fractal Rough Surfaces with the Spectral

Representation Method." *Proceedings of the Institution of Mechanical Engineers, Part J: Journal of Engineering Tribology* 235, no. 12 (2021): 2640–2653.

46. Pandey, Darpan, Kamal Niwaria, and Bharti Chourasia. "Machine learning algorithms: a review." *International Research Journal of Engineering and Technology (IRJET)* 6.2 (2019): 916–922.
47. Zhou, Zhi Hua. *Machine Learning*. Springer, 2021.
48. Jordan, Michael I., and Tom M. Mitchell. "Machine Learning: Trends, Perspectives, and Prospects." *Science* 349, no. 6245 (2015): 255–260.
49. Goodfellow, Ian, Yoshua Bengio, and Aaron Courville. *Deep Learning*. MIT Press, 2016.
50. LeCun, Yann, Yoshua Bengio, and Geoffrey Hinton. "Deep Learning." *Nature* 521, no. 7553 (2015): 436–444.
51. Ronneberger, Olaf, Philipp Fischer, and Thomas Brox. "U-Net: Convolutional Networks for Biomedical Image Segmentation." In *Medical Image Computing and Computer Assisted Intervention — MICCAI 2015: 18th International Conference*, pp. 234–241. Springer, 2015.
52. Dieleman, Sander, Kyle W. Willett, and Joni Dambre. "Rotation Invariant Convolutional Neural Networks for Galaxy Morphology Prediction." *Monthly Notices of the Royal Astronomical Society* 450, no. 2 (2015): 1441–1459.
53. Mousavi, S. Mostafa, and Gregory C. Beroza. "Machine Learning in Earthquake Seismology." *Annual Review of Earth and Planetary Sciences* 51, no. 1 (2023): 105–129.
54. Wang, Han, and Jie Zhang. "A Deep Learning Framework for Suppressing Prestack Seismic Random Noise without Noise Free Labels." *Energy Geoscience* 5, no. 3 (2024): 100297.
55. Jumper, J., Evans, R., Pritzel, A. et al. "Highly Accurate Protein Structure Prediction with AlphaFold." *Nature* 596, no. 7873 (2021): 583–589.
56. Baek, Minkyung et al. "Accurate Prediction of Protein Structures and Interactions Using a Three Track Neural Network." *Science* 373, no. 6557 (2021): 871–876.
57. E, Weinan. "Machine Learning and Computational Mathematics." arXiv preprint arXiv:2009.14596 (2020).
58. Chong, Sue Sin, Yi Sheng Ng, Hui-Qiong Wang, and Jin-Cheng Zheng. "Advances of Machine Learning in Materials Science: Ideas and Techniques." *Frontiers of Physics* 19, no. 1 (2024): 13501.

59. Pyzer-Knapp, E. O., J. W. Pitera, P. W. Staar, S. Takeda, T. Laino, D. P. Sanders, ... and A. Curioni. "Accelerating Materials Discovery Using Artificial Intelligence, High Performance Computing and Robotics." *npj Computational Materials* 8, no. 1 (2022): 84.

60. Arunachalam, Ayush, S. Novia Berriel, Parag Banerjee, Kanad Basu. "Machine Learning Enhanced Efficient Spectroscopic Ellipsometry Modeling." arXiv preprint arXiv:2201.04933 (2022).

61. Wang, Ziyang, Yuxuan Cosmi Lin, Kunyan Zhang, Wenjing Wu, Shengxi Huang. "EllipsoNet: Deep Learning Enabled Optical Ellipsometry for Complex Thin Films." arXiv preprint arXiv:2210.05630 (2022).

62. Xie, Tian, and Jeffrey C. Grossman. "Crystal Graph Convolutional Neural Networks for an Accurate and Interpretable Prediction of Material Properties." *Physical Review Letters* 120, no. 14 (2018): 145301.

63. Stanev, Vladan, Corey Oses, A. G. Kusne, Efrain Rodriguez, Johnpierre Paglione, Stefano Curtarolo, and Ichiro Takeuchi. "Machine Learning Modeling of Superconducting Critical Temperature." *npj Computational Materials* 4, no. 1 (2018): 29.

64. Honrao, Shreyas J. et al. "Discovery of Novel Li Solid State Electrolyte and Anode Coatings Using Interpretable Machine Learning and High Throughput Multi Property Screening." *Scientific Reports* 11, no. 1 (2021): 16484.

65. Khakurel, H., M. F. N. Taufique, A. Roy et al. "Machine Learning Assisted Prediction of the Young's Modulus of Compositionally Complex Alloys." *Scientific Reports* 11, no. 1 (2021): 17149.

66. Esterhuizen, Jacques A., Bryan R. Goldsmith, and Suljo Linic. "Interpretable Machine Learning for Knowledge Generation in Heterogeneous Catalysis." *Nature Catalysis* 5, no. 3 (2022): 175–184.

67. Karniadakis, George Em, Ioannis G. Kevrekidis, Lu Lu, Paris Perdikaris, Sifan Wang, and Liu Yang. "Physics Informed Machine Learning." *Nature Reviews Physics* 3, no. 6 (2021): 422–440.

68. Zhang, Linfeng, Jiequn Han, Han Wang, Roberto Car, and Weinan E. "Deep Potential Molecular Dynamics: A Scalable Model with the Accuracy of Quantum Mechanics." *Physical Review Letters* 120, no. 14 (2018): 143001.

69. Vurtur Badarinath, Poojitha, Maria Chierichetti, and Fatemeh Davoudi Kakhki. "A Machine Learning Approach as a Surrogate for a Finite Element Analysis: Status of Research and Application to One Dimensional Systems." *Sensors* 21, no. 5 (2021): 1654.

70. Peurifoy, Jonathan, Yichen Shen, Li Jing, Yi Yang, Francisco Cano Rentería, John D. Joannopoulos, Max Tegmark, and Marin Soljačić. "Nanophotonic Particle Simulation and Inverse Design Using Artificial Neural Networks." *Science Advances* 4, no. 6 (2018): eaar4206.

71. LeCun, Yann. "A Theoretical Framework for Back-Propagation." In *Proceedings of the 1988 Connectionist Models Summer School,* D. S. Touretzky, G. Hinton, T. Sejnowski (eds.), pp. 21-28. Morgan Kaufmann, 1988.

72. Butler, Keith T., David W. Davies, Hugh Cartwright, Olexandr Isayev, and Aron Walsh. "Machine Learning for Molecular and Materials Science." *Nature* 559, no. 7715 (2018): 547–555.

73. Ghojogh, Benyamin, and Mark Crowley. "The Theory behind Overfitting, Cross Validation, Regularization, Bagging, and Boosting: Tutorial." arXiv preprint arXiv:1905.12787 (2019).

74. Prechelt, Lutz. "Early Stopping—but When?" In *Neural Networks: Tricks of the Trade,* G. Montavon, G. B. Orr, K. R. Müller (eds.), pp. 53–67. Springer, 2012.

75. LeCun, Yann, Léon Bottou, Yoshua Bengio, and Patrick Haffner. "Gradient Based Learning Applied to Document Recognition." *Proceedings of the IEEE* 86, no. 11 (1998): 2278–2324.

76. Zeiler, Matthew D., and Rob Fergus. "Visualizing and Understanding Convolutional Networks." In *Computer Vision — ECCV 2014: 13th European Conference,* pp. 818–833. Springer, 2014.

77. Krizhevsky, Alex, Ilya Sutskever, and Geoffrey E. Hinton. "ImageNet Classification with Deep Convolutional Neural Networks." *Advances in Neural Information Processing Systems* 25 (2012): 1097–1105.

78. Almeida, Cássio Danelon de, Thales Tozatto Filgueiras, Moisés Luiz Lagares, Jr., Bruno da Silva Macêdo, Camila Martins Saporetti, Matteo Bodini, and Leonardo Goliatt. "A CNN Based Method for Quantitative Assessment of Steel Microstructures in Welded Zones." *Fibers* 13, no. 5 (2025): 66.

79. Sun, Yixuan, Imad Hanhan, Michael D. Sangid, and Guang Lin. "Predicting Mechanical Properties from Microstructure Images in Fiber Reinforced Polymers Using Convolutional Neural Networks." *Journal of Composites Science* 8, no. 10 (2024): 387.

80. Esteva, Andre, Brett Kuprel, Roberto A. Novoa, Justin Ko, Susan M. Swetter, Helen M. Blau, and Sebastian Thrun. "Dermatologist Level Classification of Skin Cancer with Deep Neural Networks." *Nature* 542, no. 7639 (2017): 115–118.

81. Shrestha, Prem, Nicholas Kuang, and Ji Yu. "Efficient End to End Learning for Cell Segmentation with Machine Generated Weak Annotations." *Communications Biology* 6, no. 1 (2023): 232.

82. Kondi, Alex, Efi Maria Papia, Dimitris Nioras, Ioannis Syngelakis, Chrysa Aivalioti, Elias Aperathitis, and Evangelos Gogolides. "Measurement of Thickness of Thin Coatings on Rough Substrates via Computational Analysis of SEM Images." *Micro and Nano Engineering* 28 (2025): 100315.

83. Otsu, Nobuyuki. "A Threshold Selection Method from Gray Level Histograms." *IEEE Transactions on Systems, Man, and Cybernetics* 9, no. 1 (1979): 62–66.

84. Zhang, Youshan, Brian D. Davison, Vivien W. Talghader, Zhiyu Chen, Zhiyong Xiao, and Gary J. Kunkel. "Automatic Head Overcoat Thickness Measure with NASNet Large Decoder Net." In *Proceedings of the Future Technologies Conference*, pp. 159–176. Springer, 2021.

85. Dizon-Paradis, Olivia P., Nitin Varshney, M Tanjidur Rahman, Michael Strizich, Haoting Shen, and Navid Asadizanjani. "In Situ Thickness Measurement of Die Silicon Using Voltage Imaging for Hardware Assurance." arXiv preprint arXiv:2307.13118 (2023).

86. Chien, Chen Fu, Wei Tse Hung, Chin Wei Pan, and Tran Hong Van Nguyen. "Decision Based Virtual Metrology for Advanced Process Control to Empower Smart Production and an Empirical Study for Semiconductor Manufacturing." *Computers & Industrial Engineering* 169 (2022): 108245.

87. Xie, Yunsong, and Ryan Stearrett. "Machine Learning Based CVD Virtual Metrology in Mass Produced Semiconductor Process." arXiv preprint arXiv:2107.05071 (2021).

88. Lee, Suk Ki, and Hyunwoong Ko. "Generative Machine Learning in Adaptive Control of Dynamic Manufacturing Processes: A Review." arXiv preprint arXiv:2505.00210 (2025).

89. Dey, Bappaditya, Dipam Goswami, Sandip Halder, Kasem Khalil, Philippe Leray, and Magdy A. Bayoumi. "Deep Learning Based Defect Classification and Detection in SEM Images." In *Metrology, Inspection, and Process Control XXXVI*. SPIE, 2022.

90. Ambegoda, Thanuja D., Julien N. P. Martel, Jozef Adamcik, Matthew Cook, and Richard H. R. Hahnloser. "Estimation of Z Thickness and XY Anisotropy of Electron Microscopy Images Using Gaussian Processes." arXiv preprint arXiv:2002.00228 (2020).

https://doi.org/10.1142/9789819830763_0003

CHAPTER 3

Advances in Machine Learning for Electron Beam Nanofabrication

Ampere A. Tseng
Manufacturing Institute, Arizona State University, Tempe, Arizona 85287-6106, USA
ampere.tseng@asu.edu

Abstract

This chapter reviews 16 selected studies demonstrating the integration of machine learning into electron beam nanofabrication. The studies are organized into three domains: atomic maneuverability enhancement, metrological improvement, and process predictive modeling. Collectively, they show that machine learning has progressed from supporting data interpretation to actively guiding fabrication, enabling nanoscale precision, scalability, and adaptive process optimization. In atomic maneuverability, machine learning has enhanced precision, automation, and detectability in atomic manipulation, modification, and fabrication. In metrology, machine learning models have improved the accuracy of critical dimension measurement, line edge roughness detection, and proximity effect correction. In predictive modeling, machine learning-driven approaches have optimized dose control, defect detection, and structural predictability, yielding gains in throughput and reproducibility. The chapter concludes by outlining persistent challenges such as limited datasets, lack of transferability, and insufficient physical interpretability, and recommends future efforts emphasizing standardized datasets, physics-informed machine learning frameworks, real-time adaptive control, and digital twin architectures for intelligent and sustainable nanofabrication.

Keywords: atomic manipulation, critical dimension control, electron beam lithography, fogging effect correction, line edge roughness, machine learning, multibeam electron beam lithography, proximity effect correction

3.1 Introduction

Advances in nanofabrication continue to push the boundaries of miniaturization, precision, and functional complexity, enabling the creation of ever-smaller devices, novel quantum materials, and atomically precise architectures. Among various patterning and fabrication techniques, electron beam methods, such as electron beam lithography and scanning transmission electron microscopy, have emerged as uniquely powerful tools. Their ultra-fine spatial resolution, flexibility for direct writing, and capability for atomic-level manipulation and material modification make them indispensable for next-generation nanomanufacturing. However, the full potential of electron beam nanofabrication remains constrained by several persistent challenges: the complexity of multi-parameter processes, stochastic beam–matter interactions (including scattering and backscattering), limited throughput, and lengthy optimization cycles.

In recent years, the integration of machine learning (ML) into the electron beam fabrication workflow has opened new possibilities for overcoming these challenges. ML, broadly defined as data-driven algorithms that learn patterns, correlations, and predictive models from training data, has transitioned from a niche analytical technique to a core enabler of modern nanofabrication. In electron beam nanofabrication, ML facilitates solutions to previously intractable problems: predicting critical dimensions based on resist and beam parameters; correcting proximity and scattering effects; optimizing beam exposure and resist formulations; performing real-time image analysis for feedback control; and even enabling atom-by-atom manipulation and modification through *in situ* monitoring and ML-driven beam steering.

This chapter reviews 16 selected studies that exemplify the application of ML in electron beam nanofabrication. The studies were identified through comprehensive searches using Google Scholar, Engineering Village, and IEEE Xplore. Initial searches yielded more than 40 publications from the past decade; these were narrowed to 16 that focus directly on fabrication processes rather than solely on design, maintenance, inspection, or device testing. Scholarly criteria including citation impact, methodological rigor, and publication recency guided the final selection.

The reviewed works are grouped into three categories: (1) atomic maneuverability enhancement, (2) metrological improvements, and (3) process predictive modeling. The first category (four studies) includes ML-assisted investigations of atomic manipulation, modification, and fabrication. The second category (four studies) focuses on metrological enhancement in detecting carbon nanotubes, critical dimension variations, stitching errors and line roughness in both single-beam and multibeam electron beam lithography. The third (eight studies) deals with predictive modeling of dimension and property changes in fabrication processes, including the prediction of photoresist pattern dimensions, line edge roughness, device performance, fabrication variations, contact-hole accuracy, proximity effect correction, and fogging effect correction.

Ultimately, this chapter aims to clarify a rapidly evolving yet fragmented research landscape. Although many individual studies report significant advances, their findings are often isolated, making it difficult to discern overarching trends or methodological convergence. By consolidating these developments, this review maps the areas where ML has proven most impactful, such as lithography optimization, yield enhancement, precision improvement, and defect reduction, and identifies where opportunities remain underexplored.

3.2 Machine Learning-Aided Atomic Manipulation, Modification, and Fabrication Using Electron Beams

This section reviews the applications of ML for atomic maneuvers involving manipulation, modification, and fabrication using electron beams. The reviews of four selected studies illustrate how electron beams integrating ML can enhance precision, automation, and detectability of the subtle tasks in identifying and controlling atomic-scale entities such as individual atoms and quantum dots (QDs).

Atomic manipulation refers to the controlled relocation or modification of individual atoms or atomic clusters on a material surface using a focused electron beam. This technique typically employs scanning transmission electron microscopy (STEM), which enables sub-angstrom resolution imaging and beam positioning for atom-by-atom movements. In the first two studies, ML-integrated STEM was used to interpret manipulation data, identify atomic configurations, and facilitate deterministic manipulation along predefined trajectories, highlighting the potential of intelligent feedback control in atom-scale engineering.

In atomic modification presented by Boebinger et al.,[1] the focused electron beam not only assists in identifying and moving atoms but also in modifying atomic configurations by precisely controlling beam dosages to perform atom ejection, atomic reordering, defect formation, and nanopore generation using STEM.

In atomic device fabrication by Donges et al.,[2] an artificial neural network was trained to assist electron beam lithography by denoising 2D cathodoluminescence maps and identifying QD positions for integration with circular Bragg grating resonators.

Across all four studies, a unifying challenge lies in the inherent uncertainty and atomic scale size of the targeted entities, whether atoms or QDs. Accurate positional identification is thus the critical first step before executing specific manipulation, modification, or fabrication tasks.

For readers' convenience, the key parameters and findings of these studies are consolidated in Table 3.1, where each row summarizes the process context, ML models employed, data used for training and testing, experimental configurations, and performance outcomes. This comparative overview highlights the growing impact of ML in atomic-scale fabrication domains.

3.2.1 *Machine Learning-Enhanced Atomic Manipulation of Single Si Atoms in Graphene*

Maxim et al. developed a ML-enhanced STEM system for the deterministic manipulation of single silicon (Si) atoms along predefined trajectories within a graphene lattice.[3] By integrating two ML algorithms, deep fully convolutional neural networks (DFCNNs) and the Gaussian mixture model (GMM), the system analyzed changes in the Si atom's local environment during manipulation, providing statistical insights into possible defect configurations and enabling the investigation of symmetry-breaking phenomena and material evolution at the atomic scale.

Table 3.1. ML-assisted atom manipulation, modification, and fabrication.*

Section. Author (Year)[Ref.]	Subject and Process	ML Models, Algorithms, and Platforms	Training and Testing Data	Key Equipment	Key Findings for Model Performance
3.2.1. Maxim et al. (2021)[3]	STEM for atomic manipulation of Si atoms in graphene	DFCNNs, GMM, UC	3,000 simulated + experiment STEM images	Nion UltraSTEM 100, CVD	DFCNNs + GMM create bond maps in 1 s and perform atomic manipulation successfully
3.2.2. Roccapriore et al. (2022)[5]	Detection and classification in atomic manipulation by STEM	DCNNs, GMM, DELMs, ELIT, AtomAI	Simulated and experimental STEM images	Nion UltraSTEM 100	DCNNs perform better than human eye in detection and classification in atomic manipulation
3.2.3. Boebinger et al. (2024)[2]	Predicting doses for STEM atomic modification	DCNNs, ELIT, U-Net, abTEM, AtomAI	Simulated and 50 experiment STEM images	Nion UltraSTEM 100	DCNNs achieve atomic resolution at scan rates up to 100 fps and enable selective S-atom removal for forming 2D structures
3.2.4. Donges et al. (2022)[2]	Locating QDs in EBL of circular Bragg grating	ANN, Autoencoder, TensorFlow, Keras	26,880 CL emission submaps for training	JEOL JSM 840 SEM, CLS, EBL	Autoencoder improves QD detection by 69% and 60% yield in fabricating QD-based devices

*The representing words or phrases for the acronyms or abbreviations used in the above table are defined below in alphabetical order: 1) ANN = Artificial neural network, 2) CL = Cathodoluminescence, 3) CLS = Cathodoluminescence spectroscopy, 4) CVD = Chemical vapor deposition, 5) DCNNs = Deep convolutional neural networks, 6) DELMs = Deep ensemble learning models, 7) DFCNNs = Deep fully convolutional neural networks, 8) EBL = Electron beam lithography, 9) ELIT = Ensemble learning and iterative training, 10) fps = Frames per second, 11) GMM = Gaussian mixture model, 12) QD = Quantum dot, 13) STEM = Scanning transmission electron microscopy, and 14) UC = Unsupervised clustering.

Specifically, the DFCNN was employed to identify atomic positions in noisy experimental data. Using an encoder-decoder architecture, it processed raw STEM images to generate pixel-wise classification maps that indicated the probability of each pixel corresponding to an atom or background based on intensity and local neighborhood information. The GMM, combined with principal component analysis, was used to analyze local atomic neighborhoods and identify lattice distortions. The unsupervised GMM algorithm clustered graphene sublattices centered on atoms into distinct classes,[4] while principal component analysis examined atomic distortions and orthogonal displacement patterns within each class identified by GMM. This combined ML model demonstrated the potential of electron beam-based atom manipulation for precise, atom-by-atom fabrication.

The DFCNN was trained on approximately 3,000 simulated STEM images of graphene with varied configurations and impurities. These simulated images were generated using MultiSlice simulations to replicate realistic atomic structures and mimic the appearance of STEM images, with data augmentation applied to simulate noise, drift, and instrumental effects. Model weights were optimized using the Adam optimizer with a cross-entropy loss function, while dropout layers mitigated overfitting. Batch normalization was deliberately omitted, as it degraded performance when applied to real experimental data.

Experimental validation used high-resolution STEM video frames acquired with a Nion UltraSTEM 100 microscope, operating at a 60 pA beam current and an accelerating voltage of 60 kV, to image and manipulate Si atoms within the graphene lattice. The high-angle annular dark-field detector was employed

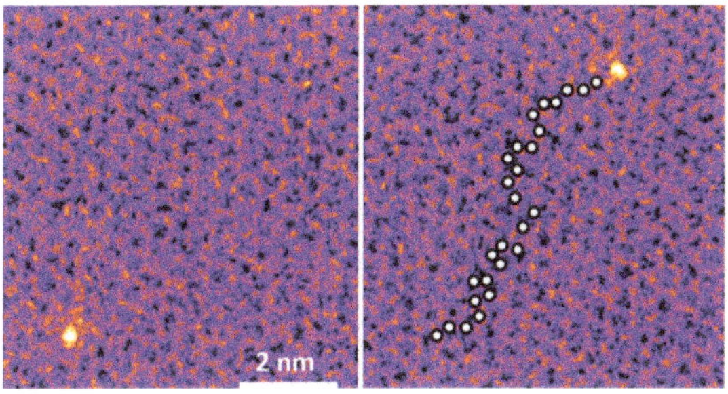

(a) (b)

Figure 3.1. Movement of a Si dopant linearly through the graphene lattice using an ML-enhanced STEM system: (a) initial configuration; (b) final configuration, with intermediate positions marked by dots. Noise was reduced using a Gaussian blur, and images were artificially colored using the 'Fire' lookup table. (Courtesy of Maxim et al.,[3] reprinted from open-access data.)

for imaging. Spurious Si dopants were linearly displaced through the lattice under electron beam control. Figure 3.1 illustrates the Si dopant movement, with images noise-filtered using a Gaussian blur and artificially colored with the 'Fire' lookup table.

Although specific quantitative performance metrics (e.g., accuracy or recall) were not reported, the combined ML model achieved rapid data analysis, generating bond maps from raw STEM data within approximately one second. The system demonstrated robust performance in identifying atomic positions and guiding manipulations even in noisy environments, enabling reliable, real-time monitoring of atomic displacements and lattice distortions. These results highlight the model's potential for efficient and precise atom-by-atom manipulation or fabrication.

3.2.2 *Machine Learning-Based Framework for Autonomous Atomic Manipulation in Scanning Transmission Electron Microscopy*

Roccapriore et al. implemented an ensemble learning and iterative training (ELIT) framework to perform real-time data analysis for guiding STEM in precise atomic manipulation and defect creation in materials like graphene and MoS_2.[5]

The ELIT framework integrated ensemble learning with three ML models: deep convolutional neural networks (DCNNs), deep ensemble learning models (DELMs), and a GMM to enhance predictive performance and experimental robustness. The DCNNs, based on a U-Net architecture, were employed to decode STEM images and extract atomic identities and coordinates, enabling real-time data stream analysis for atom-by-atom manipulation and fabrication. The DELMs consisted of multiple neural networks trained with different initializations and shuffled datasets, producing mean ensemble predictions with associated uncertainty estimates. This ensemble strategy improved robustness, helped identify out-of-distribution data, and ensured accurate predictions under varying experimental conditions. The GMM performed localized analysis at each atomic coordinate, classifying atoms and defects based on their bonding environments. Together, these models formed an automated, adaptive framework for atomic manipulation, defect creation, and closed-loop feedback during STEM experiments.

The training data consisted primarily of simulated STEM images with well-defined ground truths (atomic positions), while live experimental data were optionally incorporated during experiments for model refinement. To enhance generalization, a data augmentation pipeline introduced random zooms, rotations, contrast scaling, and noise to simulate variations between simulated and experimental data distributions. The DCNNs were trained and implemented via the AtomAI package,[6] a deep learning framework designed for STEM image and spectroscopy analysis.

Experimental validation used a Nion UltraSTEM 100, an aberration-corrected STEM system, for imaging and manipulation of graphene and MoS_2 samples. Accelerating voltages of 100 kV and 60 kV (with a 30 pA probe current) were applied for atomic imaging and beam-induced transformations. The Nion Swift interface, combined with Python scripting, enabled automated workflows directly on the microscope's control computer. The ELIT framework successfully facilitated automated atomic manipulation, such as creating single vacancy lines in MoS_2 and topological defects in graphene, while providing real-time feedback for beam control and model validation under varying imaging conditions.

Although the authors did not specify the number of simulated or experimental datasets, nor report explicit numerical performance metrics such as accuracy or error rates, they noted that the DCNNs achieved performance levels surpassing human visual recognition in atom detection and classification, even under low signal-to-noise conditions.

Overall, the ELIT framework demonstrated the integration of ML with STEM for autonomous and scalable atomic-level manipulation. Its combination of ensemble learning, uncertainty estimation, and iterative feedback represents an advancement toward real-time, intelligent control in electron beam nanofabrication.

3.2.3 *Real-Time Dose Prediction and Atomic Modification Using Machine Learning-Assisted Scanning Transmission Electron Microscopy*

Boebinger et al. employed an ELIT framework to investigate the electron beam dosages required for irregularity formation such as atom ejection, atomic reordering, defect formation, and nanopore generation in STEM experiments.[1] Combined with a custom beam control system, the ELIT framework enabled aberration-corrected STEM to precisely modify atomic structures in monolayer MoS_2 through controlled electron beam irradiation.

The ELIT framework integrated DCNNs for decoding low-quality, high-speed STEM datasets and classifying atomic positions and defect configurations. Using U-Net software for image segmentation, the DCNNs identified atomic sites (e.g., Mo, S_2) by decoding individual frames from rapid atomic-resolution STEM experiments. The framework directed the model's attention toward recognized features, quantified pixel-wise uncertainty, and adapted dynamically to changing imaging conditions, thereby improving classification accuracy and robustness.

Training data were generated from simulated crystal models of pristine MoS_2 containing random sulfur-site defects to mimic beam-induced fabrication. Simulated STEM images were created using the abTEM package[7] and augmented through cropping, rotations, noise, and blurring to replicate experimental conditions (Fig. 2.a in that paper). Additional software tools included AtomAI for deep learning microscopy analysis, STEMtool for refinement, and FaSTEM for dataset processing. These tools collectively enabled efficient decoding of high-speed, low-resolution STEM data, offering valuable insights into atomic reordering and defect evolution. The exact number of simulated training images was not specified.

Experimental data were collected using an aberration-corrected Nion UltraSTEM 100 microscope operated at 60 kV, equipped with a custom scanning feedback and beam control system for high temporal resolution imaging (up to 100 frames per second). Experimental data underwent preprocessing: resizing, cropping, denoising, and adding noise and blurring to match simulated conditions. Fifty experimental runs were performed, generating datasets for model testing and cross-validation within the ELIT framework.

The DCNN models trained under the ELIT workflow were iteratively compared until an optimal configuration was identified. Results demonstrated that the framework effectively decoded sparse datasets with resolutions as low as 64 × 64 pixels, achieving atomic resolution at scan rates up to 100 fps. The DCNNs accurately classified atomic sites with high certainty, even at 50 fps, maintaining a spatial resolution of approximately 1 Å. The study found that higher scan rates required increased electron doses to induce atomic disordering, defect formation, and nanopore generation, ranging from 1.9×10^6 e$^-$/Å^2 at 10 fps to 1.5×10^7 e$^-$/Å^2 at 33 fps (a 789% dose increase). Atomic tracking revealed consistent removal of 7 ± 9% Mo and 51 ± 9% S atoms prior to nanopore formation, which is crucial because this selective removal of S atoms aligns with the stoichiometric requirements for forming Mo_6S_6 nanowire edge structures after nanopore formation. This is essential for tailoring the electronic, optical, and magnetic properties of 2D materials like MoS_2. The ELIT framework further utilized mean ensemble predictions and uncertainty quantification to assess model confidence and detect anomalous behavior.

Overall, this study demonstrated that DCNN-assisted STEM enables accurate, high-speed tracking of atomic movements and transitions, offering deep insights into defect formation mechanisms. The ML-integrated approach significantly enhances real-time tracking, accelerates data acquisition, reduces electron dose requirements, and advances the precision and scalability of atomic-level analysis in 2D materials.

3.2.4 *Machine Learning-Enhanced In Situ Electron Beam Lithography for Deterministic Quantum Dot Integration*

Donges et al. demonstrated that ML-based image processing, specifically through an autoencoder, can significantly enhance *in situ* electron beam lithography (EBL) for quantum device fabrication.[2] The autoencoder-assisted EBL was used to deterministically integrate single InGaAs QDs into circular Bragg grating (CBG) resonators to form bright single-photon sources. A distributed Bragg reflector served as a back reflector to strengthen light confinement and complement the QD–CBG optical design.

The autoencoder, a type of artificial neural network, was trained to denoise 2D cathodoluminescence maps and identify QD positions for integration. Since QDs are inherently random in position and emission wavelength, accurate localization is essential for aligning them with nanophotonic structures such as CBG resonators. Denoising the cathodoluminescence emission maps improved the signal-to-noise ratio, enabling precise QD localization in noisy luminescence data. Such precision facilitates deterministic device fabrication, enhancing photon extraction efficiency, optical alignment, and overall yield.

The autoencoder was trained on cathodoluminescence maps with varying signal-to-noise ratios to optimize its ability to identify QD emission features in noisy images. A total of 26,880 submaps (28 × 28 pixels), augmented with spectral information, was used for training. The model was implemented and trained using TensorFlow and Keras.

The cathodoluminescence maps were acquired using a JEOL JSM 840 SEM (equipped with cathodoluminescence spectroscopy and an EBL pattern generator), a liquid-helium cryostat, a spectrometer (Si charge-coupled device and InGaAs detectors), and a micro-photoluminescence setup. Measurements were conducted at low temperatures (~10 K) on QDs embedded within a GaAs capping layer grown on an n-doped GaAs substrate. QD locations appeared as bright emission spots in the cathodoluminescence maps, and a 2D Gaussian fit was applied to determine their precise x–y coordinates. The resist CSAR 62 was used for spin coating prior to lithography.

Performance metrics demonstrated the advantages of ML-enhanced *in situ* EBL. Without ML assistance, the QD localization accuracy (fit deviation) degraded with decreasing beam dose, from 32.4 ± 6.9 nm at 8 mC/cm^2 to 129.7 ± 35.4 nm at 0.5 mC/cm^2. With ML, the deviation remained below the 50 nm accuracy target across the full dose range (0.5–8 mC/cm^2); at 0.5 mC/cm^2, it was only 40.2 ± 5.1 nm, a 69% accuracy improvement for locating QDs at low doses. This enhancement also increased photon extraction efficiency and device alignment precision: the autoencoder-assisted EBL achieved 35% photon extraction efficiency at 1.0 mC/cm², whereas conventional EBL required 8 mC/cm² to reach 35.6%. Moreover, the fabrication yield of optically active QD–CBG devices rose from 50% to nearly 80% at 0.5 mC/cm², a 60% improvement enabled by ML.

Collectively, autoencoder-enhanced *in situ* EBL markedly improves QD localization accuracy, preserves photon extraction efficiency at reduced beam doses, and significantly increases the yield of optically active devices. The autoencoder-assisted EBL technique therefore represents a key step toward scalable, high-precision fabrication of QD-based photonic devices.

3.3 Machine Learning-Aided Metrology Using Single-Beam and Multibeam Electron Beam Lithography

This section explores the emerging synergy between ML and EBL using single and multiple beams for advanced metrological applications. The four reviewed studies demonstrated how ML enhances precision, efficiency, and controllability in identifying and measuring nanoscale entities such as individual carbon nanotubes (CNTs) in single-beam EBL, and critical dimension (CD) variations, stitching error detections, and line roughness measurements in multibeam EBL.

Single-beam EBL uses one focused electron beam to write patterns serially, offering ultra-high resolution but low throughput. Multibeam EBL employs thousands of beams that expose resist in parallel, greatly increasing speed while maintaining high precision. Multibeam EBL has become a leading-edge industry tool used for making advanced photomasks, particularly for leading-edge semiconductor nodes (e.g., 7 nm, 5 nm,

3 nm and beyond), including those used in extreme ultraviolet lithography.

The third-generation production-scale multibeam EBLs, MBMW-301 (by IMS Nanofabrication) and MBM-4000 (by NuFlare Technology) reported by Tomandl et al. and Matsumoto et al., respectively, use 600,000 and 500,000 beamlets.[8,9] The MAPPER's third-generation FLX platform even uses 650,000 beamlets to achieve a 40-wafers-per-hour throughput. Precise operation of so many beamlets within a resolution down to 1 nm range could introduce complex challenges in beam calibration, data handling, stitching correction, edge detection, and process uniformity across all beamlets. In this section, the first three studies reviewed provide ML-based responses to these challenges.

In the fourth study, an ML-based recognition framework was implemented to accurately determine CNT tip positions for field emitter fabrication. Since the electrodes (cathodes) patterned via EBL must be precisely aligned within an 80 nm gap of the CNT tips (anodes) to achieve efficient quantum field emission, ML significantly enhanced detection accuracy and device yield.

Table 3.2 summarizes the review outcomes, providing an overview on the metrological applications of ML on these diverse EBL-based nanofabrication domains.

3.3.1 *Machine Learning-Assisted Monitoring of Critical Dimension Variations in Multibeam Electron Beam Lithography*

Figueiro et al. investigated an ML-based scatterometry for monitoring CD variations in multibeam EBL.[10] The goal was to enhance traditional model-based scatterometry by improving

Table 3.2. ML-aided metrology using single-beam and multibeam EBL.*

Section. Author (Year)[Ref.]	Subject and Process	ML models, Algorithms, and Platforms	Training and Testing Data	Key Equipment	Key Findings for Model Performance
3.3.1. Figueiro et al. (2018)[10]	CD variations, anomalies, and beam defects quantified by scatterometry in multibeam EBL	ML-based scatterometry (no algorithm details)	575 spectra data pairs for validating; no number of training data is reported	Multibeam EBL and CD-SEM	ML-based scatterometry achieved a 77% improvement (R^2 = 0.99) in TMU for CDV measurement and detected CD shift as small as 0.9 nm
3.3.2. Rademaker et al. (2018)[11]	Monitoring and measuring stitching errors in multibeam EBL	Unspecified advanced ML	324 measurement data sets	Mapper FLX-1200, Vistec SB3054 VSB	ML models achieved AC = 94–99% in detecting stitching errors with RMSE reduction by 70–80%
3.3.3. Chaudhary (2019)[12]	Measuring line edge roughness in multibeam EBL and denoising SEM images	CNN, DnCNN, EDGENet, LineNet1, LineNet2, SEMNet	133,920 simulated image pairs for training and 17,280 pairs for validating and testing	Tesla K80 GPU, Intel Xeon E5-2680, Intel i7-2600 CPU	SEMNet, EDGENet, and LineNet achieved AC > 91% in line edge roughness estimates and were 13% higher in peak signal-to-noise ratio

3.3.4. Tadokoro et al. (2023)[16]	Recognition of CNTs in EBL of electrodes and markers	Faster R-CNN, CNN, RPN, MSGD, GitHub gdsCAD, PyTorch	1,311 SEM images for training and 93 SEM images for testing	EBL, SEM, Nvidia GPU	Faster R-CNN was highly efficient (20,000 times faster) with 60% accuracy in detecting CNTs (within 150 nm)

*The representing words or phrases for the acronyms or abbreviations used in the above table are defined below in alphabetical order: 1) AC = Accuracy, 2) CD = Critical dimension, 3) CD-SEM = Critical dimension scanning electron microscopy, 4) CDV = Critical dimension variation, 5) CNN = Convolutional neural network, 6) CNT = Carbon nanotube, 7) DnCNN = Denoising convolutional neural network, 8) EBL = Electron beam lithography, 9) ML = Machine learning, 10) MSGD = Momentum stochastic gradient descent, 11) R^2 = Coefficient of determination for regressions, 12) R-CNN = Region-based convolutional neural network, 13) RMSE = Root mean square error, 14) RPN = Region proposal network, 15) SE = Stitching error, 16) SEM = Scanning electron microscopy, and 17) TMU = Total measurement uncertainty.

correlation with reference measurements, defect detection, and sensitivity to subtle lithographic anomalies. Scatterometry, a metrology technique for characterizing periodic and non-periodic structures, was augmented by ML to achieve higher accuracy, greater robustness against the variations of secondary parameter (resist morphology or baseline dose), and faster analysis of measurement data. Multibeam EBL, also known as maskless lithography, employs multiple electron beams to directly write patterns without masks. Its high resolution makes it a promising technology for advanced semiconductor manufacturing.

Although the authors did not specify the ML algorithms used, they described the training process: scatterometry spectra were paired with reference CD values measured by critical dimension scanning electron microscopy (CD-SEM). The model learned to predict CD values from the input spectra, enabling it to detect beam-to-beam dose variations and lithographic defects. Here, CD refers to the width of the smallest feature in an EBL-patterned structure.

Experiments were performed on two wafers. The first evaluated ML-enhanced scatterometry under dose-induced CD variations, while the second targeted smaller CD shifts using an updated lithography process. The wafer structure comprised a patterned electron beam resist on an anti-reflective coating, a spin-on carbon hard mask, and a silicon substrate (Figure 3.2). Total measurement uncertainty (TMU) and coefficient of determination (R^2) were used to assess metrology performance, with CD-SEM as the reference. For correlation analysis, 197 data pairs were used for the first wafer and 378 for the second.

Performance results were reported by comparing scatterometry measurements of CD variations with CD-SEM references.

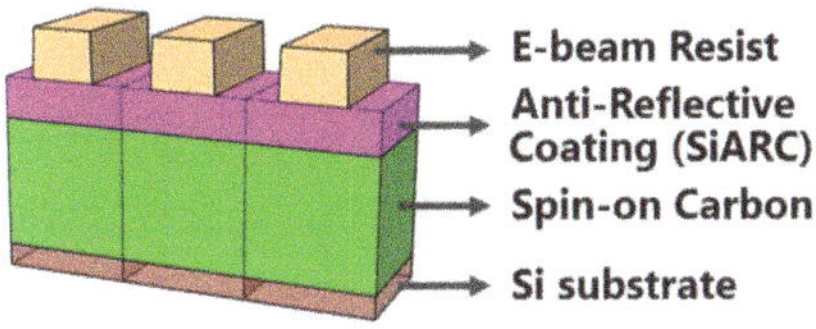

Figure 3.2. Film stack used in multiple EBL wafer structures: patterned electron beam resist atop an anti-reflective coating, spin-on carbon hard mask, and silicon substrate.

For the first wafer, the ML-based scatterometry achieved a TMU of 0.95 nm and an R^2 of 0.94, substantially outperforming model-based scatterometry (TMU = 3.65 nm, $R^2 = 0.44$), a 74% improvement in TMU and 214% improvement in R^2 comparisons. For the second wafer, ML-based scatterometry achieved TMU = 0.84 nm (77% improvement) and $R^2 = 0.99$, again superior to the model-based method (TMU = 2.3 nm, $R^2 = 0.96$), a 3% improvement.

The ML model also demonstrated higher robustness to physical anomalies and better correlation to CD-SEM across the wafer. Notably, it also detected beam defects with high sensitivity, identifying a single non-process-of-record stripe with a CD shift as small as 0.9 nm, with TMU = 0.92 at a 99.7% confidence level.

Overall, ML-based scatterometry effectively complements model-based methods for monitoring multibeam EBL. It exhibited superior correlation with CD-SEM, good sensitivity to resist and beam-related anomalies, and high precision in detecting sub-nanometer beam defects. These results underscore ML's value as an advanced metrology tool for next-generation lithography. For reference, the second row of Table 3.2 summarizes the major findings and key data discussed in this section.

3.3.2 ML-Assisted Optical Critical Dimension Metrology for Stitching Error Characterization in Multibeam Electron Beam Lithography

Rademaker et al. developed ML-assisted optical critical dimension (OCD) techniques for monitoring and measuring stitching errors in multibeam EBL.[11] Their approach incorporated predictive analytics to interpret scatterometry data and forecast stitching errors based on experimental measurements obtained using the Mapper FLX-1200 tool. The study aimed to quantify the sensitivity of OCD metrology to various types of stitching errors, with particular emphasis on displacement errors (DX, DY) and diagonal stitching errors. OCD, a non-destructive metrology technique, applies light reflection and diffraction to measure the dimensions and profiles of features on a wafer by comparing measured optical spectra to simulated models.

The authors reported that advanced ML algorithms were used to analyze scatterometry data and enhance the detection of stitching errors, although specific algorithmic details were not disclosed. The ML framework processed spectra collected from both standard and novel optical channels in the scatterometry measurements. Using labeled data with known stitching parameters, the models learned to correlate spectral features with corresponding stitching error types — DX, DY, and diagonal errors — allowing prediction of the type, number of erroneous stripes, and error magnitude directly from the scatterometry signals.

To train and validate the ML models, spectroscopic reflectometry measurements were collected from OCD targets fabricated on wafers specifically designed to simulate typical stitching errors in the Mapper FLX-1200 system. Each collected spectrum was labeled with information regarding the error type, number of beams, and error magnitude. The dataset, comprising 324 samples, was divided into a training set for model calibration and a validation set for performance evaluation. Consequently, the ML models used labeled scatterometry spectra as input and produced corresponding stitching error predictions (DX, DY, and diagonal) as output.

The ML models demonstrated varying performance across stitching error types. For the DX displacement targets, the linear fit yielded a slope of 0.94 ± 0.01 (1σ) and an intercept of 2.2 ± 0.7 a.u., with a normalized RMSE (root mean square error) of 3.9 a.u., which decreased to 2.38, 1.13, and 0.70 a.u. for 1, 3, and 6 erroneous stripes, respectively. For the DY displacement targets, the slope was 0.981 ± 0.003 (1σ) and the intercept 0.7 ± 0.2 a.u., with a normalized RMSE of 1.0, decreasing to 0.73, 0.25, and 0.16 a.u. for 1, 3, and 6 erroneous stripes, respectively. Because a perfect prediction corresponds to a slope of 1.0 and an intercept of 0.0, the models achieved approximately 94–99% predictive accuracy in detecting stitching errors with the best results observed for horizontal gratings. For the diagonal targets, the RMSE decreased by up to 70–80% as the number of erroneous stripes with stitching errors increased, indicating significant improvement in accuracy and generalization with larger datasets.

The authors also identified opportunities for improvement, such as enhancing the model's sensitivity to DX errors and stripe magnification (MX) targets, as well as investigating potential cross-correlations among different error types. Overall, the study effectively demonstrated the potential of combining scatterometry with ML to enable accurate, data-driven monitoring and quantification of stitching errors in multibeam EBL, achieving promising results for certain error categories and establishing a foundation for future metrological enhancements.

3.3.3 *Machine Learning for Improving Scanning Electron Microscopy Metrology Accuracy in Multibeam Electron Beam Lithography*

In his thesis, Chaudhary addressed challenges in multibeam EBL and SEM metrology by applying algorithms and architectures to improve efficiency and accuracy, particularly enhancing line edge and width roughness measurements and denoising noisy SEM images.[12]

The algorithms primarily included deep learning models SEMNet, EDGENet, and LineNet for image denoising and edge detection, along with data compression methods such as run-length encoding. Architectures referred to the implementation designs of these algorithms, including convolutional neural network (CNN) structures with convolutional, batch normalization, and dropout layers, and the multibeam datapath architecture for efficient data communication. The roles of these ML algorithms and architectures form the focus of this review.

SEMNet, EDGENet, and LineNet were CNN-based models comprising convolutional, batch normalization, and dropout layers for effective image processing. SEMNet performed Poisson denoising of SEM images to preserve fine details and improve the accuracy of line edge roughness (LER) and line width roughness (LWR) measurements. This denoising enabled more precise feedback for lithographic process optimization. EDGENet directly estimated edge geometries from noisy SEM images, predicting line edge positions without additional filtering or parameter tuning. This direct estimation enhanced defect detection and roughness correction, improving overall mask writing yield. LineNet combined denoising and edge estimation into a single process with LineNet1 for single-line and LineNet2 for multi-line SEM images for saving time and memory while maintaining accuracy.

These deep learning algorithms addressed key challenges in multibeam EBL such as data processing bottlenecks, throughput limitations, and metrology precision. By enhancing SEM image quality and enabling faster, more accurate edge detection, SEMNet, EDGENet, and LineNet advanced efficient and cost-effective multibeam EBL systems, supporting continued semiconductor miniaturization under Moore's law.

Training and testing data for these models were simulated using the Thorsos method and ARTIMAGEN library due to the lack of public rough-line SEM datasets. Two datasets were generated.[13,14] For the single-line dataset, each original image was corrupted with 10 different levels of Poisson noise resulting in a total of 100,800 pairs of noisy SEM images and edge images, with 89,280 image pairs for training, 2,880 image pairs for validating, and 8,640 pairs for testing. For the multiple-line dataset, each original image was split into smaller images of dimension 256 $\times$ 256 pixels, resulting in a total of 50,400 noisy SEM images and edge images, with 44,640 pairs for training, 1,440 pairs for validating, and 4,320 pairs for testing. Here, the Thorsos method (Monte Carlo spectral method) simulates rough surfaces with Gaussian-distributed heights and slopes, while the ARTIMAGEN library, developed by the U.S. National Institute of Standards and Technology, generates realistic SEM images of nanostructures with features such as random backgrounds and Gaussian blur.[15]

Performance metrics were reported for SEMNet, EDGENet, LineNet, and the benchmark denoising convolutional neural network (DnCNN).[15] SEMNet showed superior denoising, achieving a peak signal-to-noise ratio (PSNR) of 28.66 dB for Poisson noise of 10 electrons per pixel, surpassing DnCNN (25.24 dB, +13.5%), Wavelet (20.91 dB, +37.1%), and Total Variation (22.39 dB, +28.0%). By comparing with the simulation data using the Thorsos method, SEMNet achieved 91.9% LER accuracy (0.68 nm vs. simulated value 0.74 nm) and low mean absolute percentage error (MAPE) of 0.22 for high-noise images. Note that these simulated edges were used as ground truth for training and evaluating SEMNet.

EDGENet achieved 98.1% LWR accuracy (1.04 nm vs. simulated 1.06 nm) with a MAPE of 0.22 pixels. LineNet1 demonstrated comparable denoising to SEMNet (PSNR = 28.26 dB) and accurate LER/LWR estimates (0.75 nm/1.04 nm, matching simulated 0.75 nm/1.06 nm). Smaller LineNet1 variants required less computation but performed worse under high noise.

Overall, this study provided effective solutions to enhance the efficiency and accuracy of multibeam EBL and SEM metrology in integrated circuit manufacturing, demonstrating the strong potential of deep learning and data compression in advancing lithographic process precision.

3.3.4 *Faster Region-Based Convolutional Neural Network-Assisted Carbon Nanotube Recognition for Rapid Nanocantilever Fabrication in Electron Beam Lithography*

Tadokoro et al. developed a ML-assisted recognition system to enable high-throughput fabrication of nanocantilevers and electrodes using EBL.[16] Each nanocantilever, composed of a single CNT, needed precise pairing with an electrode. The CNT-based nanocantilevers served as field emitters, where quantum field emission occurred across an ultra-small gap (~80 nm) between the CNT tip (anode) and 110-nm-thick Ti/Au electrodes (cathode) fabricated via EBL and metal deposition. These nanoscale field emitters exhibit efficient low-voltage field emission, producing high output currents suitable for applications in displays, X-ray computed tomography, and optical systems. CNTs were used for nanocantilevers due to their exceptional electrical, mechanical, and thermal properties, which make them highly suitable for field emitters and other applications such as reprogrammable universal logic gates for neuromorphic computing, mentioned by the authors.

CNTs were synthesized via arc discharge, dispersed in isopropyl alcohol through ultrasonic treatment, and deposited onto a heated substrate, resulting in randomly distributed single CNTs. To overcome the positional randomness, Ti/Au markers were patterned on the substrate using EBL and deposition. These served as spatial references for the ML-assisted recognition system, which analyzed SEM images to accurately locate CNT tips suitable for clamping with electrodes. Guided by the recognition results, electrodes were subsequently patterned by EBL and deposition to form the desired quantum-scale gaps.

The recognition system addressed the critical challenge of accurately identifying the positions of randomly placed CNTs, a prerequisite for precise electrode fabrication. At its core, the system employed a faster region-based convolutional neural network (Faster R-CNN), consisting of a CNN for feature extraction, a region proposal network for locating candidate regions, and a classifier for object recognition and labeling.

Training and testing were conducted using a custom dataset of 1,404 manually labeled SEM images containing 2,080 CNTs and associated markers. Each image was annotated with bounding boxes and labels specifying the preferred electrode attachment edges. The bounding box outputs were automatically linked to computer-aided design software to generate electrode designs for subsequent EBL and deposition. The dataset was divided into 1,311 images for training and 93 for testing. The model was trained for 500 epochs using momentum stochastic gradient descent, with a learning rate reduced from 0.001 to 0.0001 after 375 epochs. Training was executed on an NVIDIA Tesla K80 GPU using PyTorch, while electrode pattern design employed the gdsCAD library.

The ML-assisted system achieved dramatic efficiency gains. Recognition and measurement tasks were completed in 2 seconds, compared to 12 hours manually, more than 20,000 times improvement. In recognition accuracy, the model detected approximately 60% of CNTs within 150 nm of their true position and 90% within 200 nm, with an average recognition error of 567.4 nm (SD = 1161.8 nm). Larger errors were attributed to CNT deformation or particle contamination.

Despite moderate spatial accuracy, this R-CNN-assisted recognition approach markedly streamlined the EBL process, enabling consistent formation of CNT-based nanocantilevers with controlled gap and length specifications, thus ensuring optimal field emission performance and demonstrating the potential of ML-driven metrology and automation in nanoscale device fabrication.

3.4 Machine Learning-Assisted Predictive Modeling in Electron Beam Lithography

This section presents eight reviews on ML applications to predictive modeling for EBL. The predictive modeling considers process and material effects for predicting CDs of photoresist patterns, edge roughness in resist patterning, fabricated device performance, fabrication variations, fabricated contact-hole dimensions, proximity effect corrections and fogging effect corrections in EBL.

EBL stands as the most powerful electron-based technique for patterning and fabricating nanoscale structures, achieving resolutions down to sub-10 nm.[17] Despite its high resolution, EBL traditionally suffers from low throughput. Enhanced versions, such as multibeam EBL, address this limitation by employing multiple parallel beams to achieve higher throughput.[8–11] While EBL is frequently used for patterning resists (acting as a temporary mask), fabricating a complete nanostructure or device typically involves numerous subsequent processes.

Table 3.3 summarizes major findings of the reviews presented in this section, according to chronological order of the articles' publication dates.

3.4.1 *Regression-Based Machine Learning Modeling for Chemical Gradient Prediction in Electron Beam Lithography*

Jin, Kozawa, and Tamura developed a ML framework to predict the chemical gradient, a key indicator of LER, in resist patterning using EBL.[18] The framework analyzed how variations in resist material properties and EBL process parameters influence the

Table 3.3. ML-assisted predictive modeling in electron beam lithography.*

Section. Author (Year)[Ref.]	Subject and Process	ML Models, Algorithms, and Platforms	Training and Testing Data	Key Equipment	Key Findings for Model Performance
3.4.1. Jin et al. (2021)[18]	Line edge roughness prediction for EBL	LR, LSR, RR	47,870 data sets for training and 11,968 data sets for testing	EBM-9000 EB mask writer	LSR predicted line edge roughness (RMSE = 0.00555 nm^{-1}, 44.5% less than acceptable threshold) better than RR and LR
3.4.2. Yao et al. (2022)[19]	Predicting proximity effect correction for EBL	ANN, MLPNN	Simulation data for training and testing	Unspecified	MLPNN achieved AC > 99% in predicting correct exposure doses and was 7–10 times faster
3.4.3. Hajijafari et al. (2022)[21]	Predicting FE correction for EBL	CNN, DNN, DQL, MP, RL	Simulated and experimental data (number unspecified)	Unspecified	RL was 13 times faster in training, 200 times faster in running, and achieved a 92% reduction in FE variations, a 7.5% improvement
3.4.4. Gostimirovic et al. (2022; 2023)[25,26]	Predicting fabrication variations in EBL and dry etching	ANNs, DCNNs, Ensemble learning	50,680 slices split evenly for training and testing	EBL, SEM, IC production line	DCNNs reduced fabrication deviations by 259% and errors in predicting fabrication variations by 32–40%
3.4.5. Getman (2023)[27]	Inverse design and performance prediction for flat optics	ALFRED, Autoencoder, EfficientNet, k-means, MLPs, SS, t-SNE, ResNet	10,000 simulation data points for training, 4,000 data points for testing	PECVD, EBL system	ALFRED was 10,000 times faster in computation with high accuracy (MSE = 0.0011) in predicting spectral responses

3.4.6. Zhao et al. (2024)[29]	Predating and classifying contact hole layer images for EBL	LSTM network, SVM	145 data sets from SEM for training and validating; 35 for testing	Raith150-Two EBL, FE-SEM	LSTM achieved AC > 93% and $R^2 \approx 1$ in CH size prediction; SVM achieved AC = 86% in classifying CH layer
3.4.7. Zhao et al. (2024)[30]	Predicting CDs in MONP by EBL	LSTM network, MFNNs	300 datasets from SEM for training and validating; 69 sets for testing	NB5 EBL, FE-SEM	LSTM achieved MAPE = 0.089, $R^2 = 0.94$ in CD prediction with minimum CD dimension = 26.05 nm
3.4.8. Huang et al. (2024)[31]	Predicting proximity effect corrections in EBL	TransUNet, MLP, U-Net, Transformer	Simulated pre- and post-CDM in 8:2 ratio for training and testing	EBL	TransUNet achieved 99% accuracy in PEC prediction with a correction time of 0.13 s and DP time of 0.081 s

*The representing words or phrases for the acronyms or abbreviations used in the above table are defined below in alphabetical order: 1) AC = Accuracy, 2) ALFRED = Autonomous learning framework for rule-based evolutionary design, 3) ANN = Artificial neural network, 4) CD = Critical dimension, 4) CDM = Correction dose matrices, 5) CH = Contact hole, 6) CNN = Convolutional neural network, 7) DCNNs = Deep convolutional neural networks, 8) DNN = Deep neural network, 9) DP = Data processing, 10) DQL = Deep Q-learning, 11) EBL = Electron beam lithography, 12) FE = Fogging effect, 13) FE-SEM = Field-emission scanning electron microscopy, 14) IC = Integrated circuit, 15) LR = Lasso regression, 16) LSR = Least squares regression, 17) LSTM = Long short-term memory, 18) MAPE = Mean absolute percentage error, 19) MFNNs = Multilayer feed-forward neural networks, 20) MLP = Multilayer perceptron, 21) MLPNN = Multilayer perceptron neural network, 22) MONP = Metal oxide nanoparticle photoresist, 23) MP = Max pooling, 24) MSE = Mean squared error, 25) PEC = Proximity effect correction, 26) PECVD = Plasma-enhanced chemical vapor deposition, 27) R^2 = Coefficient of determination for regressions, 28) RL = Reinforcement learning, 29) RR = Ridge regression, 30) SEM = Scanning electron microscopy, 31) SS = Semantic segmentation, and 32) SVM = Support vector machine.

resulting LER, thereby identifying critical factors for optimizing resist formulations and lithographic conditions in nanoscale device fabrication. The resist parameters considered included the concentrations of photoacid generator, photo-decomposable quencher (PDQ), and total sensitizer, while the EBL parameters encompassed dose sensitivity, half-pitch, exposure pattern width, and beam blur.

The ML framework incorporated three regression algorithms — least squares regression (LSR), ridge regression (RR), and lasso regression (LR) — to model the relationship between resist/EBL parameters (inputs) and the chemical gradient (output). LSR was used to minimize the error function and evaluate the RMSE across training and validation datasets, effectively capturing the nonlinear relationship between the chemical gradient and six key parameters. However, it exhibited overfitting at higher polynomial degrees and required a large dataset (over 5,000 data points) for stable predictions. RR addressed this limitation by adding a regularization term to constrain polynomial feature weights, maintaining high model complexity without overfitting. It demonstrated superior generalization performance compared to LR. Similarly, LR employed a regularization term but also performed feature selection by eliminating less significant parameters; however, it yielded slightly lower fitting accuracy than RR in this study.

The training and testing datasets were generated through simulation using a previously established model describing the chemical and physical interactions in EBL. The simulation accounted for six main parameters, sensitivity, photoacid generator concentration, photo-decomposable quencher concentration, half-pitch, exposure pattern width, and beam blur, as well as other variables such as reaction radii, diffusion constants, and deprotonation efficiencies. In the model, the Monte Carlo method was employed to trace secondary electron trajectories and their reactions with photoacid generator and PDQ, while reaction–diffusion equations were solved to model catalytic chain deprotection during post-exposure baking. This approach produced 59,838 data points, of which 80% (47,870) were used for training and 20% (11,968) for testing. Normalization was applied to standardize the feature scales, enabling efficient model convergence. This simulation-based data generation strategy provided a practical alternative to time-intensive experimental data collection.

Experimental validation was conducted using an EBM-9000 electron beam mask writer, incorporating its specifications (50 kV acceleration voltage and beam blur parameters) into the simulation to closely approximate real exposure conditions. The resist materials included poly(4-hydroxystyrene) and triphenylsulfonium triflate. Following electron beam exposure and post-exposure baking, SEM and atomic force microscopy were used to characterize resist patterns and measure the concentration gradient of protected polymer units across the line edge. These experimental measurements were compared with ML predictions to validate model performance. The protected units refer to chemical groups on the polymer side chains in chemically amplified resists that are removed during lithographic deprotection, enabling pattern formation during development.

The validation results demonstrated that the LSR model achieved the highest predictive accuracy, with an RMSE of 0.00555 nm^{-1} at the 8th polynomial degree, though overfitting was observed. At the 5th degree, the RMSE increased slightly to 0.00735 nm^{-1}, providing a better balance between accuracy and generalization. RR achieved an RMSE of 0.00826 nm^{-1} and a PDQ prediction error of 0.0165, indicating strong generalization capability. LR exhibited an RMSE of 0.00982 nm^{-1} and a PDQ error of 0.0319, performing slightly worse than RR. All RMSE values ranged from 0.00555–0.00982 nm^{-1}, 44.5% to 1.8% below the acceptable threshold of 0.01 nm^{-1}, confirming the high predictive reliability of the models. The RR model's accurate PDQ prediction further underscored its robustness in parameter optimization.

Overall, the study demonstrated that ML techniques can significantly streamline the analysis, prediction, and optimization of resist material properties and EBL process conditions. Both LSR and RR accurately identified key determinants of LER, guiding future resist design and process development for improved nanoscale pattern fidelity in EBL.

3.4.2 *Machine Learning-Assisted Prediction for Proximity Effect Correction in Electron Beam Lithography*

Yao et al. employed a ML algorithm, specifically a multilayer perceptron neural network (MLPNN), to predict corrected exposure doses for proximity effect correction (PEC) in EBL.[19] This approach aimed to enhance computational efficiency and eliminate the costly iterative calculations required by conventional PEC methods. PEC compensates for electron forward and backscattering, which can otherwise cause unintended energy deposition and pattern distortion, thereby ensuring uniform exposure and high-fidelity nanoscale fabrication.

The MLPNN, a type of artificial neural network comprising multiple layers of interconnected neurons, was trained for predictive dose correction. Its training data were generated using the traditional model-based PEC method and further diversified through data augmentation techniques such as flipping and translation to expand the dataset. In the model-based PEC framework, electron scattering is characterized by a two-component model: forward scattering in the resist and backscattering from the substrate.[20] The deposited energy is expressed by a double Gaussian point spread function (PSF), consisting of a narrow Gaussian for short-range forward scattering and a broad Gaussian for long-range backscattering, enabling iterative dose and geometry optimization. The point spread function's energy deposition distribution served as the input, while the corrected exposure doses were used as the output.

Although the model-based PEC approach achieves high accuracy, its iterative optimization is computationally intensive, especially for complex nanoscale layouts. The MLPNN effectively addressed this limitation by bypassing iterative computation while maintaining correction precision. The ML-based MLPNN model achieved a PEC inference error below 1%, comparable to traditional PEC accuracy, and delivered a computational efficiency of seven to ten times faster. This demonstrates that ML-based PEC can retain high correction accuracy while significantly accelerating processing.

3.4.3 *Reinforcement Learning-Based Placement for Reducing Fogging Effects in Mixed-Signal Integrated Circuits*

Hajijafari et al. proposed a reinforcement learning-based placement technique for mixed-signal integrated circuits to minimize fogging effect variations (FEVs) on sensitive analog devices.[21] The approach employed deep Q-learning to train an agent that intelligently predicted the optimal placement of integrated circuit modules, mitigating fogging effects caused by re-scattered electrons during EBL. The fogging effect, a long-range and low-intensity exposure from backscattered electrons, induces pattern distortion and performance degradation in analog circuits.[22]

The placement technique involved four ML algorithm/models: reinforcement learning, deep Q-learning, deep neural network, and max pooling. The reinforcement learning served as the overarching framework through the implementation using deep Q-learning. The deep neural network acted as the reinforcement learning agent, learning optimal integrated circuit placement by interacting with the reinforcement learning environment and minimizing FEVs. The deep neural network received input states representing current integrated circuit layout configurations encoded in the B*-tree topological representation and selected actions (rotate, move, swap, or no move) to optimize module arrangements. The B*-tree structure efficiently encoded spatial relationships and positions of modules within the layout. Deep Q-learning approximated the Q-value function to guide the deep neural network in selecting actions that maximized cumulative rewards. The first two layers of the deep neural network employed max pooling, a common technique in CNNs, to reduce input dimensionality and improve processing efficiency. This architecture enhanced the agent's ability to handle complex layout data and make accurate placement decisions.

Collectively, the ML framework leveraged the B*-tree representation to predict optimal placements of mixed-signal integrated circuit modules, thereby reducing FEVs, improving circuit performance, and achieving faster placement than existing analytical reinforcement learning-based methods.[23,24] Although the authors did not explicitly specify their data sources, the training and testing datasets were likely derived from netlists and layout configurations of mixed-signal integrated circuits. Fogging effect maps could be simulated using Gaussian point spread functions to model energy distribution. Synthetic data may be generated from simulation tools, while experimental data from EBL systems can serve as validation. Thus, the framework's input is the integrated circuit layout encoded in the B*-tree representation, and the output is the predicted integrated circuit placement with optimized module arrangements.

Performance evaluations showed that the B*-tree reinforcement learning placement technique significantly outperformed the analytical reinforcement learning approach across all metrics. The B*-tree reinforcement learning model achieved an average runtime of 8.4 seconds versus 1695.5 seconds (a 202× reduction) and reduced training time from 1027 minutes to 77 minutes (a 13.3× reduction). It also minimized layout area (99 μm^2 vs. 323.77 μm^2, a 327% reduction) and wirelength (98.72 nm vs. 300.5 nm, a 304% reduction). Additionally, FEV was reduced by 92% versus 85.6% (7.5% improvement), and direct current gain improved to 52.01 dB compared to 51.45 dB (1.1% improvement), demonstrating substantial gains in efficiency and performance.

Overall, the B*-tree reinforcement learning placement technique effectively minimized fogging-induced variations in sensitive analog devices within mixed-signal integrated circuits. By employing a topological representation to reduce the search space, it achieved faster training, superior layout quality, and enhanced circuit performance metrics — including area, wirelength, fogging variation, and direct current gain, representing a significant improvement over state-of-the-art analytical reinforcement learning placement methods.

3.4.4 *Machine Learning-Driven Prediction and Correction of Fabrication Errors by Electron Beam Lithography and Etching*

In 2022, Gostimirovic et al. applied a ML approach to predict fabrication-induced structural dimension variations in silicon photonic devices.[25] They developed a DCNN to predict imperfections such as over- and under-etching, corner rounding, and channel filling, common issues arising from EBL and dry etching. The primary goal was to validate designs prior to fabrication, thereby reducing prototyping costs and time while improving device performance.

DCNNs, a subclass of artificial neural networks tailored for grid-structured data such as images, were trained to analyze detailed features in both graphic design system layouts and corresponding SEM images. By learning correlations between design layouts and fabricated outcomes, the DCNN could accurately predict fabrication effects and assist designers in validating and optimizing photonic layouts before committing to costly physical prototyping.

The dataset used for model training and testing was obtained from an integrated photonics foundry employing EBL for fabrication and SEM imaging at 1.5 nm/pixel resolution. Thirty randomly generated design patterns (2.8×2.0 μm^2), created using Fourier transform-based filters, simulated structural variations characteristic of topologically optimized photonic devices.

These SEM images were preprocessed — binarized, cropped, and aligned in resolution — to match the graphic design system layouts. The processed images were divided into overlapping 128×128 pixel segments with 32-pixel steps, yielding 50,680 augmented data slices, evenly split between training and testing datasets. Model optimization used the Adam optimizer and binary cross-entropy loss.

To enhance generalization, ten identical DCNNs were trained independently with randomized weight initializations and shuffled training data, forming an ensemble DCNN model implemented in TensorFlow. When evaluated on unseen SEM test images, the ensemble achieved a 0.5% lower error than the best individual model and a 6% lower error than the worst-performing one, demonstrating improved accuracy and robustness through ensemble learning.

The DCNN model was benchmarked against a traditional uniform erosion method for predicting fabrication variations in two subwavelength grating couplers and a topologically optimized wavelength demultiplexer. Using mean squared error (MSE) as the evaluation metric, the DCNN reduced prediction errors by 40% for the 2D subwavelength grating coupler, 37% for the focusing subwavelength grating coupler, and 32% for the demultiplexer. The DCNN effectively captured fabrication effects such as corner rounding, channel filling, and edge smoothing, substantially outperforming conventional prediction methods. Its accuracy and versatility make it a valuable tool for photonic design validation across both grating and subwavelength structures.

Building on these results, Gostimirovic et al. extended their work by developing a "corrector model," a DCNN-based system that automatically adjusts photonic device layouts prior to fabrication to ensure closer conformity between designed and realized structures.[26] Validated through experiments and simulations, the corrector model significantly enhanced design fidelity and optical performance. For a topologically optimized wavelength-division multiplexer, it reduced fabrication deviations from 2,133 to 825 error pixels (a 259% improvement) and decreased peak wavelength shifts from over 10 nm to approximately 1 nm.

Together, these two studies demonstrate the power of ML-driven DCNNs in improving the reliability and performance of silicon photonic devices. By addressing fabrication variability and inverse design challenges inherent to EBL, these models enable more accurate, efficient, and miniaturized photonic component fabrication within existing technological constraints.

3.4.5 *Machine Learning-Enhanced Performance Prediction and Inverse Design for Flat Optics*

Getman developed the Autonomous Learning Framework for Rule-based Evolutionary Design (ALFRED), a software platform integrating ML and simulation techniques for performance prediction and inverse design of high-efficiency flat-optics devices operating in the visible spectrum.[27] Using EBL for precise nanoscale patterning, ALFRED leverages neural network predictors and inverse techniques to address challenges such as material loss, fabrication tolerances, and deformation robustness, enabling advances in flat optics, nanooptics, and flexible electronics.

ALFRED features two ML-based modules, the neural predictor unit as well as clustering and post-training analysis, each supporting the prediction and design for flat-optics devices.

The neural predictor unit employed the EfficientNet architecture to process binary images of device geometries, extracting low-dimensional structural features that feed into MLPs to predict spectral responses for both transverse electric and transverse magnetic polarizations. This significantly accelerates evaluation, reducing reliance on time-intensive first-principles simulations.[28] The clustering and post-training analysis module improves prediction accuracy and robustness by identifying high-failure clusters using k-means and t-distributed stochastic neighbor embedding analysis. These clusters were used to generate additional data for post-training, enhancing model reliability and performance. Together, these models enabled ALFRED to employ user-defined desired optical response for a flat-optics device (as the input) to predict and design the optimized geometry of the flat-optics device that meets the desired optical response (the output).

Training and testing datasets were derived from first-principles FDTD (finite-difference time-domain) simulations. The training set included 4,000 samples with a 500 nm period and thicknesses from 50–300 nm in 25 nm steps. An additional 10,000 spectra from high-failure clusters were generated for post-training refinement, resulting in 10,000 training and 4,000 testing samples.

ALFRED evaluated objective functions about 10,000 times faster than FDTD simulations. By comparing spectral responses predictions with FDTD simulations, EfficientNet, serving as the convolutional base, achieved an average MSE of 0.0011 across 4,000 test samples, surpassing models such as VGG and ResNet. Over 96% of predictions were below the MSE threshold of 0.004. Pre-training (via autoencoders and semantic segmentation) accelerated convergence and reduced error, while post-training lowered MSE from 0.0086 to <0.001, demonstrating high predictive precision. Since normalized MSE values ranged from 0 to 1, an MSE of 0.0011 corresponds to an average deviation of ~0.033 per prediction, indicating excellent accuracy.

To validate ALFRED's predictions, two flexible flat-optics devices — polarizing beam splitters operating between 600–800 nm — were designed, fabricated, and tested. The fabricated devices achieved >80% polarization and transmission efficiency over a 200 nm bandwidth, with polarization and transmission efficiencies of 99% and 98.3% at 600 nm. The dichroic mirror exhibited 90% transmission in the visible range, and transverse electric polarization transmission reached 85% for flexible flat-optics polarizers. Experimental spectra closely matched FDTD predictions, with deviations of only 10–15% beyond 650 nm, confirming high fabrication and design accuracy.

Both ML model comparisons and experimental results validated ALFRED's efficiency and robustness, with polarization and transmission rates often exceeding 90%, outperforming many state-of-the-art devices. Overall, this work demonstrates the feasibility of ML-driven inverse design for scalable, efficient, and flexible flat-optics fabrication.

As shown in the SEM micrograph in Figure 3.3, the flexible flat-optics structure consists of 250-nm-high patterned silicon nanostructures on 30-μm Kapton tape. Fabrication steps involved: (a) adhering the Kapton to a 200-μm glass substrate; (b) depositing a 250-nm amorphous silicon layer via plasma-enhanced chemical vapor deposition; (c) spin-coating ZEP520A resist; (d) patterning with EBL and depositing 22-nm chromium; (e) forming a Cr mask via resist liftoff; (f) etching silicon through the Cr mask; (g) removing the mask; and (h) detaching the final flexible flat optic from glass.

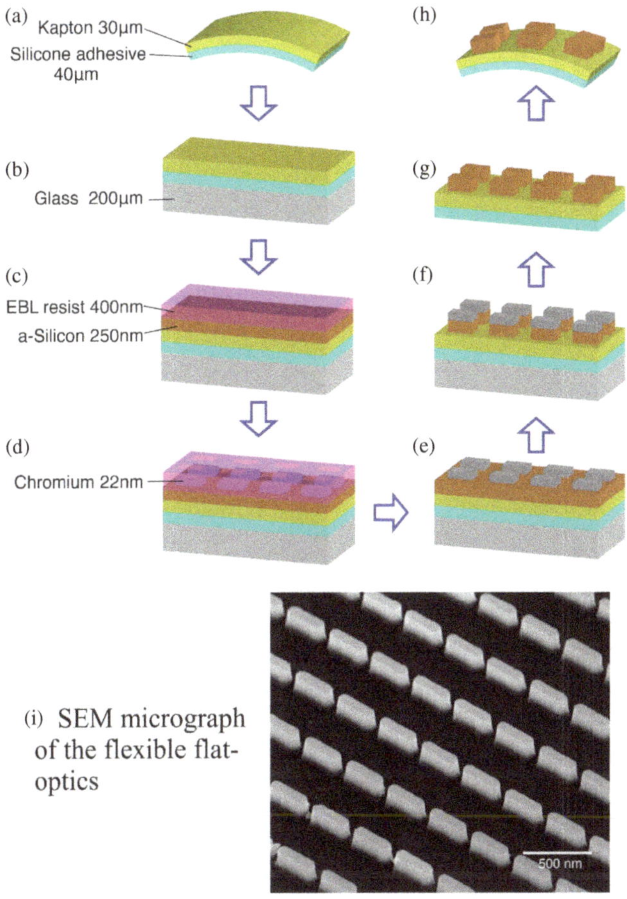

Figure 3.3. Fabrication sequence and SEM micrograph of the completed flat-optics device: (a) Kapton tape, (b) adhering the Kapton to glass substrate, (c) depositing silicon layer and spin-coating resist, (d) patterning with EBL and depositing 22-nm chromium, (e) forming a Cr mask via resist liftoff, (f) etching silicon through the mask, (g) removing the mask, (h) detaching the flexible flat optic from glass, and (i) photo of the completed device. (Courtesy of Getman,[27] reprinted from open-access data.)

3.4.6 Machine Learning-Aided Prediction and Classification for Optimizing Contact Hole Imaging in Electron Beam Lithography

Zhao et al. utilized ML to predict contact hole (CH) imaging and classify process conditions, enabling precise and efficient fabrication of CH layers in photoresists, critical in EBL.[29] Their ML models effectively captured the long-term influence of EBL parameters on the dimensional accuracy and uniformity of the fabricated CHs.

CH layers contain small openings patterned into a photoresist spun onto a silicon wafer, allowing conductive materials to connect transistors or interconnects within a chip. These holes demand exceptional precision and uniformity; deviations can seriously degrade device performance. Oversized holes may cause short circuits, while undersized ones increase contact resistance. Thus, dimension precision and uniformity of CHs are essential for reliable chip functionality as feature sizes continue to shrink.

The authors employed a long short-term memory (LSTM) network to predict CH imaging and a Support Vector Machine (SVM) to classify process conditions. The LSTM analyzed CH images to predict average horizontal, vertical, and overall contact width under various process conditions, while the SVM identified conditions meeting local critical dimension uniformity (LCDU) tolerance for optimal imaging. SVM excels in binary classification tasks, such as classifying process conditions based on whether they meet the LCDU tolerance requirements. Combined, the models formed an optimization framework to screen process conditions satisfying both width and LCDU requirements.

Data for training and testing were obtained through EBL experiments using the Raith150-Two system. Process conditions including soft bake temperature, beam current, exposure dose, and development time were systematically varied for dense square CHs (10–100 nm). After experiment, hole dimensions were measured by high-resolution SEM, and LCDU was derived based on the standard deviation of the CDs of the SEM-associated imaging. A total of 180 SEM data sets were collected; 145 were used for training and validation, and 35 for testing.

For the LSTM, 180 data sets were collected, with 145 sets used for training and validation, and 35 sets for testing. The LSTM model took process conditions (soft bake temperature, beam current, exposure dose, and development time) as inputs and predicted imaging for gauging horizontal, vertical, and overall contact width. The SVM also used the same 180 SEM imaging data sets for training and testing with the process conditions and LCDU values as inputs to classify conditions that meet the LCDU tolerance threshold for optimal lithography imaging.

The LSTM, using process parameters as inputs, accurately predicted horizontal, vertical, and overall contact width. During training, RMSE values were 4.832, 4.834, and 4.293 nm, with MAPE values of 0.052, 0.053, and 0.044, and R^2 near 1. Testing yielded RMSE values of 3.934, 5.348, and 4.298 nm, with MAPE values of 0.057, 0.073, and 0.058, and R^2 still close to 1, confirming robust predictive performance. The SVM achieved 86% accuracy in classifying process conditions based on LCDU tolerance. Together, the models identified process windows meeting both LCDU and horizontal contact width specifications with relative errors under 7%, demonstrating their effectiveness in lithography optimization.

This ML-assisted EBL approach enhances imaging dimension quality, reduces experimental effort, and offers a scalable pathway for high-resolution and reliable patterning in advanced chip manufacturing.

3.4.7 *Long Short-Term Memory-Enhanced Prediction of Critical Dimensions in Metal Oxide Nanoparticle Photoresists*

Zhao et al. employed a LSTM network to enhance the prediction accuracy of CDs based on photoresist composition and process conditions in EBL.[30] Leveraging its ability to process sequential data, the LSTM model effectively captured complex relationships between metal oxide nanoparticle photoresist (MONP) components and process parameters, enabling efficient optimization of resist formulations to meet CD requirements. This approach replaced conventional trial-and-error methods that are both time-consuming and costly.

The datasets for training and testing were obtained from EBL experiments conducted with an 80-kV NB5 system. MONP films were prepared with varying compositions of photoacid generators and quenchers under different process conditions, including beam current, exposure dose, and development time. The CDs of the resulting resist patterns were measured using a field-emission SEM. A total of 369 field-emission SEM datasets were collected, of which 300 were used for training and validation, and 69 for model testing.

The LSTM-based model successfully screened photoresist formulations achieving CDs below 29 nm, with a minimum CD of 26.05 nm. Predicted CDs closely matched experimental values, exhibiting a maximum relative error of 9.24%. The model achieved a MAPE of 0.089 and an R^2 of 0.94, confirming its high predictive accuracy and robustness. Compared with multilayer feed-forward neural networks previously used for similar purposes, the LSTM model demonstrated superior performance and reliability.

Overall, the integration of LSTM with EBL significantly improved prediction accuracy, accelerated photoresist development, reduced experimental costs, and enhanced precision in EBL high-resolution patterning.

3.4.8 *Machine Learning-Assisted TransUNet Model for Efficient Proximity Effect Corrections in Electron Beam Lithography*

Huang et al. introduced a ML-assisted PEC method for EBL to overcome the limitations of traditional PEC techniques, such as high computational cost and complex data preparation.[31] The intrinsic precision of EBL is often degraded by the proximity effect, primarily caused by electron backscattering within the resist and substrate, leading to unintended energy deposition and feature broadening.[17] Numerous PEC techniques have been developed to compensate for these scattering effects. Conventional PEC methods, such as Monte Carlo simulations based on the point spread function, are accurate but computationally intensive.

To address these challenges, the authors implemented an ML model based on the TransUNet architecture for end-to-end PEC in EBL. TransUNet integrated the Transformer framework, which captured global contextual features, with the U-Net architecture, which preserved local spatial details. The model directly mapped the layout pattern to the corrected dose distribution, simplifying the correction process, reducing data preparation, and improving computational efficiency while maintaining accuracy.

For training and testing, datasets were generated by producing pre-correction (input) and post-correction dose matrices (output). Monte Carlo simulations were first used to determine point spread function parameters and simulate the energy deposition distribution for the target layout. The layout was discretized into square pixels with uniform initial exposure doses. Traditional iterative PEC methods then produced post-correction dose matrices, which were normalized and used as training labels. The implementation of mixed-precision training reduced memory consumption and accelerated convergence, while a learning-rate scheduling mechanism enhanced training stability. A high-resolution EBL system was used for direct writing on polymethyl methacrylate resists coated on silicon substrates. The dataset was divided into training, validation, and testing sets in an 8:1:1 ratio.

Model performance was evaluated using MSE, PEC time, and data processing time. Comparing the TransUNet-predicted dose patterns (after PEC) with target dose distributions, the model achieved an error of 0.76%, comparable to traditional Monte Carlo PEC (0.6%) and MLP-based PEC (0.67%). Importantly, TransUNet reduced correction time to 0.13 s, far faster than traditional PEC (5.63 s), though slightly slower than MLP-based PEC (0.01 s). Its greatest advantage was in data processing time, only 0.081 s, compared with 2,465 s for MLP-based PEC. Moreover, TransUNet preserved fine pattern features, including sharp corners and uniform linewidths, demonstrating superior pattern fidelity. Experimental results confirmed that TransUNet achieved accuracy comparable to both traditional iterative and MLP-based PEC methods while substantially reducing computational cost and simplifying the correction workflow.

Note that Yao et al. employed a MLP-based PEC model to predict corrected dose patterns, as discussed in Section 3.4.2. Although the MLP-based PEC achieved comparable accuracy and higher computational efficiency than traditional PEC, it required transforming 2D input matrices into 1D sequences, complicating data preparation and significantly increasing data processing time with finer grids or larger layouts.

In conclusion, the TransUNet-based PEC framework represents a significant advancement in EBL. Its exceptional computational efficiency, high accuracy, and ability to preserve pattern integrity make it a highly promising solution for addressing proximity effect challenges.

3.5 Summary and Conclusions

This review surveyed a wide range of ML algorithms and models applied to electron beam nanofabrication, highlighting their transformative roles and performance outcomes. The studies reviewed demonstrate that ML has evolved from a complementary analytical tool into an indispensable enabler of electron beam nanofabrication. Across atomic manipulation, metrology, and predictive modeling, ML-driven frameworks have consistently enhanced pattern fidelity, process throughput, and device yield. Nonetheless, realizing the full potential of ML-assisted nanofabrication requires overcoming several foundational challenges that span data infrastructure, model generalization, interpretability, and system integration.

First, the performance of ML models remains strongly data-dependent. High-quality, annotated datasets representative of diverse process conditions are scarce due to the proprietary nature of industrial data, experimental cost, and sensitivity to instrument configurations. Insufficient or biased datasets can limit the ability of models to generalize beyond narrowly defined conditions, leading to overfitting and reduced robustness when transferred between lithography platforms or fabrication systems.

Second, model generalization and cross-domain transferability remain unresolved. ML models trained on one lithographic tool or material stack often fail to perform accurately on another due to variations in beam characteristics, resist chemistry, or environmental parameters. Achieving universal applicability requires domain adaptation strategies, multi-task learning architectures, and standardized data representations that encode physical priors.

Third, interpretability and physics integration are critical yet underdeveloped. Many deep learning architectures operate as "black boxes," providing accurate but opaque predictions that lack mechanistic transparency. Bridging data-driven and physics-based modeling through hybrid physical-ML frameworks, explainable artificial intelligence, and uncertainty quantification would enhance trust, reproducibility, and process control reliability.

Fourth, hardware–software integration poses practical barriers. Embedding ML algorithms into *in situ* control loops for real-time dose correction, defect detection, or beam steering demands low-latency computation, high-bandwidth data pipelines, and close coupling between model inference and instrument feedback. Current electron beam systems lack standardized interfaces for ML deployment.

Finally, computational costs, reproducibility, and standardization remain systemic issues. The lack of open benchmarks and reproducible software environments hinders cross-laboratory comparison and algorithm validation. Large-scale training and inference also impose substantial computational and energy costs, raising concerns about sustainability and accessibility for smaller research facilities.

Overcoming these challenges calls for a convergence of disciplines, combining nanofabrication science, data science, and computational engineering to establish a standardized, interpretable, and scalable ML frame system for next-generation nanofabrication.

To accelerate the maturation of ML-integrated electron beam nanofabrication, future research should adopt a multi-tiered strategy that emphasizes data infrastructure, algorithmic innovation, system integration, and ethical-sustainability considerations.

First, developing large, open-access, and standardized nanofabrication datasets is paramount. These datasets should capture variability in beam parameters, resist morphologies, and environmental conditions to support model generalization. Incorporating federated learning architectures would enable collaborative model training across academic and industrial institutions without compromising data privacy or intellectual property.

Second, future frameworks should embed physical constraints and process knowledge within ML architectures. Hybrid models, combining neural networks with differential equations or Monte

Carlo simulations, can enforce physical plausibility while maintaining computational efficiency. Such approaches bridge the gap between empirical prediction and mechanistic understanding.

Third, integrating ML with reinforcement learning and real-time control algorithms can enable self-optimizing systems capable of autonomous tuning of beam parameters, dose modulation, and exposure sequencing for autonomous and adaptive manufacturing. The convergence of ML with digital twin architectures and advanced probes (e.g., 4D-STEM, *in situ* spectroscopy) will facilitate continuous learning and adaptive manufacturing at the nanoscale.

Fourth, leveraging quantum computing, neuromorphic hardware, or edge-AI accelerators could substantially reduce training and inference times, enabling near-real-time process correction and multibeam synchronization. This integration with emerging computational paradigms will be essential for scaling ML applications to industrial throughput levels.

Fifth, the community should establish standardized metrics for evaluating ML performance in lithographic accuracy, stability, and computational efficiency. Transparent model documentation, open-source code dissemination, and peer-reviewed benchmark datasets are necessary for reproducibility and scientific credibility.

Finally, future developments should balance automation with human oversight. Incorporating explainable interfaces and operator-in-the-loop feedback mechanisms ensures safety, interpretability, and accountability in automated nanofabrication workflows.

In summary, ML has proven transformative in advancing the precision, speed, and adaptability of electron beam nanofabrication. The next phase of progress depends on bridging data-driven intelligence with physical insight, standardizing infrastructure across facilities, and embedding ML into autonomous yet interpretable manufacturing systems. These efforts will enable scalable, reliable, and sustainable ML-driven nanofabrication, ushering in a new paradigm of intelligent process-based materials design and nanoscale manufacturing.

Acknowledgements

The author gratefully acknowledges financial support from Pacific Technology, LLC (Phoenix, USA), which enabled the author to spend time to prepare substantial portions of this manuscript. In addition, the author used Gemini, an artificial intelligence language model developed by Google, to assist in addressing technical questions arising from the review of related literature and to help refine the language of the manuscript. All AI-assisted content was critically evaluated, verified, and edited by the author, who takes full responsibility for the accuracy, originality, and integrity of the final manuscript.

References

1. Boebinger, Matthew G., Ayana Ghosh, Kevin M. Roccapriore, Sudhajit Misra, Kai Xiao, Stephen Jesse, Maxim Ziatdinov, Sergei V. Kalinin, and Raymond R. Unocic. "Exploring electron-beam induced modifications of materials with machine-learning assisted high temporal resolution electron microscopy." *npj Computational Materials* 10, no. 1 (2024): 260.
2. Donges, Jan, Marvin Schlischka, Ching-Wen Shih, Monica Pengerla, Imad Limame, Johannes Schall, Lucas Bremer, Sven Rodt, and Stephan Reitzenstein. "Machine learning enhanced in situ electron beam lithography of photonic nanostructures." *Nanoscale* 14, no. 39 (2022): 14529–14536.
3. Maxim, Ziatdinov, Stephen Jesse, Bobby G. Sumpter, Sergei V. Kalinin, and Ondrej Dyck. "Tracking atomic structure evolution during directed

electron beam induced Si-atom motion in graphene via deep machine learning." *Nanotechnology* 32, no. 3 (2021): 035703.

4. Yu, Guoshen, Guillermo Sapiro, and Stéphane Mallat. "Solving inverse problems with piecewise linear estimators: From Gaussian mixture models to structured sparsity." *IEEE Transactions on Image Processing* 21, no. 5 (2011): 2481–2499.
5. Roccapriore, Kevin M., Matthew G. Boebinger, Ondrej Dyck, Ayana Ghosh, Raymond R. Unocic, Sergei V. Kalinin, and Maxim Ziatdinov. "Probing electron beam induced transformations on a single-defect level via automated scanning transmission electron microscopy." *ACS Nano* 16, no. 10 (2022): 17116–17127.
6. Ziatdinov, Maxim, Ayana Ghosh, Chun Yin Wong, and Sergei V. Kalinin. "AtomAI framework for deep learning analysis of image and spectroscopy data in electron and scanning probe microscopy." *Nature Machine Intelligence* 4, no. 12 (2022): 1101–1112.
7. Madsen, Jacob, and Toma Susi. "The abTEM code: transmission electron microscopy from first principles." *Open Research Europe* 1 (2021): 24.
8. Tomandl, Mathias, Christoph Spengler, Peter Hudek, Christof Klein, Hans Loeschner, and Elmar Platzgummer. "Multi-beam mask writing opens up new fields of application, including curvilinear mask pattern for high numerical aperture extreme ultraviolet lithography." *Journal of Micro/Nanopatterning, Materials, and Metrology* 23, no. 1 (2024): 011205–011205.
9. Matsumoto, Hiroshi, Haruyuki Nomura, Jumpei Yasuda, Shunsuke Isaji, Kenichi Yasui, Hayato Kimura, and Yoshinori Kojima. "Multi-beam mask writer MBM-4000 for high-NA EUV era." In *Novel Patterning Technologies 2025*, vol. 13427, edited by Ricardo Ruiz, Richard A. Farrell, p. 1342706. SPIE, 2025.
10. Figueiro, Nivea, Francisco Sanchez, Roy Koret, Michael Shifrin, Yoav Etzioni, Shay Wolfling, Matthew Sendelbach et al. "Application of scatterometry-based machine learning to control multiple electron beam lithography: AM: Advanced metrology." In *2018 29th Annual SEMI Advanced Semiconductor Manufacturing Conference (ASMC)*, pp. 328–333. IEEE, 2018.
11. Rademaker, Guido, Yoann Blancquaert, Thibault Labbaye, Lucie Mourier, Nivea Figueiro, Francisco Sanchez, Roy Koret et al. "Feasibility of monitoring a multiple e-beam tool using scatterometry and machine learning: stitching error detection." In *34th European Mask and Lithography Conference*, vol. 10775, pp. 22–28. SPIE, 2018.

12. Chaudhary, Narendra. "Algorithms and architectures for some problems in multibeam electron beam lithography and SEM metrology." Ph.D. Thesis, Texas A&M University, 2019.

13. Cizmar, Petr, András E. Vladár, Bin Ming, Michael T. Postek, and National Institute of Standards and Technology. "Simulated SEM images for resolution measurement." *Scanning* 30, no. 5 (2008): 381–391.

14. Cizmar, Petr, András E. Vladár, and Michael T. Postek. "Optimization of accurate SEM imaging by use of artificial images." In *Scanning Microscopy 2009*, vol. 7378, pp. 208–213. SPIE, 2009.

15. Zhang, Kai, Wangmeng Zuo, Yunjin Chen, Deyu Meng, and Lei Zhang. "Beyond a gaussian denoiser: Residual learning of deep CNN for image denoising." *IEEE Transactions on Image Processing* 26, no. 7 (2017): 3142–3155.

16. Tadokoro, Yukihiro, Keita Funayama, Keisuke Kawano, Atsushi Miura, Jun Hirotani, Yutaka Ohno, and Hiroya Tanaka. "Artificial-intelligence-assisted mass fabrication of nanocantilevers from randomly positioned single carbon nanotubes." *Microsystems & Nanoengineering* 9, no. 1 (2023): 32.

17. Tseng, Ampere A., Kuan Chen, Chii D. Chen, and Kung J. Ma. "Electron beam lithography in nanoscale fabrication: recent development." *IEEE Transactions on Electronics Packaging Manufacturing* 26, no. 2 (2003): 141–149.

18. Jin, Yuqing, Takahiro Kozawa, and Takao Tamura. "Analysis of mitigating factors for line edge roughness generated during electron beam lithography using machine learning." *Japanese Journal of Applied Physics* 60, no. 7 (2021): 076509.

19. Yao, Wenze, Yujie Yang, Jie Liu, Hongcheng Xu, Siyuan Zhang, Haojie Zhao, and Xin Zhang. "Accurate and Efficient Proximity Effect Correction for Electron Beam Lithography Based on Multilayer Perceptron Neural Network." In *2022 International Workshop on Advanced Patterning Solutions (IWAPS)*, pp. 1–4. IEEE, 2022.

20. Liu, Chun-Hung, Pei-Lin Tien, Philip CW Ng, Yu-Tian Shen, and Kuen-Yu Tsai. "Model-based proximity effect correction for electron-beam direct-write lithography." In *Alternative Lithographic Technologies II*, vol. 7637, pp. 430–437. SPIE, 2010.

21. Hajijafari, Mohammad, Mehrnaz Ahmadi, Zhenxin Zhao, and Lihong Zhang. "Fogging-effect-aware mixed-signal IC placement with reinforcement learning." In *2022 IEEE International Symposium on Circuits and Systems (ISCAS)*, pp. 2895–2899. IEEE, 2022.

22. Chang, Yao-Wen, Ru-Gun Liu, and Shao-Yun Fang. "EUV and e-beam manufacturability: Challenges and solutions." In *Proceedings of the 52nd Annual Design Automation Conference*, pp. 1–6. ACM-SIGDA and IEEE-CEDA, 2015.

23. Ahmadi, Mehrnaz, and Lihong Zhang. "Analog layout placement for FinFET technology using reinforcement learning." In *2021 IEEE International Symposium on Circuits and Systems (ISCAS)*, pp. 1–5. IEEE, 2021.

24. Hudek, Peter, Ulrich Denker, Dirk Beyer, Nikola Belic, and Hans Eisenmann. "Fogging effect correction method in high-resolution electron beam lithography." *Microelectronic Engineering* 84, no. 5–8 (2007): 814–817.

25. Gostimirovic, Dusan, Dan-Xia Xu, Odile Liboiron-Ladouceur, and Yuri Grinberg. "Deep learning-based prediction of fabrication-process-induced structural variations in nanophotonic devices." *ACS Photonics* 9, no. 8 (2022): 2623–2633.

26. Gostimirovic, Dusan, Yuri Grinberg, Dan-Xia Xu, and Odile Liboiron-Ladouceur. "Improving fabrication fidelity of integrated nanophotonic devices using deep learning." *ACS Photonics* 10, no. 6 (2023): 1953–1961.

27. Getman, Fedor. "Fabrication of metasurfaces operating in the visible via nanotechnology and artificial intelligence." Ph.D. Thesis, King Abdullah University of Science and Technology, Thuwal, Saudi Arabia, 2023. (Accessed 12/10/25: https://repository.kaust.edu.sa/bitstreams/61669e63-0016-42b7-aa23-9e6f131b0e2f/download.)

28. Segall, M. D., Philip J. D. Lindan, M. J. al Probert, Christopher James Pickard, Philip James Hasnip, S. J. Clark, and M. C. Payne. "First-principles simulation: ideas, illustrations and the CASTEPcode." *Journal of Physics: Condensed Matter* 14, no. 11 (2002): 2717–2744.

29. Zhao, Rongbo, Xiaolin Wang, Yayi Wei, Xiangming He, and Hong Xu. "Machine learning applied to electron beam lithography to accelerate process optimization of a contact hole layer." *ACS Applied Materials & Interfaces* 16, no. 17 (2024): 22465–22470.

30. Zhao, Rongbo, Xiaolin Wang, Hong Xu, Yayi Wei, and Xiangming He. "Machine learning in electron beam lithography to boost photoresist formulation design for high-resolution patterning." *Nanoscale* 16, no. 8 (2024): 4212–4218.

31. Huang, Hongwei, Haolan Wang, Yuyang Liu, and Sikun Li. "TransUNet-based end-to-end proximity effect correction for electron beam lithography." In *Eighth International Workshop on Advanced Patterning Solutions (IWAPS 2024)*, vol. 13423, pp. 262–269. SPIE, 2024.

https://doi.org/10.1142/9789819830763_0004

CHAPTER 4

Machine Learning in Emerging Nanofabrication Processes: From Imprinting and Tip-Based to Ion-Beam, Laser, and Capillary Approaches

Ampere A. Tseng[1,*], Miroslav Raudensky[2], Jun-ichi Shirakashi[3]

[1]Manufacturing Institute, Arizona State University, Tempe, Arizona 85287-6106, USA
[2]Faculty of Mechanical Engineering, Brno University of Technology, 616 69 Brno, Czech Republic
[3]Department of Electrical Engineering and Computer Science, Tokyo University of Agriculture and Technology, Tokyo 184-8588, Japan
*Corresponding author: ampere.tseng@asu.edu

Abstract

Machine learning is increasingly revolutionizing emerging nanofabrication processes by enabling predictive modeling, process optimization, and autonomous control at the nanoscale. This review systematically examines machine learning applications across five representative techniques — nanoimprinting, tip-based nanofabrication, focused ion beam nanofabrication, nanocapillary force lithography, and laser nanofabrication. Fifteen key studies employing 26 machine learning algorithms were analyzed and demonstrated significant improvements in accuracy, defect reduction, and process adaptability compared with conventional methods. The review identifies artificial neural network, support vector regression, and general regression neural network as top performers in precision and efficiency, while highlighting critical challenges such as limited datasets, interpretability, and scalability. Future research directions emphasize data-efficient hybrid models, explainable artificial intelligence, and real-time adaptive control. Overall, machine learning-assisted nanofabrication represents a paradigm shift toward intelligent, sustainable, and high-precision nanofabrication, bridging computational intelligence with experimental nanoscale engineering.

Keywords: atomic force microscopy, ion beam, machine learning, nanocapillary force lithography nanofabrication, nanoimprint, scanning probe microscopy, tip-based nanofabrication

4.1 Introduction

Machine learning (ML), a subfield of artificial intelligence, has emerged as a transformative paradigm across science and engineering. Its primary goal is to develop systems that learn from data to perform prediction, classification, and decision-making tasks without explicit rule-based programming.[1] By leveraging large datasets, ML algorithms automatically uncover latent patterns and correlations, enabling data-driven modeling of highly nonlinear and complex systems that are often intractable through analytical or empirical approaches. A typical ML workflow — comprising data collection, preprocessing, model selection, training, and validation — provides a systematic framework for improving accuracy, adaptability, and scalability across diverse applications.[2]

In recent years, ML has been increasingly integrated into nanofabrication, the science of creating materials, patterns, and devices with dimensions below 100 nanometers. Traditional nanofabrication methods, governed by coupled physical and chemical interactions, often suffer from high experimental cost, limited throughput, and unpredictable process variability. ML offers a powerful computational framework to overcome these limitations through predictive modeling, process optimization, adaptive control, anomaly detection, and autonomous operation.[3,4] These data-driven approaches not only enhance precision and efficiency but also enable self-optimizing and adaptive fabrication platforms that integrate experimental feedback into intelligent processing platforms.

While semiconductor manufacturing remains the most industrially optimized nanofabrication technique, numerous emerging or specialized nanofabrication methods — such as nanoimprinting lithography, tip-based nanofabrication, focused ion beam nanofabrication, nanocapillary force lithography, and laser nanofabrication — offer unique advantages in resolution, material selectivity, cost effectiveness, or pattern versatility. However, these processes have yet to achieve the yield, reproducibility, and scalability required for high-volume manufacturing. ML integration has become a key enabler in accelerating their transition toward practical and industrial relevance by providing tools for real-time prediction, process tuning, and defect minimization.

This review focuses on these five emerging nanofabrication processes, highlighting how ML accelerates their technical maturation and broadens their application potential. Specifically, it analyzes 15 representative articles, categorized into three main sections: (1) nanoimprinting (seven articles), (2) tip-based nanofabrication (four articles), and (3) other processes (four articles). Each article is reviewed in terms of (1) motivation and objectives for ML integration, (2) algorithms and models employed, (3) dataset generation and training strategies, and (4) quantitative performance metrics.

These articles were identified through systematic searches across Google Scholar, Engineering Village, and IEEE Xplore, employing diverse keyword combinations including ML, artificial intelligence, nanoimprinting, tip-based nanofabrication, focused ion beam, laser nanofabrication, nanomanufacturing, and atomic force microscopy. From over 100 initial results, a multi-stage screening process was applied based on topical relevance, methodological depth, and publication recency. The final corpus of 15 articles was selected for their direct emphasis on fabrication processes, excluding studies focused solely on design, inspection, or device characterization, and for meeting scholarly criteria such as citation impact and technical rigor.

By bridging computational intelligence with nanoscale engineering, ultimately, this review aims to provide a state-of-the-art perspective on ML-assisted nanofabrication, emphasizing demonstrated advances, prevailing challenges, and future directions.

4.2 Machine Learning-Assisted Nanoimprinting

Two principal setups have been developed for nanoimprinting: 1) Stationary mold (non-rotating), typical for batch nanoimprinting, relatively slow but precise; and 2) Rotating roll mold, used in roll-to-roll nanoimprinting (R2R-NI), which enables high-throughput replication over large patterns.

Figure 4.1 illustrates a stationary mold nanoimprinting process, consisting of four major steps: a) Fabrication of a

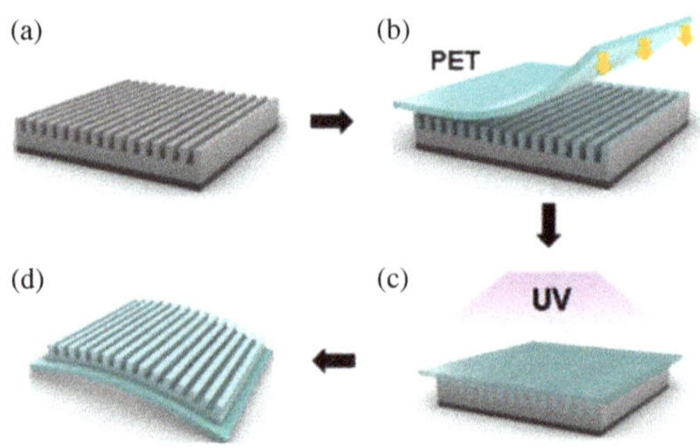

Figure 4.1. Schematic of a stationary mold nanoimprinting process: (a) stationary nanograting mold, (b) spin-coating of polyurethane acrylate resin and covering with polyethylene terephthalate film, (c) Pressing and UV curing, and (d) peeling off the polymer nanograting film. (Courtesy of Jeong et al.,[5] reprinted from open-access data.)

nano-patterned mold; b) Coating a soft resin in a rubbery state (e.g., polyurethane acrylate) onto a thin plastic sheet (e.g., polyethylene terephthalate). The resin-coated sheet is placed onto the stationary mold; c) Pressing the resin into the mold, followed by UV curing; d) Peeling off the hardened resin film (light blue in Figure 4.1d), with or without the plastic backing (dark blue), leaving the desired nanostructure.[5]

As shown in Figure 4.2, an R2R-NI system, similar to plastic calendering, consists of a hard patterned roll and an elastomeric roll.[6] A soft film passes through the gap between the two rolls. The elastomeric roll, pressed against the patterned roll by actuators at its ends, transfers the nanoscale pattern onto the film. However, the applied load can induce bending and deflection of the elastomeric roll (the exaggerated "bending curve" shown in Figure 4.2), which results in non-uniform pressure, deformation distributions and degraded pattern uniformity. It is worth noting that the patterned roll, typically made of steel alloys, is much larger and stiffer than the elastomeric roll; therefore, its bending during operation is minimal and can be neglected compared with that of the elastomeric roll.

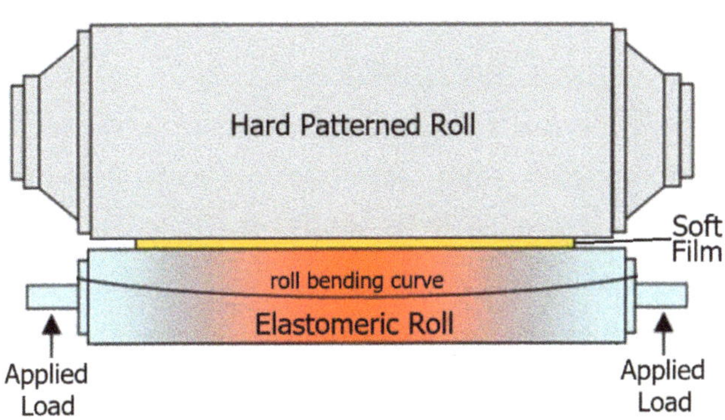

Figure 4.2. Schematic of a roll-to-roll nanoimprinting setup.

The following subsections review articles that apply ML techniques to improve nanoimprinting processes. Notably, one article specifically applies ML to minimize pattern non-uniformity in R2R-NI. Finally, for readers' convenience, the key data and major findings of each study are consolidated into Table 4.1, which provides a chronological, comparative overview of the reviewed literature.

4.2.1 *Machine Learning-Assisted Quality Rank Prediction of Nanoimprint Lithography*

Akter and Desai developed an ML-assisted framework for predicting imprint quality ranks and minimizing defects in nanoimprinting using static molds (Figure 4.1).[7] Three ML models — general regression neural network (GRNN), back-propagation neural network (BPNN), and probabilistic neural network (PNN) — were implemented to predict imprint quality using datasets compiled from both experiments and literature. GRNN offers fast training without iterative convergence issues, making it highly efficient for imprint quality prediction. BPNN provides flexibility in handling linear and nonlinear relationships but requires extensive iterative training and large datasets. PNN, a Bayes–Parzen classifier, delivers faster training and better generalization for classification problems.

Datasets for model training and testing were derived from two primary sources. First, experiments were conducted using a thermal nanoimprint lithography (NIL) system (Nanonex 2000) under varied temperature, pressure, and imprinting time conditions, with imprint quality assessed via scanning electron microscopy (SEM). Second, additional data were collected from published experimental and computational studies,[8,9]

Table 4.1. ML-assisted nanoimprinting.*

Section. Author (Year)[Ref.]	Subject	ML Models, Algorithms, and Platforms	Training and Testing Data	Key Equipment	Key Findings for Model Performance
4.2.1. Akter & Desai (2018)[7]	Quality prediction for nanoimprinting	BPC, GRNN, LM, BPNN, PNN	375 experimental and simulated data sets: training 75% and testing 25%	Nanonex 2000 NI, SEM	GRNN achieved AC = 92.37%, R = 0.921 in quality predictions and outperformed BPNN and PNN
4.2.2. Hirai et al. (2022)[10]	Predicting material formability in nanoimprinting	BPNN, deep learning	43 experimental and 120 simulated data sets: 70–85% for training	Stationary mold nanoimprint equipment	BPNN achieved ~10% error (SSE = 0.05) in predicting pattern height
4.2.3. Sayed et al. (2023)[11]	Predicting φ, Erec and Fmax for metallic glasses nanoimprinting	MELM, ELM	Simulated data sets (no number specified)	Unspecified	MELM achieved $R^2 = 0.959$, 0.976 and 0.971 in predicting Erec, φ, and Fmax, respectively
4.2.4. Nguyen et al. (2023)[12]	Improving large imprint uniformity in R2R-NI using backup roll	ANN, TensorFlow	233 simulated data sets: training 75% and test 25%; experimental data sets for validating	R2R-NI lithography system, thin-film sensors	ANNs achieved AC = 97.5% in non-uniformity prediction comparing with experimental, and AC = 99.99% in comparing with simulated
4.2.5. Jeong et al. (2024)[5]	Improving optical performance of nanoimprint-fabricated nanograting	ANN, GR, KNN, SVM, SVR	216 simulated data sets for training and testing; experimental data for validating	Experimental scale nanoimprinting, KrF lithography	SVR achieved $R^2 = 0.995$, MSE = 0.004 in transmittance predictions, outperforming KNN, GR and ANN, with 10–57% improvement in device performance

(Continued)

Table 4.1. (*Continued*)

Section. Author (Year)[Ref.]	Subject	ML Models, Algorithms, and Platforms	Training and Testing Data	Key Equipment	Key Findings for Model Performance
4.2.6. Sun et al. (2025)[15]	Predicting replication quality (CFP) in nanoimprint	BPNN, CNN, MFNN, PNN	100 experimental data points: training 70%, testing 14%, validating 14%	Experimental nanoimprinting setup	BPNNs achieved AC = 95.5%, MSE = 0.00065, $R^2 = 0.955$ in CFP prediction, outperforming CNN and PNN
4.2.7. Chien & Lee (2025)[16]	Layout-to-SEM image reconstruction for nanoimprint lithography	CNN, CQR, U-Net, BM, TFTM	240 simulated SEM images for training, validation, calibration, and testing	Not applicable	TFTM reduced MAE by 30–34% and increased 2.4–3.2% in predicting ICR, with 80% less data usage and 72% training time

*The representing words or phrases for the acronyms or abbreviations used in the above table are defined below in alphabetical order: 1) φ = Strain localization, 2) AC = Accuracy, 3) ANN = Artificial neural network, 4) BM = Baseline model, 5) BPC = Bayes–Parzen classifier, 6) BPNN = Back-propagation neural network, 7) CFP = Cavity filling proportion, 8) CNN = Convolutional neural network, 9) CQR = Conformalized quantile regression, 10) ELM = Extreme learning machine, 11) Erec = Elastic recovery, 12) Fmax = Maximum imprint force, 13) GR = Gaussian regression, 14) GRNN = General regression neural network, 15) ICR = Interval coverage rate, 16) KNN = k-nearest neighbor, 17) LM = Levenberg–Marquardt, 18) MELM = Meta-learning extreme learning machine, 19) MFNN = Multi-layer feedforward neural network, 20) ML = Machine learning, 21) MSE = Mean squared error, 22) NI = Nanoimprinting, 23) PNN = Probabilistic neural network, 24) R = Correlation coefficient, 25) R^2 = Coefficient of determination, 26) R2R= Roll-to-roll, 27) SEM = Scanning electron microscopy, 28) SSE = Sum of squared error, 29) SVM = Support vector machine, 30) SVR = Support vector regression, and 31) TFTM = Transfer fine-tuning model.

encompassing variations in NIL parameters, stamp geometries, and material types.

Data compilation proceeded in two stages. In Stage 1, 75 datasets containing six input factors were collected. In Stage 2, the dataset was expanded to 300 entries with nine input factors — including process, material, and geometry parameters — while maintaining a single output variable — the *quality of imprint* — ranked from 1 (low) to 3 (high) based on SEM images. The data were divided into 75% (281 sets) for training and 25% (94 sets) for testing. All three ML models were implemented in MATLAB using the Levenberg–Marquardt algorithm, with robustness evaluated through tenfold randomized cross-validation.

Model performance was assessed using mean squared error (MSE) and regression analysis by predicting imprint quality ranks. In Stage 1, GRNN achieved the highest accuracy (58.22%) with an overall correlation coefficient (R) of 0.46871, indicating weak prediction accuracy. In Stage 2, all models showed marked improvement: GRNN achieved the best performance with 92.37% accuracy (R = 0.92089), followed by PNN (91.19%, R = 0.91) and BPNN (84.79%, R = 0.91). GRNN consistently outperformed the other models, demonstrating superior robustness and predictive accuracy for NIL imprint quality. Increased dataset size and inclusion of additional input factors significantly enhanced model performance. The R-value here represents the correlation coefficient, quantifying the linear relationship between predicted and actual values, ranging from −1 (perfect negative correlation) to +1 (perfect positive correlation).

Overall, this study illustrates that artificial neural network (ANN)-assisted frameworks — particularly GRNN — can serve as effective predictive tools for assessing NIL process quality and reducing fabrication defects.

4.2.2 Experimental and Simulation-Based Learning for Formability Prediction in Nanoimprinting

Hirai et al. developed a hybrid deep learning-based system for optimizing materials and process parameters in stationary-mold nanoimprinting by predicting imprinting outcomes for various materials and process conditions using both experimental and simulation data for model training and testing.[10] The study focused on predicting the formability of polyvinyl alcohol films with glycerol additives in a low-temperature nanoimprint process and using polymethyl methacrylate as a reference material in simulation-based learning.

A multi-layer BPNN served as the core algorithm of the hybrid system. This multi-layer BPNN was a deep learning model comprising an input layer, two hidden layers, and an output layer, where neurons were interconnected through adjustable weights and trained using the gradient descent method to minimize prediction errors. The BPNN employed the sigmoid activation function to transform inputs into nonlinear outputs.

The hybrid system combined two complementary components: experimental learning and simulation-based learning. The experimental learning model used data from 43 nanoimprinting experiments to predict pattern height, a measure of formability, based on glycerol concentration and process temperature. The simulation-based model relied on 120 numerically generated datasets involving polymethyl methacrylate to predict pattern height as a function of mold geometry, polymer topography, and press pressure. Approximately 70–85% of the datasets were used for training, with the remainder reserved for testing and validation.

The input parameters of the models included pressure, temperature, glycerol concentration, and topographical features, while the output was the predicted pattern height after imprinting. Model performance was evaluated using the sum of squared error. The optimal configuration, two hidden layers with ten nodes each, yielded a sum of squared error below 0.05 μm^2, corresponding to roughly 10% average error for pattern heights in the 2–3 μm range. As shown in Figure 8 of the original paper, the predicted pattern heights were in good agreement with experimental results, although slight overestimations occurred at higher temperatures (e.g., 140 °C) and low glycerol concentrations (<10%).

Overall, while the BPNN predictions were not flawless, the model effectively captured key trends and provided practical process design guidelines for nanoimprint optimization. The authors noted that increasing the volume of training data could further enhance accuracy but would require significant experimental effort.

4.2.3 *Machine Learning-Assisted Prediction of Elastic Recovery, Strain Localization, and Imprint Force in Nanoimprinting*

Sayed et al. developed an ML-assisted framework to predict key parameters in the nanoimprinting of CuZr metallic glasses, including elastic recovery (*Erec*), strain localization (φ), and maximum imprint force (*Fmax*).[11] Here, *Erec* represents the extent to which the material returns to its original shape after unloading; a lower *Erec* indicates higher imprinting accuracy due to reduced deformation errors. The parameter φ quantifies strain concentration, where a larger value signifies sharper localized deformation that can degrade pattern fidelity and cause non-uniform imprinting.

The proposed predictive framework employed the meta-learning extreme learning machine (MELM) algorithm, which integrated both structural features of metallic glasses and imprinting process parameters to predict φ, *Erec*, and *Fmax* in the nanoimprinting of CuZr metallic glasses. MELM improved upon the conventional extreme learning machine by partitioning the training dataset into multiple subsets for meta-learning. Each subset was trained independently to enhance robustness, adaptability, and weight initialization. This approach enabled MELM to map input descriptors (comprising structural and process features) to output targets with fast convergence, strong generalization, reduced overfitting, and high predictive accuracy, making it particularly suitable for modeling the nanoimprinting behavior of CuZr metallic glasses.

Training and testing datasets were generated through molecular dynamics simulations of the nanoimprinting process under varied conditions. Atomic boxes with CuZr alloy compositions ($Cu_{64-50}Zr_{36-50}$) were prepared and subjected to filling temperatures ranging from 300–700 K and filling displacement rates from 2×10^{-5} to 5×10^{-5} nm per time step (1 femtosecond). The CuZr metallic glasses were produced by heating to 2,500 K, quenching at 10^9 K/s to form amorphous structures, followed by relaxation to remove defects. Input descriptors included both structural features (coordination number, free volume, flexibility volume) and process parameters (imprinting temperature, filling rate). Outputs were the three target responses: φ, *Erec*, and *Fmax*. Data were normalized using min–max rescaling and split into training (80%) and testing (20%) sets with fivefold cross-validation. However, the authors did not report the total number of simulated datasets or whether experimental data were used for validation.

Regression analyses showed that the MELM model achieved strong correlations between predicted and simulated results, with R^2 values of 0.959 for *Erec*, 0.976 for φ, and 0.971 for *Fmax*. The model also exhibited minimal variation, with mean standard deviations of 1.9% for *Erec*, 0.015 for φ, and 1.05 μN for *Fmax*. Error ratio distributions for all outputs were tightly centered around zero, confirming the model's high predictive reliability.

These results highlight the MELM model's strong performance in predicting key parameters of the nanoimprinting process, accurately capturing the complex relationships between structural and process descriptors with minimal error. Overall, the authors presented a robust and efficient MELM-assisted framework for characterizing critical parameters in the nanoimprinting of CuZr metallic glasses.

4.2.4 *Machine Learning-Assisted Prediction of Imprint Uniformity in Roll-to-Roll Nanoimprinting with Backup Roll*

Nguyen et al. developed an ML-assisted framework to enhance imprint uniformity in large-area R2R-NI lithography using a backup roll.[12] The framework integrated ANN modeling, finite element simulations, and grey relational analysis to minimize non-uniform pressure distribution, a key factor for high-fidelity pattern replication. The study investigated the backup roll's role in compensating bending deformation and experimentally validated the approach.

As shown in Figure 4.2, in R2R-NI, a thin soft film passes through the roll gap between a hard pattern roll and an elastomeric roll. Roll bending often causes imprint non-uniformity, which can be mitigated using a backup roll, crowned rolls, or through roll crossing.[13] The authors focused on the backup roll strategy, employing an additional roll to back up the elastomeric roll to compensate for bending deformation.

Two ANN models were implemented in Python using TensorFlow and Keras. The first predicted imprint non-uniformity based on input parameters including elastomeric roll length, backup roll length, and force ratio, the ratio of backup roll force (F_2) to elastomeric roll force (F_1). The second ANN predicted the distributed force (DF) curve along the roll gap using inputs such as the average applied force. Model precision was enhanced by predicting the residual error of the intercept (C) in the quadratic DF curve modeled by the equation of $DF = Ax^2 + C$.

Grey relational analysis was applied to identify optimal roll parameters (length, diameter, thickness) under varying applied forces, speeds, and alignments to minimize non-uniformity. As a multi-attribute optimization method, grey relational analysis effectively quantifies the relationships among influencing variables.

Finite element simulations in ANSYS generated datasets for training and testing ANN models by modeling the effects of roll dimensions and force ratios on pressure distributions. For non-uniformity prediction, 125 datasets were created using three inputs: elastomeric roll length (L), backup roll length (B), with f fixed at 3 N/mm. For DF curve prediction, 108 datasets were generated with four inputs (f, L, B, R). These datasets were normalized, scaled, and split into 75% training and 25% testing sets.

Experimental validation employed a large-scale R2R-NI system with a 2,650-mm steel hard roll, 1,400-mm elastomeric roll, and 600-mm backup roll. A Tekscan Pressure Mapping Sensor 5555 measured pressure distributions, showing excellent agreement among ANN predictions, finite element simulations, and experiments. A representative case reported non-uniformity values of 2.048% (ANN prediction), 2.0479% (finite element simulation), and 2.10% (experiment), corresponding to 99.99% and 97.5% accuracy relative to simulation and experiment, respectively. Here, non-uniformity quantifies variation in DF along the roll contact area, with lower values indicating improved uniformity.

The ANN model's accuracy was further confirmed by MSE analysis against finite element data. For non-uniformity prediction, the MSE was 0.41. For DF curve prediction, data refinement improved accuracy significantly, reducing MSE from 0.395 to 0.14 for coefficient A (a 64.6% reduction) and from 0.056 to 0.003 for intercept C (a 94.6% reduction). The refinement involved recognizing a linear relationship between the f and the intercept C of the DF distribution curve. These results validate the model's strong capability to predict force distribution and deformation behavior.

Overall, this study demonstrated that ML can accurately predict and enhance imprint uniformity in R2R-NI with a backup roll, enabling real-time process control and optimization. The approach adopted advances imprint reproducibility, uniformity, and scalability for large-area nanoimprinting.

4.2.5 Support Vector Regression-Assisted Prediction of Nanoimprinted Nanograting Structures for Enhanced Organic Light-Emitting Diodes and Solar Cell Performance

Jeong et al. introduced a Support Vector Regression (SVR) model to predict the transmittance of nanograting structures by capturing nonlinear relationships among structural parameters: pitch, duty cycle, height, and optical performance.[5] Two nanograting devices, organic light-emitting diodes and organic solar cells, were fabricated using stationary mold NIL and analyzed via finite difference simulations to develop an efficient ML model for enhancing light manipulation, reducing reflection and scattering, and improving device efficiency.[14]

Four ML algorithms — SVR, k-nearest neighbor (KNN), Gaussian Regression (GR), and ANN — were adopted for predicting the performance of nanograting structures. SVR served as the primary model, trained with limited data to predict continuous transmittance and enable contour mapping for structure design. It extends support vector machines (SVMs) to regression tasks by identifying a hyperplane that fits the data within a defined error margin. KNN, GR, and ANN were used as benchmarks. KNN predicts values based on nearby data averages, GR outputs distributions probabilistically as Gaussian processes, and ANN learns complex patterns through interconnected nodes.

Finite-difference time-domain simulations generated 216 transmittance data points across pitch (200–400 nm), duty cycle (0.4–0.8), and height (200–400 nm) at $\lambda = 850$ nm. An optimized radial basis function kernel was tuned by grid search to effectively map nonlinear input–output relationships.

Experimental validation involved fabricating three nanograting structures: (1) P = 200 nm, D = 0.5, H = 250 nm; (2) P = 200 nm, D = 0.5, H = 350 nm; and (3) P = 400 nm, D = 0.625, H = 350 nm. Polyurethane acrylate and polyethylene terephthalate substrates were patterned via stationary mold nanoimprinting lithography through four steps as shown in Figure 4.1: (a) silicon mold fabrication, (b) polyurethane acrylate resin spin-coating, (c) UV curing at 365 nm (50 mJ/cm^2), and (d) mold detachment. The transmittance of these three nanograting structures was measured using UV-visible spectroscopy in the visible light range (400–1000 nm), showing closely matched SVR predictions.

When integrated into solution-processed organic light emitting diodes and organic solar cells, the nanograting substrates markedly improved performance over flat substrate-based reference devices: solution-processed organic light emitting diodes exhibited a 22.9% gain in external quantum efficiency (6.86% vs. 5.58%) and a 57.9% luminance increase (17,038 cd/m^2 vs. 10,792 cd/m^2), while organic solar cells showed a 10.7% rise in power conversion efficiency (8.57% vs. 7.74%). Although the paper reported improvements of 17% and 16% for external quantum efficiency and luminance, respectively, the higher values calculated here (from data on page 7 in the reviewed article) likely reflect minor discrepancies or typographical errors.

By comparing with finite-difference time-domain simulations, model performance across 450, 650, and 850 nm confirmed SVR's superiority, achieving coefficient of determination (R^2) = 0.995 and MSE = 0.004, surpassing KNN (R^2 = 0.971, MSE = 0.022), GR (R^2 = 0.833, MSE = 0.130), and ANN (R^2 = 0.909, MSE = 0.071, training time 840.97 s vs. 8.38 s for SVR). The SVR also accurately predicted transmittance for untrained nanostructures, enabling efficient contour mapping and fabrication tolerance analysis.

Overall, the study demonstrated that SVR provides a robust and data-efficient framework for optimizing nanograting-based optoelectronic devices. By integrating ML prediction with scalable nanoimprinting fabrication, Jeong et al. offered a practical pathway to enhance light management and device efficiency while addressing the design and manufacturing challenges of complex nanostructures.

4.2.6 *Machine Learning-Assisted Prediction of Replication Quality for Thermal Nanoimprinting*

Sun et al. applied BPNNs to prediction models for evaluating replication quality in thermal NIL.[15] The model used four key imprinting parameters — initial resist thickness, imprint temperature, imprint duration, and average pressure — as inputs to predict replication quality, quantified by the cavity filling proportion (CFP). CFP measures the extent to which mold cavities are filled with resist material after imprinting; higher CFP values indicate better replication fidelity with fewer defects.

Three ML algorithms — BPNN, convolutional neural network (CNN), and PNN — were implemented to predict CFP. BPNN served as the primary model, while CNN and PNN were used as benchmarks. BPNN, a multi-layer feedforward neural network trained via the error back-propagation algorithm, adjusts weights based on the error gradient to minimize prediction errors. CNN leverages convolutional layers to extract spatial and hierarchical features, making it suitable for image-like data, whereas PNN employs kernel-based probabilistic classification for pattern recognition tasks.

The dataset comprised 36 thermal NIL experiments involving four process parameters and corresponding CFP outputs. To overcome data scarcity, data augmentation (via linear interpolation and Gaussian noise addition) expanded the dataset to 100 samples. Preprocessing included normalization and outlier removal using the three-sigma rule, followed by fivefold cross-validation to enhance generalization. The dataset was divided into training (70%), testing (14%), and validation (14%) subsets, with 2% unallocated.

By comparing CFP prediction with the experimental augmented data, BPNN achieved the best predictive performance with 95.5% accuracy, outperforming CNN (93.33%) and PNN (90.0%). BPNN yielded an MSE of 0.00065, $R^2 = 0.955$, superior to CNN (MSE = 0.00075, $R^2 = 0.948$) and PNN (MSE = 0.00080, $R^2 = 0.940$). In classification metrics, BPNN attained 96% precision and 95% recall, indicating strong reliability in identifying replication quality patterns. Moreover, BPNN demonstrated remarkable computational efficiency, with training and prediction times of 0.002 s and 0.0015 s per sample, respectively, much faster than CNN (training: 3.42 s, prediction: 0.0401 s) and comparable to PNN (training: 0.001 s, prediction: 0.001 s). The model's mean relative error of 9.85% and average relative error of 4.37% further underscored its predictive robustness.

Overall, the study demonstrates that BPNNs can accurately and efficiently predict NIL replication quality, outperforming CNN and PNN models across accuracy, MSE, R^2, and average relative error. The BPNN approach not only reduces experimental costs but also enhances research efficiency, offering significant potential for high-throughput nanoimprinting.

4.2.7 *Data-Efficient Deep Learning for Uncertainty-Aware Scanning Electron Microscopy Reconstruction in Nanoimprinting*

Chien and Lee developed a data-efficient deep learning framework to predict post-imprint SEM images and pixel-level uncertainty maps from binary design layouts in NIL.[16] The predicted SEM images were used to identify defects, assess pattern fidelity, and analyze nanoscale features, facilitating optical proximity correction optimization, precise patterning, and enhanced process robustness. Pixel-wise uncertainty maps complemented this by highlighting error-prone regions, enabling targeted corrections and defect-aware optimization.

Two deep learning algorithms were designed for layout-to-SEM image reconstruction and uncertainty quantification. A U-Net-based CNN predicted grayscale SEM images from binary mask layouts. A shallow U-Net variant was employed to avoid overfitting and ensure data efficiency, suitable for the limited dataset. Its encoder–decoder structure with skip connections preserved spatial features and captured intricate nanoscale patterns critical to NIL. Conformalized quantile regression was integrated to generate statistically valid prediction intervals for each pixel, supporting reliable uncertainty quantification. Transfer learning further refined the model by freezing the encoder and fine-tuning the decoder on high-uncertainty regions, improving accuracy and robustness.

A custom physics-based simulation model generated training and testing data, modeling NIL through three stages — electron beam lithography, reactive ion etching, and nanoimprinting — to produce realistic, noisy SEM-like grayscale images. The electron beam lithography and reactive ion etching stages created the master imprint mold. The input to the ML models consisted of binary design layouts (256 × 256 pixels), and the output was grayscale SEM-like images (256 × 256 pixels) reconstructing post-imprint topographies. A total of 240 labeled images were simulated and divided into training (60%), validation (18%), calibration (12%), and testing (10%) subsets.

Two models were developed and compared: the baseline model and the transfer fine-tuning model (TFTM). The baseline model combined the shallow U-Net with conformalized quantile regression and used the full dataset, requiring 36 minutes of graphics processing unit training. The TFTM used the same architecture but froze the encoder to retain learned hierarchical features while fine-tuning the decoder and output layers. It trained on only 48 labeled images in 10 minutes, focusing on high-error regions identified by conformal prediction.

Performance metrics showed that TFTM substantially outperformed baseline model in accuracy, reliability, and computational efficiency when compared with simulated SEM images. On the calibration set, the mean absolute error from TFTM dropped from 0.0355 to 0.0235 (a 34% reduction), while the prediction interval coverage rate rose from 90.2% to 93.1% (a 3.2% increase). On the test set, the mean absolute error decreased from 0.0365 to 0.0255 (a 30% reduction), and coverage improved from 90.4% to 92.6% (a 2.4% gain). The TFTM reduced data usage by 80% (48 vs. 240 images) and graphics processing unit time by 72% (10 vs. 36 minutes).

Overall, the study presented a highly data-efficient transfer fine-tuning framework for layout-to-SEM image reconstruction, achieving superior accuracy, reliability, and computational efficiency. This approach enables robust, uncertainty-aware prediction for advanced NIL processes.

4.3 Machine Learning in Tip-Based Nanofabrication

The applications of ML to tip-based nanofabrication (TBN) presented in four selected articles are reviewed in this section. TBN encompasses techniques that use a sharp nanoscale tip to manipulate, remove, or deposit materials at the atomic or molecular level, enabling the creation of complex structures for electronics, bioengineering, and materials science.[17]

TBN techniques derive from scanning probe microscopy (SPM), particularly atomic force microscopy (AFM) and scanning tunneling microscopy (STM), developed in the 1980s and 1990s.[18] AFM and STM operate on distinct physical principles. AFM measures interatomic forces between a sharp tip and the sample surface, allowing imaging and material modification on both insulating and conductive substrates through mechanical, electrical, or thermal interactions. In contrast, STM relies on quantum tunneling current between a conductive tip and the sample, limiting its use to conductive or semiconductive surfaces but offering true atomic-scale spatial resolution. Consequently, AFM provides broader material versatility, while STM enables unparalleled atomic precision in nanofabrication.

TBN is valued for its high precision, relatively low cost, and flexibility to operate in diverse environments, air, vacuum, and liquid. This section reviews four articles demonstrating how ML enhances TBN capabilities, improving control, reproducibility, and scalability.

The four reviews are presented in chronological order in the following subsections. The key data and major findings of each review are presented in Table 4.2 for clarity and comparison.

4.3.1 *Machine Learning-Assisted Autonomous Molecular Manipulation Using Scanning Probe Microscopy*

Leinen et al. developed an ML-assisted framework for autonomous molecular manipulation using an ultra-high-vacuum, low-temperature non-contact SPM on a supramolecular substrate composed of a compact perylene-tetracarboxylic dianhydride (PTCDA) monolayer on an Ag(111) surface.[19] The framework employed reinforcement learning (RL) as a software agent to autonomously manipulate single molecules with high precision, speed, and adaptability, enabling the construction of complex supramolecular structures and paving the way for quantum matter design beyond traditional, human-controlled approaches. PTCDA, an organic semiconductor, was selected for its potential to advance quantum devices through custom molecular configurations.

The RL agent employed a neural network with a dueling architecture to autonomously control the SPM. The neural network, consisting of a three-neuron input layer and one hidden layer, approximated Q-values through separate value and advantage

Table 4.2. ML-assisted tip-based nanofabrication.*

Section. Author (Year)[Ref.]	Subject	ML Models, Algorithms, and Platforms	Training and Testing Data	Key Equipment	Key Findings for Model Performance
4.3.1. Leinen et al. (2020)[19]	Autonomous molecular manipulation by SPM	RL, NNs, DRL	Over 165 full simulated and experiment data sets	Ultra-high-vacuum SPM, PtIr tip, qPlus	DRL reduced 84% in failure rate; finding strong tips enabled faster task success
4.3.2. Jha & Agarwal (2021)[20]	Predicting L-D curves and AFM indent images for tip-based nanoindentation	ANN, LM algorithm	23,420 experiment data sets: 70% training, 15% testing, and 15% validation	Nanoinductor, AFM, MATLAB's Deep Learning Toolbox™	ANN achieved R^2 values of 0.995–0.9999 with MSE ~ 1.0 to 10.0 nm^2 in L-D curve predictions
4.3.3. Liu et al. (2023)[22]	Accelerating AFM scratch process for large-area patterning	HR-C-OCR, SSN	576 AFM images: 80% training, 10% validation, and 10% testing	NT-MDT AFM instrument	HR-C-OCR increased throughput by 48%, in AFM scratching of stitchless 1×1 mm^2 line pattern
4.3.4. Wu et al. (2024)[24]	Autonomous on-surface synthesis by STM atomic manipulation	AutoOSS, BO, DRL, MTarget, ResNet18, MInd, MDiss, MTriple	1,350 measured STM images	Createc LT-STM, with Au-coated PtIr tip	AutoOSS achieved >94% accuracy and >98% AUC in image classification, and 80% success rate in C–Br bond dissociation

*The representing words or phrases for the acronyms or abbreviations used in the above table are defined below in alphabetical order: 1) AFM = Atomic force microscopy, 2) ANN = Artificial neural network, 3) AUC = Area under the curve, 4) AutoOSS = Automated on-surface synthesis, 5) BO = Bayesian optimization, 6) DRL = Deep reinforcement learning, 7) HR-C-OCR = High-resolution convolutional object-contextual representation, 8) L-D = Load displacement, 9) LM = Levenberg–Marquardt, 10) NNs = Neural networks, 11) RL = Reinforcement learning, 12) SPM = Scanning probe microscopy, 13) SSN = Semantic segmentation network, and 14) STM = Scanning tunneling microscopy.

streams. The value function represented the benefit of the current tip position, while the advantage function evaluated how much better a specific movement was relative to other alternatives. The double Q-learning, a deep reinforcement learning (DRL) algorithm, mitigated overestimation of Q-values and enabled efficient determination of optimal manipulation trajectories.

Training data were obtained from both simulations and experiments. Simulated data were generated using synthetic bond-breaking criteria derived from prior studies, while experimental data were collected during molecular manipulation of the PTCDA monolayer using an ultra-high-vacuum, low-temperature SPM that monitored tunneling current and force gradients. Experiments were performed in non-contact mode with an SPM equipped with a qPlus tuning-fork sensor at low temperature, where the PTCDA monolayer was stabilized by a network of hydrogen bonds (right inset, Figure 4.3). During manipulation, PTCDA molecules could spontaneously attach to the PtIr SPM tip and be removed from the surface upon tip retraction along a suitable trajectory. Bond formation and rupture caused significant tunneling current fluctuations.[32] A sudden drop in current during tip retraction, indicative of bond rupture, was designated as a "failure reward" (left inset, Figure 4.3). The RL agent controlled the SPM tip via a voltage source connected to piezo actuators and could select among five predefined movement directions (Figure 4.3), optimizing trajectories based on the probability of bond-breaking events learned during training.

Each molecular removal took approximately 5–10 minutes. The RL agent stored state-action-reward-state tuples in an experience memory for training, with 10% of samples sourced from experience and 90% generated through simulation. Training

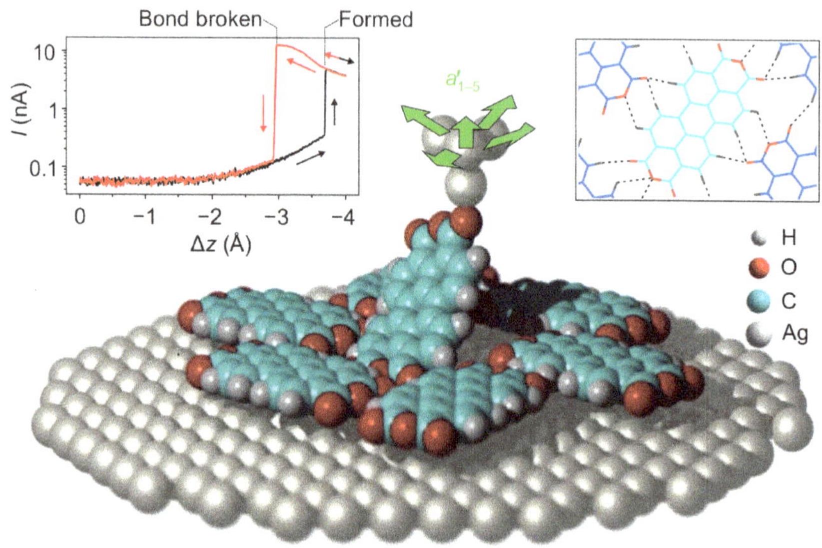

Figure 4.3. Manipulation of a PTCDA molecule on an Ag(111) surface using SPM. PTCDA molecules can spontaneously bind to the SPM tip and be removed from the monolayer upon tip retraction along a suitable trajectory. Bond formation and rupture cause pronounced tunneling current fluctuations (left inset). The task is challenging because PTCDA molecules are stabilized by hydrogen bond networks (dotted lines, right inset). The RL agent selects among five movement options (green arrows): (1) +x (+0.3 Å), (2) −x (−0.3 Å), (3) +y (+0.3 Å), (4) −y (−0.3 Å), and (5) no lateral movement ($\Delta z = 0.1$ Å). These actions allow the RL agent to adaptively adjust tip positions for optimal manipulation trajectories. (Courtesy of Leinen et al.,[19] reprinted from open-access data.)

employed an off-policy Expected SARSA algorithm with temporal difference learning, while model-based RL (Dyna algorithm) generated additional training samples using deterministic state transitions. A rupture-avoidance mechanism with negative training temperature further accelerated learning by discouraging failure states.

The RL agent generally required more than 150 episodes in simulations to successfully learn and complete the molecular removal task. Each episode represented one full attempt to lift a molecule through a sequence of actions. Although the total number of data points was not specified, the RL agent used Cartesian coordinates (x, y, z) of the SPM tip as input and determined one of five possible actions: moving the tip by +0.01 nm in z, ±0.03 nm in x or y, or remaining stationary.

Performance was evaluated based on the number of episodes required for successful manipulation and failure rates. Conventional agents exhibited about 70% failure after more than 150 episodes. Incorporating model-based RL and rupture avoidance reduced failure rates to 11% and accelerated convergence, corresponding to an 84% improvement. Pretrained agents outperformed randomly initialized ones, achieving faster learning and demonstrating transferable knowledge across different tip configurations. Tip-specific variations were also observed: strong tips achieved quicker success, whereas weak tips required more episodes and exhibited narrower success trajectories. These results highlight the RL agent's adaptability, efficiency, and robustness in nanoscale manipulation tasks.

Overall, the RL-based framework demonstrated efficient learning, robustness under varying experimental conditions, and enhanced performance through model-based planning. This study represents a significant step toward autonomous molecular manipulation and nanoscale fabrication using ML-assisted SPM.

4.3.2 *ML-Assisted Simulation of Atomic Force Microscopy Nanoindentation for Material Characterization*

Jha and Agarwal developed an ML-assisted platform, termed Graphical User Interface/Application, which leveraged ANNs to simulate the nanoindentation process performed by an AFM-like nanoindentation tester equipped with an AFM for providing indent images.[20] The platform predicted relationships between load–displacement curves and AFM indent images, thereby reducing the number of physical experiments required for material testing. The nanoindentation testing process involves pressing an indenter tip against a sample surface with a load cell measuring the depth of the indentation and force and displacement during loading and unloading. Mechanically, the nanoindentation process is one of the AFM-based modification processes, i.e., using a nanoscale tip to deform a surface. For example, Kim et al. applied AFM to successfully perform nanotip indentation lithography for fabricating plasmonic nanostructures.[21]

The ANN was implemented using MATLAB's Deep Learning Toolbox™. Two types of ANN models were developed: indentation depth models and AFM image models. Indentation depth models were used for predicting indentation depth and generating load–displacement curves based on input parameters such as maximum applied load, holding time, load, and time extracted from tester-generated text files. AFM image models were used for predicting AFM indent images based on data from the hierarchical data format files, which contained information about the surface profile in gradient and topography modes generated by AFM. The ANNs were chosen for their ability to model complex, nonlinear relationships in large, high-dimensional datasets like nanoindentation and AFM imaging data. ANNs excel at capturing intricate patterns between experimental parameters (e.g., load, holding time) and outputs (e.g., indentation depth, AFM images).

Training and testing data were obtained from the nanoindentation experiment performed on three materials/coatings: cold-sprayed aluminum-based bulk metallic glass coating, carbon nanotube-reinforced aluminum composite coating, and spark plasma-sintered hydroxyapatite samples. A total of 24 text files were produced, each containing approximately 1,034 rows of data. Each row included the maximum applied load, holding time, load–displacement curve, and the associated AFM indent image. After preprocessing, 23 files (23,420 rows) were used, 70% for training (16,394 samples), 15% for validation (3,513 samples), and 15% for testing (3,513 samples), while one file was reserved for independent evaluation. The ANNs were trained using the Levenberg–Marquardt algorithm, with optimal performance achieved using 15 hidden neurons in the hidden layer. The models were designed to simulate experimental outcomes for new test conditions, substantially reducing the need for additional physical experiments.

The accuracy of the two ANN models was evaluated using MSE and R^2 values. For the indentation depth models, predicted and experimental load–displacement curves showed excellent agreement. For cold-sprayed aluminum-based bulk metallic glass coatings, the ANN model achieved an MSE of ~4.5 nm^2 and an R^2 value near 0.999, indicating high predictive fidelity. Similarly, AFM image models successfully reproduced system-generated AFM indent images, effectively capturing experimental morphology. Comparable accuracy was achieved for other materials: for carbon nanotube-reinforced aluminum composite coatings, predictions yielded an MSE of ~1.0 and an R^2 of ~0.9999, while spark plasma-sintered hydroxyapatite samples achieved an MSE of ~10.0 and an R^2 of ~0.995. These results confirmed the model's robustness and generalizability across different materials.

In summary, the ANN-assisted platform demonstrated outstanding accuracy in predicting both load–displacement curves and AFM indent images. By enabling reliable virtual simulations of the nanoindentation process, this ML framework significantly reduces experimental workload, time, and cost in materials characterization.

4.3.3 *Machine Learning-Driven Framework for High Throughput in Stitchless Atomic Force Microscopy Nanopatterning*

Liu et al. proposed an ML-assisted framework to optimize and accelerate scanning probe lithography (SPL), focusing on three direct writing techniques: mechanical SPL, oxidation SPL, and bias-induced SPL.[22] In their NT-MDT system, the authors used AFM mode for both nano-lithography and *in situ* characterization. However, the oxidation SPL and bias-induced SPL processes described were essentially identical to AFM oxidation.[18] Likewise, in mechanical SPL, the AFM probe operated in tapping mode for dynamic plowing lithography, more commonly known as AFM scratching.[23] For clarity, the term SPL may therefore be replaced by AFM, consolidating the three techniques into two: AFM scratching and AFM oxidation. The AFM probes used featured long cantilevers with conical silicon tips, 15 μm in height and ~10 nm in apex radius (Tap190Al-G, BudgetSensors®, Sofia, Bulgaria).

Nanopatterning by AFM direct writing is highly sensitive to process conditions and requires labor-intensive characterization.[17,23] The proposed ML-assisted framework addresses this challenge by integrating *in situ* AFM topographic imaging with automated feature segmentation and parameter optimization. This enables faster, more reliable fabrication and direct extraction of key structural parameters such as full width at half maximum and pitch.

The authors employed a high-resolution convolutional object-contextual representation (HR-C-OCR) model, a supervised ML algorithm incorporating semantic segmentation networks, for pixel-level AFM image analysis. This approach enabled real-time statistical evaluation of nanostructure features, improving fabrication quality through data-driven insights. Data augmentation was applied to enhance robustness and reduce overfitting.

Training and testing data were experimentally generated by fabricating nanostructures and capturing AFM images. Silicon wafers were coated with a ~20 nm polymethyl methacrylate thin film via spin coating and baking. AFM scratching was then performed in tapping mode under ambient conditions (25 °C, ~50% humidity). The same probe conducted *in situ* semi-contact AFM characterization at 1 Hz with a resolution of 256 × 256 pixels. Flattened topographic images, processed with NOVA_PX software, served as input for the HR-C-OCR model. A total of 121 parameter pairs yielded 1,089 nanostructure lines; after augmentation, 576 grayscale AFM images were used (80% training, 10% validation, 10% testing). Training was implemented in PyTorch, with preprocessing in MATLAB. The resulting segmentation maps effectively segmented nanostructures for subsequent analysis, with false positives and negatives primarily at edges.

The framework's capability was demonstrated by fabricating a stitchless 1 × 1 mm^2 line array, the largest area achieved, featuring a 500 nm pitch and 22.99 nm critical dimension in 86 minutes. Compared to conventional stitched methods, which divide and rejoin sub-patterns, the ML-assisted approach improved throughput by 48% and eliminated stitching errors, enabling efficient large-area patterning.

Nevertheless, certain aspects remain insufficiently addressed. First, the authors provided limited information on conventional stitched AFM methods, hindering a full comparison of performance gains. Given that typical AFM scanners have lateral ranges of only 50–100 µm, producing millimeter-scale stitchless patterns poses significant technical challenges. Second, although segmentation performance metrics, pixel accuracy, mean intersection over union, and frequency-weighted intersection over union were reported, direct numerical comparisons between ML-predicted results and experimental AFM measurements were lacking. Such validation would have strengthened claims regarding fabrication accuracy.

Overall, Liu et al. effectively demonstrated that ML can accelerate the process operation in AFM scratching.[22] Their study represents an advancement toward integrating ML with AFM direct writing to enhance throughput and accuracy in TBN.

4.3.4 *Deep Learning-Driven Autonomous on-Surface Synthesis by Scanning Tunneling Microscopy*

Wu et al. developed an efficient and autonomous software infrastructure, AutoOSS (automated on-surface synthesis), to automate STM manipulations for precise and large-scale chemical transformations from the STM tip to the surface.[24] The studied transformation involved the dissociation of C–Br bonds in $ZnBr_2Me_4DPP$ molecules adsorbed on an Au(111) surface. The STM tip induced specific chemical reactions (e.g., C–Br bond cleavage) by applying optimized manipulation parameters, enabling on-surface synthesis for constructing complex molecular architectures with potential applications in optoelectronics and spintronics.

AutoOSS integrated ResNet18-based neural networks, DRL, and Bayesian optimization to detect $ZnBr_2Me_4DPP$ molecules, interpret STM outputs, and optimize manipulation parameters for accurate chemical reactions. Four ResNet18-based models — MTarget, MInd, MDiss, and MTriple — were trained to detect targets and classify outcomes as successful dissociation, intact, or indeterminate. A DRL model based on the soft actor-critic algorithm optimized STM parameters such as tip position, bias voltage, and tunneling current, while Bayesian optimization guided 3D structure search and reaction mechanism exploration. A signal classifier further analyzed voltage–topography curves to confirm dissociation success, together enabling efficient autonomous synthesis.

Training and testing data were derived from STM images of $ZnBr_2Me_4DPP$ molecules and their dissociation products, obtained using a Createc LT-STM system with a gold-coated PtIr tip. $ZnBr_2Me_4DPP$ molecules were synthesized, characterized, and evaporated onto Au(111) at 7 K. STM scans (3.5–100 nm range) were captured at 1000 Å/s and 128 pixels per frame (~42 s/image). Images were preprocessed into grayscale and binary form, with the Otsu algorithm used for contour detection and cluster exclusion. A labeled dataset of 1,350 images was manually constructed, including 270 true molecules and 1,080 non-true molecules (clusters or fragments), improving the models' discrimination accuracy.

AutoOSS achieved strong performance in autonomous molecular manipulation. Its image classifiers (ResNet18) reached >94% accuracy and area under the curve >98%, ensuring reliable reactant and product recognition. Area under the curve is the ability of ML to distinguish between different classes. Using random action, AutoOSS achieved a 34% success rate for C–Br bond dissociation, which increased to 62% after 60 DRL training episodes and further to 80% following parameter optimization (higher voltage, lower current).

This work demonstrates a powerful and scalable framework for automated, high-precision STM-based molecular synthesis, paving the way for versatile on-surface chemical manipulation across diverse molecular systems.

4.4 Machine Learning-Assisted Nanofabrication Using Ion Beam, Nanocapillary Force and Laser

Focused ion beam (FIB) nanofabrication is a direct-write process, using a tightly focused ion beam for nanoscale material removal (milling, etching) or addition (implantation, deposition). Applications include semiconductor device modification, mask repair, and fabrication of complex 2D and 3D nanostructures.[25] Successful FIB operation requires careful control of multiple parameter groups: 1) Five beam parameters: current, accelerating voltage, ion species, and spot size; 2) Four scan parameters: dwell time, pixel spacing/overlap, number of passes, and ion dose; and 3) Four material parameters: angle of incidence, atomic mass, crystallographic orientation, redeposition behavior. The interdependence of these parameters (variables) makes predictive modeling particularly challenging. One article reviewed applied ML to model the FIB milling process, illustrating the potential of ML to improve predictive accuracy and optimize nanoscale fabrication outcomes.

Two articles on nanocapillary force lithography (NFL) were selected for review. This process exploits capillary forces to fill nanoscale cavities with a photopolymeric melt, enabling patterned polymer films after UV-induced cross-linking. At the nanoscale, however, capillary rise is complex. Sudden drops or incomplete filling may occur if driving forces are insufficient. Understanding nanoscale capillarity remains challenging.

Laser nanofabrication (LN) employs tightly focused laser beams, often femtosecond pulses, to induce localized melting, ablation, or photochemical reactions for creating nanoscale structures. It offers high precision, flexibility, and non-contact processing across diverse materials.

The reviewed studies demonstrate how ML has been applied to FIB, LN and NFL processes, highlighting the breadth of applications enabled by nanoscale processes considered. In chronological order, Table 4.3 consolidates the key findings from each study.

4.4.1 *Machine Learning-Assisted Prediction of Beam Dosage and Post-Fabrication Outcomes in Focused Ion Beam Nanofabrication*

Buchnev et al. explored the use of ML to accurately predict the dosage required and post-fabrication appearance of nanostructures

Table 4.3. ML-assisted nanofabrication using ion beam, nanocapillary force, and laser.*

Section. Author (Year)[Ref.]	Subject and Process	ML Models, Algorithms, and Platforms	Training and Testing Data	Key Equipment	Key Findings for Model Performance
4.4.1. Buchnev et al. (2022)[26]	Predicting dosage required and fabricated SEM images for FIB nanofabrication	CNN, cGAN	680 SEM images and data pairs	FEI Helios Nanolab 600 FIB/SEM	CNN achieved a ~7.4% error (RMSE = 0.52 mC/cm^2) in predicting dose; cGAN achieved AC = 96–98% in predicting SEM images
4.4.2. Cho et al. (2023)[27]	Identifying rules behind nanocapillary rise in NFL	BE, DL, RC, RNN	Experiment data (no quantity specified)	AFM, NFL equipment	BE identified the forbidden gap (rise height stop) at a UV dose near 35 mJ/cm^2
4.4.3. Chapagain et al. (2024)[29]	Predicting UV dose needed for inverse design in NFL	BE, BDL	22,100 experiment and simulated data pairs	AFM, NFL equipment	BDL achieved $R^2 = 0.995$ in predicting required dose based on input heights
4.4.4. Zhang et al. (2025)[30]	Focal position detection in laser nanofabrication	RQGPR, KALS, QSVM, TNN	1,289 CCD images for training and 227 for testing	Innofocus NanoPrint, CMOS CCD camera	TNN achieved best RMSE = 0.2566–0.3427 μm, MAE = 0.1857– 0.2437 μm, $R^2 = 0.97$–0.98, MPS = 1,000) in predicting focal distance

*The representing words or phrases for the acronyms or abbreviations used in the above table are defined below in alphabetical order: 1) AC = Accuracy, 2) AFM = Atomic force microscopy, 3) BDL = Backward deep learning, 4) BE = Bayesian evolutionary, 5) CNN = Convolutional neural network, 6) cGAN = Conditional generative adversarial network, 7) CMOS CCD = complementary metal oxide semiconductor charge-coupled detector, 8) DL = Deep learning, 9) FIB = Focused ion beam, 10), KALS = Kernel approximation least squares, 11) MAE = Mean absolute error, 12) MPS = Mean prediction speed, 13) NFL = Nanocapillary force lithography, 14) QSVM = Quadratic support vector machine, 15) R^2 = Coefficient of determination, 16) RC = Reservoir computing, 17) RMSE = Root mean square error, 18) RNN = Recurrent neural network, 19) RQGPR = Rational quadratic Gaussian process regression, 20) SEM = Scanning electron microscopy, 21) TNN = Trilayered neural network, and 22) UV = Ultraviolet.

produced by FIB milling.[26] FIB milling is a complex nanofabrication technique involving multiple interdependent parameters that traditionally require extensive trial-and-error optimization.[25] In this context, "appearance" refers to the morphological features visible in SEM images, including the shape, dimensions, and defects of milled patterns.

Two ML models, a CNN and a conditional generative adversarial network (cGAN), were developed for predictive modeling. The CNN analyzed SEM images of simple geometric designs (e.g., chevrons) to predict ion beam dosage with high accuracy based on pixel intensity patterns. The cGAN, by contrast, simulated FIB milling outcomes for arbitrary patterns, learning from paired binary design images and SEM results to predict post-fabrication appearances, including artifacts, for unseen designs.

Training and testing datasets were obtained using a FEI Helios Nanolab 600 DualBeam FIB/SEM system. For the CNN, chevron structures were milled into a gold thin film at 70 distinct ion beam dosages with a fixed current of 9.8 pA, producing 630 SEM images. Half of the dataset (315 images) was used for training and half for testing. For the cGAN, 59 binary design–SEM image pairs were generated using various ion beam currents on gold films, including straight-line and circular features with diverse orientations, lengths, and widths.

The CNN demonstrated strong predictive accuracy, achieving a root mean squared error (RMSE) of 0.52 mC/cm^2 for dosage prediction. Considering a typical FIB dosage of approximately 7 mC/cm^2, this corresponds to a 7.4% prediction error. The cGAN achieved outstanding performance in reproducing SEM appearances of the *EPSRC* and *Light* logos, reaching over 96% and 98% accuracy, respectively, with mean pixel intensity errors between

1.5% and 2.6%. Predictions were generated within milliseconds, allowing near-real-time process feedback and optimization.

The ML-assisted predictive framework proposed by the authors represents a transformative step toward intelligent FIB nanofabrication. It enhances precision, reproducibility, and process efficiency while significantly reducing experimental overhead. This approach holds substantial promise for broader applications for nanoscale device prototyping.

4.4.2 Interpretable Machine Learning Framework for Uncovering Hidden Physical Rules in Nanocapillary Force Lithography

Cho, Ji, and Kim developed an ML-assisted framework to uncover the hidden physical rules governing irregularities in NFL, particularly focusing on the "forbidden gap" phenomenon that restricts achievable capillary rise heights.[27] The framework served primarily as a theoretical model for elucidating nonlinear physical interactions underlying pattern irregularities during nanoscale capillary rise.

NFL is a soft lithography technique in which liquid polymers are drawn into nanoscale cavities of a patterned mold through capillary action, driven by surface tension, adhesion, and cohesion forces. At the nanoscale, capillary rise can exhibit irregular behaviors, such as sudden drops or unreachable height ranges, when the driving force becomes insufficient. These unachievable height ranges, termed the forbidden gap, have remained poorly understood. The ML framework was designed to reveal the underlying mechanisms responsible for this phenomenon. As shown in Figure 4.4, the NFL process involves four principal steps: (a) a

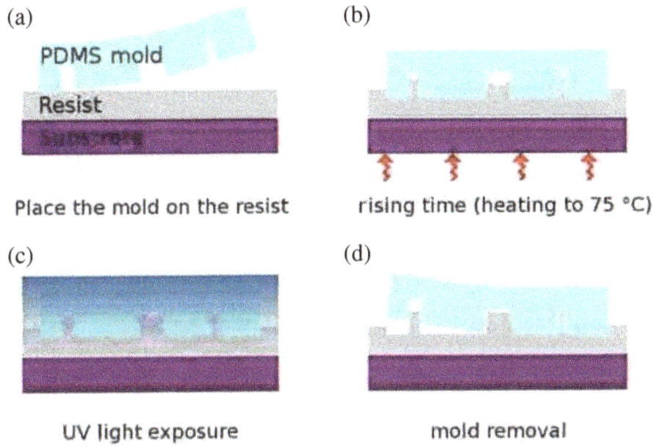

Figure 4.4. Schematic of the NFL process: (a) mold placement, (b) capillary rise of resist with optional heating, (c) UV cross-linking, and (d) mold removal. (Courtesy of Hansen et al.,[28] reprinted from open-access data.)

polydimethylsiloxane mold is placed on a resist-coated substrate; (b) the resist rises into mold grooves by capillary action, often enhanced by mild heating; (c) the resist is cross-linked under UV exposure; and (d) the mold is detached, leaving a patterned surface.[28]

The authors integrated a Bayesian evolutionary (BE) algorithm with an interpretable ML approach grounded in physics-based insights. Initial rule structures, represented as link functions, were designed to retain physical transparency. The BE algorithm employed two-parameter exponential link functions to model nonlinear dependencies among key parameters such as air diffusivity, dynamic viscosity, and surface tension. Through iterative genetic operations (mutation and selection), the BE algorithm refined these relationships, continuously updating prior knowledge with experimental data. This hybrid strategy effectively balanced exploration of new relationships with exploitation of existing physical knowledge, yielding interpretable, physics-consistent rules that successfully explained the forbidden gap phenomenon.

Alternative ML methods, including deep learning, reservoir computing, and recurrent neural networks using cubic regression spline link functions, were also tested. However, these approaches failed to identify interpretable rules or exhibited poor generalization due to the limited dataset size, highlighting the advantage of the BE-based framework in small-data regimes.

Experimental data for model training were obtained from nanocapillary rise studies using cylindrical array molds spin-coated with photocurable polymer (NOA73) and cured under UV light. The resulting nanopillar structures were characterized via AFM, generating topographical datasets capturing rise heights and UV dose effects. Input features included UV dose, air mass flux rate, and physical parameters such as pressure, temperature, and mass, while outputs consisted of air diffusivity, dynamic viscosity, surface tension, and final capillary rise height. Although the total number of samples was not specified, the dataset was divided into training and validation subsets for model development.

The BE-based model predictions closely matched experimental observations. Specifically, it successfully identified the forbidden gap near a UV dose of 35 mJ/cm^2, with predicted rise heights of 292 nm and 232 nm above and below the gap, respectively, values confirmed by AFM measurements. The top ML-derived rules reproduced experimental trends with strong consistency, demonstrating the model's capacity to generalize physical relationships. Roughly speaking, as the UV dose increases, the temperature rises, leading to a decrease in the surface tension of the photopolymer. When the UV dose reaches approximately

35 mJ/cm^2, the rise height of the liquid photopolymer experiences a sudden drop, referred to by the authors as the "forbidden gap".

Overall, this study illustrated how a physics-informed ML framework can elucidate complex nanoscale phenomena that remain inaccessible to first-principles modeling. By integrating BE with interpretable rule discovery, the approach provided a transparent and extensible foundation for understanding and predicting nonlinear effects in NFL, paving the way toward data-driven, physically grounded nanoscale process modeling.

4.4.3 Deep Learning-Assisted Inverse Design for Capillary Force Lithography

Chapagain and Cho developed an ML-based framework to predict optimal UV doses from target grating and attained heights, enabling inverse design and practical fabrication of multifunctional nano-patterned surfaces using capillary force lithography.[29]

The framework integrated two complementary ML approaches. First, a hybrid intelligence approach combined human expertise, such as physical rules and experimental observations, with ML through a BE algorithm to identify hidden rules and optimize parameters for modeling UV-controlled capillary force lithography, as discussed in Section 4.4.2.[27] This stage enriched the dataset by establishing quantitative relationships among UV dose, grating height, and attained height. Second, a backward deep learning model was trained on the enriched datasets to predict the necessary UV dose for a desired grating and attained height, thus enabling inverse design of nano-patterned surfaces.

Training data were derived from capillary force lithography experiments involving nanogratings and nanopillars of varying heights to establish correlations between UV dosage and fabricated structure dimensions. The BE algorithm further expanded this dataset, generating 22,100 data points for three key variables (UV dose, grating height, and attained height), where 85 grating height values ranged from 1.02 μm to 3.35 μm.

The backward deep learning model exhibited excellent predictive accuracy, achieving R^2 of 0.995, confirming its reliability in estimating UV doses for targeted geometries and enabling high-precision inverse process control.

In summary, this study extended prior work by integrating interpretable, physics-informed rules into a data-driven framework for the design and fabrication of multifunctional nanopatterned surfaces with tunable properties.

4.4.4 *Machine Learning-Based Real-Time Position Recognition in Laser Nanofabrication*

Zhang et al. developed an ML-based method for real-time focal position recognition in LN, achieving sub-half-wavelength precision without requiring additional optical components.[30] This approach significantly improves detection accuracy, efficiency, and adaptability in laser-based systems, addressing critical challenges such as uneven surfaces and surface tilt that often degrade focal spot positioning during fabrication.

Four ML models were evaluated: rational quadratic Gaussian process regression (RQGPR), kernel approximation least squares (KALS), quadratic support vector machine (QSVM), and a trilayered neural network (TNN). RQGPR, a probabilistic regression model, predicted focal distances by mapping relationships between focal spot features and positions. KALS, a regression model, applied kernel functions to approximate focal distances but demonstrated slower processing and lower precision. QSVM utilized a quadratic kernel for regression, offering improved accuracy. The TNN model used a deep learning approach with three fully connected layers and applied rectified linear unit activation functions for these layers, which help introduce nonlinearity and improve learning capabilities. It also included layer biases and weights to optimize its performance to achieve the highest accuracy, speed, and reliability for real-time focal distance detection in laser nanofabrication.

Training and testing data were obtained using a NanoPrint3D laser nanoprinting system (Innofocus, Australia) equipped with a femtosecond laser ($\lambda = 800$ nm, PRF = 1,000 Hz, PW = 400 fs). The experiment fabricated Archimedean spiral patterns consisting of 499 points on a 200 nm graphene oxide film inclined at 7°, generating a range of focal distances. A total of 1516 charge-coupled detector images were recorded, of which 499 had known focal distances. The remaining data were labeled via linear interpolation. Each image (205 × 205 pixels) was processed with a 51 × 51 region of interest capturing the laser focal spot intensity profile.

The dataset was divided into 85% training (1,289 images) and 15% testing (227 images) subsets. The trained models were evaluated using RMSE, mean absolute error, R^2, and mean prediction speed. The TNN model outperformed all others, achieving the lowest RMSE (0.3427 μm training, 0.2566 μm testing), lowest mean absolute error (0.2437 μm training, 0.1857 μm testing), highest R^2 (0.97 training, 0.98 testing), and fastest mean prediction speed (~1,000 observations/s). KALS exhibited the poorest performance, while RQGPR and QSVM achieved moderate accuracy and speed.

Overall, this study demonstrated a robust ML framework for real-time focal distance detection in laser nanofabrication, achieving sub-half-wavelength accuracy and eliminating the need for auxiliary optical components. The demonstrated TNN model represented a significant step toward intelligent, self-correcting laser nanofabrication systems with enhanced precision, speed, and operational simplicity.

4.5 Summary and Conclusions

This review underscores the transformative impact of ML on five emerging nanofabrication processes: nanoimprinting, TBN, FIB nanofabrication, NFL, and LN. Across 15 representative studies, 26 distinct ML algorithms and models were implemented to address process-specific challenges. Collectively, these works demonstrate that ML consistently enhances predictive accuracy, process efficiency, and adaptability beyond conventional modeling and empirical optimization. The integration of ML has enabled precise control of nanoscale parameters, reduction of fabrication defects, and optimization of structural fidelity, key advances toward intelligent nanomanufacturing.

4.5.1 *Overview of Machine Learning Algorithms and Models*

The 26 algorithms examined in this review, summarized in Table 4.4, encompass a wide range of model types — from classical regression and neural networks to advanced physics-informed and autonomous learning frameworks. Each table provides a comparative analysis of its key features, applications,

Table 4.4. Twenty-six ML models/algorithms reviewed.*

No. Model/ Algorithm	Basic Features	Applications and Section Located	Performance Metrics
1. Artificial neural network (ANN)	Inspired by the human brain, consisting of interconnected nodes (neurons) designed to process information and learn patterns from data.	Used in Sections 4.2.4, 4.2.5, and 4.3.2 for predicting imprint quality rank, improving imprint uniformity, optimizing nanograting design, and simulating nanoindentation.	Achieved 97.5–99.99% accuracy in predicting uniformity (Section 4.2.4) and R^2 = 0.995–0.9999 in predicting load–displacement curves in nanoindentation (Section 4.3.2).
2. Back-propagation neural network (BPNN)	Using back-propagation algorithm to minimize prediction errors by adjusting weights based on error gradients.	Applied in Sections 4.2.1, 4.2.2, and 4.2.6 for predicting imprint quality and formability, and replication quality in nanoimprinting.	Achieved 10% error in predicting pattern height (Section 4.2.2) and 95.5% accuracy with R^2 = 0.955 in predicting replication quality (CFP) (Section 4.2.6).
3. Backward deep learning (BDL)	Using enriched datasets to predict input parameters required to achieve desired outputs, enabling inverse design.	Used in Section 4.4.3 to predict optimal UV doses for inverse design and fabricating nanopatterned surfaces using capillary force lithography.	Achieved R^2 = 0.995 in predicting UV doses for desired grating and attained heights.

(*Continued*)

Table 4.4. (*Continued*)

No. Model/ Algorithm	Basic Features	Applications and Section Located	Performance Metrics
4. Bayesian evolutionary (BE)	Combining Bayesian inference with evolutionary algorithms to discover interpretable rules and optimize parameters.	Applied in Sections 4.4.2 and 4.4.3 to uncover hidden physical rules in NFL and predict UV doses for inverse design in capillary force lithography.	Identified the forbidden gap near a UV dose of 35 mJ/cm^2 (Section 4.4.2) and achieved $R^2 = 0.995$ in predicting UV doses for inverse design (Section 4.4.3).
5. Bayesian optimization (BO)	Using a probabilistic algorithm for exploring and exploiting parameter spaces to identify optimal solutions.	Integrated in AutoOSS framework (Section 4.3.4), BO aiding 3D structure search and optimizing reaction mechanisms for chemical transformations during STM molecular synthesis.	In Section 4.3.4, BO helped to improve the success rate of C–Br bond dissociation from 62% to 80% after parameter optimization.
6. Conditional generative adversarial network (cGAN)	Generating synthetic data by learning the mapping between input conditions and output images.	Used in Section 4.4.1 for predicting post-fabrication shape, dimensions, and defects of nanostructures in FIB milling.	Achieved 96–98% accuracy in predicting shape, dimensions, and defects in SEM images of fabricated nanostructures.

Table 4.4. (*Continued*)

No. Model/ Algorithm	Basic Features	Applications and Section Located	Performance Metrics
7. Conformalized quantile regression (CQR)	Statistically generating prediction intervals for regression tasks for reliable uncertainty quantification.	Used in Section 4.2.7 for layout-to-SEM image reconstruction to identify defects, assess pattern fidelity, and analyze nanoscale features in nanoimprinting.	Reduced MAE by 30–34% and increased interval coverage rate by 2.4–3.2% with 80% less data usage and 72% less training time.
8. Convolutional neural network (CNN)	Applying deep learning for processing grid-like data (images) by using convolutional layers to extract spatial features.	Applied in Sections 4.2.6 and 4.4.1 for predicting replication quality in nanoimprinting and analyzing SEM images in FIB milling.	Achieved $R^2 = 0.955$ in predicting replication quality (Section 4.2.6) and a ~7.4% error in predicting dosage for FIB milling (Section 4.4.1).
9. Deep reinforcement learning (DRL)	Using deep neural networks to approximate value functions and optimize decision-making.	Applied in Sections 4.3.1 and 4.3.4 for autonomous molecular manipulation using SPM and optimizing STM parameters for on-surface synthesis.	Reduced failure rates by 84% in molecular manipulation (Section 4.3.1) and achieved an 80% success rate in C–Br bond dissociation (Section 4.3.4).

(*Continued*)

Table 4.4. (*Continued*)

No. Model/ Algorithm	Basic Features	Applications and Section Located	Performance Metrics
10. Extreme learning machine (ELM)	Using a single hidden layer feedforward neural network with random weights for fast training and strong generalization.	Used in Section 4.2.3 to predict elastic recovery, strain localization, and imprint force in metallic glasses during nanoimprinting.	Achieved R^2 values of 0.959, 0.976, and 0.971 for elastic recovery, strain localization, and imprinting force predictions, respectively.
11. Gaussian regression (GR)	Probabilistically predicting outcomes based on Gaussian processes, providing uncertainty estimates for predictions	Used in Section 4.2.5 as a benchmark for predicting nanograting transmittance.	Achieved $R^2 = 0.833$ (lower than SVR and ANN) and MSE = 0.130 (higher than SVR and ANN) in predicting transmittance.
12. General regression neural network (GRNN)	Providing fast training and generalization for regression tasks without iterative convergence.	Applied in Section 4.2.1 for predicting imprint quality ranks in nanoimprinting.	Achieved 92.37% accuracy and R = 0.92089 in predicting imprint quality ranks, outperforming BPNN and PNN.
13. High-resolution convolutional object-contextual representation (HR-C-OCR)	A supervised ML algorithm using semantic segmentation networks for pixel-level image analysis.	Used in Section 4.3.3 to analyze AFM images and optimize the AFM scratching process.	Increased throughput by 48% by AFM scratching of stitchless 1 × 1 mm^2 line pattern, enabling efficient large-area patterning.

Table 4.4. (*Continued*)

No. Model/ Algorithm	Basic Features	Applications and Section Located	Performance Metrics
14. Kernel approximation least squares (KALS)	Using kernel functions to approximate relationships between input and output variables, often for nonlinear mapping.	Used in Section 4.4.4 for real-time focal position recognition in laser nanofabrication.	Demonstrated slower processing and lower precision compared to other models (TNN, RQGPR and QSVM) in focal distance prediction.
15. k-nearest neighbor (KNN)	Predicting outcomes based on the average of the nearest data points in the feature space.	Used in Section 4.2.5 as a benchmark for predicting the transmittance of nanograting structures fabricated by nanoimprinting.	Achieved R^2 = 0.971 and MSE = 0.022, which was worse than the performance of SVR but better than that of GR and ANN models.
16. Levenberg–Marquardt (LM)	An optimization algorithm used to minimize the error in nonlinear least squares problems, often for training neural networks.	Used in Sections 4.2.1 and 4.3.2 to train neural networks for predicting imprint quality ranks in nanoimprinting and simulating nanoindentation processes.	Enabled robust training of models, such as GRNN achieving 92.37% accuracy in imprint quality rank prediction (Section 4.2.1) and ANN achieving R^2 values of 0.995–0.9999 in nanoindentation predictions (Section 4.3.2).

(*Continued*)

Table 4.4. (*Continued*)

No. Model/ Algorithm	Basic Features	Applications and Section Located	Performance Metrics
17. Meta-learning extreme learning machine (MELM)	An advanced version of the extreme learning machine that uses meta-learning to partition training datasets into subsets for enhanced robustness and adaptability.	Applied in Section 4.2.3 to predict elastic recovery, strain localization, and imprinting force in the nanoimprinting of metallic glasses.	Achieved R^2 values of 0.959, 0.976, and 0.971 for elastic recovery, strain localization, and imprinting force predictions, respectively, demonstrating high predictive accuracy.
18. Probabilistic neural network (PNN)	Using a probabilistic approach for classification tasks, leveraging Bayes–Parzen classifiers for fast training and generalization.	Used in Sections 4.2.1 and 4.2.6 for predicting imprint quality ranks and replication quality in nanoimprinting.	Achieved 91.19% accuracy in predicting imprint quality ranks (Section 4.2.1) and 90% accuracy in replication quality prediction (Section 4.2.6), outperformed by GRNN and BPNN.
19. Quadratic support vector machine (QSVM)	A variant of SVM that uses a quadratic kernel for regression tasks, improving accuracy in nonlinear data mapping.	Applied in Section 4.4.4 for real-time focal position recognition in laser nanofabrication.	Achieved moderate accuracy and speed in focal distance prediction, but was outperformed by TNN.

Table 4.4. *(Continued)*

No. Model/ Algorithm	Basic Features	Applications and Section Located	Performance Metrics
20. Rational quadratic Gaussian process regression (RQGPR)	A probabilistic regression model that uses rational quadratic kernels to predict outcomes with uncertainty quantification.	Used in Section 4.4.4 for real-time focal position recognition in laser nanofabrication.	Achieved moderate accuracy in focal distance prediction, but was outperformed by TNN.
21. Recurrent neural network (RNN)	Processing sequential data by maintaining a memory of previous inputs, making it suitable for time-series and sequential prediction tasks.	Discussed in Section 4.4.2 as an alternative approach to uncover hidden physical rules in NFL.	RNN was tested but failed to identify interpretable rules or exhibited poor generalization due to the limited dataset size, highlighting the advantage of the BE algorithm in small-data regimes.
22. ResNet18	A deep convolutional neural network architecture that uses residual learning to improve training efficiency and accuracy, particularly for image classification tasks.	Applied in Section 4.3.4 as part of the AutoOSS framework to detect molecules and classify outcomes in STM-based molecular synthesis.	Achieved over 94% accuracy and an area under the curve greater than 98% in image classification, ensuring reliable recognition of reactants and products during molecular manipulation.

(Continued)

Table 4.4. (*Continued*)

No. Model/ Algorithm	Basic Features	Applications and Section Located	Performance Metrics
23. Semantic segmentation network (SSN)	For pixel-level classification in images, enabling the identification and segmentation of specific features within high-resolution images.	Used in Section 4.3.3 as part of the HR-C-OCR framework to analyze AFM images and extract structural parameters for optimizing AFM scratching processes.	Enabled real-time segmentation of nanostructures, improving fabrication quality and throughput by 48% in large-area (1×1 mm^2) AFM patterning.
24. Support vector machine (SVM)	Finding the optimal hyperplane that separates data points into different classes for classification and regression tasks.	Mentioned in Section 4.2.5 as a benchmark for predicting the transmittance of nanograting structures fabricated by nanoimprinting.	Achieved $R^2 = 0.995$ and MSE = 0.004 in transmittance predictions, outperforming KNN, GR, and ANN, and contributed to a 10–57% improvement in device performance.
25. Support vector regression (SVR)	An extension of SVM for regression tasks, which identifies a hyperplane that fits the data within a defined error margin, making it suitable for predicting continuous variables.	Applied in Section 4.2.5 to predict the transmittance of nanograting structures fabricated by nanoimprinting and optimize their optical performance.	Achieved $R^2 = 0.995$ and MSE = 0.004 in transmittance predictions, outperforming KNN, GR, and ANN models, and contributed to a 10–57% improvement in device performance.

Table 4.4. (*Continued*)

No. Model/ Algorithm	Basic Features	Applications and Section Located	Performance Metrics
26. Trilayered neural network (TNN)	A deep learning model with three fully connected layers using rectified linear unit activation functions and optimized weights and biases for high accuracy and speed in predictions.	Applied in Section 4.4.4 for real-time focal position recognition in laser nanofabrication.	Outperformed KALS, RQGPR and QSVM, achieving the lowest RMSE (0.2566–0.3427 μm), lowest MAE (0.1857–0.2437 μm), highest R^2 (0.97–0.98), and fastest mean prediction speed (~1,000 observations/s) in focal distance detection.

*The representing words or phrases for the acronyms or abbreviations used in the above table are defined below in alphabetical order: 1) AFM = Atomic force microscopy, 2) MAE = Mean absolute error, 3) MSE = Mean squared error, 4) R = Correlation coefficient, 5) R^2 = Coefficient of determination, 6) RMSE = Root mean square error, 7) SEM = Scanning electron microscopy, and 8) STM = Scanning tunneling microscopy.

and performance metrics, allowing a systematic understanding of how different ML paradigms contribute to process innovation.

These models include neural architectures such as GRNN, BPNN, PNN, ANN, and ResNet18; regression-based methods such as SVR and QSVM; and specialized frameworks including MELM, HR-C-OCR, AutoOSS, and BE algorithms. Their applications span predictive modeling, optimization, and real-time control across NIL, TBN, FIB, LN, and NFL.

While models like ANN, SVR, and GRNN demonstrated exceptional predictive accuracy and efficiency, physics-informed approaches such as BE algorithms offered interpretability and mechanistic insights into complex nanoscale phenomena. Despite challenges, most notably limited datasets and scalability, these ML techniques collectively signify a paradigm shift toward data-driven, adaptive, and intelligent nanofabrication.

4.5.2 *Top Three Machine Learning Performers in Nanofabrication*

Among the 26 reviewed algorithms, three emerged as the most effective in terms of predictive power, adaptability, and generalizability:

1) Artificial Neural Networks (ANNs): ANNs demonstrated exceptional flexibility in learning complex datasets for prediction, classification, and pattern recognition. In nanoimprinting, ANNs achieved 97.5–99.99% accuracy in uniformity prediction, while in tip-based nanoindentation, ANNs achieved R^2 values approaching 0.9999 in predicting load–displacement curves and AFM indent images. Their ability to generalize across domains underscores their central role in ML-driven nanofabrication.
2) Support Vector Regression (SVR): SVR models nonlinear dependencies with high precision in continuous-variable predictions. Applied to nanograting transmittance prediction, SVR achieved $R^2 = 0.995$ and MSE = 0.004, surpassing KNN, GR, and ANN models. Its predictive precision contributed to a 10–57% enhancement in optoelectronic device performance.

3) General Regression Neural Network (GRNN): GRNN effectively captures nonlinear relationships with minimal training data and computational cost. In NIL applications, it achieved 92.37% accuracy (R = 0.92089) for predicting imprint quality and minimizing defects, outperforming BPNN and PNN. Its robustness and fast convergence make it ideal for real-time fabrication optimization.

Together, GRNN, SVR, and ANNs exemplify the current state-of-the-art in integrating ML with nanoscale process control, bridging experimental physics and computational intelligence.

4.5.3 *Challenges and Knowledge Gaps*

Despite its transformative potential, ML in nanofabrication faces several challenges. Limited datasets are a significant hurdle, as nanoscale experiments often produce small amounts of data, hindering robust model training and generalization. Computational resource constraints pose difficulties, especially for complex models like deep learning algorithms that require substantial processing power. Additionally, the "black box" nature of many ML models limits interpretability, making it challenging to understand the physical principles behind predictions. Scalability and real-time adaptive control systems remain underdeveloped, restricting dynamic adjustments during fabrication to minimize defects and optimize throughput.

Moreover, ML should be viewed as an augmentative tool rather than a substitute for scientific reasoning. Fundamental discovery, hypothesis generation, and creative insight remain uniquely human endeavors. Overreliance on ML for optimization alone risks narrowing inquiry to parameter tuning rather than fostering conceptual advances in nanoscience. Hybrid, physics-informed ML approaches grounded in first-principles understanding are essential for interpretable, generalizable, and scientifically meaningful outcomes.

4.5.4 *Recommendations and Future Outlook*

To address these challenges, future ML development in these nanofabrication processes should focus on several key areas. First, efforts should prioritize data-efficient algorithms, such as transfer learning and hybrid models that integrate physics-based approaches. Second, enhancing computational efficiency through optimized algorithms and hardware acceleration will be crucial for scaling ML applications. Third, incorporating explainable artificial intelligence techniques, such as SHAP (SHapley Additive exPlanations) or interpretable models, can improve understanding of process-property relationships and foster trust in ML predictions.[31] Fourth, interdisciplinary collaboration among artificial intelligence researchers, engineers, and scientists is essential to embed domain knowledge into ML frameworks. Lastly, advancing real-time autonomous control systems will enable dynamic adjustments, reduce defects and improve precision in nanofabrication.

Looking ahead, ML's role in nanofabrication will expand from prediction and optimization toward autonomous experimentation, self-correcting manufacturing, and generative nano-system design. The convergence of ML, advanced metrology, and physics-based modeling promises a new era of intelligent nanomanufacturing marked by precision, scalability, and sustainability.

Acknowledgements

The authors gratefully acknowledge financial support from the Ministry of Education of the Czech Republic under the programme INTER-ACTION, No. LUAUS24006. The authors also employed Gemini (Google, Mountain View, CA) to address technical questions arising from the review of related literature and to help refine the language of the manuscript. All artificial intelligence-assisted content was critically evaluated, verified, and edited by the authors, who take full responsibility for the accuracy, originality, and integrity of the final manuscript.

References

1. Mahesh, B. "Machine learning algorithms-a review." *International Journal of Science and Research* 9, no. 1 (2020): 381–386.
2. Tseng, Ampere A. "Advancing nanoparticle synthesis with machine learning: fundamentals and insights from an umbrella review." In *Nanofabrication in the Era of Machine Learning: Advances and Emerging Frontiers,* edited by Ampere A. Tseng, pp. 452–513. Singapore: World Scientific, 2026.
3. Fahle, Simon, Christopher Prinz, and Bernd Kuhlenkötter. "Systematic review on machine learning (ML) methods for manufacturing processes–Identifying artificial intelligence (AI) methods for field application." *Procedia CIRP* 93 (2020): 413–418.
4. Chen, Ying-Lin, Sara Sacchi, Bappaditya Dey, Victor Blanco, Sandip Halder, Philippe Leray, and Stefan De Gendt. "Exploring machine learning for semiconductor process optimization: a systematic review." *IEEE Transactions on Artificial Intelligence* 5, no. 12 (2024): 5969- 5989.
5. Jeong, Won-Kyeong, Ki-Hoon Kim, Chaehyun Park, Dae Geun Song, Myungkwan Song, and Min-Ho Seo. "Highly accurate, efficient, and fabrication tolerance-aware nanostructure prediction for high-performance optoelectronic devices." *Scientific Reports* 14, no. 1 (2024): 1–11.
6. Tseng, Ampere A., S.R. Wang, and J. Hashemi. "On-line modeling for gage control of calendering." *ASME Manufacturing Review* 7 (1994): 72–85.
7. Akter, Tahmina, and Salil Desai. "Developing a predictive model for nanoimprint lithography using artificial neural networks." *Materials & Design* 160 (2018): 836–848.

8. Cook, Deborah F., and Chin-Chou Chiu. "Predicting the internal bond strength of particleboard, utilizing a radial basis function neural network." *Engineering Applications of Artificial Intelligence* 10, no. 2 (1997): 171–177.
9. Hirai, Yoshihiko, Yuki Onishi, Toshiaki Tanabe, Mayuko Shibata, Takuya Iwasaki, and Yasuroh Iriye. "Pressure and resist thickness dependency of resist time evolutions profiles in nanoimprint lithography." *Microelectronic Engineering* 85, no. 5–6 (2008): 842–845.
10. Hirai, Yoshihiko, Sou Tsukamoto, Hidekatsu Tanabe, Kai Kameyama, Hiroaki Kawata, and Masaaki Yasuda. "Smart systems for material and process designing in direct nanoimprint lithography using hybrid deep learning." *Nanomaterials* 12, no. 15 (2022): 2571.
11. Sayed, Biju Theruvil, Mahmoud M. Al-Sakhnini, Asaad AH Alzubaidi, Ahmed HR Alawadi, Ahmed Jaber Ibrahim, and Shavan Askar. "Assessment of nano-imprinting process in CuZr amorphous films through combination of machine learning and molecular dynamics." *Journal of Electronic Materials* 52, no. 10 (2023): 6943–6958.
12. Nguyen, Truong Sinh, Anton Nailevich Gafurov, Jeongdai Jo, Taik-Min Lee, Seung-Hyun Lee, and Kyunghoon Kim. "Multi-geometry parameters optimization of large-area roll-to-roll nanoimprint module using grey relational analysis and artificial neural network." *Polymers* 15, no. 13 (2023): 2909.
13. Tseng, Ampere A., W. H. Shih, C. L. Thomas, and S. J. Chen. "Intelligent processing of polymer sheets for calendering." *Advances in Polymer Technology: Journal of the Polymer Processing Institute* 12, no. 4 (1993): 361–378.
14. Park, Jongwoon, Yoichi Kawakami, and Seoung-Hwan Park. "Numerical analysis of multilayer organic light-emitting diodes." *Journal of Lightwave Technology* 25, no. 9 (2007): 2828–2836.
15. Sun, Hong-Wen, Yan-Chun Huang, Tian-Hua Tang, Jie Shen, Li-Jun Gu, and Jing-Sheng Wang. "A prediction model for the replication quality of nanoimprinting patterns based on BP neural network." *Microsystem Technologies* 31 (2025): 2385–2395.
16. Chien, Jean, and Eric Lee. "Deep-CNN-based layout-to-SEM image reconstruction with conformal uncertainty calibration for nanoimprint lithography in semiconductor manufacturing." *Electronics* 14, no. 15 (2025): 2973.
17. Tseng, Ampere A. "Constraints and challenges in tip-based nanofabrication." In *Tip-Based Nanofabrication: Fundamentals and Applications*, edited by Ampere A. Tseng, pp. 401–443. Springer, 2011.

18. Tseng, Ampere A., Andrea Notargiacomo, and T. P. Chen. "Nanofabrication by scanning probe microscope lithography: A review." *Journal of Vacuum Science & Technology B* 23, no. 3 (2005): 877–894.

19. Leinen, Philipp, Malte Esders, Kristof T. Schütt, Christian Wagner, Klaus-Robert Müller, and F. Stefan Tautz. "Autonomous robotic nanofabrication with reinforcement learning." *Science Advances* 6, no. 36 (2020): eabb6987.

20. Jha, Rajesh, and Arvind Agarwal. "Software (GUI/APP) for developing AI-based models capable of predicting load-displacement curve and AFM image during nanoindentation." *Coatings* 11, no. 3 (2021): 299.

21. Kim, Jongwoo, Jeong Seop Lee, Ji-woong Kim, Peter De Wolf, Seunghyun Moon, Dong Hwan Kim, Joo-Hyun Song et al. "Fabrication of plasmonic arrays of nanodisks and nanotriangles by nanotip indentation lithography and their optical properties." *Nanoscale* 13, no. 8 (2021): 4475–4484.

22. Liu, Yijie, Xuexuan Li, Ben Pei, Lin Ge, Zhuo Xiong, and Zhen Zhang. "Towards smart scanning probe lithography: a framework accelerating nano-fabrication process with in-situ characterization via machine learning." *Microsystems & Nanoengineering* 9, no. 1 (2023): 128.

23. Tseng, Ampere A. "Nanoscale scratching with single and dual sources using atomic force microscopes" In *Tip-Based Nanofabrication: Fundamentals and Applications*, edited by Ampere A. Tseng, pp. 1–64. Springer, 2011.

24. Wu, Nian, Markus Aapro, Joakim S. Jestila, Robert Drost, Miguel Martínez García, Tomás Torres, Feifei Xiang et al. "Precise large-scale chemical transformations on surfaces: deep learning meets scanning probe microscopy with interpretability." *Journal of the American Chemical Society* 147, no. 1 (2024): 1240–1250.

25. Tseng, Ampere A. "Recent developments in nanofabrication using focused ion beams," *Small*, 1, no. 10 (2005): 924–939.

26. Buchnev, Oleksandr, James A. Grant-Jacob, Robert W. Eason, Nikolay I. Zheludev, Ben Mills, and Kevin F. MacDonald. "Deep-learning-assisted focused ion beam nanofabrication." *Nano Letters* 22, no. 7 (2022): 2734–2739.

27. Cho, In Ho, Myung Gi Ji, and Jaeyoun Kim. "Pursuit of hidden rules behind the irregularity of nano capillary lithography by hybrid intelligence." *Scientific Reports* 13, no. 1 (2023): 13649.

28. Hansen, M., M. Ziegler, H. Kohlstedt, A. Pradana, M. Raedler, and M. Gerken. "UV capillary force lithography for multiscale structures." *Journal of Vacuum Science & Technology B* 30, no. 3 (2012): 031601.

29. Chapagain, Ashish, and In Ho Cho. "Multiphysics machine learning framework for on-demand multi-functional nano pattern design by light-controlled capillary force lithography." *Communications Physics* 7, no. 1 (2024): 213.

30. Zhang, Hao, Jinchuan Zheng, Guiyuan Cao, Han Lin, and Baohua Jia. "Real-time machine learning-based position recognition in laser nanofabrication with sub-half-wavelength precision." *Engineering* (2025). (Available online 16 June 2025) https://doi.org/10.1016/j.eng.2025.03.037

31. Yumeng, Yu Cai, Shunyuan Lou, and Yining Chen. "Explainable prediction of deposited film thickness in IC fabrication with CatBoost and SHapley Additive exPlanations (SHAP) models." *Applied Intelligence* 54, no. 1 (2024): 246–263.

32. Tseng, Ampere A., and Zhuang Li. "Manipulations of atoms and molecules by scanning probe microscopy." *Journal of Nanoscience and Nanotechnology* 7, no. 8 (2007): 2582–2595.

https://doi.org/10.1142/9789819830763_0005

CHAPTER 5

Nanofabrication Through Superplastic Nanomolding: From Basic Principles to Nanophotonic Applications

Jianxin Liu[1], Hui Fang[1], Yangyang Pan[1], Ze Liu[1,*], Yong Li[2], Shiyu Fan[2], Tao Ding[2,*]

[1]Department of Engineering Mechanics, School of Civil Engineering, Wuhan University, Wuhan 430072, People's Republic of China

[2]Key Laboratory of Artificial Micro/Nano Structure of Ministry of Education, School of Physics and Technology, Wuhan University, Wuhan 430072, People's Republic of China

*Corresponding authors: t.ding@whu.edu.cn (T. D.); ze.liu@whu.edu.cn (Z. L.)

Abstract

This chapter introduces recent progress on superplastic nanomolding (or thermomechanical nanomolding) technique, a powerful way towards ordered nanoarrays/nanostructures potentially in any metals/alloys with exceptional crystallinity, precision and high efficiency. We begin by elucidating the fundamental principles of nanoimprinting, stamping and embossing, detailing the mechanisms that enable transferring nanopatterns in soft materials at elevated temperatures. The chapter then transitions to the core concept of superplastic nanomolding, explaining how diffusion-based deformation mechanism is harnessed to replicate intricate patterns from a master mold onto target crystalline materials at the nanoscale. We discuss various superplastic nanomolding processing parameters, including temperature, pressure, and strain rate, and their influence on the final product's quality and fidelity. Finally, this chapter highlights the significant impact of superplastic nanomolding on nanophotonic applications. We showcase how superplastic nanomolding enables the fabrication of high-performance surface-enhanced Raman scattering substrates with precisely controlled plasmonic "hot spots" for enhanced molecular detection. We also detail the use of superplastic nanomolding for creating complex plasmonic waveguides with tailored geometries for efficient light manipulation at the nanoscale. Furthermore, we explore the fabrication of chiral nanostructures via superplastic nanomolding, demonstrating their application in chiroptical sensing and the generation of circularly polarized luminescence. This chapter concludes by discussing current limitations and future research directions for superplastic nanomolding, particularly in advancing surface-enhanced Raman scattering, plasmonic waveguide technology, and the expanding field of chiral nanophotonics.

Keywords: superplastic nanomolding; nanoarrays; nanophotonics; nanomechanics; chiral sensing

5.1 Introduction

Nanofabrication is a fundamental technology of modern science. In the past few decades, the development and application of nanofabrication techniques have attracted widespread attention and great advances have been achieved. Some examples include electron beam lithography (EBL), nanoimprint lithography (NIL), self-assembly/colloidal lithography (SAL), glancing angle deposition (GLAD), laser direct writing (LDW), and dip-pen nanolithography (DPN). In general, nanofabrication methods can be classified into "top-down" or "bottom-up" processing strategies. The former is usually subtractive manufacturing or net-forming, while the latter mostly corresponds to additive manufacturing. Different nanofabrication methods have their own advantages and disadvantages. In terms of nanophotonic applications, ultrafine feature size and precise controllability is the key for a nanofabrication technique. Based on developed nanofabrication methods, many optical devices including frequency,[1,2] amplitude,[3–5] chromatic dispersion,[6] surface plasmon polariton,[7,8] phase delay,[9] and resonance[10–13] devices have been developed and used for practical applications, such as metamaterial lens,[14–19] metaholograms,[20–31] color filters,[32–38] surface-enhanced Raman scattering (SERS),[39–41] and plasmonic waveguides.[42–45] Publications utilizing diverse fabrication methods for nanophotonic applications are summarized in Figure 5.1. It is obvious that NIL is the most popular technique due to its facileness and scalability, while LDW is growing fast in the last decade due to its improved precision, cost-effectiveness and easy operation.

In the following section, we will briefly elucidate different fabrication methods and make a comparative analysis to highlight the importance of NIL.

EBL is a fundamental nanofabrication technology that enables the direct fabrication of any 2D nanostructures in most materials, when combined with the techniques of chemical/physical vapor deposition or electrodeposition. The core principle of this method is the change of chemical properties of a photosensitive resist by irradiation of focused electron beams. Usually, for positive resists after electron beam irradiation, they

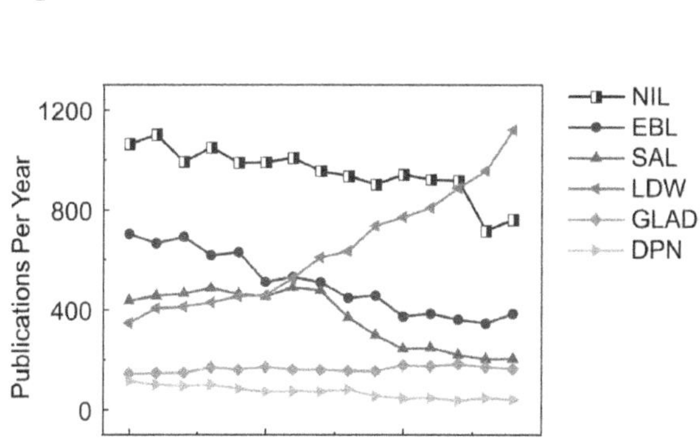

Figure 5.1. Publications with different methods for nanophotonic applications.

will decompose into smaller and more easily soluble fragments. For negative resists, the irradiation causes resists to undergo cross-linking reactions, combining smaller polymers into larger polymers with lower solubility, thus reducing the solubility.[46–54] EBL exhibits sub-10 nanometer resolution; the main limitation is the contradiction between high resolution and high efficiency (large-area patterning). SAL represents a convenient and economic processing method, which usually uses colloid self-assembly to form a layer of microparticle arrays as a mask template, and obtains patterns through subsequent etching or deposition processes. It has the advantages of simplicity and low cost, but is only effective for simple periodic patterns.[55–59] GLAD originates from the chemical/physical vapor deposition technique, where the substrate is placed at a certain angle relative to the direction of the vapor flow, therefore the growth of columnar nanostructures can be controlled by manipulating the substrate such as rotation, and the shadow effect combined with the surface diffusion mechanism finally leads to batch fabrication of 3D nanostructures.[60–66]

LDW utilizes a focused laser beam to scan and expose photosensitive material on a substrate surface. Through a developing process, a desired pattern is formed on the material surface. Then, this pattern can be transferred to the substrate through subsequent etching or deposition.[67–71] In recent years, two-/multiple-photon polymerization has garnered considered interest, which utilizes the characteristic of two-/multiple-photon absorption of materials to achieve the polymerization of irradiated regions in the photosensitive material. Similar to the LDW technique, the two-photon polymerization technique also requires processes such as developing and deposition to obtain designed

3D objects.[72–78] Owing to its simplicity and 3D processibility, LDW as well as the two-photon polymerization show obvious rising trends in publications from the 2010s to 2020s (Figure 5.1). This acceleration is driven by the rapid technological iteration (e.g., faster lasers and parallelization) that transformed LDW into a versatile 3D nanoprinting tool for micro-optics, microfluidics, and metasurfaces, attracting widespread interests in academia and industry, positioning LDW as a prevalent method for rapid prototyping of complex photonic structures.

DPN is a nanofabrication technology based on scanning probe microscopy. This technology combines the direct writing of soft materials with the high resolution of atomic force microscopy. It employs an atomic force microscopy probe "pen" coated with "ink"; when the pen is in contact with a substrate, the ink spreads through the substrate and is readily deposited.[79–85] The DPN technique exhibits very high processing resolution because of sharp atomic force microscopy tip-based inking, but suffers from low throughput and scalability issues.

Among the developed nanofabrication techniques, NIL possesses a unique combination of high resolution, low cost, and high throughput. Therefore, the NIL technique has emerged as one of the most important and commercially available techniques for the large-scale fabrication of nanophotonic devices. The mechanism of NIL relies on contact deformation of moldable materials by hard molds to replicate nanostructures, therefore it requires the pre-fabrication of hard templates. The traditional NIL is divided into thermal curing and UV light curing. Thermal curing takes advantage of the low viscosity of materials at relatively low temperatures. When a master template is pressed onto a polymer coating-substrate system, the polymer coating is thermally cured and the template is subsequently separated after cooling. The patterns in the template are then transferred to the polymer surface. This process is very simple and can be directly used for fabricating superstructures.[86–90] The UV light curing method involves applying a liquid polymer onto a substrate by spin coating. During the imprinting process, the polymer is cured by UV radiation. Therefore, the template needs to be transparent to UV light. This method eliminates the need for heating and cooling, thereby increasing production efficiency and enabling the imprinting to be carried out at room temperature and under low pressure.[21,91–95]

In the view of industrial applications, especially nanophotonic applications, a fabrication method with high efficiency (parallel processability) and low cost is highly desirable. NIL is exactly the technique in need: a single, high-resolution master mold containing the desired nanopatterns is pressed into a thin layer of resist spun onto a substrate. In a single imprint step lasting seconds to minutes, all patterns in the master mold are replicated across the substrate. This stamping process reduces the fabrication time from hours to minutes, slashing the cost per device by orders of magnitude compared to optical lithography systems. The cost of NIL systems can be as little as 10–20% of the cost of a conventional lithography technique,[96] which makes it capable of bridging the gap from the laboratory to the factory for high-volume nanophotonic products. Furthermore, NIL is not limited by the diffraction limit of light or the physical constraints of self-assembly. Because it is a mechanical deforming process, NIL can achieve resolutions below 10 nm, limited only by the quality of the master mold. This is critical for next-generation nanophotonic devices that rely on deep sub-wavelength features to manipulate light with unprecedented precision. Unlike SAL, which is restricted to close-packed arrangement of spheres, or GLAD, which produces specific columnar morphologies, NIL can replicate any arbitrary 2D or 3D patterns that are defined on a master mold. This freedom is essential for creating complex metasurfaces, photonic integrated circuits, and other sophisticated optical components. Besides, NIL allows one to obtain nanopatterns in a wide variety of materials, including metals, silicon, glass, polymers, and III-V semiconductors when combined with deposition or etching methods. From the perspective of the demand for low cost and large-scale production of complex nano-optical components, NIL may become a fundamental technology for unlocking the widespread commercialization of nanophotonics, impacting a wide range of applications from data centers and 6G communication to healthcare and quantum computing.

In the following section, we will briefly introduce the developmental history of NIL (Figure 5.2). The concept of NIL was first proposed by Stephen Y. Chou in 1996.[97,98] In this pioneering work, they used a rigid mold with nanopatterns to perform hot pressing on a polymethyl methacrylate (PMMA) film at a temperature above its glass transition temperature, successfully fabricating patterns with a resolution as high as 25 nm. It demonstrated the potential for low-cost, high-throughput, and high-resolution fabrication, opening a new way for the mass production of nanodevices. In the meantime, George M. Whitesides's group proposed the concept of soft lithography in 1997.[99] Unlike the hot embossing technique that uses rigid molds, soft lithography employs an elastic polymer (like polydimethylsiloxane) as a stamp or template. The advent of soft

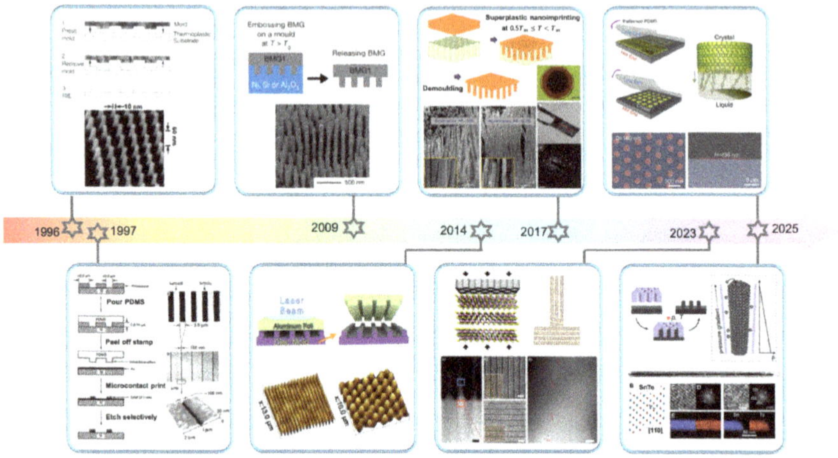

Figure 5.2. The developmental history of the nanoimprinting technique, starting from polymer hot pressing in 1996[97,98] and soft lithography with polydimethylsiloxane in 1997,[99] through metallic glasses in 2009,[100] crystalline metals with laser shock imprinting in 2014[101,102] and superplastic nanomolding in 2017,[103] to layered materials in 2023,[104] topological nanowires via thermomechanical epitaxy[105] and organic semiconductor crystals via nanoimprint crystallithography in 2025.[106] Reprinted with permission from [98] by Elsevier. Reprinted with permission from [99] by Royal Society of Chemistry. Reprinted with permission from [100] by Springer Nature. Reprinted with permission from [102] by John Wiley and Sons. Reprinted under the terms of CC-BY Creative Commons Attribution 4.0 International License from [103]. Reprinted with permission from [104] by American Chemical Society. Reprinted with permission from [105] by Elsevier. Reprinted under the terms of CC-BY-NC-ND Creative Commons Attribution NonCommercial NoDerivatives 4.0 International License from [106].

lithography greatly enhances the adaptability of nanoimprinting to non-planar surfaces.

Early nanoimprint technology primarily focuses on polymer materials. With the development of nanoscience and technology, especially the miniaturization of microsystem/

optical components, the demand for directly fabricating nanostructures on inorganic materials like metals becomes increasingly urgent. A key challenge was that traditional crystalline metals lack the thermo-induced rheology of polymers, making them difficult to shape through hot pressing at relatively low temperatures and pressures. Noting that amorphous metals (also known as metallic glasses) exhibit similar thermoplasticity in their supercooled liquid region (i.e., above the glass transition temperature T_g, but below the crystallization temperature T_x), Jan Schroers' group from Yale University successfully demonstrated the direct molding of metallic glass nanorods with aspect ratios exceeding 50 in 2009.[100] More importantly, because metallic glasses at temperature below T_g possess extremely high strength and hardness, the imprinted metallic glass structures could also be used as new molds to imprint other soft materials. Unlike silicon molds, which is intrinsically brittle, the great mechanical properties of metallic glass molds allow them to be used for thousands of times, significantly reducing the processing costs.

On the contrary, crystalline metals usually exhibit limited formability at the nanoscale, resulting from the size effect in strength/deformation (the smaller the stronger[107]) and friction.[108] To overcome the size effect in strength, Gary J. Cheng's group developed a laser shock imprinting technique,[101,102] which uses a nanosecond laser pulse to generate an immense shockwave pressure (on the order of gigapascal) on the material's surface. This shockwave generates extremely high strain rate (10^6–10^7 s^{-1}), deriving a crystalline metal sheet flowing into a nanomold at room temperature. The mechanism explores the high strain rate-activated superplastic deformation, which can overcome the size effect in strength. However, the aspect ratio of formed metallic structures is usually below 5.

Considering that diffusion creep may dominate at nanoscale, Ze Liu from Wuhan University proposed a superplastic nanomolding (SPNM) technique to directly shape crystalline metals in the diffusion-based temperature ranges (typically about ~0.5 T_m, T_m being the melting temperature) at nanoscale.[103] It shows that at the nanoscale, atomic diffusion-based deformation mechanism endows crystalline metals with excellent mold-filling capabilities, enabling the fabrication of nanowires arrays with feature size as small as ~8 nm and aspect ratio of ~2,000. Because metals and alloys exhibit similar diffusivity in the diffusion deformation-dominated temperature ranges, the SPNM technique could be applicable for most metals/alloys. Indeed, Judy J. Cha's group successfully applied thermomechanical nanomolding to directly shape layered materials (Te, In_2Se_3, Bi_2Te_3 Bi_2Se_3, GaSe, and Sb_2Te_3) at the nanoscale in 2023.[104] Recently, Jan Schroers and co-workers further developed a thermomechanical epitaxy technique,[105] which uses pressure-driven diffusion process to epitaxially grow single-crystal topological nanowires, where a theoretical framework for predicting the imprintability of materials is established. Besides, Hong-Bo Sun's group developed nanoimprint crystalithography for fragile organic semiconductor crystals.[106] The technique controls crystallization kinetics by precisely managing temperature gradients, achieving the fabrication of high-quality, low-defect, and residue-free organic semiconductor arrays that can be used for high-performance flexible optoelectronic devices.

The developmental history of NIL suggests that NIL has evolved from simple molding of polymers into a powerful platform capable of precisely structuring a wide range of advanced materials, including metals, semiconductors, layered materials, topological materials, and even organic crystals. The core driving force behind this evolution lies in the deep understanding and utilization of the physical deformation mechanisms of different materials at the nanoscale. Either through thermoplasticity, superplasticity, atomic diffusion, or controlled crystallization, the fundamental principles have vastly fascinated the development of nanofabrication. Furthermore, as a perspective on NIL for nanophotonics rightly suggests, "advances in nanoimprint lithography have moved the technology out of the laboratory and onto the production floor for use in a wide variety of photonic applications".[96]

This chapter will first introduce the fundamentals of SPNM and then showcase its latest applications in fabricating high-performance nanophotonic devices. We start with the core principles of SPNM and discuss how key processing parameters can be precisely tuned to control the geometry and aspect ratio of the resulting nanostructures. Then we will review state-of-the-art applications based on the SPNM technique, including SERS substrates, plasmonic waveguides and chiral nanostructures for chiroptical applications. Finally, we conclude this chapter by discussing the remaining challenges and offering a future perspective on how SPNM will continue to drive innovation in nanophotonic applications.

5.2 Principles of Superplastic Nanomolding

Unlike conventional NIL, which is primarily applied to amorphous materials like polymers or metallic glasses, the core advantage of SPNM lies in its ability to directly shape crystalline materials at temperatures well below the melting points. This section will first introduce the processing of SPNM and clarify its distinction from conventional nanoimprinting. Subsequently, we will discuss the mechanisms of SPNM, as well as the key processing parameters such as temperature, pressure, and frequency, and their influence on molding results. The discussion will then focus on the interfacial effects, introducing engineering strategies aimed at enhancing molding efficiency. Finally, we will introduce the essential elements of mold design and fabrication, which are fundamental to the application of SPNM.

5.2.1 *Superplastic Nanomolding Process and the Role of Superplasticity*

For a typical SPNM experiment, a crystal ingot (such as a metal foil) is firstly placed onto a hard mold with prefabricated nanocavities. Subsequently, under the action of specific pressure and temperature (typically $0.4\,T_m \leq T < T_m$, where T_m is the melting temperature in Kelvin), the crystal would fill the nanocavities of the mold by creep flow. A typical experimental setup for SPNM of crystals is shown in Figure 5.3a. After processing, the mold is removed through chemical etching to yield a nanostructure array that is the inverse of the mold. The key to this process is the utilization of superplasticity under controlled temperature and stress conditions. Superplasticity is usually related to the phenomenon where a crystalline material under tension can undergo enormous plastic deformation without fracture (with elongation of larger than ~400%).[109] Similarly, during the SPNM process, we may define superplastic deformation when the extrusion of metal nanorods leads to aspect ratio larger than 5, which is usually challenging in the temperature range dominated by

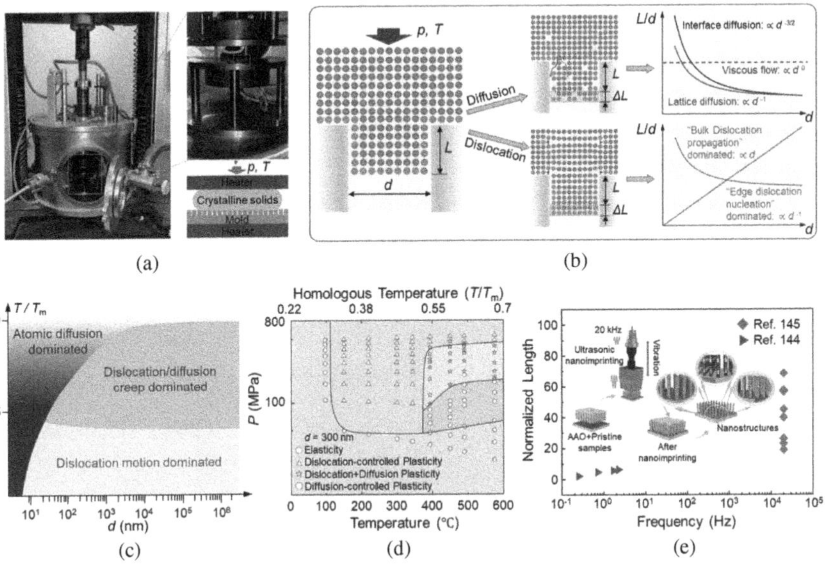

Figure 5.3. The process and mechanism of nanomolding. (a) Typical experimental set-up for SPNM.[133] Reprinted under the terms of CC-BY Creative Commons Attribution 4.0 International License from [133]. (b) Deformation mechanisms in SPNM: the molding of crystals is based on diffusion transport mechanism at high homologous temperature ($T > 0.4\ T_m$) and dislocation slip mechanism at low homologous temperature ($T < 0.4\ T_m$).[128] Reprinted with permission from [128] by American Physical Society. (c) Sketch of the temperature- and size-dependent deformation mechanism map in crystalline metals.[133] Reprinted under the terms of CC-BY Creative Commons Attribution 4.0 International License from [133]. (d) Experimentally constructed deformation mechanism map of Ag.[134] Reprinted with permission from [134] by Elsevier. (e) Effect of superimposed micro-vibration on SPNM; the length is normalized by that molded without vibration.[144,145] Reprinted under the terms of CC-BY Creative Commons Attribution 4.0 International License from [145].

dislocation motion as the interface between the metal and cavity wall hinders dislocation movement. One possible way is to increase the processing temperature (usually larger than $0.4\ T_m$) to activate atomic diffusion.

Although SPNM can be considered as a vital branch of NIL, they differ significantly in terms of the materials processed, the underlying mechanism, and the morphology of the final products. Key distinguishing metrics are the aspect ratio and minimum feature size. Conventional nanoimprinting was primarily used to create shallow 2D or quasi-3D patterns on the surface of materials such as polymers. The aspect ratio is typically less than 5, and the minimum feature size is limited by the flow unit of polymers. This process usually correlates with the viscous flow of (supercooled) liquids, and focuses more on "imprinting" to create surface patterns in resist coating. In contrast, SPNM was specifically designed to fabricate nanostructures with pronounced 3D features. Leveraging a unique atomic diffusion mechanism, SPNM can achieve exceptionally high aspect ratios, typically ranging from 10:1 to 1000:1. A key and unprecedented advantage of this technique, which we will elaborate on later, is its "the smaller the easier" molding characteristic. Fundamentally, SPNM operates through the diffusional flow of crystalline materials within a confined space, prioritizing a "molding" approach to create high-aspect-ratio nanostructures.

5.2.2 *Processing and Mechanism of Superplastic Nanomolding*

The SPNM technique requires only a universal testing machine equipped with a heater (Figure 5.3a). When the bulk metal/nanomold is heated up to the diffusion-dominated temperature range, the applied pressure from the universal testing machine will create a pressure gradient within the metal, which drives atoms to migrate directionally from high- (the base) to low-pressure regions (the tip of partially filled nanorods) until the entire mold cavities are fully filled. This diffusion-dominated flow behavior enables the precise replication of the mold's fine features, even down to a few nanometers, without the limitation of grain size.[103]

The SPNM technique is intrinsically related to moldability of crystals. In general, the moldability of a material is associated with its flowability. Such flowability is high in thermoplastics, gels, and some glasses but low in crystalline metals.[110,111] When the load exceeds the flow stress, crystals will plastically deform under the interaction of several independent mechanisms, such as nucleation and propagation of dislocations, diffusional flow, and viscous flow. To identify the processing conditions for SPNM, Frost and Ashby have plotted a phase map of materials deformation and mechanisms,[112] which have already proven to be very useful in a qualitative way for predicting superplasticity,[113,114] selecting materials for engineering applications, and investigating the flow mechanism.[115,116] It should be noted that the prevalent deformation maps generally cannot be applied to nanostructured materials since they are usually obtained based on empirical relations.[114,115,117] A considerable number of experiments and simulations on nanostructured materials show that the mechanical properties of materials are size-dependent.[118,119] For instance, it is found that the strength or hardness of polycrystalline metals generally increases with the decrease of grain sizes and follows the well-known Hall-Petch behavior.[120,121] However, the Hall-Petch effect will break down at elevated temperatures due to the onset of diffusional creep and enhanced grain boundary migration/sliding.[122–125] It is crucial to investigate the temperature effect on the mechanical properties of materials at small scales, which is not only important for the application of the diffusion-based mechanism for facile processing of microparts at economical high temperatures,[126,127] but also crucial for the material selection of high-strength components in a certain temperature range.[112,114] In recent works, it has been recognized that identifying the boundaries of dislocation and diffusion-dominant mechanisms is crucial in SPNM as it affects the efficiency and quality of nanoforming.[103,128–134]

During the nanomolding process, the deformation mechanism of a crystal can be dominated by diffusion or dislocation movement (Figure 5.3b), which depends on temperature and pressure. Considering that the cavity size is usually far smaller than the grain size of bulk metals/alloys,[103] the nanorod can be regarded as growing from a single crystal. For high-temperature molding (e.g., $T > 0.4\ T_m$), the molding process is dominated by diffusion, and the thermal equilibrium of a metal will introduce defects such as vacancies in the crystal, thus the matter can transport through vacancies by diffusion.[135–137] For low-temperature conditions (e.g., $T < 0.4\ T_m$), the deformation mechanism will be taken over by dislocation movement, and the growth of metallic nanorods can be considered as the nucleation and propagation of dislocations. For the simplest case, the edge dislocation loops nucleate and emit at the edge of a nanocavity.[128] A similar mechanism is used to explain the surface indent by a stiff indenter.[138] The prominent deformation mechanism is a function of the sample size and temperature (Figure 5.3c). With decreasing length scale, dislocation slip[139] ceases to operate either due to the reduction of required space to form dislocations (dislocation starvation),[140] or due to the lack of dislocations.[118,141,142] As a consequence, the feasibility of nanomolding based on plasticity and dislocation slip is limited on the nanoscale.[139,142] Considering that the atomic diffusion is controlled by the exponential temperature dependence of vacancy concentration, the deformation can be dominated by diffusion at relatively high temperatures. Therefore, it can be expected that bulk diffusion (Nabarro-herring creep)[128] and interface diffusion (Coble creep)[103,135,143] will play a dominant role in the deformation mechanism for high-temperature nanomolding of metals/alloys.

Considering that both deformation mechanisms based on dislocation motion and atomic diffusion are thermally activated physical processes, in principle any method that can help overcome or reduce the energy barrier can improve the molding efficiency. Recently, researchers found that the forming efficiency of the molded nanostructures can be improved by approximately one order of magnitude when superimposed micro-vibration is applied to the molding pressure (Figure 5.3e).[144,145] This enhanced molding efficiency at low temperatures is attributed to the alternating stress field-assisted releasing of accumulated dislocations.

5.2.3 *Interfacial Effect*

As dimensions of a material shrink to the nanoscale, its specific surface area will increase dramatically, making interfacial effects such as friction, adhesion, surface energy barriers, and chemical reactions dominant. As a result, interfacial resistance may drastically reduce forming efficiency or even cause complete forming failure. Consequently, the precise control of the feedstock-mold interface, namely a practice termed interfacial engineering, has emerged as a core research direction to overcome this challenge. This section will systematically elucidate, with reference to four exemplary cases shown in Figure 5.4, how different interfacial

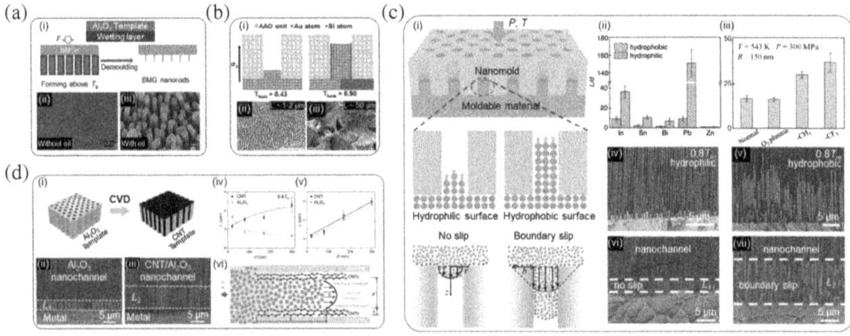

Figure 5.4. Interfacial engineering strategies for enhancing the efficiency of SPNM. (a) Wetting layer lubrication: (i) Schematic of using a liquid wetting layer to facilitate the molding of reactive materials. (ii) and (iii) Scanning electron microscopy (SEM) images of molding a $Zr_{35}Ti_{30}Cu_{8.25}Be_{26.75}$ bulk metallic glass (BMG) without and with a high-temperature oil lubricant, respectively.[146] Reprinted with permission from [146] by IOP Publishing. (b) Eutectic interface diffusion-enhanced nanomolding: (i) Schematic of diffusion enhancement by a eutectic interface. (ii) and (iii) are SEM images of molded pure Au and an $Au_{81}Si_{19}$ eutectic alloy at 300 °C, respectively.[147] Reprinted under the terms of CC-BY-NC-ND Creative Commons Attribution NonCommercial NoDerivatives 4.0 International License from [147]. (c) Boundary slip-enhanced nanomolding via wettability control: (i) Schematic of nanomolding with different surface chemistries (hydrophilic vs. hydrophobic) and the corresponding velocity profiles for no-slip vs. boundary slip conditions. (ii) and (iii) show the length difference of molded metal/alloy nanowires when using molds with hydrophilic and hydrophobic surface, respectively. (iv) and (v) are SEM images of Bi nanowires molded in hydrophilic and hydrophobic nanochannels, respectively, and (vi) and (vii) show similar comparison for molded Pt-BMG nanorods.[148,149] Reprinted under the terms of CC-BY Creative Commons Attribution 4.0 International License from [148] and with permission from [149] by American Chemical Society. (d) Solid lubricant-enhanced nanomolding: (i) Schematic for fabricating carbon nanotube nanochannels from an Al_2O_3 template. (ii) and (iii) are SEM images of Ag nanorods molded in pristine Al_2O_3 and carbon nanotube-coated Al_2O_3 nanochannels, respectively. (iv) and (v) show the lubricant effect in nanomolding of crystalline Ag and metallic glass liquids, respectively. (vi) A schematic of the two competing transport mechanisms during nanomolding.[150] Reprinted with permission from [150] by American Physical Society.

engineering strategies can be employed to effectively overcome the interface challenges in SPNM, enabling high-efficiency, low-temperature, and high-precision nanomanufacturing.

5.2.3.1 *Overcoming Forming Barriers with a Liquid Wetting Layer*

For chemically reactive materials, such as zirconium- and magnesium-based bulk metallic glasses (BMGs) and active crystalline metals, they can readily react with atmospheric oxygen at high temperatures and form a hard and brittle oxide film in the surface. This nanoscale oxide film, together with the capillary effects caused by poor wettability between the material and the mold, creates a significant barrier for forming, particularly during the filling of small nanopores (<1 μm). To overcome this challenge, a simple yet highly effective strategy has been proposed: the introduction of a liquid wetting layer, such as a high-temperature lubricating oil, onto the mold surface prior to forming (as shown in Figure 5.4a). This oil film can wet both the BMG and the aluminum oxide (Al_2O_3) mold surface, taking effect via two mechanisms: reducing interfacial energy and providing boundary lubrication. Experimental results have confirmed the efficiency of this strategy. As shown in Figure 5.4a, an attempt to mold $Zr_{35}Ti_{30}Cu_{8.25}Be_{26.75}$ BMG without a wetting layer failed completely (with no nanostructures formed in the surface). In contrast, under the same processing conditions but with the addition of the oil wetting layer, nanorods arrays with an aspect ratio exceeding 5 were successfully fabricated.[146] This strategy has demonstrated broad applicability through its successful application to the nanoforming of various highly reactive Zr- and Mg-based BMGs.

5.2.3.2 *Enhancing Diffusion via Eutectic Interfaces for Low-Temperature Molding*

Another major limitation of conventional SPNM is its high processing temperature requirement. For crystalline metals, it is typically necessary to heat them to approximately half of their absolute melting point (~0.5 T_m) to activate atomic diffusion mechanism. The "eutectic interface" strategy offers a paradigm-shifting solution for low-temperature and even room-temperature nanomolding. The core concept is to utilize the unique properties of the interface between two phases in a eutectic alloy system. In a system with a deep eutectic point (such as the Au-Si system, shown in Figure 5.4b), the interfacial region between the Au and Si phases exhibits an effective melting temperature far below that of either pure component, corresponding instead to the eutectic composition (364 °C for Au-Si). When the processing temperature ($T_{process}$) is well below the melting point of pure gold but near the eutectic temperature, the local homologous temperature ($T_{hom} = T_{process}/T_{eutectic}$) at this interface becomes very high (approaching 1), which leads to an increase in the diffusion rate of interfacial atoms by several orders of magnitude.[147] As shown in Figure 5.4b, the forming efficiency of an $Au_{81}Si_{19}$ eutectic alloy at 300 °C is about 50 times that of pure gold at the same processing conditions. Furthermore, at room temperature (30 °C), pure gold cannot be molded, whereas the eutectic alloy readily forms nanorods with an aspect ratio of ~4. Interestingly, this enhancement is not strictly dependent on the bulk material's composition, as even adding as little as 0.5 at.% of silicon to pure gold is sufficient to form the critical "eutectic interfaces" and achieve a comparable forming efficiency. This principle can be further extended by creating an artificial eutectic interface through direct matching of the feedstock and mold, such as by molding a pure gold feedstock into a silicon mold. This approach has likewise enabled high-efficiency, room-temperature nanomanufacturing.

5.2.3.3 *Inducing Boundary Slip by Controlling Wettability*

In a confined nanoscale space, friction at the interface between the flowing material (fluid or fluid-like, such as metals at high temperature) and the cavity wall can be the dominated source of forming resistance. Classical fluid dynamics assumes a "no-slip boundary condition", where the fluid velocity at the wall is zero. However, it has been shown that this fundamental assumption can be broken by altering the interface's chemical properties at the micro/nanoscale. The core of this strategy is to control the surface wettability of the cavity walls (as shown in Figure 5.4c). Typically, the surfaces of ceramic molds such as aluminum oxide are hydrophilic. Through chemical modification methods such as silanization, the surface can be rendered hydrophobic. For a BMG flowing in its highly viscous state (viscosity up to $\sim 10^8$ Pa · s) or a crystalline metal undergoing diffusion creep, a hydrophobic mold surface significantly weakens interactions and friction with the filled materials, thereby inducing a phenomenon known as "boundary slip". In the slip regime, the material has a non-zero velocity at the wall, causing the overall velocity profile to approach that of a "plug flow", which dramatically increases the average flow rate and forming efficiency. As per typical experimental results shown in Figure 5.4c,[148] for a wide range of crystalline metals (e.g., Bi, Pb, In, Sn) and for Pt-based BMG, the length of nanorods molded in hydrophobic nanochannels is far greater than in hydrophilic nanochannels under the same processing conditions, with the molding efficiency increased by more than one order of magnitude. For highly viscous Pt-BMG, studies have quantified the slip length on the hydrophobic interface to be in the range of 50 to 500 nm.[149] It has also been found that the slip length correlates with the processing pressure, which can be rationalized from shear-dependent viscosity. The experiments demonstrate that surface chemistry is a powerful tool for controlling confined flow at the nanoscale.

5.2.3.4 *Controlling Transport Mechanisms with Solid Lubricants*

During the SPNM process, material transport can involve a competition between multiple physical mechanisms, primarily atomic-scale diffusion along the feedstock-mold interface (interface diffusion) and collective plastic flow within the material (e.g., creep). Interface diffusion is highly sensitive to surface defects (such as atomic steps and vacancies), which act as "pinning sites" that severely impede atomic migration. The use of a solid lubricant can not only reduce friction but also fundamentally control the dominant transport mechanism (as shown in Figure 5.4d).[150] It is recently shown that carbon nanotubes (CNTs) grown in the cavity wall surface can serve as a solid lubricating layer though they contain a large number of defects. Owing to the high surface defects and low adhesion energy of chemically inert surfaces, CNT coatings play a dual role in suppressing interface diffusion and promoting boundary slip. Typical results are shown in Figure 5.4d, where the length of Ag nanorods molded with CNT-coated mold is approximately three times that molded with a pristine alumina mold, demonstrating the excellent lubricating performance of CNTs even at high temperatures (up to ~470 °C). More importantly, this strategy reverses the size dependence of the forming process. In the pristine mold, the forming efficiency follows "the smaller the easier" behavior, which is consistent with an interface diffusion model. In the CNT-coated mold, however, this trend is inverted to "the larger the more favorable", which aligns with a model of boundary slip-assisted collective flow. This strategy shows that by designing the solid lubricant interface, it is possible to actively select and control the matter transport at the nanoscale, thereby significantly improving the nanoforming efficiency.

5.2.4 *Mold Design and Fabrication*

The controlled fabrication of templates plays an important role in the structural diversity of prepared nanostructures by nanoimprinting/nanomolding. This section will review the design and fabrication of advanced nanomolds with a focus on anodic aluminum oxide (AAO) templates. AAO templates have become one of the most commonly used and extensively studied hard templates in nanoimprint technology due to its low cost, ease of preparation, ability to form large-area uniform nanopore arrays, and highly controllable structural parameters. We will first discuss methods for achieving precise control over 2D planar structures, including the formation of highly ordered arrays and the engineering of non-traditional pore shapes and gradients. Next, we will explore the construction of complex 3D nanochannel structures, such as axially modulated and branched nanopore networks. Finally, we will introduce advanced techniques for fabricating hierarchical templates that integrate features from the nanometer to the centimeter scale, enabling the creation of artificial hierarchical devices.

5.2.4.1 *Precise Control of 2D Planar Nanostructures*

Normally the anodization of aluminum can form nanopores, but their arrangement is disordered. To obtain the highly ordered structures required for nanoimprinting, Hideki Masuda's group pioneered the "two-step anodization" method.[158] This method utilizes a pre-anodization step, where the interaction between pores leads to the formation of a concave array with short-range order on the aluminum surface. After removing this initial oxide layer, the remaining ordered concavities on the Al surface can serve as nucleation sites for the formation of a long-range ordered nanopore array during the second anodization. The ordered structure formed through this self-organization process typically exhibits a hexagonal close-packed arrangement, which has the lowest energy. The two-step anodization method is a foundational technique for preparing high-quality AAO templates.

Similar to the two-step anodization method, concavities can also be pre-imprinted into aluminum foil surface by using a master mold prepared by techniques such as EBL, interference lithography, or step-and-flash NIL. These artificially defined, perfectly arranged concavities can precisely guide the nucleation site of each pore during the subsequent anodization process, resulting in nearly perfect ordered AAO nanochannel arrays on a wafer scale (Figure 5.5a).[159]

Interestingly, the final shape of the AAO nanopores is not directly determined by the shape of the pre-imprinted concavities but rather by the arrangement pattern of the concavities on the aluminum surface. During anodization, a "cell" forms around

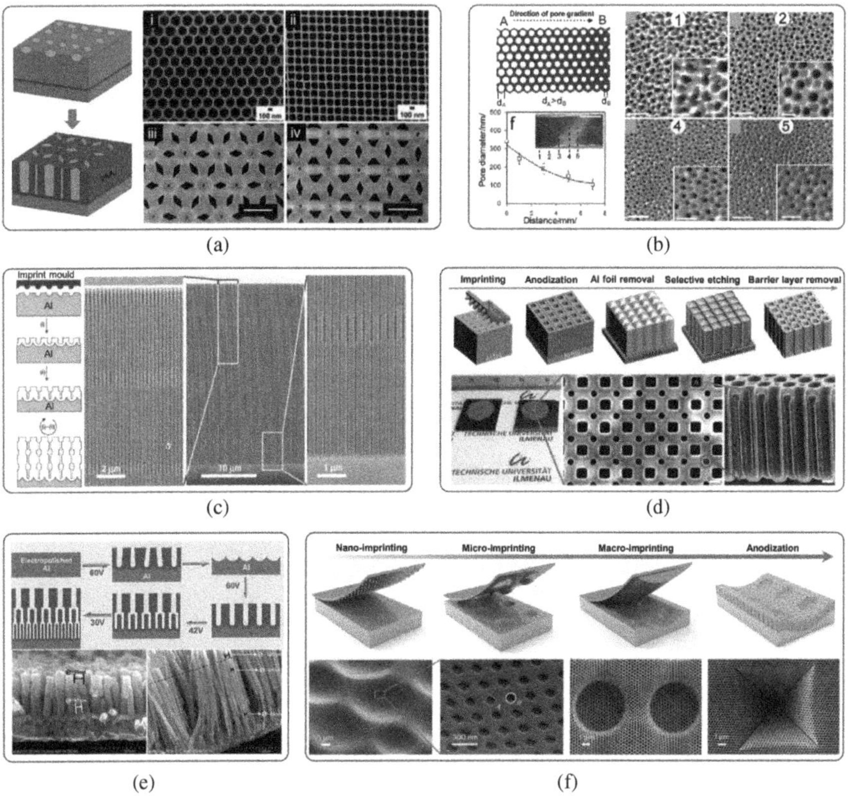

Figure 5.5. Diverse anodic aluminum oxide template. (a) Various nanopore shapes achieved by controlling the arrangement of pre-imprinted patterns.[151,152] Reprinted with permission from [151] by American Chemical Society and with permission from [152] by AIP Publishing. (b) Anodic aluminum oxide template with a horizontal pore diameter gradient.[153] Reprinted with permission from [153] by American Chemical Society. (c) Anodic aluminum oxide template with vertical diameter modulation.[154] Reprinted with permission from [154] by Springer Nature. (d) Double-sided heterostructured templates.[155] Reprinted with permission from [155] by Springer Nature. (e) Branched nanopore networks.[156] Reprinted under the terms of CC-BY Creative Commons Attribution 4.0 International License from [156]. (f) Cross-scale (nano-micro-macro) multilevel structural templates.[157] Reprinted with permission from [157] by Springer Nature.

each pore, and the pore's shape is the result of compression and competition between these cells. By leveraging this principle, one can break the traditional hexagonal symmetry by designing different arrangement patterns. For example, the orthogonal arrangement of pre-imprinted patterns can produce square holes.[151,160] By selectively omitting certain imprinting sites in a hexagonal lattice, surrounding cells are forced to stretch or compress, leading to the formation of diamond-shaped pores or hybrid triangle-diamond pores.[152]

In addition to pores of uniform diameter, it is also possible to create a gradient of pore diameters in the horizontal direction of the template.[153] By employing a non-parallel electrode setup during anodization (e.g., tilting the anode and cathode), a non-uniform gradient electric field and current density can be generated across the aluminum substrate surface. Since the AAO pore diameter is positively correlated with the anodization voltage/current density, this non-uniform anodization condition leads to the formation of a gradient structure with continuously varying pore diameters from one end of the template to the other (Figure 5.5b).

5.2.4.2 *Construction of 3D Channel Nanostructures*

Nanochannel control in the vertical dimension of the AAO template (i.e., along the pore depth) has opened up the possibilities for fabricating 3D nanostructures and heterostructures. This subsection will elaborate further on this aspect.

1) *Axially Varying Pore Diameter*

   By dynamically varying the voltage during anodization, it is possible to achieve periodic changes in pore diameter in the vertical direction. For example, by alternating between "mild anodization" (low voltage) and "hard anodization" (high voltage), "bamboo-like" structures with alternating diameters can be created within the same channel (Figure 5.5c).[154] This axially modulated channel structure provides a template for preparing diameter-modulated nanowires.

2) *Branched Nanochannel Networks*
Through a stepwise reduction of the voltage during anodization, nanochannels with branched structures can be fabricated.[156] When the voltage is lowered, the bottom of the existing single nanochannel re-nucleates and splits into multiple new, smaller nanochannels that continue to grow, thus forming Y-shaped or even higher-order dendritic branched nanostructures (Figure 5.5e). Such branched nanopores are a crucial foundation for constructing 3D nanoscale networks and complex nanodevices.

3) *Janus-Typed Templates*
Through sophisticated process design, it is possible to fabricate pores of different shapes and sizes on the opposite face of the same AAO template. For instance, square pores arrays can be formed on one side through pre-imprinting and anodization, and then through selective etching at specific junctions on the other side (the original aluminum substrate side), a different set of round pores can be generated (Figure 5.5d).[155] This results in a unique "Janus" template, where the openings of the two sets of pores face opposite sides, allowing for independent filling and modification of each set of nanopore structures.

5.2.4.3 *Fabrication of Multiscale Templates*

Bridging the gap between nanoscale features and macroscopic devices requires templates with hierarchical structures that span multiple length scales. A novel technique, termed aluminum-based 3D lithography, has been developed by combining sequential imprinting with anodization to create true multiscale AAO templates (Figure 5.5f).[157]

The process begins with sequential nano-micro-macro-imprinting on a single aluminum substrate. First, a nanostructured mold is used to create nanopatterns. Then, a mold with microscale patterns is imprinted on top of the nanopatterned surface. Finally, a mold with macroscale patterns can be used to shape the entire substrate. The key to preserving the smaller features during subsequent larger-scale imprinting originates from the hardening of aluminum. Each imprinting step plastically deforms the aluminum surface, creating a high density of dislocations, which increases the material's local hardness and strength, making the previously formed patterns robust enough to withstand the pressure of the next imprinting step. This process permits high-fidelity transfer of patterns across scales from nanometers to centimeters.

Once the sequential imprinting is completed, the multiscale patterned aluminum surface is anodized. The final step converts the entire hierarchical structure into a multi-scale AAO template, where high-aspect-ratio nanopores are formed directly on top of the micro-patterns (e.g., micropyramids) and macro-patterns (e.g., curved surfaces). This method enables the creation of templates with precisely defined features across at least seven orders of magnitude in length scale.

5.3 Superplastic Nanomolding for Nanophotonic Applications

Recent advances in nanophotonics have driven the demand for scalable, high-precision fabrication techniques capable of producing complex metallic nanostructures with tailored optical properties. SPNM has emerged as a transformative approach, enabling the rapid prototyping of plasmonic architectures with unprecedented design flexibility and reproducibility. This section explores the application of SPNM in three critical areas of nanophotonics: (1) SERS substrates, where controlled nanoscale roughness and plasmonic hotspots amplify molecular signatures; (2) plasmonic waveguides, which leverage SPNM's ability to form low-loss, high-confinement optical pathways; and (3) chiral nanostructures for chiroptical applications, where 3D asymmetry at the nanoscale enables polarization-dependent light-matter interactions.

5.3.1 *Surface-Enhanced Raman Scattering Substrates*

The development of high-performance SERS substrates has undergone significant technological evolution, progressing from conventional lithography techniques to advanced nanomolding approaches. This section systematically examines the fabrication methodology, performance characteristics, and limitations of SERS substrates prepared by EBL, NIL, and SPNM, with particular emphasis on their structural control, enhancement factors, and practical applications.

Traditional EBL-fabricated SERS substrates, while offering exceptional precision, present substantial limitations in scalability and cost-effectiveness. As demonstrated by Gutierrez-Rivera et al., optimized EBL processes can produce Au nanodot arrays with tunable interparticle gaps (10–25 nm) on fused silica substrates (Figure 5.6a), achieving sensitive detection of protein A in aqueous solutions.[161] The small nanogaps with size of

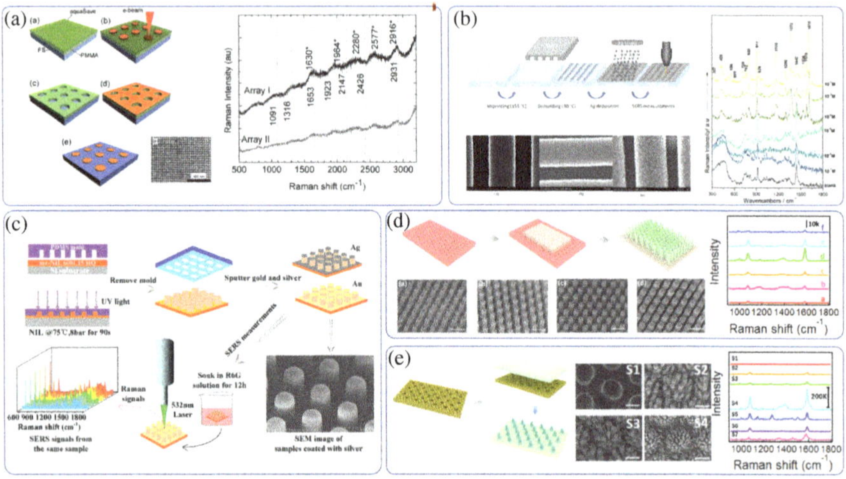

Figure 5.6. SERS substrates fabricated by EBL, NIL and SPNM. (a) Fabrication of Au nanodot arrays on fused silica via EBL for SERS detection of protein A.[161] Reprinted with permission from [161] by American Vacuum Society. (b) Schematic of the flexible SERS substrate fabricated by NIL and the SERS performance.[162] Reprinted with permission from [162] by Elsevier. (c) Highly ordered Ag/Au nanoparticle arrays fabricated by UV-NIL for SERS applications.[163] Reprinted with permission from [163] by Elsevier. (d) Fabrication of highly ordered Ag sub-micrometer pillars via SPNM using AAO template.[164] Reprinted with permission from [164] by John Wiley and Sons. (e) Hierarchical Ag nanostructures fabricated by SPNM using hierarchical AAO template.[165] Reprinted with permission from [165] by American Chemical Society.

10 nm exhibited strong SERS signals due to intense plasmonic coupling effects. However, this technique requires sophisticated anti-charging schemes during the high-resolution patterning of dielectric substrates, and the serial writing process fundamentally limits the throughput. Furthermore, the dependence on semiconductor-grade cleanroom facilities significantly increases the fabrication costs, rendering EBL impractical for large-area SERS substrate production.

NIL has emerged as a promising alternative, combining nanoscale resolution with scalable manufacturing capabilities. Recent advances in NIL-based SERS substrates have demonstrated remarkable performance metrics. Colniță et al. developed flexible SERS platforms featuring 3D silver-metallized nanotrenches (400 nm width/800 nm pitch) on plastic substrates (Figure 5.6b), where the 25 nm Ag coating achieved an enhancement factor of 10^7 and a detection limit of 10 pM for crystal violet.[162] The periodic nanostructure geometry facilitates synergistic plasmonic coupling between adjacent features, while the polymer substrate enables conformal contact with irregular surfaces which is a critical advantage for practical sensing applications. Similarly, Cai et al. reported highly ordered Ag/Au nanoparticle arrays fabricated via UV-NIL and plasma sputtering (Figure 5.6c), which exhibit exceptional enhancement factors (0.8×10^{11}) and reproducibility (RSD ~3.72%).[163] These substrates benefit from the inherent parallel processing nature of NIL, allowing wafer-scale patterning with sub-50 nm resolution. Nevertheless, NIL-based approaches still face challenges related to master mold fabrication (typically requiring EBL), limited 3D design flexibility, and potential degradation of soft molds during repeated use.

The advent of SPNM has addressed several fundamental limitations of conventional lithographic techniques for SERS substrate fabrication. Xiang et al. use the SPNM technique with AAO templates to create highly ordered metallic nanostructures.[164] In this innovative approach, BMG molds prepattern aluminum sheets prior to anodization, enabling the fabrication of Ag sub-micrometer pillars with tunable diameters (137–292 nm) and aspect ratios (~2). These substrates demonstrate strong SERS performance with enhancement factors of ~10^6 for 4-mercaptobenzoic acid detection (Figure 5.6d). The SPNM technique offers three key advantages over traditional methods. First, it eliminates the photolithography steps through direct metal forming. Second, it holds the capability to produce complex hierarchical architectures in a single molding process. Third, it shows exceptional mold durability (>50 uses without degradation).

Further developments in SPNM have yielded hierarchical Ag nanostructures consisting of microscale cavities decorated with dense nanopillars (30–200 nm diameter).[165] These two-level architectures achieve enhancement factors of $\sim 7.0 \times 10^5$ with high reproducibility (RSD = 7%), enabling detection of biomolecules (bovine serum albumin) and dyes (rhodamine 6G) at concentrations as low as 10^{-10} M (Figure 5.6e). The superior performance stems from the combined effects of increased surface area for molecular adsorption and optimized plasmonic coupling between the nanostructures. Importantly, SPNM's compatibility with various metals and alloys provides additional degrees of freedom for tailoring plasmonic responses across the visible to near-infrared region.

Comparative analysis of these fabrication technologies reveals distinct performance trade-offs. While EBL achieves the highest resolution (<10 nm gaps) and NIL offers excellent balance between performance and scalability, SPNM provides unique advantages in terms of structural complexity, material versatility, and cost-effectiveness for large-area applications. The transition from conventional lithography to nanomolding approaches represents a paradigm shift in SERS substrate manufacturing, particularly for applications requiring 3D nanostructures or flexible substrates. The combination of design flexibility, performance consistency, and manufacturing scalability positions SPNM as a transformative technology for next-generation SERS platforms.

5.3.2 *Plasmonic Waveguides*

Plasmonic waveguides have emerged as essential components for integrated photonic circuits due to their unique capability of subwavelength light confinement. The fabrication of high-performance plasmonic waveguides requires precise nanoscale patterning techniques that balance resolution, throughput, and material compatibility. This section compares three typical approaches that have been developed for the fabrication of plasmonic waveguides, namely EBL,[166] NIL,[167] and SPNM.[168] We highlight their advantages and limitations for photonic applications.

EBL has been widely employed for prototyping plasmonic waveguide structures owing to its exceptional patterning resolution. Martiradonna et al. demonstrated the fabrication of colloidal nanocrystal (CdSe/ZnS)-based ridge waveguides with integrated distributed Bragg reflectors (DBRs) and distributed feedback gratings using EBL.[42] As shown in Figure 5.7a, the PMMA/nanocrystals blend waveguide exhibited efficient optical confinement ($n = 1.62$ at 640 nm) without requiring additional etching steps. While EBL enables sub-10 nm feature sizes, its serial writing process results in low throughput and high operational costs, making it impractical for large-scale device fabrication. Furthermore, EBL faces challenges in processing certain materials, particularly brittle crystalline metals that are crucial for low-loss plasmonic applications.

Thermal NIL offers a high-throughput alternative by replicating master patterns through mechanical deformation. Fehrman

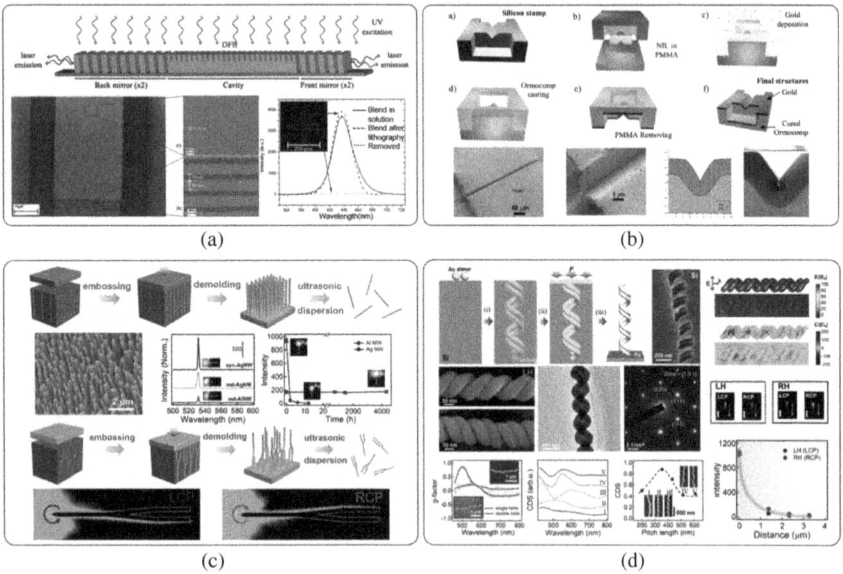

Figure 5.7. Plasmonic waveguides fabricated via EBL, NIL and SPNM. (a) Schematic and SEM images of a colloidal nanocrystal (CdSe/ZnS)-based waveguide fabricated using EBL. The structure consists of a PMMA/nanocrystals ridge waveguide with integrated distributed Bragg reflectors and a distributed feedback grating. The spectra shows emission detected on the patterned PMMA/nanocrystals sample.[42] Reprinted with permission from [42] by Elsevier. (b) Schematic illustration of the fabrication process for V-groove plasmonic waveguides using NIL. Results of simulations and SEM images showing smooth Au grooves with oxidation-sharpened angles (72°→50°) for enhanced plasmon confinement.[169] Reprinted with permission from [169] by American Vacuum Society. (c) Schematic of the SPNM process for fabricating single-crystalline Al NWs and Y-shaped NWs using AAO templates. SEM images waveguiding performance and simulations.[44] Reprinted with permission from [44] by American Chemical Society. (d) Fabrication of Ag double helices via SPNM. Left panel: Schematic of the fabrication process; SEM images of the spiral pores double helical enantiomer and their transmission electron microscopy image along with electron diffraction pattern; chiroptic response and its structural correlation. Right panel: Near-field simulations and waveguiding performance.[45] Reprinted with permission from [45] by American Chemical Society.

Cory et al. successfully implemented NIL to directly pattern inverted ridge waveguides into DNA-CTMA (hexadecyltrimethyl-ammonium chloride) biopolymer films. The solvent-free imprinting process at 120 °C and 500 psi produced 0.2-μm-deep channels with straight sidewalls and minimal roughness, achieving electro-optic coefficients twice as high as conventional materials.[43] Fernandez-Cuesta et al. developed a NIL process for creating V-groove plasmonic waveguides in Au films.[169] As shown in Figure 5.7b, their approach utilized silicon stamps with anisotropically etched V-grooves (72 ° angle) that were replicated into PMMA, followed by Au deposition. The technique achieved smooth sidewalls and enabled simultaneous patterning of micron-scale alignment channels for optical fiber integration. Subsequent thermal oxidation of silicon stamps at 1150 °C for six hours further improved groove sharpness to 50 °, enhancing plasmon confinement characteristics. However, NIL suffers from inherent limitations including template wear, residual layer issues, and the requirement for multistep processing. The demolding process often induces structural defects, especially for high-aspect-ratio nanostructures, and the technique remains constrained by the availability of high-quality master stamps.

Recent advances in SPNM have addressed these fundamental challenges through a solid-state forming approach. Yao, et al. developed centimeter-scale single-crystalline aluminum nanowires via SPNM using AAO templates.[44] These waveguides exhibited broadband guiding (400–700 nm) with exceptional long-term stability (>6 months) and thermal modulation capability (61.4% depth at 4 mW power), as illustrated in Figure 5.7c. The SPNM process, conducted at 500 °C (0.63 T_m) with ~300 MPa pressure, leverages diffusion-dominated deformation to achieve atomic-level precision without grain boundary defects. In contrast to the chemically synthesized Ag nanowires (NWs), the SPNM-produced monocrystalline Al NWs show fivefold weaker waveguiding efficiency initially but stabilize over the long term. A key innovation is the fabrication of Y-shaped Al NWs via SPNM for polarization-dependent light routing and splitting, achieving high efficiency with minimal junction loss, which is a significant improvement over lithographically patterned or mechanically assembled structures. Furthermore, the thermal-optical modulation capability of Al NWs enables reversible and durable optical switching, making them promising for applications in nanophotonic circuits, optical encryption, and on-chip information processing. This work establishes SPNM as a versatile platform for functional nanophotonic devices and integrated optical chips.

The unique capability of SPNM extends to 3D chiral plasmonic structures. Li et al. fabricated Ag double helices with sub-2 nm nanocrevices by compressing bulk silver into pre-etched helical Si templates.[45] As depicted in Figure 5.7d, the three-step process involves catalytic etching of double helical nanopores, superplastic molding at 500 °C, and demolding, which produced enantiomeric nanostructures with g-factors reaching ±0.8. These structures enabled ultrasensitive chiral detection with detection limit of 10^{-19} mole and generation of circularly polarized luminescence with $g_{lum} \approx 0.5$. When the circularly polarized luminescence travels along the double helical waveguide, its g_{lum} further increases to 0.8. This anisotropy arises from the chiral interaction between the surface plasmon polaritons and helical plasmonic nanocrevices, which boosts the luminescence dissymmetry. However, the propagation loss (~2 μm decay length) remains an issue due to plasmon damping. The work pioneers the integration of chiral waveguiding with ultrasensitive sensing, suggesting potential for on-chip chiral photonic circuits.

The transition from conventional lithography to SPNM represents a paradigm shift in nanophotonic device fabrication. While EBL provides unmatched resolution and NIL offers high throughput, SPNM uniquely combines atomic-level precision with material generality and structural versatility. This is achieved through thermally activated solid-state diffusion that enables direct molding of bulk materials into complex nanostructures without grain boundaries or interfacial defects. The technique's ability to produce monocrystalline metallic waveguides with tailored optoelectronic properties opens new avenues for robust, high-performance plasmonic circuits. Future developments may focus on expanding the library of moldable materials and integrating SPNM with hybrid photonic-plasmonic architectures for on-chip applications.[44,88,92]

5.3.3 *Chiral Nanostructures for Sensing*

The fabrication of chiral nanostructures represents a frontier in nanophotonics, where the intrinsic handedness of 3D architectures enables unprecedented control over circularly polarized light-matter interactions. Unlike conventional planar metamaterials that rely on extrinsic chirality through oblique illumination, true 3D chiral nanostructures exhibit intrinsic optical activity independent of incident angle, making them ideal for robust biosensing applications. The development of scalable fabrication techniques for these complex geometries has evolved from serial writing methods[170] to advanced molding approaches,[167] with SPNM emerging as a particularly promising route for achieving both structural fidelity and material versatility.[168]

Traditional approaches to chiral nanostructure fabrication have relied heavily on direct laser writing[174–177] and focused ion beam milling,[178–180] which suffer from lengthy fabrication time and limited material choices, although they are capable of producing arbitrary 3D geometries. The advent of NIL brings significant improvements in throughput, though early implementations were restricted to quasi-2D chiral patterns. Karimullah et al. pioneered the use of injection molding to create shuriken-shaped plasmonic resonators with engineered superchiral fields (Figure 5.8a).[172] These 100 nm-thick templated

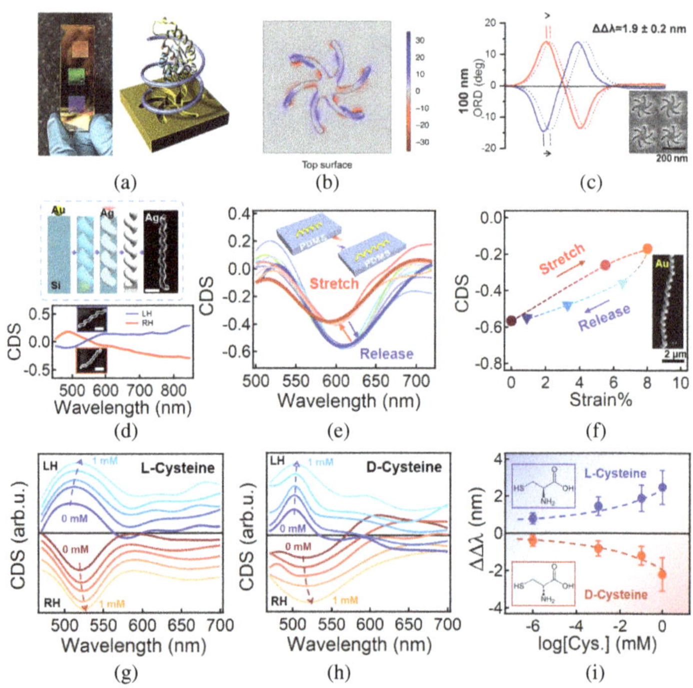

Figure 5.8. Chiral nanophotonic structures fabricated by SPNM and their biosensing applications. (a) Photo image and schematic illustration of a shuriken-shaped nanostructure used for protein sensing and detection.[171] Reprinted under the terms of CC-BY Creative Commons Attribution 4.0 International License from [171]. (b) Superchiral near fields at the top

plasmonic substrates generate optical chirality densities exceeding 30 times that of circularly polarized light, creating highly localized regions of enhanced chiral light-matter interaction (Figure 5.8b). When applied to protein sensing, these substrates demonstrated remarkable sensitivity to conformational changes, detecting as small as 251 pg of Concanavalin A through differential wavelength shifts of 1.9±0.2 nm in optical rotatory dispersion measurements (Figure 5.8c). The mechanism relies on

Figure 5.8. (*Continued*) surface of the shuriken structures, highlighting the chiral field distribution.[172] Reprinted under the terms of CC-BY Creative Commons Attribution 4.0 International License from [172]. (c) Optical rotatory dispersion spectra measured with 100 nm-thick templated plasmonic substrates. Red lines correspond to measurements on left-handed substrates, while blue lines represent right-handed substrates. Solid lines indicate spectra collected in buffer solution, and dotted lines show spectra in the presence of Concanavalin A solution. Vertical bars mark the positions of the peaks under consideration, with arrows indicating the direction and relative magnitude of the spectral shift. Reprinted under the terms of CC-BY Creative Commons Attribution 4.0 International License from [171]. (d) Schematic of the fabrication process for helical nanowires using the SPNM technique. Circular differential scattering (CDS) spectra of helical Ag nanowire on silicon substrate (bottom), with inset showing the corresponding SEM images.[173] (e) CDS spectra of single Au helix-on-polydimethylsiloxane system with reversible strains. (f) Variation of CDS intensity during a single cycle of stretching and releasing. The inset shows a SEM image of the helical Au nanowire. (d-f) Reprinted with permission from [173] by John Wiley and Sons. (g, h) CDS spectra of double helical Ag NW enantiomers coated with a monolayer of (g) L-cysteine and (h) D-cysteine molecules, incubated with different solution concentrations.[45] (i) Change in the spectral dissymmetric factor ($\Delta\Delta\lambda$) as a function of L-/D-cysteine solution concentration. Error bars represent standard deviation. (g-i) Reprinted with permission from [45] by American Chemical Society.

the co-localization of superchiral hotspots with protein binding sites, where the overlap of nearly parallel electric and magnetic fields produces maximum chiral asymmetry.

A recent breakthrough towards true 3D chiral nanostructure is based on the adaptation of the SPNM technique to the fabrication of (double) helical NWs. Li et al. developed a distinctive approach combining metal-assisted chemical etching with high-temperature compression molding to produce single-crystalline metallic helices with unprecedented structural perfection (Figure 5.8d).[173] The process begins with the creation of helical nanopores in silicon templates using Au nanoparticle catalysts, followed by superplastic deformation of bulk metals (Au, Ag, Cu) into these templates at 0.6–0.7 T_m. The resulting structures exhibit circular differential scattering signals with high g-factors, far exceeding those achievable through bottom-up synthesis or conventional lithography. Critically, the monocrystalline nature of SPNM-fabricated helices eliminates grain boundary scattering, resulting in plasmonic losses 5–10 times lower than polycrystalline counterparts.

The integration of SPNM-fabricated chiral nanostructures with flexible substrates has opened new avenues for dynamic biosensing. By embedding Au helices in polydimethylsiloxane matrices, we achieved mechanically tunable chiroptical responses with remarkable reproducibility over six cycles of stretching and releasing (Figures 5.8e–f). This mechanical modulation enables real-time calibration and background subtraction, addressing a long-standing challenge in chiral biosensing where environmental fluctuations often mask molecular signals.[45] The platform's versatility extends to surface functionalization, where enantiomeric Ag helices modified with L- or D-cysteine monolayers exhibit outstanding chiral sensing performance with detection limits reaching 10^{-19} mole, which is several orders of magnitude higher than conventional circular dichroism spectroscopy (Figures 5.8g–i).[181–183] Recent advances have pushed SPNM-based chiral biosensing into clinically relevant applications. The fabrication of hierarchical chiral assemblies, combining microscale spiral architectures with nanoscale surface textures, has enabled multiplexed detection of disease biomarkers. For instance, arrays of helical nanowires with systematically varied pitch and handedness can simultaneously discriminate multiple chiral analytes through their distinct spectral signatures.[184,185] The technique's compatibility with various metals allows optimization of plasmonic resonances across the visible to near-infrared spectrum, matching specific molecular absorption bands. Furthermore, the exceptional structural reproducibility of SPNM (RSD $< 5\%$ for critical dimensions) ensures reliable quantitative analysis, a prerequisite for diagnostic applications.

The future of SPNM-fabricated chiral nanostructures lies in their integration with advanced photonic platforms. Current research explores the combination of chiral plasmonic elements with photonic crystals and metasurfaces to create hybrid devices with enhanced functionality.[186–188] These include chiral photonic circuits for on-chip enantiomer separation,[189,190] circularly polarized single-photon sources for quantum communication,[191–193] and biomimetic sensors inspired by natural photonic structures.[194–196] The ability to mold complex 3D architectures in diverse materials, from noble metals to magnetic alloys, positions SPNM as an enabling technology for next-generation chiroptical devices that bridge fundamental science and practical applications.[45,103,128,129,173]

5.4 Conclusion and Future Perspective

This chapter has comprehensively examined SPNM as a transformative fabrication technology for nanophotonic applications, demonstrating its unique capabilities in creating high-performance SERS substrates, plasmonic waveguides, and chiral sensing. Through systematic comparison with established techniques like EBL and NIL, we have shown that SPNM offers an unprecedented combination of atomic-level precision, material versatility, and scalable manufacturing potential. The technique's ability to produce monocrystalline metallic nanostructures with controlled 3D geometries addresses fundamental limitations in conventional nanofabrication, enabling optical devices with superior performance metrics, including enhancement factors exceeding 10^{11} for SERS, propagation loss below 0.1 dB/μm for plasmonic waveguides, and chiroptical g-factors up to 0.8.

The impact of SPNM extends beyond incremental improvements in device performance. By enabling direct molding of bulk materials through thermally activated superplastic deformation, this technique eliminates many processing steps associated with traditional lithography, reducing fabrication complexity and cost. The demonstrated applications such as ultrasensitive molecular detection and dynamic chiroptical sensing highlight SPNM's potential to accelerate the translation of nanophotonic concepts into practical technologies. Particularly noteworthy is the technique's compatibility with diverse material systems, from noble metals to metallic glasses,[45,173,197,198] opening new design spaces for photonic devices with tailored optical, mechanical, and chemical properties.

Despite these remarkable achievements, several critical challenges must be addressed to fully realize SPNM's potential in industry-scale nanophotonic manufacturing (Figure 5.9). Demolding issues remain a persistent concern, particularly for high-aspect-ratio structures where adhesion forces can cause catastrophic failure during the demolding process.[199,200] Current

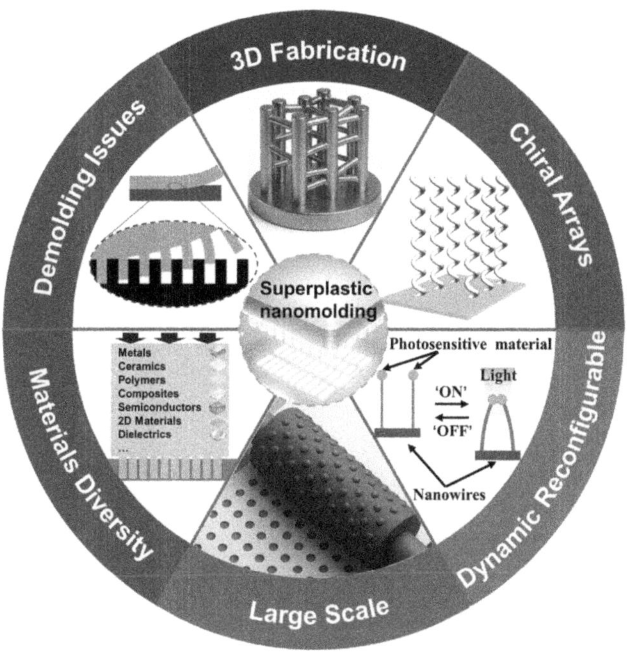

Figure 5.9. Overview of key challenges in SPNM for nanophotonic applications. The circular diagram illustrates six critical challenges in SPNM technology: a) Demolding Issues, which are critical for preventing structural damage and adhesion during mold separation; b) 3D Fabrication, which is crucial for achieving complex 3D photonic architectures through multi-step molding; c) Chiral Arrays, which involves the fabrication of helical and twisted nanostructures with controlled handedness; d) Dynamic Reconfigurable, meaning creating adaptable photonic structures with tunable properties; e) Large Scale, which extends nanoscale precision to wafer-level manufacturing; f) Materials Diversity, which enables molding of various optical materials from polymers to semiconductors and metallic nanostructures.

solutions involving surface modifications and controlled release mechanisms show promise but require further optimization for different material combinations.[201,202] The fabrication of complex 3D photonic architectures presents another frontier, where multi-step molding processes must maintain nanoscale accuracy of alignment while avoiding cumulative errors. Recent demonstrations of layer-by-layer SPNM with *in situ* alignment markers suggest viable pathways,[203] though significant engineering challenges remain.

The creation of large-area chiral nanostructure arrays demands precise control over structural handedness across centimeter-scale substrates, where even minor temperature or pressure gradients can induce unwanted variations. Dynamic reconfigurable systems, while demonstrated at the proof-of-concept level, require integration with robust actuation mechanisms and feedback control to enable adaptive photonic devices.[173,204,205] Perhaps most critically, scaling SPNM from laboratory demonstrations to wafer-level manufacturing necessitates fundamental advances in equipment design,[157] including uniform heating systems, precision pressure control, and automated handling capabilities that maintain nanoscale fidelity across large areas.

Materials diversity represents both an opportunity and a challenge for future SPNM development. While the technique has proven successful for various metals and alloys,[133] extending its applicability to semiconductors, oxides, and hybrid organic-inorganic systems would dramatically expand the scope of achievable photonic functionalities. This requires deep understanding of material-specific deformation mechanisms and the development of tailored processing conditions for each system.[110,206] The integration of multiple materials within a single device demands innovative approaches such as sequential molding, selective area processing, or hybrid fabrication schemes combining SPNM with complementary techniques.

Looking forward, the convergence of SPNM with emerging fields such as quantum photonics,[207,208] neuromorphic computing,[209,210] and synthetic biology[211,212] presents exciting opportunities. The ability to fabricate deterministic single-photon sources through precisely positioned quantum emitters in SPNM-defined cavities could enable scalable quantum information processing.[213,214] Similarly, the creation of complex plasmonic neural networks through 3D molding of interconnected waveguides offers a path toward optical computing architectures with unprecedented speed and energy efficiency.[215,216] In the biological realm, SPNM-fabricated biomimetic photonic structures could lead to new classes of implantable sensors and therapeutic devices that interface seamlessly with living systems.[217,218]

The successful development of SPNM for next-generation nanophotonics will require coordinated efforts across multiple disciplines. Materials scientists must continue exploring new superplastic systems and understanding their deformation mechanisms at the nanoscale. Mechanical engineers need to develop innovative mold designs and demolding strategies that preserve structural integrity. Optical engineers should push the boundaries of device design, leveraging SPNM's unique capabilities to create functionalities impossible with conventional fabrication. Finally, manufacturing engineers must bridge the gap between laboratory demonstrations and industrial implementation, developing the tools and processes necessary for reliable, high-volume production.

In conclusion, SPNM represents a paradigm shift in nanophotonic device fabrication, offering solutions to long-standing challenges while opening new avenues for innovation.[168] As the field advances from fundamental research toward practical implementation, SPNM is poised to play a central role in realizing the full potential of nanophotonics across applications ranging from ultrasensitive biosensing to quantum information processing. The continued development of this technology, guided by the challenges and opportunities outlined in this chapter, will be essential for the next generation of photonic technologies that will shape our increasingly connected and data-driven world.

Acknowledgments

We acknowledge the support of the National Natural Science Foundation of China (12374356, 12172260, 92477138, 124B2036), and the National Key Research and Development Program of China (2020YFA0211300).

References

1. Basudeb, Sain, Meier Cedrik, and Zentgraf Thomas, "Nonlinear optics in all-dielectric nanoantennas and metasurfaces: A review." *Advanced Photonics* 1, no. 2 (2019): 024002.
2. Deng, Juan, Yan Yang, Jin Tao, Liangui Deng, Daoqun Liu, Zhiqiang Guan, Gongfa Li, Zile Li, Shaohua Yu, Guoxing Zheng, Zhongyang Li, and Shuang Zhang, "Spatial frequency multiplexed meta-holography and meta-nanoprinting." *ACS Nano* 13, no. 8 (2019): 9237–9246.
3. Kumar, Karthik, Huigao Duan, Ravi S. Hegde, Samuel C. W. Koh, Jennifer N. Wei, and Joel K. W. Yang, "Printing colour at the optical diffraction limit." *Nature Nanotechnology* 7, no. 9 (2012): 557–561.
4. Liu, Lixiang, Xueqian Zhang, Mitchell Kenney, Xiaoqiang Su, Ningning Xu, Chunmei Ouyang, Yunlong Shi, Jiaguang Han, Weili Zhang, and Shuang Zhang, "Broadband metasurfaces with simultaneous control of phase and amplitude." *Advanced Materials* 26, no. 29 (2014): 5031–5036.

5. Lee, Gun-Yeal, Gwanho Yoon, Seung-Yeol Lee, Hansik Yun, Jaebum Cho, Kyookeun Lee, Hwi Kim, Junsuk Rho, and Byoungho Lee, "Complete amplitude and phase control of light using broadband holographic metasurfaces." *Nanoscale* 10, no. 9 (2018): 4237–4245.
6. Zang, Wenbo, Quan Yuan, Run Chen, Lin Li, Tianyue Li, Xiujuan Zou, Gaige Zheng, Zhuo Chen, Shuming Wang, Zhenlin Wang, and Shining Zhu, "Chromatic dispersion manipulation based on metalenses." *Advanced Materials* 32, no. 27 (2020): 1904935.
7. Zayats, Anatoly V., Igor I. Smolyaninov, and Alexei A. Maradudin, "Nano-optics of surface plasmon polaritons." *Physics Reports* 408, no. 3 (2005): 131–314.
8. Berini, Pierre, and Israel De Leon, "Surface plasmon–polariton amplifiers and lasers." *Nature Photonics* 6, no. 1 (2012): 16–24.
9. Kim, Jaekyung, Junhwa Seong, Younghwan Yang, Seong-Won Moon, Trevon Badloe, and Junsuk Rho, "Tunable metasurfaces towards versatile metalenses and metaholograms: A review." *Advanced Photonics* 4, no. 2 (2022): 024001.
10. Lee, Dohyeon, Younghwan Yang, and Junsuk Rho, "Lithography-free tailoring of thin-film nanomorphology for enhanced real-time refractive-index sensing with tunable resonance peaks in the visible spectrum." *Journal of Applied Physics* 135, no. 9 (2024): 093102.
11. Jung, Chunghwan, Soo-Jung Kim, Jaehyuck Jang, Joo Hwan Ko, Doa Kim, Byoungsu Ko, Young Min Song, Sung-Hoon Hong, and Junsuk Rho, "Disordered-nanoparticle–based etalon for ultrafast humidity-responsive colorimetric sensors and anti-counterfeiting displays." *Science Advances* 8, no. 10 (2022): eabm8598.
12. Badloe, Trevon, Joohoon Kim, Inki Kim, Won-Sik Kim, Wook Sung Kim, Young-Ki Kim, and Junsuk Rho, "Liquid crystal-powered Mie resonators for electrically tunable photorealistic color gradients and dark blacks." *Light: Science & Applications* 11, no. 1 (2022): 118.
13. Lee, Taejun, Joohoon Kim, Ishwor Koirala, Younghwan Yang, Trevon Badloe, Jaehyuck Jang, and Junsuk Rho, "Nearly perfect transmissive subtractive coloration through the spectral amplification of Mie scattering and lattice resonance." *ACS Applied Materials & Interfaces* 13, no. 22 (2021): 26299–26307.
14. Schlickriede, Christian, Naomi Waterman, Bernhard Reineke, Philip Georgi, Guixin Li, Shuang Zhang, and Thomas Zentgraf, "Imaging through nonlinear metalens using second harmonic generation." *Advanced Materials* 30, no. 8 (2018): 1703843.

15. Kim, Joohoon, Yeseul Kim, Wonjoong Kim, Dong Kyo Oh, Dohyun Kang, Junhwa Seong, Jeong Woo Shin, Dohyun Go, Chanwoong Park, Hyoin Song, Jihwan An, Heon Lee, and Junsuk Rho, "8" wafer-scale, centimeter-sized, high-efficiency metalenses in the ultraviolet." *Materials Today* 73, (2024): 9–15.

16. Badloe, Trevon, Yeseul Kim, Joohoon Kim, Hyemi Park, Aleksandr Barulin, Yen N. Diep, Hansang Cho, Won-Sik Kim, Young-Ki Kim, Inki Kim, and Junsuk Rho, "Bright-field and edge-enhanced imaging using an electrically tunable dual-mode metalens." *ACS Nano* 17, no. 15 (2023): 14678–14685.

17. Barulin, Aleksandr, Yeseul Kim, Dong Kyo Oh, Jaehyuck Jang, Hyemi Park, Junsuk Rho, and Inki Kim, "Dual-wavelength metalens enables epi-fluorescence detection from single molecules." *Nature Communications* 15, no. 1 (2024): 26.

18. Badloe, Trevon, Junhwa Seong, and Junsuk Rho, "Trichannel spin-selective metalenses." *Nano Letters* 23, no. 15 (2023): 6958–6965.

19. Badloe, Trevon, Inki Kim, Yeseul Kim, Joohoon Kim, and Junsuk Rho, "Electrically tunable bifocal metalens with diffraction-limited focusing and imaging at visible wavelengths." *Advanced Science* 8, no. 21 (2021): 2102646.

20. Kim, Inki, Won-Sik Kim, Kwan Kim, Muhammad Afnan Ansari, Muhammad Qasim Mehmood, Trevon Badloe, Yeseul Kim, Junho Gwak, Heon Lee, Young-Ki Kim, and Junsuk Rho, "Holographic metasurface gas sensors for instantaneous visual alarms." *Science Advances* 7, no. 15 (2021): eabe9943.

21. So, Sunae, Joohoon Kim, Trevon Badloe, Chihun Lee, Younghwan Yang, Hyunjung Kang, and Junsuk Rho, "Multicolor and 3D holography generated by inverse-designed single-cell metasurfaces." *Advanced Materials* 35, no. 17 (2023): 2208520.

22. Kim, Joohoon, Dongmin Jeon, Junhwa Seong, Trevon Badloe, Nara Jeon, Gyeongtae Kim, Jaekyung Kim, Sangwon Baek, Jong-Lam Lee, and Junsuk Rho, "Photonic encryption platform via dual-band vectorial metaholograms in the ultraviolet and visible." *ACS Nano* 16, no. 3 (2022): 3546–3553.

23. Kim, Inki, Jaehyuck Jang, Gyeongtae Kim, Jihae Lee, Trevon Badloe, Jungho Mun, and Junsuk Rho, "Pixelated bifunctional metasurface-driven dynamic vectorial holographic color prints for photonic security platform." *Nature Communications* 12, no. 1 (2021): 3614.

24. Kim, Joohoon, Younghwan Yang, Trevon Badloe, Inki Kim, Gwanho Yoon, and Junsuk Rho, "Geometric and physical configurations of meta-atoms for advanced metasurface holography." *InfoMat* 3, no. 7 (2021): 739–754.

25. Kim, Joohoon, Jun-Hyung Im, Sunae So, Yeongseon Choi, Hyunjung Kang, Bogyu Lim, Minjae Lee, Young-Ki Kim, and Junsuk Rho, "Dynamic hyperspectral holography enabled by inverse-designed metasurfaces with oblique helicoidal cholesterics." *Advanced Materials* 36, no. 24 (2024): 2311785.

26. Latif, Sabiha, Joohoon Kim, Hafiz Saad Khaliq, Nasir Mahmood, Muhammad Afnan Ansari, Xianzhong Chen, Jehan Akbar, Trevon Badloe, Muhammad Zubair, Yehia Massoud, Muhammad Qasim Mehmood, and Junsuk Rho, "Spin-selective angular dispersion control in dielectric metasurfaces for multichannel meta-holographic displays." *Nano Letters* 24, no. 2 (2024): 708–714.

27. Naeem, Taimoor, Joohoon Kim, Hafiz Saad Khaliq, Junhwa Seong, Muhammad Tariq Saeed Chani, Tauseef Tauqeer, Muhammad Qasim Mehmood, Yehia Massoud, and Junsuk Rho, "Dynamic chiral metasurfaces for broadband phase-gradient holographic displays." *Advanced Optical Materials* 11, no. 5 (2023): 2202278.

28. Mehmood, Muhammad Qasim, Junhwa Seong, Muhammad Ashar Naveed, Joohoon Kim, Muhammad Zubair, Kashif Riaz, Yehia Massoud, and Junsuk Rho, "Single-cell-driven tri-channel encryption meta-displays." *Advanced Science* 9, no. 35 (2022): 2203962.

29. Khaliq, Hafiz Saad, Joohoon Kim, Taimoor Naeem, Kashif Riaz, Trevon Badloe, Junhwa Seong, Jehan Akbar, Muhammad Zubair, Muhammad Qasim Mehmood, Yehia Massoud, and Junsuk Rho, "Broadband chiro-optical effects for futuristic meta-holographic displays." *Advanced Optical Materials* 10, no. 22 (2022): 2201175.

30. Javed, Isma, Joohoon Kim, Muhammad Ashar Naveed, Dong Kyo Oh, Dongmin Jeon, Inki Kim, Muhammad Zubair, Yehia Massoud, Muhammad Qasim Mehmood, and Junsuk Rho, "Broad-band polarization-insensitive metasurface holography with a single-phase map." *ACS Applied Materials & Interfaces* 14, no. 31 (2022): 36019–36026.

31. Naveed, Muhammad Ashar, Joohoon Kim, Isma Javed, Muhammad Afnan Ansari, Junhwa Seong, Yehia Massoud, Trevon Badloe, Inki Kim, Kashif Riaz, Muhammad Zubair, Muhammad Qasim Mehmood, and Junsuk Rho, "Novel spin-decoupling strategy in liquid crystal-integrated metasurfaces for interactive metadisplays." *Advanced Optical Materials* 10, no. 13 (2022): 2200196.

32. Lee, Chihun, Seokho Lee, Junhwa Seong, Dong Yong Park, and Junsuk Rho, "Inverse-designed metasurfaces for highly saturated transmissive colors." *Journal of the Optical Society of America B* 41, no. 1 (2024): 151–158.

33. Jang, Jaehyuck, Trevon Badloe, Younghwan Yang, Taejun Lee, Jungho Mun, and Junsuk Rho, "Spectral modulation through the hybridization of Mie-scatterers and quasi-guided mode resonances: Realizing full and gradients of structural color." *ACS Nano* 14, no. 11 (2020): 15317–15326.

34. Sun, Shang, Zhenxing Zhou, Chen Zhang, Yisheng Gao, Zonghui Duan, Shumin Xiao, and Qinghai Song, "All-dielectric full-color printing with TiO_2 metasurfaces." *ACS Nano* 11, no. 5 (2017): 4445–4452.

35. Kim, Inki, Juyoung Yun, Trevon Badloe, Hyuk Park, Taewon Seo, Younghwan Yang, Juhoon Kim, Yoonyoung Chung, and Junsuk Rho, "Structural color switching with a doped indium-gallium-zinc-oxide semiconductor." *Photonics Research* 8, no. 9 (2020): 1409–1415.

36. Jeong, Heonyeong, Younghwan Yang, Hanlyun Cho, Trevon Badloe, Inki Kim, Ren-Min Ma, and Junsuk Rho, "Emerging advanced metasurfaces: Alternatives to conventional bulk optical devices." *Microelectronic Engineering* 220 (2020): 111146.

37. Jang, Jaehyuck, Trevon Badloe, Young Chul Sim, Younghwan Yang, Jungho Mun, Taejun Lee, Yong-Hoon Cho, and Junsuk Rho, "Full and gradient structural colouration by lattice amplified gallium nitride Mie-resonators." *Nanoscale* 12, no. 41 (2020): 21392–21400.

38. Kang, Hyunjung, Dohyeon Lee, Younghwan Yang, Dong Kyo Oh, Junhwa Seong, Jaekyung Kim, Nara Jeon, Dohyun Kang, and Junsuk Rho, "Emerging low-cost, large-scale photonic platforms with soft lithography and self-assembly." *Photonics Insights* 2, no. 2 (2023): R04.

39. Lim, Dong-Kwon, Ki-Seok Jeon, Hyung Min Kim, Jwa-Min Nam, and Yung Doug Suh, "Nanogap-engineerable Raman-active nanodumbbells for single-molecule detection." *Nature Materials* 9, no. 1 (2010): 60–67.

40. Lee, Jung-Hoon, Jwa-Min Nam, Ki-Seok Jeon, Dong-Kwon Lim, Hyoki Kim, Sunghoon Kwon, Haemi Lee, and Yung Doug Suh, "Tuning and maximizing the single-molecule surface-enhanced Raman scattering from DNA-tethered nanodumbbells." *ACS Nano* 6, no. 11 (2012): 9574–9584.

41. Lee, Haemi, Jung-Hoon Lee, Seung Min Jin, Yung Doug Suh, and Jwa-Min Nam, "Single-molecule and single-particle-based correlation studies between localized surface plasmons of dimeric nanostructures with ~1 nm gap and surface-enhanced Raman scattering." *Nano Letters* 13, no. 12 (2013): 6113–6121.

42. Martiradonna, L., A. Qualtieri, T. Stomeo, L. Carbone, R. Cingolani, and M. De Vittorio, "Lithographic nano-patterning of colloidal nanocrystal emitters for the fabrication of waveguide photonic devices." *Sensors and Actuators B: Chemical* 126, no. 1 (2007): 116–119.

43. Emily, M. Fehrman-Cory, S. Aga Roberto, Jr., P. Lombardi Jack, III, M. Bartsch Carrie, M. Sarangan Andrew, and M. Heckman Emily, "Nanoimprint lithography of deoxyribonucleic acid biopolymer films." *Journal of Micro/Nanolithography, MEMS, and MOEMS* 12, no. 4 (2013): 040501.

44. Yao, Jiacheng, Hui Fang, Yong Li, Ze Liu, Hongxing Xu, and Tao Ding, "Superplastic nanomolding of aluminum waveguides for subwavelength light routing, splitting, and encryption." *ACS Nano* 17, no. 17 (2023): 17342–17349.

45. Li, Yong, Hui Fang, Ze Liu, and Tao Ding, "Helical plasmonic nanocrevices formed by metallic double helices for ultrasensitive chiral detection and high-performance chiral photoluminescence." *ACS Photonics* 12, no. 5 (2025): 2787–2794.

46. Ni, Xingjie, Naresh K. Emani, Alexander V. Kildishev, Alexandra Boltasseva, and Vladimir M. Shalaev, "Broadband light bending with plasmonic nanoantennas." *Science* 335, no. 6067 (2012): 427–427.

47. Avayu, Ori, Euclides Almeida, Yehiam Prior, and Tal Ellenbogen, "Composite functional metasurfaces for multispectral achromatic optics." *Nature Communications* 8, no. 1 (2017): 14992.

48. Chen, Xianzhong, Lingling Huang, Holger Mühlenbernd, Guixin Li, Benfeng Bai, Qiaofeng Tan, Guofan Jin, Cheng-Wei Qiu, Shuang Zhang, and Thomas Zentgraf, "Dual-polarity plasmonic metalens for visible light." *Nature Communications* 3, no. 1 (2012): 1198.

49. Huang, Lingling, Holger Mühlenbernd, Xiaowei Li, Xu Song, Benfeng Bai, Yongtian Wang, and Thomas Zentgraf, "Broadband hybrid holographic multiplexing with geometric metasurfaces." *Advanced Materials* 27, no. 41 (2015): 6444–6449.

50. Luo, Xuhao, Yueqiang Hu, Xin Li, Yuting Jiang, Yasi Wang, Peng Dai, Qing Liu, Zhiwen Shu, and Huigao Duan, "Integrated metasurfaces with microprints and helicity-multiplexed holograms for real-time optical encryption." *Advanced Optical Materials* 8, no. 8 (2020): 1902020.

51. Yu, Ping, Jianxiong Li, Xin Li, Gisela Schütz, Michael Hirscher, Shuang Zhang, and Na Liu, "Generation of switchable singular beams with dynamic metasurfaces." *ACS Nano* 13, no. 6 (2019): 7100–7106.

52. Deng, Zi-Lan, Mingke Jin, Xuan Ye, Shuai Wang, Tan Shi, Junhong Deng, Ningbin Mao, Yaoyu Cao, Bai-Ou Guan, Andrea Alù, Guixin Li, and Xiangping Li, "Full-color complex-amplitude vectorial holograms based on multi-freedom metasurfaces." *Advanced Functional Materials* 30, no. 21 (2020): 1910610.

53. Kim, Joohoon, Junhwa Seong, Wonjoong Kim, Gun-Yeal Lee, Seokwoo Kim, Hongyoon Kim, Seong-Won Moon, Dong Kyo Oh, Younghwan Yang, Jeonghoon Park, Jaehyuck Jang, Yeseul Kim, Minsu Jeong, Chanwoong Park, Hojung Choi, Gyoseon Jeon, Kyung-il Lee, Dong Hyun Yoon, Namkyoo Park, Byoungho Lee, Heon Lee, and Junsuk Rho, "Scalable manufacturing of high-index atomic layer–polymer hybrid metasurfaces for metaphotonics in the visible." *Nature Materials* 22, no. 4 (2023): 474–481.

54. Moon, Seong-Won, Joohoon Kim, Chanwoong Park, Wonjoong Kim, Younghwan Yang, Jaekyung Kim, Seokho Lee, Minseok Choi, Hansang Sung, Jaein Park, Hyoin Song, Heon Lee, and Junsuk Rho, "Wafer-scale manufacturing of near-infrared metalenses." *Laser & Photonics Reviews* 18, no. 4 (2024): 2300929.

55. Zhang, Xu A., Jonathan Elek, and Chih-Hao Chang, "Three-dimensional nanolithography using light scattering from colloidal particles." *ACS Nano* 7, no. 7 (2013): 6212–6218.

56. Qiu, Tengfei, Eser Metin Akinoglu, Bin Luo, Muxina Konarova, Jung-Ho Yun, Ian R. Gentle, and Lianzhou Wang, "Nanosphere lithography: A versatile approach to develop transparent conductive films for optoelectronic applications." *Advanced Materials* 34, no. 19 (2022): 2103842.

57. Go, Myeongcheol, Dasol Lee, Sanghoon Kim, Junho Jang, Keon-Woo Kim, Jaeyong Lee, Sangmin Shim, Jin Kon Kim, and Junsuk Rho, "Facile fabrication of titanium nitride nanoring broad-band absorbers in the visible to near-infrared by shadow sphere lithography." *ACS Applied Materials & Interfaces* 15, no. 2 (2023): 3266–3273.

58. Mayer, Martin, Max J. Schnepf, Tobias A. F. König, and Andreas Fery, "Colloidal self-assembly concepts for plasmonic metasurfaces." *Advanced Optical Materials* 7, no. 1 (2019): 1800564.

59. Vinnacombe-Willson, Gail A., Ylli Conti, Andrei Stefancu, Paul S. Weiss, Emiliano Cortés, and Leonardo Scarabelli, "Direct bottom-up in situ growth: A paradigm shift for studies in wet-chemical synthesis of gold nanoparticles." *Chemical Reviews* 123, no. 13 (2023): 8488–8529.

60. Young, Niels O., and Jakym Kowal, "Optically active fluorite films." *Nature* 183, no. 4654 (1959): 104–105.

61. Robbie, K., D. J. Broer, and M. J. Brett, "Chiral nematic order in liquid crystals imposed by an engineered inorganic nanostructure." *Nature* 399, no. 6738 (1999): 764–766.

62. Tan, Shawn J., Michael J. Campolongo, Dan Luo, and Wenlong Cheng, "Building plasmonic nanostructures with DNA." *Nature Nanotechnology* 6, no. 5 (2011): 268–276.

63. Saito, Koichiro, and Tetsu Tatsuma, "Chiral plasmonic nanostructures fabricated by circularly polarized light." *Nano Letters* 18, no. 5 (2018): 3209–3212.

64. Wang, Ling, Augustine M. Urbas, and Quan Li, "Nature-inspired emerging chiral liquid crystal nanostructures: From molecular self-Assembly to DNA mesophase and nanocolloids." *Advanced Materials* 32, no. 41 (2020): 1801335.

65. Goerlitzer, Eric S. A., Reza Mohammadi, Sergey Nechayev, Kirsten Volk, Marcel Rey, Peter Banzer, Matthias Karg, and Nicolas Vogel, "Chiral surface lattice resonances." *Advanced Materials* 32, no. 22 (2020): 2001330.

66. Lingstädt, Robin, Fatemeh Davoodi, Kenan Elibol, Masoud Taleb, Hyunah Kwon, Peer Fischer, Nahid Talebi, and Peter A. van Aken, "Electron beam induced circularly polarized light emission of chiral gold nanohelices." *ACS Nano* 17, no. 24 (2023): 25496–25506.

67. Komlenok, M. S., S. G. Tikhodeev, T. Weiss, S. P. Lebedev, G. A. Komandin, and V. I. Konov, "All-carbon diamond/graphite metasurface: Experiment and modeling." *Applied Physics Letters* 113, no. 4 (2018): 041101.

68. Yan, Chao, Xiong Li, Mingbo Pu, Xiaoliang Ma, Fei Zhang, Ping Gao, Yinghui Guo, Kaipeng Liu, Zuojun Zhang, and Xiangang Luo, "Generation of polarization-sensitive modulated optical vortices with all-dielectric metasurfaces." *ACS Photonics* 6, no. 3 (2019): 628–633.

69. Berzinš, Jonas, Simonas Indrišiūnas, Koen van Erve, Arvind Nagarajan, Stefan Fasold, Michael Steinert, Giampiero Gerini, Paulius Gečys, Thomas Pertsch, Stefan M. B. Bäumer, and Frank Setzpfandt, "Direct and high-throughput fabrication of Mie-resonant metasurfaces via single-pulse laser interference." *ACS Nano* 14, no. 5 (2020): 6138–6149.

70. Li, Bohong, Xiaowei Li, Ruizhe Zhao, Guocui Wang, Weina Han, Bingquan Zhao, Lingling Huang, Yan Zhang, Yongfeng Lu, and Lan Jiang, "Polarization multiplexing terahertz metasurfaces through spatial femtosecond laser-shaping fabrication." *Advanced Optical Materials* 8, no. 12 (2020): 2000136.

71. Wang, Ziyu, Tieshan Yang, Yupeng Zhang, Qingdong Ou, Han Lin, Qianhui Zhang, Huanyang Chen, Hui Ying Hoh, Baohua Jia, and Qiaoliang Bao, "Flat lenses based on 2D perovskite nanosheets." *Advanced Materials* 32, no. 30 (2020): 2001388.

72. Zipfel, Warren R., Rebecca M. Williams, and Watt W. Webb, "Nonlinear magic: Multiphoton microscopy in the biosciences." *Nature Biotechnology* 21, no. 11 (2003): 1369–1377.

73. Faniayeu, Ihar, and Vygantas Mizeikis, "Realization of a helix-based perfect absorber for IR spectral range using the direct laser write technique." *Optical Materials Express* 7, no. 5 (2017): 1453–1462.

74. Zhao, Wenjing, Weixuan Zhang, Rong-Yao Wang, Yinglu Ji, Xiaochun Wu, and Xiangdong Zhang, "Photocontrollable chiral switching and selection in self-assembled plasmonic nanostructure." *Advanced Functional Materials* 29, no. 20 (2019): 1900587.

75. Wang, Hao, Yejing Liu, Qifeng Ruan, Hailong Liu, Ray J. H. Ng, You Sin Tan, Hongtao Wang, Ying Li, Cheng-Wei Qiu, and Joel K. W. Yang, "Off-axis holography with uniform illumination via 3D printed diffractive optical elements." *Advanced Optical Materials* 7, no. 12 (2019): 1900068.

76. Wang, Hao, Hongtao Wang, Wang Zhang, and Joel K. W. Yang, "Toward near-perfect diffractive optical elements via nanoscale 3D printing." *ACS Nano* 14, no. 8 (2020): 10452–10461.

77. Bürger, Johannes, Vera Schalles, Jisoo Kim, Bumjoon Jang, Matthias Zeisberger, Julian Gargiulo, Leonardo de S. Menezes, Markus A. Schmidt, and Stefan A. Maier, "3D-nanoprinted antiresonant hollow-core microgap waveguide: An on-chip platform for integrated photonic devices and sensors." *ACS Photonics* 9, no. 9 (2022): 3012–3024.

78. Johannes, Bürger, Kim Jisoo, Weiss Thomas, A. Maier Stefan, and A. Schmidt Markus, "On-chip twisted hollow-core light cages: Enhancing planar photonics with 3D nanoprinting." *Advanced Photonics* 7, no. 4 (2025): 046002.

79. Piner, Richard D., Jin Zhu, Feng Xu, Seunghun Hong, and Chad A. Mirkin, ""Dip-pen" nanolithography." *Science* 283, no. 5402 (1999): 661–663.

80. Salaita, Khalid, Yuhuang Wang, and Chad A. Mirkin, "Applications of dip-pen nanolithography." *Nature Nanotechnology* 2, no. 3 (2007): 145–155.

81. Biswas, Soma, Falko Brinkmann, Michael Hirtz, and Harald Fuchs, "Patterning of quantum dots by dip-pen and polymer pen nanolithography." *Nanofabrication* 2, no. 1 (2015): 19–26.

82. Liu, Guoqiang, Sarah Hurst Petrosko, Zijian Zheng, and Chad A. Mirkin, "Evolution of dip-pen nanolithography (DPN): From molecular patterning to materials discovery." *Chemical Reviews* 120, no. 13 (2020): 6009–6047.

83. Li, Ze-Qi, Ping Guo, and Yi-Ge Zhou, "Dip-pen nanolithography enabled functional nanomaterials and their applications." *Advanced Materials Technologies* 6, no. 4 (2021): 2000897.

84. Berganza, Eider, Evgeniy Boltynjuk, George Mathew, Fabio Fernando Vallejo, Roland Gröger, Torsten Scherer, Sylwia Sekula-Neuner, and Michael Hirtz, "3D nanolithography by means of lipid ink spreading inhibition." *Small* 19, no. 10 (2023): 2205590.

85. Nam, Tae Won, Min-Jae Choi, and Yeon Sik Jung, "Ultrahigh-resolution quantum dot patterning for advanced optoelectronic devices." *Chemical Communications* 59, no. 19 (2023): 2697–2710.

86. Makarov, Sergey V., Valentin Milichko, Elena V. Ushakova, Mikhail Omelyanovich, Andrea Cerdan Pasaran, Ross Haroldson, Balasubramaniam Balachandran, Honglei Wang, Walter Hu, Yuri S. Kivshar, and Anvar A. Zakhidov, "Multifold emission enhancement in nanoimprinted hybrid perovskite metasurfaces." *ACS Photonics* 4, no. 4 (2017): 728–735.

87. Chen, Wenxiang, Mykhailo Tymchenko, Prashanth Gopalan, Xingchen Ye, Yaoting Wu, Mingliang Zhang, Christopher B. Murray, Andrea Alu, and Cherie R. Kagan, "Large-area nanoimprinted colloidal Au nanocrystal-based nanoantennas for ultrathin polarizing plasmonic metasurfaces." *Nano Letters* 15, no. 8 (2015): 5254–5260.

88. Yao, Yuhan, and Wei Wu, "All-dielectric heterogeneous metasurface as an efficient ultra-broadband reflector." *Advanced Optical Materials* 5, no. 14 (2017): 1700090.

89. Park, Chanwoong, Wonjoong Kim, Yeseul Kim, Hansang Sung, Jaein Park, Hyoin Song, Joohoon Kim, Dong Kyo Oh, Hyunjung Kang, Nara Jeon, Junsuk Rho, and Heon Lee, "High-throughput fabrication of large-scale metaholograms via one-step printing." *Advanced Optical Materials* 12, no. 9 (2024): 2301562.

90. Yoon, Gwanho, Kwan Kim, Se-Um Kim, Seunghoon Han, Heon Lee, and Junsuk Rho, "Printable nanocomposite metalens for high-contrast near-infrared imaging." *ACS Nano* 15, no. 1 (2021): 698–706.

91. Miyazaki, Hideki T., Takeshi Kasaya, Hirotaka Oosato, Yoshimasa Sugimoto, Bongseok Choi, Masanobu Iwanaga, and Kazuaki Sakoda, "Ultraviolet-nanoimprinted packaged metasurface thermal emitters for

infrared CO_2 sensing." *Science and Technology of Advanced Materials* 16, no. 3 (2015): 035005.

92. Yao, Yuhan, He Liu, Yifei Wang, Yuanrui Li, Boxiang Song, Richard P. Wang, Michelle L. Povinelli, and Wei Wu, "Nanoimprint-defined, large-area meta-surfaces for unidirectional optical transmission with superior extinction in the visible-to-infrared range." *Optics Express* 24, no. 14 (2016): 15362–15372.
93. Finn, Andreas, Bo Lu, Robert Kirchner, Xaver Thrun, Karola Richter, and Wolf-Joachim Fischer, "High aspect ratio pattern collapse of polymeric UV-nano-imprint molds due to cleaning." *Microelectronic Engineering* 110 (2013): 112–118.
94. Einck, Vincent J., Mahsa Torfeh, Andrew McClung, Dae Eon Jung, Mahdad Mansouree, Amir Arbabi, and James J. Watkins, "Scalable nanoimprint lithography process for manufacturing visible metasurfaces composed of high aspect ratio TiO_2 meta-atoms." *ACS Photonics* 8, no. 8 (2021): 2400–2409.
95. Yoon, Gwanho, Kwan Kim, Daihong Huh, Heon Lee, and Junsuk Rho, "Single-step manufacturing of hierarchical dielectric metalens in the visible." *Nature Communications* 11, no. 1 (2020): 2268.
96. Kreindl, Gerald, Thomas Glinsner, and Ron Miller, "Making a good impression." *Nature Photonics* 4, no. 1 (2010): 27–28.
97. Chou, Stephen Y., Peter R. Krauss, and Preston J. Renstrom, "Imprint lithography with 25-nanometer resolution." *Science* 272, no. 5258 (1996): 85–87.
98. Chou, Stephen Y., and Peter R. Krauss, "Imprint lithography with sub-10 nm feature size and high throughput." *Microelectronic Engineering* 35, no. 1–4 (1997): 237–240.
99. Zhao, Xiao-Mei, Younan Xia, and George M. Whitesides, "Soft lithographic methods for nano-fabrication." *Journal of Materials Chemistry* 7, no. 7 (1997): 1069–1074.
100. Kumar, Golden, Hong X. Tang, and Jan Schroers, "Nanomoulding with amorphous metals." *Nature* 457, no. 7231 (2009): 868–872.
101. Gao, Huang, Yaowu Hu, Yi Xuan, Ji Li, Yingling Yang, Ramses V. Martinez, Chunyu Li, Jian Luo, Minghao Qi, and Gary J. Cheng, "Large-scale nanoshaping of ultrasmooth 3D crystalline metallic structures." *Science* 346, no. 6215 (2014): 1352–1356.
102. Jin, Shengyu, Zhiguang Zhou, Enas Said Attia Sakr, Maithilee Motlag, Xinyu Huang, Lei Tong, Peter Bermel, Lei Ye, and Gary J. Cheng, "Scalable

nanoshaping of hierarchical metallic patterns with multiplex laser shock imprinting using soft optical disks." *Small* 15, no. 18 (2019): 1900481.

103. Liu, Ze, "One-step fabrication of crystalline metal nanostructures by direct nanoimprinting below melting temperatures." *Nature Communications* 8, no. 1 (2017): 14910.

104. Sam, Quynh P., Qishuo Tan, Christian D. Multunas, Mehrdad T. Kiani, Ravishankar Sundararaman, Xi Ling, and Judy J. Cha, "Nanomolding of Two-Dimensional Materials." *ACS Nano* 18, no. 1 (2024): 1110–1117.

105. Liu, Naijia, Yi-Xiang Yang, Cai Lu, Sebastian A. Kube, Arindam Raj, Sungwoo Sohn, Xiaoyu Zhang, Miguel B. Costa, Ze Liu, and Jan Schroers, "Realizing one-dimensional single-crystalline topological nanomaterials through thermomechanical epitaxy." *Matter* 8, no. 9 (2025): 102128.

106. Li, Shun-Xin, Guan-Yao Huang, Hong Xia, Tairan Fu, Xiao-Jie Wang, Xin Zeng, Xinfeng Liu, Yan Hao Yu, Qi-Dai Chen, Linhan Lin, and Hong-Bo Sun, "Nanoimprint crystalithography for organic semiconductors." *Nature Communications* 16, no. 1 (2025): 3636.

107. Uchic, Michael D., Dennis M. Dimiduk, Jeffrey N. Florando, and William D. Nix, "Sample dimensions influence strength and crystal plasticity." *Science* 305, no. 5686 (2004): 986–989.

108. Nosonovsky, Michael, and Bharat Bhushan, "Multiscale friction mechanisms and hierarchical surfaces in nano- and bio-tribology." *Materials Science and Engineering: R: Reports* 58, no. 3–5 (2007): 162–193.

109. Langdon, Terence G., "Seventy-five years of superplasticity: Historic developments and new opportunities." *Journal of Materials Science* 44, no. 22 (2009): 5998–6010.

110. Guo, L. Jay, "Nanoimprint lithography: Methods and material requirements." *Advanced Materials* 19, no. 4 (2007): 495–513.

111. Schroers, Jan, Thomas M. Hodges, Golden Kumar, Hari Raman, Anthony J. Barnes, Quoc Pham, and Theodore A. Waniuk, "Thermoplastic blow molding of metals." *Materials Today* 14, no. 1–2 (2011): 14–19.

112. Frost, Harold J., and M. F. Ashby, "Deformation-mechanism maps: The plasticity and creep of metals and ceramics."

113. McFadden, S. X., R. S. Mishra, R. Z. Valiev, A. P. Zhilyaev, and A. K. Mukherjee, "Low-temperature superplasticity in nanostructured nickel and metal alloys." *Nature* 398, no. 6729 (1999): 684–686.

114. Yamakov, V., D. Wolf, S. R. Phillpot, A. K. Mukherjee, and H. Gleiter, "Deformation-mechanism map for nanocrystalline metals by molecular-dynamics simulation." *Nature Materials* 3, no. 1 (2004): 43–47.

115. Ashby, Michael F.A., "A first report on deformation-mechanism maps." *Acta Metallurgica* 20, no. 7 (1972): 887–897.

116. Darling, K. A., M. Rajagopalan, M. Komarasamy, M. A. Bhatia, B. C. Hornbuckle, R. S. Mishra, and K. N. Solanki, "Extreme creep resistance in a microstructurally stable nanocrystalline alloy." *Nature* 537, no. 7620 (2016): 378–381.

117. Zhu, Ting, and Ju Li, "Ultra-strength materials." *Progress in Materials Science* 55, no. 7 (2010): 710–757.

118. Greer, Julia R., and Jeff Th M. De Hosson, "Plasticity in small-sized metallic systems: Intrinsic versus extrinsic size effect." *Progress in Materials Science* 56, no. 6 (2011): 654–724.

119. El-Awady, Jaafar A., "Unravelling the physics of size-dependent dislocation-mediated plasticity." *Nature Communications* 6, no. 1 (2015): 5926.

120. Hall, E. O., "The deformation and ageing of mild steel: III discussion of results." *Proceedings of the Physical Society. Section B* 64, no. 9 (1951): 747.

121. Petch, Norman J., "The cleavage strength of polycrystals." *Journal of the Iron and Steel Institute, London* 174 (1953): 25–28.

122. Schneibel, Joachim H., and Martin Heilmaier, "Hall–Petch breakdown at elevated temperatures." *Materials Transactions* 55, no. 1 (2014): 44–51.

123. Chokshi, Atul Haris, Abraham R. Rosen, J. Karch, and Herbert Gleiter, "On the validity of the Hall-Petch relationship in nanocrystalline materials." *Scripta Metallurgica* 23, no. 10 (1989): 1679–1683.

124. Zhu, Ting, Ju Li, Amit Samanta, Austin Leach, and Ken Gall, "Temperature and strain-rate dependence of surface dislocation nucleation." *Physical Review Letters* 100, no. 2 (2008): 025502.

125. Kim, Dongchoul, and Wei Lu, "Creep flow, diffusion, and electromigration in small scale interconnects." *Journal of the Mechanics and Physics of Solids* 54, no. 12 (2006): 2554–2568.

126. Geiger, M., M. Kleiner, R. Eckstein, N. Tiesler, and U. Engel, "Microforming." *CIRP Annals* 50, no. 2 (2001): 445–462.

127. Vollertsen, Frank, H. Schulze Niehoff, and Z. Hu, "State of the art in micro forming." *International Journal of Machine Tools and Manufacture* 46, no. 11 (2006): 1172–1179.

128. Liu, Ze, "Investigation of temperature and feature size effects on deformation of metals by superplastic nanomolding." *Physical Review Letters* 122, no. 1 (2019): 016101.

129. Liu, Ze, Guoxing Han, Sungwoo Sohn, Naijia Liu, and Jan Schroers, "Nanomolding of crystalline metals: the smaller the easier." *Physical Review Letters* 122, no. 3 (2019): 036101.

130. Liu, Naijia, Yujun Xie, Guannan Liu, Sungwoo Sohn, Arindam Raj, Guoxing Han, Bozhao Wu, Judy J. Cha, Ze Liu, and Jan Schroers, "General nanomolding of ordered ohases." *Physical Review Letters* 124, no. 3 (2020): 036102.

131. Liu, Baishan, Qingliang Liao, Xiankun Zhang, Junli Du, Yang Ou, Jiankun Xiao, Zhuo Kang, Zheng Zhang, and Yue Zhang, "Strain-engineered van der Waals interfaces of mixed-dimensional heterostructure arrays." *ACS Nano* 13, no. 8 (2019): 9057–9066.

132. Liu, Naijia, Guannan Liu, Arindam Raj, Sungwoo Sohn, Mayra Daniela Morales-Acosta, Jingbei Liu, and Jan Schroers, "Unleashing nanofabrication through thermomechanical nanomolding." *Science Advances* 7, no. 47 (2021): eabi4567.

133. Liu, Ze, Naijia Liu, and Jan Schroers, "Nanofabrication through molding." *Progress in Materials Science* 125 (2022): 100891.

134. Wu, Yupeng, Yujie Zhang, Langquan Shui, Jinsong Wu, and Ze Liu, "Experimental decoding of grain boundary-based plastic deformation." *Acta Materialia* 225 (2022): 117534.

135. Coble, R. L., "A model for boundary diffusion controlled creep in polycrystalline materials." *Journal of Applied Physics* 34, no. 6 (1963): 1679–1682.

136. Herring, Conyers, "Diffusional viscosity of a polycrystalline solid." *Journal of Applied Physics* 21, no. 5 (1950): 437–445.

137. Nabarro, F. R. N., "Deformation of crystals by motion of single ions." *Report of a Conference on the Strength of Solids, Physical Society, London* (1948): 75-90.

138. Seitz, Frederick, "On the generation of vacancies by moving dislocations." *Advances in Physics* 1, no. 1 (1952): 43–90.

139. Fang, T. H., W. L. Li, N. R. Tao, and K. Lu, "Revealing extraordinary intrinsic tensile plasticity in gradient nano-grained copper." *Science* 331, no. 6024 (2011): 1587–1590.

140. Li, Xiaoyan, Yujie Wei, Lei Lu, Ke Lu, and Huajian Gao, "Dislocation nucleation governed softening and maximum strength in nano-twinned metals." *Nature* 464, no. 7290 (2010): 877–880.

141. Greer, Julia R., and William D. Nix, "Nanoscale gold pillars strengthened through dislocation starvation." *Physical Review B* 73, no. 24 (2006): 245410.

142. Shan, Z. W., Raja K. Mishra, S. A. Syed Asif, Oden L. Warren, and Andrew M. Minor, "Mechanical annealing and source-limited deformation in submicrometre-diameter Ni crystals." *Nature Materials* 7, no. 2 (2008): 115–119.

143. Zhu, T. T., A. J. Bushby, and D. J. Dunstan, "Materials mechanical size effects: A review." *Materials Technology* 23, no. 4 (2008): 193–209.

144. Zhang, Yujie, Bozhao Wu, Enlai Gao, Langquan Shui, and Ze Liu, "Observation of speeding growth of metal nanowires by ultra-low frequency micro-vibration assisted superplastic nanomolding." *Materials Letters* 283, no. 15 (2021): 128890.

145. Ge, Junyu, Bin Ding, Shuai Hou, Manlin Luo, Donguk Nam, Hongwei Duan, Huajian Gao, Yee Cheong Lam, and Hong Li, "Rapid fabrication of complex nanostructures using room-temperature ultrasonic nanoimprinting." *Nature Communications* 12, no. 1 (2021): 3146.

146. Liu, Ze, and Jan Schroers, "General nanomoulding with bulk metallic glasses." *Nanotechnology* 26, no. 14 (2015): 145301.

147. Raj, Arindam, Naijia Liu, Guannan Liu, Sungwoo Sohn, Junxiang Xiang, Ze Liu, and Jan Schroers, "Nanomolding of gold and gold–silicon heterostructures at room temperature." *ACS Nano* 15, no. 9 (2021): 14275–14284.

148. Xiang, Jun-Xiang, and Ze Liu, "Observation of enhanced nanoscale creep flow of crystalline metals enabled by controlling surface wettability." *Nature Communications* 13, no. 1 (2022): 7493.

149. Xiang, Jun-Xiang, and Ze Liu, "Observation of a large slip effect in the nanoscale flow of highly viscous supercooled liquid metals." *Langmuir* 39, no. 32 (2023): 11224–11230.

150. Xiang, Jun-Xiang, Yuanpeng Yao, Hui Fang, Cai Lu, and Ze Liu, "Interplay between interface diffusion and boundary slip underlying graphite-lubricated metal nanoforming." *Physical Review Letters* 134, no. 17 (2025): 176204.

151. Kustandi, Tanu Suryadi, Wei Wei Loh, Han Gao, and Hong Yee Low, "Wafer-scale near-perfect ordered porous alumina on substrates by step and flash imprint lithography." *ACS Nano* 4, no. 5 (2010): 2561–2568.

152. Smith, Joshua T., Qingling Hang, Aaron D. Franklin, David B. Janes, and Timothy D. Sands, "Highly ordered diamond and hybrid triangle-diamond patterns in porous anodic alumina thin films." *Applied Physics Letters* 93, no. 4 (2008): 043108.

153. Kant, Krishna, Suet P. Low, Asif Marshal, Joseph G. Shapter, and Dusan Losic, "Nanopore gradients on porous aluminum oxide generated by nonuniform anodization of aluminum." *ACS Applied Materials & Interfaces* 2, no. 12 (2010): 3447–3454.

154. Lee, Woo, Ran Ji, Ulrich Gösele, and Kornelius Nielsch, "Fast fabrication of long-range ordered porous alumina membranes by hard anodization." *Nature Materials* 5, no. 9 (2006): 741–747.

155. Wen, Liaoyong, Rui Xu, Yan Mi, and Yong Lei, "Multiple nanostructures based on anodized aluminium oxide templates." *Nature Nanotechnology* 12, no. 3 (2017): 244–250.

156. Zaraska, Leszek, Elżbieta Kurowska, Grzegorz D. Sulka, and Marian Jaskuła, "Porous alumina membranes with branched nanopores as templates for fabrication of Y-shaped nanowire arrays." *Journal of Solid State Electrochemistry* 16, no. 11 (2012): 3611–3619.

157. Wang, Lang, Hangtong Li, Chen Zhao, Liqiang Zhang, Jiye Li, Salah Ud Din, Zichen Wang, Jiachen Sun, Sergio Andres Galindo Torres, Zhiyong Fan, and Liaoyong Wen, "Aluminium surface work hardening enables multi-scale 3D lithography." *Nature Materials* 24, no. 1 (2025): 39–47.

158. Masuda, Hideki, and Kenji Fukuda, "Ordered metal nanohole arrays made by a two-step replication of honeycomb structures of anodic alumina." *Science* 268, no. 5216 (1995): 1466–1468.

159. Masuda, Hideki, Haruki Yamada, Masahiro Satoh, Hidetaka Asoh, Masashi Nakao, and Toshiaki Tamamura, "Highly ordered nanochannel-array architecture in anodic alumina." *Applied Physics Letters* 71, no. 19 (1997): 2770–2772.

160. Lee, Woo, Ran Ji, Caroline A Ross, Ulrich Gösele, and Kornelius Nielsch, "Wafer-scale Ni imprint stamps for porous alumina membranes based on interference lithography." *Small* 2, no. 8–9 (2006): 978–982.

161. Gutierrez-Rivera, Luis, Robert F. Peters, Steven K. Dew, and Maria Stepanova, "Application of EBL fabricated nanostructured substrates for surface enhanced Raman spectroscopy detection of protein A in aqueous solution." *Journal of Vacuum Science & Technology B, Nanotechnology and Microelectronics: Materials, Processing, Measurement, and Phenomena* 31, no. 6 (2013): 06F901.

162. Colniţă, Alia, Daniel Marconi, Nicoleta Elena Dina, Ioana Brezeştean, Diana Bogdan, and Ioan Turcu, "3D silver metallized nanotrenches fabricated by nanoimprint lithography as flexible SERS detection platform."

Spectrochimica Acta Part A: Molecular and Biomolecular Spectroscopy 276 (2022): 121232.

163. Cai, Jieyi, Ruihua Liu, Siyi Jia, Zhuohong Feng, Lin Lin, Zhiqiang Zheng, Shaofan Wu, and Zhezhe Wang, "SERS hotspots distribution of the highly ordered noble metal arrays on flexible substrates." *Optical Materials* 122, no. Part B (2021): 111779.

164. Xiang, Junxiang, Yunxia Wang, Yupeng Wu, Qi Peng, Langquan Shui, Wengen Ouyang, Tao Ding, and Ze Liu, "Superplastic nanomolding of highly ordered metallic sub-micrometer pillars arrays for surface enhanced Raman scattering." *Advanced Materials Technologies* 7, no. 3 (2021): 2100891.

165. Xiang, Junxiang, Yunxia Wang, Yupeng Wu, Hui Fang, Langquan Shui, Ze Liu, and Tao Ding, "Ordered hierarchical Ag nanostructures as surface-enhanced raman scattering platforms for (bio)chemical sensing and pollutant monitoring." *ACS Applied Nano Materials* 4, no. 11 (2021): 11644–11650.

166. Chen, Yifang, "Nanofabrication by electron beam lithography and its applications: A review." *Microelectronic Engineering* 135 (2015): 57–72.

167. Cox, Lewis M., Alina M. Martinez, Adrienne K. Blevins, Nancy Sowan, Yifu Ding, and Christopher N. Bowman, "Nanoimprint lithography: Emergent materials and methods of actuation." *Nano Today* 31 (2020): 100838.

168. Fang, Hui, Yangyang Pan, Cai Lu, Jianxin Liu, Tao Ding, and Ze Liu, "In situ nanomechanics: Opportunities based on superplastic nanomolding." *ACS Nano* 17, no. 24 (2023): 24479–24486.

169. Fernandez-Cuesta, Irene, Rasmus Bundgaard Nielsen, Alexandra Boltasseva, Xavier Borrise, Francesc Perez-Murano, and Anders Kristensen, "V-groove plasmonic waveguides fabricated by nanoimprint lithography." *Journal of Vacuum Science & Technology B: Microelectronics and Nanometer Structures Processing, Measurement, and Phenomena* 25, no. 6 (2007): 2649–2653.

170. Balena, Antonio, Marco Bianco, Ferruccio Pisanello, and Massimo De Vittorio, "Recent advances on high-speed and holographic two-photon direct laser writing." *Advanced Functional Materials* 33, no. 39 (2023): 2211773.

171. Kakkar, Tarun, Chantal Keijzer, Marion Rodier, Tatyana Bukharova, Michael Taliansky, Andrew J. Love, Joel J. Milner, Affar S. Karimullah, Laurence D. Barron, Nikolaj Gadegaard, Adrian J. Lapthorn, and Malcolm Kadodwala, "Superchiral near fields detect virus structure." *Light: Science & Applications* 9, no. 1 (2020): 195.

172. Karimullah, Affar S., Calum Jack, Ryan Tullius, Vincent M. Rotello, Graeme Cooke, Nikolaj Gadegaard, Laurence D. Barron, and Malcolm Kadodwala, "Disposable plasmonics: Plastic templated plasmonic metamaterials with tunable chirality." *Advanced Materials* 27, no. 37 (2015): 5610–5616.

173. Li, Yong, Hui Fang, Longfei Miao, Yueming Zhai, Ze Liu, and Tao Ding, "Mechano-tunable chiroptics of single crystalline metal helices made by superelastic nanomolding technique." *Advanced Optical Materials* 13, no. 14 (2025): 2403486.

174. Gansel, Justyna K., Michael Thiel, Michael S. Rill, Manuel Decker, Klaus Bade, Volker Saile, Georg von Freymann, Stefan Linden, and Martin Wegener, "Gold helix photonic metamaterial as broadband circular polarizer." *Science* 325, no. 5947 (2009): 1513–1515.

175. Lu, Jiafeng, Jing Tian, Bertrand Poumellec, Enrique Garcia-Caurel, Razvigor Ossikovski, Xianglong Zeng, and Matthieu Lancry, "Tailoring chiral optical properties by femtosecond laser direct writing in silica." *Light: Science & Applications* 12, no. 1 (2023): 46.

176. Kaschke, Johannes, Leonard Blume, Lin Wu, Michael Thiel, Klaus Bade, Zhenyu Yang, and Martin Wegener, "A helical metamaterial for broadband circular polarization conversion." *Advanced Optical Materials* 3, no. 10 (2015): 1411–1417.

177. Lu, Xiaolin, Xujie Wang, Shuangshuang Wang, and Tao Ding, "Polarization-directed growth of spiral nanostructures by laser direct writing with vector beams." *Nature Communications* 14, no. 1 (2023): 1422.

178. Esposito, Marco, Vittorianna Tasco, Massimo Cuscunà, Francesco Todisco, Alessio Benedetti, Iolena Tarantini, Milena De Giorgi, Daniele Sanvitto, and Adriana Passaseo, "Nanoscale 3D chiral plasmonic helices with circular dichroism at visible frequencies." *ACS Photonics* 2, no. 1 (2015): 105–114.

179. Esposito, Marco, Vittorianna Tasco, Francesco Todisco, Massimo Cuscunà, Alessio Benedetti, Mario Scuderi, Giuseppe Nicotra, and Adriana Passaseo, "Programmable extreme chirality in the visible by helix-shaped metamaterial platform." *Nano Letters* 16, no. 9 (2016): 5823–5828.

180. Mao, Yifei, Yun Zheng, Can Li, Lin Guo, Yini Pan, Rui Zhu, Jun Xu, Weihua Zhang, and Wengang Wu, "Programmable bidirectional folding of metallic thin films for 3D chiral optical antennas." *Advanced Materials* 29, no. 19 (2017): 1606482.

181. Miles, Andrew J., and B. A. Wallace, "Synchrotron radiation circular dichroism spectroscopy of proteins and applications in structural and functional genomics." *Chemical Society Reviews* 35, no. 1 (2006): 39–51.

182. Kelly, Sharon M., Thomas J. Jess, and Nicholas C. Price, "How to study proteins by circular dichroism." *Biochimica et Biophysica Acta (BBA) — Proteins and Proteomics* 1751, no. 2 (2005): 119–139.

183. Miles, Andrew J., and B. A. Wallace, "Circular dichroism spectroscopy of membrane proteins." *Chemical Society Reviews* 45, no. 18 (2016): 4859–4872.

184. Schäferling, Martin, Daniel Dregely, Mario Hentschel, and Harald Giessen, "Tailoring enhanced optical chirality: Design principles for chiral plasmonic nanostructures." *Physical Review X* 2, no. 3 (2012): 031010.

185. Tan, Yong, Xiaolin Lu, and Tao Ding, "Trace-amount detection of chiral molecules based on plasmonic racemic arrays fabricated via direct laser writing." *ACS Sensors* 9, no. 6 (2024): 3290–3295.

186. Valev, Ventsislav K., Jeremy J. Baumberg, Concita Sibilia, and Thierry Verbiest, "Chirality and chiroptical effects in plasmonic nanostructures: Fundamentals, recent progress, and outlook." *Advanced Materials* 25, no. 18 (2013): 2517–2534.

187. Li, Zhancheng, Yifan Jiang, Wenwei Liu, Yuebian Zhang, Hua Cheng, Junjie Li, Jianguo Tian, and Shuqi Chen, "Hybrid bilayer plasmonic metasurfaces with intrinsic chiral optical responses." *Applied Physics Letters* 122, no. 18 (2023): 181702.

188. Shen, Zhonglei, Shuting Fan, Wei Yin, Shengnan Li, Yafei Xu, Liuyang Zhang, and Xuefeng Chen, "Chiral metasurfaces with maximum circular dichroism enabled by out-of-plane plasmonic system." *Laser & Photonics Reviews* 16, no. 12 (2022): 2200370.

189. Liu, Ya, Weixuan Zhang, Lu He, and Xiangdong Zhang, "All-optical separation of chiral nanoparticles on silicon-based microfluidic chips with vector exceptional points." *APL Photonics* 8, no. 3 (2023): 036112.

190. Zhao, Yang, Amr A. E. Saleh, and Jennifer A. Dionne, "Enantioselective optical trapping of chiral nanoparticles with plasmonic tweezers." *ACS Photonics* 3, no. 3 (2016): 304–309.

191. Couteau, Christophe, Stefanie Barz, Thomas Durt, Thomas Gerrits, Jan Huwer, Robert Prevedel, John Rarity, Andrew Shields, and Gregor Weihs, "Applications of single photons to quantum communication and computing." *Nature Reviews Physics* 5, no. 6 (2023): 326–338.

192. Kan, Yinhui, Sebastian K. H. Andersen, Fei Ding, Shailesh Kumar, Changying Zhao, and Sergey I. Bozhevolnyi, "Metasurface-enabled generation of circularly polarized single photons." *Advanced Materials* 32, no. 16 (2020): 1907832.

193. Liu, Xujing, Yinhui Kan, Shailesh Kumar, Danylo Komisar, Changying Zhao, and Sergey I. Bozhevolnyi, "On-chip generation of single-photon circularly polarized single-mode vortex beams." *Science Advances* 9, no. 32 (2023): eadh0725.

194. Long, Zhenghao, Yu Zhou, Yucheng Ding, Xiao Qiu, Swapnadeep Poddar, and Zhiyong Fan, "Biomimetic optoelectronics with nanomaterials for artificial vision." *Nature Reviews Materials* 10, no. 2 (2025): 128–146.

195. Potyrailo, Radislav A., Helen Ghiradella, Alexei Vertiatchikh, Katharine Dovidenko, James R. Cournoyer, and Eric Olson, "Morpho butterfly wing scales demonstrate highly selective vapour response." *Nature Photonics* 1, no. 2 (2007): 123–128.

196. Parker, Andrew R., and Helen E. Townley, "Biomimetics of photonic nanostructures." *Nature Nanotechnology* 2, no. 6 (2007): 347–353.

197. Ketkaew, Jittisa, Ze Liu, Wen Chen, and Jan Schroers, "Critical crystallization for embrittlement in metallic glasses." *Physical Review Letters* 115, no. 26 (2015): 265502.

198. Wu, Yupeng, Yongping Kang, Yiling Xie, Xiaojian Xiang, Langquan Shui, Jan Schroers, and Ze Liu, "Fabrication of 3D metallic glass architectures by a mold-strain-set method." *Materials & Design* 218 (2022): 110668.

199. Xu, Lihan, Langquan Shui, Yujie Zhang, Qi Peng, Longjian Xue, and Ze Liu, "Robust and reproducible fabrication of large area aluminum (Al) micro/nanorods arrays by superplastic nanomolding at room temperature." *Applied Physics Express* 13, no. 3 (2020): 036503.

200. Shui, Langquan, Weidong Yan, Yujie Zhang, Lihan Xu, Enlai Gao, Ze Liu, and Quanshui Zheng, "Peeling mechanics of film-substrate system with mutually embedded nanostructures in the interface." *International Journal of Solids and Structures* 251 (2022): 111737.

201. Chen, Shuo, Yu Shu, Hao Cao, Deming Wang, Hui Wan, and Chengqun Gui, "Nondestructive demolding of structure-designable high-aspect-ratio nanoimprint template." *Langmuir* 40, no. 38 (2024): 20057–20061.

202. Li, Mingjie, Yulong Chen, Wenxin Luo, and Xing Cheng, "Demolding force dependence on mold surface modifications in UV nanoimprint lithography." *Microelectronic Engineering* 236, (2021): 111470.

203. Brenner, Philipp, Ofer Bar-On, Tobias Siegle, Tobias Leonhard, Raz Gvishi, Carsten Eschenbaum, Heinz Kalt, Jacob Scheuer, and Uli Lemmer, "3D whispering-gallery-mode microlasers by direct laser writing and subsequent soft nanoimprint lithography." *Applied Optics* 56, no. 13 (2017): 3703–3708.

204. Deng, Fangfang, Juntao Chen, Junxiang Xiang, Yong Li, Yan Qiao, Ze Liu, and Tao Ding, "Light-programmed bistate colloidal actuation based on photothermal active plasmonic substrate." *Research* 6 (2023): 0020.

205. Deng, Fangfang, Junxiang Xiang, Ze Liu, and Tao Ding, "Dynamic and reversible tuning of pixelated plasmonic cluster arrays." *Journal of Materials Chemistry C* 9, no. 44 (2021): 15927–15931.

206. Schift, Helmut, "Nanoimprint lithography: 2D or not 2D? A review." *Applied Physics A* 121, no. 2 (2015): 415–435.

207. Zhang, Zimo, Zhongxiao Xu, Ran Huang, Xingda Lu, Fengbo Zhang, Donghao Li, Şahin K. Özdemir, Franco Nori, Han Bao, Yanhong Xiao, Bing Chen, Hui Jing, and Heng Shen, "Chirality-induced quantum non-reciprocity." *Nature Photonics* 19, no. 8 (2025): 840–846.

208. Suárez-Forero, D. G., M. Jalali Mehrabad, C. Vega, A. González-Tudela, and M. Hafezi, "Chiral quantum optics: Recent developments and future directions." *PRX Quantum* 6, no. 2 (2025): 020101.

209. Liu, Qi, Qi Wei, Hui Ren, Luwei Zhou, Yifan Zhou, Pengzhi Wang, Chenghao Wang, Jun Yin, and Mingjie Li, "Circular polarization-resolved ultraviolet photonic artificial synapse based on chiral perovskite." *Nature Communications* 14, no. 1 (2023): 7179.

210. Dan, Soirik, Subham Paramanik, and Amlan J. Pal, "Introducing chiro-optical activities in photonic synapses for neuromorphic computing and in-memory logic operations." *ACS Nano* 18, no. 22 (2024): 14457–14468.

211. Xu, Liguang, Xiuxiu Wang, Weiwei Wang, Maozhong Sun, Won Jin Choi, Ji-Young Kim, Changlong Hao, Si Li, Aihua Qu, Meiru Lu, Xiaoling Wu, Felippe M. Colombari, Weverson R. Gomes, Asdrubal L. Blanco, Andre F. de Moura, Xiao Guo, Hua Kuang, Nicholas A. Kotov, and Chuanlai Xu, "Enantiomer-dependent immunological response to chiral nanoparticles." *Nature* 601, no. 7893 (2022): 366–373.

212. Ma, Baojin, and Alberto Bianco, "Regulation of biological processes by intrinsically chiral engineered materials." *Nature Reviews Materials* 8, no. 6 (2023): 403–413.

213. Lu, Xiaolin, Yong Tan, Zhiyuan Gao, and Tao Ding, "Deterministic placement of achiral emitters in superchiral fields for superior chiral photoluminescence." *Nano Letters* 25, no. 17 (2025): 7137–7143.

214. Liang, Kaixiang, Yong Li, Shiyu Fan, and Tao Ding, "Room-temperature circularly polarized single photon emission from Eu^{3+}/organic complexes coupled to chiral plasmonic nanocavity." *Nano Letters* 25, no. 41 (2025): 14825–14831.

215. Han, Jeong Hyun, Yae-Chan Lim, Ryeong Myeong Kim, Jiawei Lv, Nam Heon Cho, Hyeohn Kim, Seok Daniel Namgung, Sang Won Im, and Ki Tae Nam, "Neural-network-enabled design of a chiral plasmonic nanodimer for target-specific chirality sensing." *ACS Nano* 17, no. 3 (2023): 2306–2317.

216. Li, Yu, Youjun Xu, Meiling Jiang, Bowen Li, Tianyang Han, Cheng Chi, Feng Lin, Bo Shen, Xing Zhu, Luhua Lai, and Zheyu Fang, "Self-learning perfect optical chirality via a deep neural network." *Physical Review Letters* 123, no. 21 (2019): 213902.

217. Kolle, Mathias, Pedro M. Salgard-Cunha, Maik R. J. Scherer, Fumin Huang, Pete Vukusic, Sumeet Mahajan, Jeremy J. Baumberg, and Ullrich Steiner, "Mimicking the colourful wing scale structure of the Papilio blumei butterfly." *Nature Nanotechnology* 5, no. 7 (2010): 511–515.

218. Sharma, Vivek, Matija Crne, Jung Ok Park, and Mohan Srinivasarao, "Structural origin of circularly polarized iridescence in Jeweled Beetles." *Science* 325, no. 5939 (2009): 449–451.

https://doi.org/10.1142/9789819830763_0006

CHAPTER 6

Atomic Force Microscopy-Based Nanofabrication: Fundamentals, Advances, and Prospects

Ampere A. Tseng
Manufacturing Institute, Arizona State University, Tempe, Arizona 85287, USA
ampere.tseng@asu.edu

Abstract

Over the past decades, atomic force microscopy-based nanofabrication has emerged as a versatile, high-precision platform, prized for its low cost and nanoscale accuracy. Atomic force microscopy has evolved from a characterization tool into a multifunctional fabrication system capable of interacting with materials mechanically, thermally, electrically, and chemically. This chapter reviews atomic force microscopy-based techniques employing single or multiple energy modalities, highlighting their capabilities for material removal, modification, and addition. While atomic force microscopy-based nanofabrication has enabled a wide range of nanoscale devices, components, and structures, its broader adoption is limited by three key constraints: reproducibility, achievable resolution, and throughput. We discuss these challenges and outline potential strategies, including process optimization and machine learning-assisted control, to enhance reliability, precision, and productivity, advancing atomic force microscopy-based nanofabrication toward practical, next-generation nanofabrication applications.

Keywords: atomic force microscopy, atomic force microscopy-based nanofabrication, deposition, dual sources, electrochemical-assisted, local anodic oxidation, machine learning, multiple probes, scratch, thermal-assisted

6.1 Introduction

The continued scaling of semiconductor devices and systems has driven rapid advances in nanofabrication, enabling nanoscale components to achieve their intended functionalities and performance. These advances have also catalyzed the emergence of diverse research fields and applications, including nanoelectronics, nanophotonics, nanobiology, nanomedicine, and related disciplines. Unlike conventional semiconductor technologies, many of these emerging areas demand novel materials, unconventional device architectures, and fabrication strategies capable of delivering high spatial resolution, structural complexity, and functional diversity.

Despite this growing diversity, nanofabrication techniques widely adopted by the semiconductor industry are often neither cost-effective nor practical for emerging technologies. Extreme ultraviolet photolithography, for example, is optimized for high-volume manufacturing and entails prohibitively high infrastructure and mask costs, with individual photomasks frequently costing several million U.S. dollars. Such capital-intensive approaches are poorly suited for low-volume production, rapid design iteration, or exploratory prototyping, all of which are essential during the early stages of device and system development.

Many emerging nanoscale devices remain at an early stage of technological maturity and require multiple development cycles to optimize performance, reduce fabrication cost, and improve process flexibility and reconfigurability. These requirements are particularly critical for fabricating small numbers of devices with high precision and design agility. Atomic force microscopy (AFM)-based nanofabrication has therefore emerged as a powerful enabling technology for low-volume, high-resolution prototyping. It employs microcantilevers equipped with functionalized nanoscale tips, evolved from conventional AFM probes, to remove, modify, or deposit materials with nanometer-scale accuracy. As demonstrated in earlier work by Tseng,[1] AFM-based approaches enable the fabrication of complex nanoscale patterns and structures with high precision and relatively low operational cost. Beyond purely mechanical interactions, AFM has evolved into a multifunctional nanofabrication platform capable of coupling mechanical, thermal, electrical, and chemical energy at the nanoscale.

AFM-based nanofabrication can be performed in a wide range of environments, including ambient air, controlled atmospheres, and liquid solutions, and can incorporate mechanical loading, electrical bias, and chemical or biological reagents. This review surveys AFM-based nanofabrication techniques in which single or multiple energy modalities are delivered through the probe, with emphasis on their capabilities for material removal, modification, and addition, as well as their achievable resolution and process reliability.

The review begins with AFM scratching for material removal using purely mechanical means, focusing on fundamental mechanisms, material responses, and theoretical models (Section 6.2). Representative applications of scratching-based nanofabrication across different materials and device systems are then discussed to illustrate the versatility and limitations of force-driven processes (Section 6.3). Techniques in which AFM scratching is augmented by vibrational, thermal, or electrical energies are subsequently reviewed, with particular attention to the fabrication of high-aspect-ratio 3D nanostructures and complex device architectures (Section 6.4). The use of multiple or parallel probes to enhance fabrication throughput is also examined, along with the influence of electrical bias and chemical environments on scratching behavior.

AFM-based material modification processes are addressed next, with a focus on the fundamentals and recent advances of local anodic oxidation in Section 6.5. In local anodic oxidation, a biased AFM probe induces localized electrochemical reactions under ambient conditions, enabling the selective formation of nanoscale oxide patterns with high spatial precision. Recent progress in high-resolution, photoresist-free local anodic oxidation is reviewed, along with gasification-based oxidation processes that enable clean, sub-nanometer-scale material removal for patterning materials such as graphite and graphene.

The third major class of AFM-based nanofabrication, material addition, is presented in Sections 6.6 and 6.7. The fundamentals of AFM-assisted deposition are discussed, including ink-coated tip and hollow tip approaches, in which materials are transferred controllably from the probe to the substrate. AFM tips functionalized with chemical or biological species enable localized reactions for nanoscale patterning, while additional techniques such as AFM-assisted chemical vapor deposition and electron beam-induced deposition further expand the range of depositable materials. Advances in these methods are reviewed with emphasis on coupling effects, process controllability, and application potential.

Although AFM-based nanofabrication has demonstrated remarkable versatility and precision, several fundamental and practical challenges continue to limit its broader adoption, particularly in achieving reliable sub-10-nm resolution, high throughput, and long-term reproducibility. The concluding section critically assesses these challenges and outlines future research directions, including machine learning-assisted control, multi-tip architectures, and hybrid fabrication strategies, that may enable AFM-based nanofabrication to evolve from a laboratory-scale prototyping tool into a robust and scalable nanomanufacturing platform.

6.2 Material Removal by Atomic Force Microscopy Scratching: Fundamentals

AFM scratching, also known as AFM machining, encompasses processes in which nanoscale material removal is achieved purely through mechanical interactions between an AFM probe and the substrate, with material removed by tip-induced scratching or plowing.

As illustrated in Figure 6.1, a typical AFM system comprises a microcantilever with a sharp tip, a piezoelectric scanner, and a detection system. The scanner uses piezoelectric actuators for precise positioning and scanning under electronic control, while

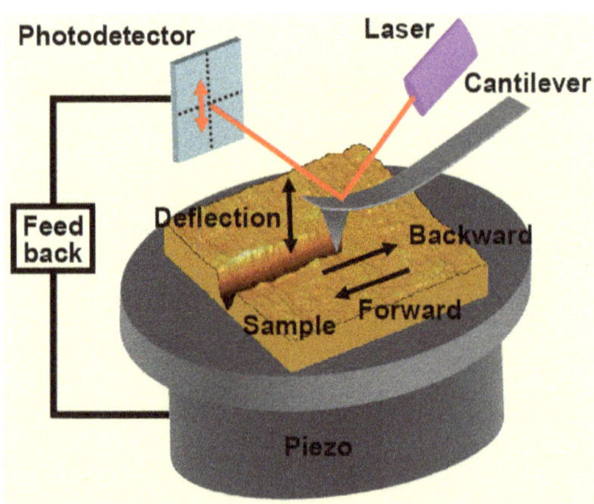

Figure 6.1. Schematic of AFM-based nanoscale scratching, illustrating material removal by a tip under an applied normal force during scanning.

the detection system commonly employs a laser–photodiode optical lever to measure vertical and lateral cantilever deflections during surface interaction.

During scratching, a controlled normal force is applied to the tip via the piezoelectric actuator, inducing cantilever deflection as the tip scans across the surface. Scratch depth is governed primarily by the tip–substrate normal force, typically maintained at a constant value through closed-loop feedback. Using this approach, trenches or grooves with widths from several tens to several hundreds of nanometers and depths of a few to several tens of nanometers have been produced on a wide range of materials, including metals, oxides, semiconductors, and soft materials.[2]

The AFM tip is characterized by its material properties and geometric parameters, such as apex radius, shape, height, aspect ratio, hardness, and stiffness. At the nanoscale, material removal depends sensitively on these tip characteristics as well as on key scratching parameters. Without a clear understanding of these factors, achieving high-fidelity and reproducible nanofabrication remains challenging.

This section reviews recent progress in understanding the parameters governing AFM scratching. Emphasis is placed on single-scratch or single-scan experiments, which most directly isolate the effects of applied normal force and scratching speed on groove geometry, while additional factors such as multiple scratching and scratchability are also discussed.

6.2.1 *Effects of Scratch Forces on Scratch Geometry*

Tseng and co-workers systematically investigated the scratching behavior of Si, Ni, and $Ni_{80}Fe_{20}$ thin films by varying the applied normal force (F_n) from 1 to 9 μN. They demonstrated that both the groove depth, d, and groove width, w_f, increase with F_n according to logarithmic relationships:[2,3]

$$d(F_n) = \alpha_1 \text{Ln } (F_n/F_{t1}) \tag{6.1a}$$

and

$$w_f (F_n) = \alpha_2 \text{Ln } (F_n/F_{t2}) \tag{6.1b}$$

where α_1 and α_2 are the depth and width scratch coefficients, respectively, and F_{t1} and F_{t2} are the corresponding threshold forces. The coefficient α characterizes the scratching efficiency of a given tip–material combination, while F_t represents the minimum normal force required to produce a measurable groove. At $F_n = F_t$, no observable scratching occurs. The groove depth d is defined as the vertical distance between the original surface and the groove minimum, whereas w_f corresponds to the full width at half maximum of the groove profile. Comparison across several metallic systems confirmed that this logarithmic dependence provides an accurate and reliable description of AFM scratching behavior.[2]

As shown in Figure 6.2, scratching can be performed in four directions: upward, forward, downward, and backward. The upward and downward configurations are physically equivalent. Experimental results show that differences between upward (or downward) and forward scratching are small and generally indistinguishable within experimental uncertainty. Consequently, only the upward and backward scratching directions are considered here.

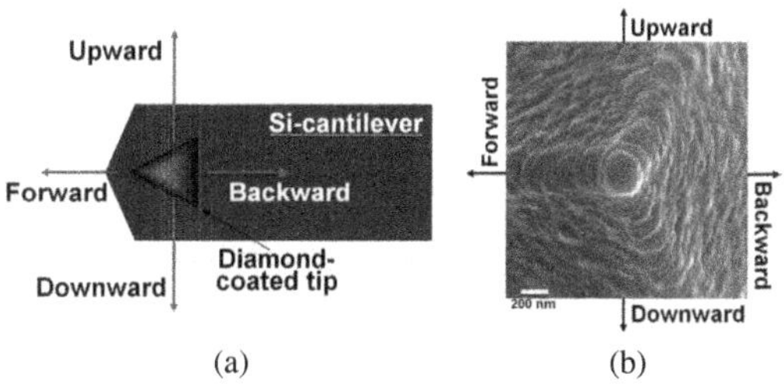

Figure 6.2. Diamond-coated AFM probe: (a) rectangular microcantilever with a triangular pyramid tip indicating four scratching directions; (b) scanning electron microscopy (SEM) top-view image of the tip showing scratching orientations.

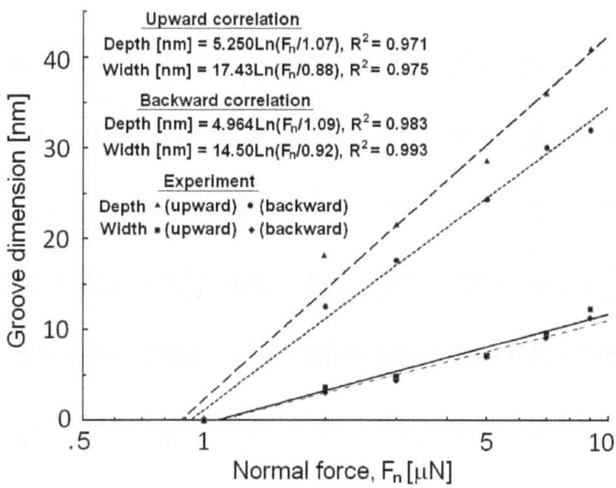

Figure 6.3. Groove depth (d) and width (w_f) as functions of applied normal force (F_n) for upward and backward scratching on a NiFe thin film at 100 nm/s. Symbols represent experimental data; solid lines are logarithmic fits.

Figure 6.3 presents experimental data and corresponding logarithmic fits for grooves scratched on a NiFe thin film in the upward and backward directions, with F_n ranging from 1 to 9 μN at a scratching speed of 100 nm/s. Both the groove depth d and width w_f closely follow Eq. (6.1). Each data point represents the mean of ten measurements taken at randomly selected locations along a groove, with standard deviations below 4% of the mean.

The coefficients of determination (R^2) for depth and width are 0.971 and 0.975, respectively, for upward scratching and 0.983 and 0.993, respectively, for backward scratching, indicating mean deviations of less than 3% and 2%, respectively. From the fits in Figure 6.3, the extracted scratch coefficients α_1 and α_2 are 5.250 nm and 17.43 nm for upward scratching, and 4.964 nm and 14.50 nm for backward scratching. On average, grooves produced by upward scratching are approximately 8% deeper than those formed in the backward direction. The corresponding threshold forces F_{t1} and F_{t2} are 1.07 and 0.88 μN for upward scratching, and 1.09 and 0.92 μN for backward scratching.

Because no measurable groove depth was observed at F_n = 1.0 μN, the true threshold force must exceed this value. The depth-based estimates ($F_{t1} \approx 1.07$–1.09 μN) are therefore consistent with experimental observations, whereas the width-based estimates ($F_{t2} \approx 0.88$–0.92 μN) underestimate the threshold force by approximately 10%, primarily due to uncertainties in width measurements. To avoid ambiguity and improve correlation reliability, Tseng et al. selected F_{t1} as the definitive scratch threshold force and recommended depth-based correlations.[2,3]

Chip or debris formation occurs once the applied load exceeds approximately 5 μN. Chip formation involves shearing along a shear plane when the applied normal force generates stresses exceeding the material's shear strength, initiating cleavage. Both continuous and discontinuous chips may form and are transported upward along the curved, semi-spherical region of the tip, often adhering to the flat or pyramid faces of the probe. In many cases, chips extending from several to tens of micrometers, often exhibiting spiral or helical morphologies, have been observed during scratching of materials such as Si,[4] Zr-based alloys,[5] and Al thin films.[6] In addition, small fragments of discontinuous chips may accumulate on one or both sides of the groove, forming symmetric or asymmetric ridges or protuberances. The height of these features generally increases with applied normal force, analogous to snow displacement by an angled snowplow blade.

At even higher applied loads (typically above ~10 μN), scratching can lead to increased surface roughness, with Ra values exceeding 1 nm, even when using a smooth diamond-coated tip. Scratching loads should therefore be carefully controlled — generally below ~25 μN — to avoid instability, which can produce unacceptable roughness and highly irregular surface morphologies.

Beyond geometric scaling with applied load, AFM scratching performance depends strongly on how efficiently a material accommodates or resists removal. To quantify this behavior, Section 6.2.2 introduces the scratch ratio and scratchability as comparative metrics across materials and scratching conditions.

6.2.2 Scratch Ratio and Scratchability

By re-arranging Eq. (6.1), the scratch coefficient, α, is expressed as:

$$\alpha = d/\mathrm{Ln}(F_n/F_t) \tag{6.2a}$$

This coefficient quantifies the scratch efficiency of an AFM tip–material system. Larger α values correspond to greater material removal per unit applied load, whereas the threshold force

F_t reflects resistance to scratching, as higher F_t values indicate that larger normal forces are required to initiate measurable material removal.

Because scratching behavior is well described by the logarithmic relationship in Eq. (6.1a), governed by the two parameters α_1 and Ft_1, their ratio, defined as the scratch ratio (SR):

$$SR = \alpha_1 / F_{t1} \tag{6.2b}$$

provides a compact descriptor of scratch behavior. In this formulation, SR represents an intrinsic nanoscale scratchability parameter that captures the combined effects of scratching efficiency and threshold force. Analogous to machinability indices in conventional machining, SR enables quantitative comparison of material removal behavior under standardized scratching conditions.[7]

In addition to the NiFe data presented in Figure 6.3, SR has been evaluated for six other materials, including three metallic thin films, one semiconductor, one metallic glass, and one polymer. As shown in Figure 6.4, all materials exhibit logarithmic correlations between groove depth and normalized normal force, resulting in nearly constant SR values over a wide range of scratching conditions. Metallic thin film data were reported for Ni,[3] Al/As_2S_3 bilayers,[8] and permalloy (NiFe),[9] while Si data were obtained from Kawasegi et al.[4] and Tseng et al.[10] Data for the metallic glass $Zr_{55}Cu_{30}Al_{10}Ni_5$ were reported by Kawasegi et al.,[5] and results for the polymer poly(glycidyl methacrylate) (PGMA) were reported by Yamamoto et al.[11]

Because these cases span a broad range of materials and scratching conditions, the observed invariance of SR indicates that it is a robust parameter applicable across diverse AFM

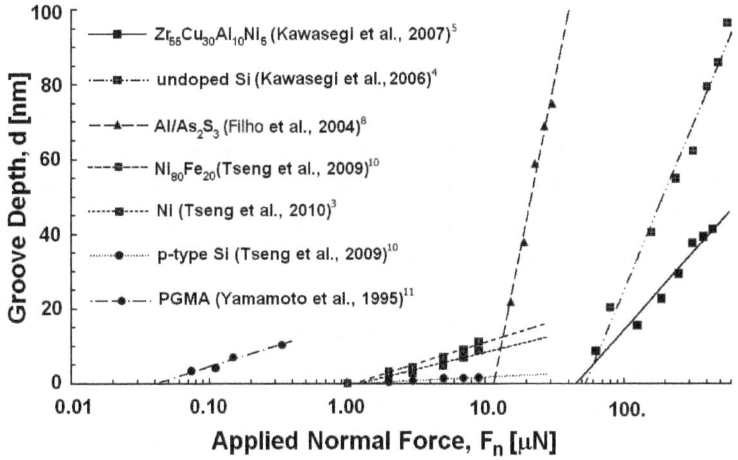

Figure 6.4. Correlation of groove depth (d) with applied normal force (F_n) for various materials under AFM scratching.

scratching regimes. Accordingly, SR may be treated as a general nanoscale scratchability metric for AFM-based mechanical material removal.

In conventional machining, material machinability is commonly assessed by comparison with a reference material under specified conditions. In AFM scratching, silicon serves as a natural reference due to its widespread use in nanofabrication studies. An AFM scratchability rating (SR*) is therefore defined as

$$SR^* \, [\%] = 100 \times SR/SR_{Si} \tag{6.2c}$$

where SR_{si} is the SR of Si and equals 0.756 nm/μN, as reported by Tseng.[2]

The scratch ratios, normalized scratchability ratings, and associated scratching conditions for the materials considered are summarized in Table 6.1. Metallic thin films such as Ni, NiFe,

Table 6.1. Comparison of scratch properties.

Parameter	p-type Si	Undoped Si	$Ni_{80}Fe_{20}$ thin film	Ni thin film	Al/As_2S_3 thin film	$Zr_{55}Cu_{30}Al_{10}Ni_5$	PGMA
Spring constant [N/m]	42.0	239.	42.0	42.0	165.	239.	0.68
Scratch speed [μm/s]	0.10	30.	0.10	0.10	5.0	30.	10.
Scratch coefficient, α_1 [nm]	0.765	38.65	4.96	3.94	80.51	18.09	4.538
Threshold force, F_{t1} [μN]	1.01	52.66	1.09	1.20	11.57	45.35	0.037
Coefficient, R^2	0.96	0.97	0.98	0.95	0.98	0.96	0.96
Scratch ratio, α_1/F_{t1} [nm/μN]	0.756	0.734	4.555	3.272	6.962	0.399	123.44
Scratchability rating, SR* [%]	100	97	603	433	921	53	16,328
Hardness, H [GPa]	10.	10.	8.0	7.35	1.00	4.90	—
Rating based on hardness	1.00	1.00	1.25	1.36	10.0	2.04	—
Reference	10	4	10	3	8	5	11

and Al/As_2S_3 exhibit scratchabilities approximately 433%, 603%, and up to 921% higher than that of Si, respectively, whereas the metallic glass $Zr_{55}Cu_{30}Al_{10}Ni_5$ is approximately 189% more difficult to scratch. As expected, the polymer PGMA exhibits a scratchability more than two orders of magnitude higher than that of Si, indicating substantially easier material removal. Notably, a conventional Si probe with a low spring constant of 0.68 N/m, typically used for routine topographic imaging, can readily be used to scratch PGMA.[11]

While the SR characterizes intrinsic material removal efficiency under single-pass scratching, its influence on cumulative material removal becomes evident when scratching is repeated over multiple cycles, as discussed in the following subsection.

6.2.3 *Scratch Cycle Number and Multiple Scratches*

Repeated scratching experiments have been widely used to investigate the influence of scratch (or scan) cycle number on AFM-based material removal behavior. When multiple scratches are performed repeatedly along the same path, the groove dimensions are expected to increase progressively with cycle number. Tseng et al. demonstrated that both the groove depth (d) and width (w_f) increase with the number of scratching cycles (N_o) according to power law relationships:[2,10]

$$d(N_o) = M_1(N_o)^{n1} \tag{6.3a}$$

and

$$w_f(N_o) = M_2(N_o)^{n2} \tag{6.3b}$$

where M_i and n_i are the multiple-scratch coefficient and exponent, respectively, and both n_i and N_o are dimensionless.

The depth and width data shown in Figure 6.5 were obtained by Tseng et al. for scratching a NiFe thin film under a constant normal load of 9 μN.[10] The fitted parameters are $M_1 = 12.711$ nm, $M_2 = 39.61$ nm, $n_1 = 0.279$, and $n_2 = 0.227$. The corresponding coefficients of determination (R^2) are 1.00 for the depth correlation and 0.96 for the width correlation, indicating an excellent fit for depth data and a very good fit for width data. Importantly, the power law model satisfies a fundamental physical requirement by predicting zero material removal at $N_o = 0$.

Similar experiments were performed by Tseng on a p-type Si(100) substrate scratched in the forward direction under a normal load of 9 μN for cycle numbers of 1, 3, and 5.[2] The fitted parameters were $M_1 = 2.694$ nm, $M_2 = 20.34$ nm, $n_1 = 0.538$, and $n_2 = 0.112$, with R^2 values of 0.98 and 0.96 for the depth

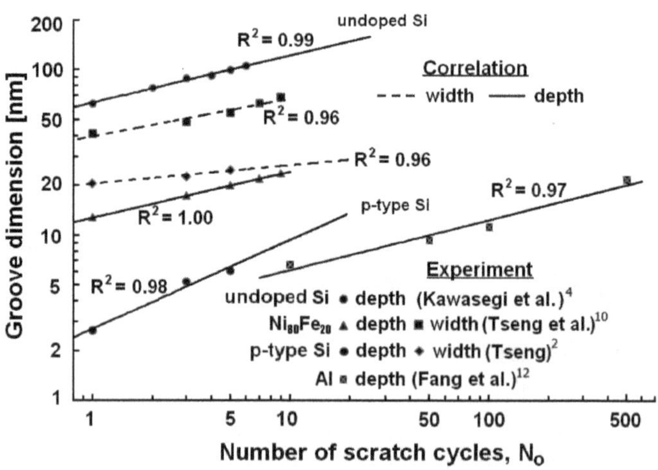

Figure 6.5. Correlation of groove dimensions with scratching cycle number (N_o) for various materials under AFM scratching.

and width correlations, respectively. Mean values of d and w_f, obtained from ten measurements at each condition, are plotted in Figure 6.5 together with their standard deviations, which are approximately twice those observed in single-cycle scratching experiments (Figure 6.4). Because closed-loop feedback is not applied to lateral (x–y) motion, this increased variability is attributed primarily to hysteresis and creep effects in the AFM piezoelectric scanner.[12]

Depth data reported by Kawasegi et al. for scratching undoped Si(100) wafers using a high-stiffness diamond-tip cantilever under a normal load of 318 μN and scratching speeds of 30 μm/s for cycle numbers ranging from 1 to 6 are also included in Figure 6.5.[4] These data are well described by the power law model, yielding M_1 = 63.06 nm, n_1 = 0.285, and R^2 = 0.99. Similarly, Fang, Weng, and Chang reported power law behavior for scribing a 500-nm-thick Al thin film using a diamond tip.[12] For cycle numbers between 10 and 500 under a tip force of 8 μN, the fitted depth correlation yielded M_1 = 3.039 nm, n_1 = 0.303, and R^2 = 0.97.

In contrast, studies by Tsuchitani et al. on carbon thin films under varying environments and by Ogino et al. on Si substrates reported an approximately linear increase in groove depth with N_0.[13,14] Although Tsuchitani et al. also observed that, in some cases, wear depths correlated better with a power law than with a linear relationship, the origin of this discrepancy remains unclear.[13]

To clarify the influence of scratch cycle number on groove geometry, further systematic studies are required to resolve these inconsistencies. It is noteworthy that a linear correlation does not satisfy the physical requirement of predicting zero material removal at $N_o = 0$, whereas the power law relationship fulfills this condition while providing excellent fits across a wide range of materials and scratching conditions, as shown in Figure 6.5. Consequently, the power law model is regarded as a reliable and physically meaningful framework for correlating scratch cycle number with scratched geometry. For comparison, the fitted parameters M_1 and n_1 for representative cases are summarized in Table 6.2.

Finally, multiple scratching shares certain similarities with wear testing used to evaluate wear coefficients and wear behavior.[15–17] In conventional wear studies, a large number of sliding cycles are typically applied without distinguishing individual scratches. In contrast, AFM scratching is primarily intended for the fabrication of high-quality, high-precision nanostructures. Achieving this objective requires careful selection of tip geometry and material, as well as precise control of normal load, scratch cycle number, and other process parameters, to remove prescribed amounts of material at predefined locations. Accordingly, each scratch must be performed under well-defined and reproducible

Table 6.2. Comparison of power law parameters for multiple-scratch experiments.

Parameter	p-type Si	Undoped Si	$Ni_{80}Fe_{20}$ thin film	Ni thin film	Al thin film
Multiple scratch coefficient, M_1	2.694	63.06	12.71	—	3.039
Multiple scratch exponent, n_1	0.538	0.285	0.279	—	0.303
R^2 for correlation of M_1 and n_1	0.98	0.99	1.00	—	0.97
Reference	2	4	10	3	12

conditions to maintain nanoscale precision and surface quality. Excessive scratching cycles should therefore be avoided, as they degrade process control and significantly reduce fabrication efficiency.

6.2.4 *Effects of Scratch Speeds*

Numerous studies have examined the influence of scratching speed on groove morphology and material removal mechanisms in AFM-based nano-scratching. Tseng et al. systematically investigated scratching on p-type Si(100) over speeds ranging from 10^1 to 10^4 nm/s under a constant normal load ($F_n = 9$ μN).[10] Both groove depth and width remained essentially invariant across this range, with only minor fluctuations. The standard deviations of depth and width from ten repeated measurements were ~4% and ~3% of their respective mean values, comparable to the variation among mean values, indicating a negligible speed effect. No measurable wear or degradation of the diamond tip was observed after a cumulative scratching length of 150 μm at 10 μm/s, confirming stable tool–surface interaction.

Similar speed-insensitive behavior for Si substrates was reported by Ogino, Nishimura, and Shirakashi over speeds from 10^1 to 10^4 nm/s,[14] and by Kawasegi et al. from 3×10^4 to 1.2×10^5 nm/s using comparable diamond tips.[4] Fang, Weng, and Chang reported analogous results for AFM scratching of Al thin films at speeds from 10^2 to 2×10^5 nm/s.[12] Collectively, these studies show that under typical AFM scratching conditions ($<10^5$ nm/s), groove geometry is governed primarily by normal load, tip geometry, and material properties rather than scratching speed. Although slightly rougher surfaces were observed at lower speeds, the dominant deformation mechanisms, elastic–plastic deformation and localized fracture, exhibit minimal rate dependence. For a diamond pyramidal tip with a radius of ~15 nm, surface roughness (R_a) decreased modestly from ~3 nm at 10^2 nm/s to ~2 nm at 2×10^4 nm/s.[12]

For amorphous or relatively soft materials, additional speed-dependent mechanisms can become significant. Rate-dependent plasticity, viscoelastic or viscoplastic responses, frictional heating, and debris formation may influence groove morphology. Tsuchitani, Kaneko, and Hirono demonstrated that during scratching of amorphous carbon films, scratch depth decreased logarithmically with increasing speed in both air and nitrogen.[13] Using a pyramidal diamond tip with a curvature radius of 70 nm, they scratched amorphous carbon films under loads of 10–80 mN and speeds from 1 to 100 μm/s. The scratch depth at 100 μm/s was ~0.1–0.2 nm shallower than at 1 μm/s, with depths ranging from ~0.3–0.7 nm in air and ~0.2–0.6 nm in nitrogen. The reduced depth at higher speeds was attributed to shortened chemical reaction times, diminished water-meniscus contact, and a reduced effective contact area.

When scratching speeds increase by two to three orders of magnitude beyond typical AFM values, speed effects become pronounced. Feng et al. scratched BK7 glass using a diamond tip with a radius of 2 μm at speeds of 1, 5, and 20 m/s and observed micro-shear fractures at ≥5 m/s due to tensile stress development and viscosity mismatch.[18] At 20 m/s, horizontal cracks replaced median cracks, indicating a transition in subsurface crack morphology and reduced subsurface damage. Consecutive scratching revealed that overlapping scratches and thermally induced expansion altered the ductile–brittle transition, underscoring the importance of thermal effects at high speeds.

At even higher effective scratching speeds, Ren et al. used 3D molecular dynamics simulations to study scratching of monocrystalline copper with and without water-layer lubrication.[19] Speeds of 50–400 m/s were simulated using LAMMPS. In dry scratching, increasing speed led to *higher* normal and tangential forces due to enhanced dislocation pile-up and frictional heating, with tangential forces increasing from ~0.8 to ~1.2 nN and normal forces from ~0.5 to ~0.8 nN. Higher speeds also promoted microcrack formation and chip generation, indicating a transition from cutting- to plowing-dominated removal. In contrast, water-lubricated scratching produced higher forces but improved surface quality by reducing atom accumulation and suppressing microcracks.

Overall, scratching speed has a limited influence on groove geometry within the low-speed range typical of AFM scratching, where geometry is dominated by load, tip geometry, and material properties. As speed increases, or when amorphous, soft, or chemically active materials are involved, rate-dependent plasticity, tribochemical reactions, and thermal effects become increasingly important. At ultra-high speeds, dynamic fracture, heat generation, and lubrication fundamentally alter material removal mechanisms.

6.3 Advancements in Atomic Force Microscopy Scratching Driven by Tip Forces (Single Loading)

In the preceding section, the basics of scratching, which were more related to metals and hard materials, were introduced. In this section, the basic AFM scratching for soft materials, including self-assembled and biomaterials, is first presented. Representative examples of fabricating more complex nanostructures, including 3D with curved surfaces and three nanodevices — including memristors, plasmonic arrays, and nanoripple channels — are then studied. Recent advances in multi-tip scratching aimed at improving process throughput are also reviewed.

6.3.1 *Scratching Self-Assembled and Biomaterials*

This subsection examines AFM scratching driven solely by mechanical tip forces for soft, self-assembled, and bio-related materials.

Grooves and pits with lateral sizes down to a few tens of nanometers could be obtained by scratching in self-assembled monolayers. Sugihara, Takahara, and Kajiyama applied a Si tip in order to dig pits with a minimum diameter of ~20 nm in a lignoceric acid self-assembled monolayer in an ambient environment.[20] The pits were made by a tip with a radius of 10 nm, a scanning rate of 2 Hz, and a tip force of 0.3 nN. The pits could be artificially distributed with different sizes and surface area densities in the organic monolayer. Zhang, Balaur, and Schmuki used a diamond-coated tip with a cantilever spring constant of 17 N/m and a scratch speed of 4 μm/s to scratch through an organic monolayer covered Si(111) surface in contact mode.[21] The scratched monolayer (1-octadecene) was used as an insulating mask for patterning a Cu-based nanostructure on the Si surface by a subsequent immersion plating process. The 1-octadecene ($C_{18}H_{36}$) layer was covalently attached to a hydrogen-terminated Si(111) surface.

Figure 6.6a shows an AFM image of four scratched grooves that were spaced by 1 μm on the 1-octadecene-covered Si surface,

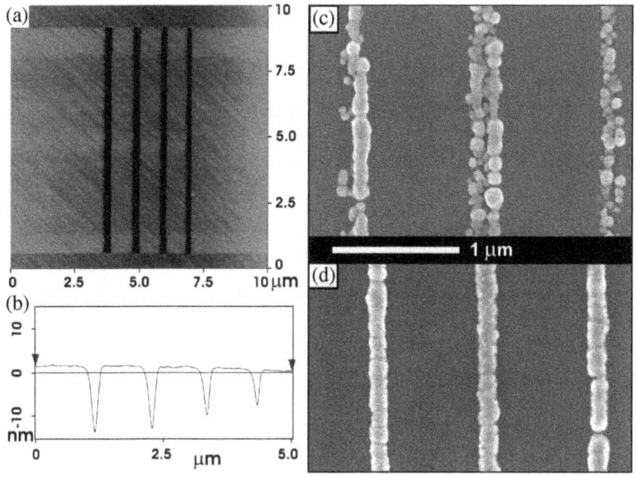

Figure 6.6. Nanopatterning of organic monolayer-covered Si surfaces by AFM scratching: (a) AFM image (5 μm × 5 μm) of four grooves (from left to right) scratched with the normal tip loads decreasing from 30 to 10 μN under contact mode on 1-octadecene ($C_{18}H_{36}$)-coated Si surface, (b) corresponding AFM measured groove profile, in which the distance between two neighboring grooves is about 1 μm, (c) SEM image of Cu deposited in scratched grooves by immersion plating in 0.05 M $CuSO_4$ + 1% HF solution with immersion time of 10 s and (d) of 15 s. (Courtesy of Professor Patrik Schmuki of University of Erlangen-Nuremberg, Germany and Dr. Eugeniu Balaur of La Trobe University, Australia.)[21]

where the four grooves shown from left to right were scratched with normal tip loads decreasing from 30 to 10 μN. Figure 6.6b is the corresponding AFM-measured groove profiles and show a uniform V shape with the depth range from 15 to 7 nm and width range from 300 to 200 nm as the normal load decreases from 30 to 10 μN. Immediately after scratching, the Cu was deposited on the scratched grooves in 0.05 M $CuSO_4$ + 1% HF solution by immersion plating. Different immersion times were used to test the selectivity of the deposition. Figure 6.6c and 6d show two SEM images that were obtained after different immersion times, 10 and 15 s, respectively. Clear effects of the immersion time on the copper deposit morphology were found: For deposition time up to 10 s single copper nuclei form selectively in the scratch, while at approximately 10–15 s these nuclei coalesce to coherent lines. It is also apparent that the longer the immersion time, the wider the deposit lines. The line width increases from 150 to 440 nm within 10–25 s. In Figure 6.6c, it is clear that the initial step of the copper deposition is the formation of globular nuclei. It was observed that for a deposition time longer than 20 s, Cu was not only deposited on the scratched grooves but also randomly on the surfaces. As shown in Figure 6.6d, an immersion time of 12–15 s can give the best results, i.e., coherently deposited lines but no deposits outside the scratched region. As concluded by the authors,[21] under an appropriate immersion time, the scratched organic layer can be used as an effective mask for patterning Cu nanostructures on Si.

In biomaterial scratching, Firtel et al. used a Si tip for nanosurgery to mechanically remove large patches of outer cell walls after appropriate medical treatment, which typically left the bacteria alive.[22] Their study can open a new direction for using AFM scratching for biological applications.

Muir et al. created nanoscale biologically active protein patterns by scratching the top low-fouling surface coating of a bilayer film, which consists of a 4 nm-thick low-fouling DGpp (diethylene glycol dimethyl ether plasma polymer) coating and a protein-fouling HApp (heptylamine plasma polymer) nonspecific binding base.[23] The material of the top DGpp coating can be selectively scratched to form a desired pattern so that the underlying HApp base, which can provide a protein-adsorbing surface, can be exposed. Incubation in a rabbit IgG protein solution leads to the adsorption of proteins onto the exposed protein-fouling patterns, while the regions with the low-fouling top coating resist protein adsorption. For example, subsequent exposure to a fluorescently labeled anti-IgG antibody results in the selective binding of the antibody to the surface-bound proteins, resulting in fluorescently active protein patterns which can be readily imaged by fluorescence as demonstrated in Figure 6.7. Figure 6.7a shows the AFM images of a scratched groove

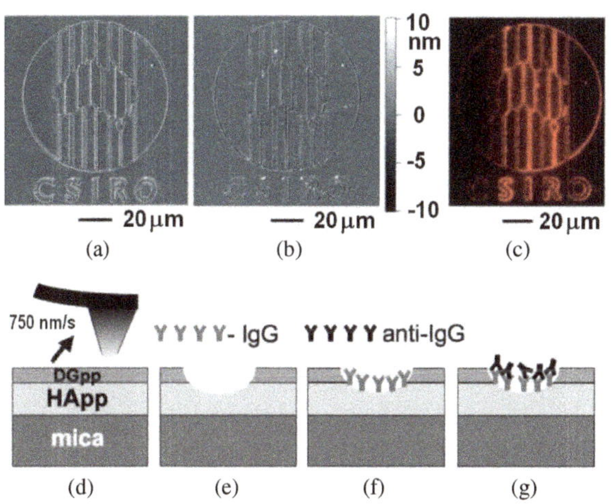

Figure 6.7. Biological protein patterning by AFM scratching: (a) AFM image of a scratched groove pattern with a typical groove depth of 7 nm and a mouth width of 300 nm, (b) AFM image after sequential adsorption of an IgG protein and a fluorescently labeled anti-IgG protein antibody, (c) fluorescence microscopy image of b, (d) DGpp surface being scratched by an AFM tip to expose the underlying HApp substrate, (e) scratched surface incubated with a rabbit IgG protein solution, (f) a fluorescently labeled anti-IgG protein that selectively binds to the rabbit IgG molecules being incubated with the surface, and (g) after rinsing, immobilized protein remains on the exposed HApp scratched regions. (Courtesy of Muir et al.,[23] reprinted from open-access data.)

pattern with a typical groove depth of 7 nm and a mouth width of 300 nm, while Figure 6.7b is the AFM image after sequential adsorption of an IgG protein and a fluorescently labeled anti-IgG protein antibody. The fluorescence microscopy image of Figure 6.7b is shown in Figure 6.7c. The schematic diagrams illustrating the steps for protein patterning are also shown, where Figure 6.7d depicts the DGpp surface being scratched by an AFM tip to expose the underlying HApp substrate; Figure 6.7e the scratched surface incubated with a rabbit IgG protein solution; Figure 6.7f a fluorescently labeled anti-IgG protein that selectively binds to the rabbit IgG molecules being incubated with the surface; and Figure 6.7g illustrates that, after rinsing, immobilized protein remains on the exposed HApp-scratched regions.

6.3.2 Atomic Force Microscopy Scratching of 3D Nanostructures and Nanodevices

A wide variety of 3D nanostructures have been fabricated for diverse applications using AFM scratching. Two examples for fabricating 3D nanostructures having various depths scratched in a control manner were first introduced in this subsection. Three different nanodevices, namely memristors, plasmonic arrays, and nanoripple channels, are then presented.

In scratching complex nanodevices or 3D structures, many researchers have integrated an AFM piezo tube scanner with a commercially available multi-axis stage to enhance the manipulability of workpiece movements. Thus, the workpiece can move or rotate in 3D and a curvilinear pattern of grooves can be singly or overlapped scratched. However, to handle relatively large dimensions, in the order of mm, the resolution of the stage is normally one to two orders of magnitude higher than that of the

AFM scanner and thus the integrated scratching system may be limited to create microscale instead of nanoscale 3D patterns. In an AFM scanner, the vertical accuracy (in the normal direction between the tip and workpiece) is normally less than 1 nm while the horizontal accuracy is on the order of a few nm. Certainly, the horizontal accuracy of the AFM scanner can be deteriorated by the effects of the nonlinearity and hysteresis of the piezo tube used, if no appropriate closed-loop control system is installed to tune the horizontal positions.

Vincenti et al. applied an antimony-doped n-type single-crystal silicon tip with a curvature radius of ~6 nm to scratch a 60-nm thick polymethyl methacrylate (PMMA) layer to create 3D PMMA nanostructures.[24] A grayscale bitmap image was used as a template for the nanostructure pattern, where grayscale values corresponded to specific tip forces applied at each pixel, with the tip scanning the surface pixel by pixel at a speed of 0.25 μm/s. The force varied from 0.171 μN to 0.630 μN, with increments ranging from 0.04 μN to 0.05 μN depending on the setpoint value. The scratched PMMA patterns were processed using a wet etching procedure to remove the debris accumulated during the scratching process. Subsequently, a dry etching step was applied to transfer the PMMA pattern, which served as a mask, onto the underlying 200 nm-thick silicon nitride (Si_xN_y) layer. Figure 6.8 illustrates four major steps in creating a Sierpinski triangle, a fractal-inspired triangular structure. Figure 6.8a shows the grayscale template, where the brightest areas (white) represent the maximum tip force (0.630 μN), while the darkest areas (black) represent the minimum force (0.171 μN). Figure 6.8b depicts an AFM topography image of the scratched triangular nanostructures. The scale bar on the right ranges from 0 nm to

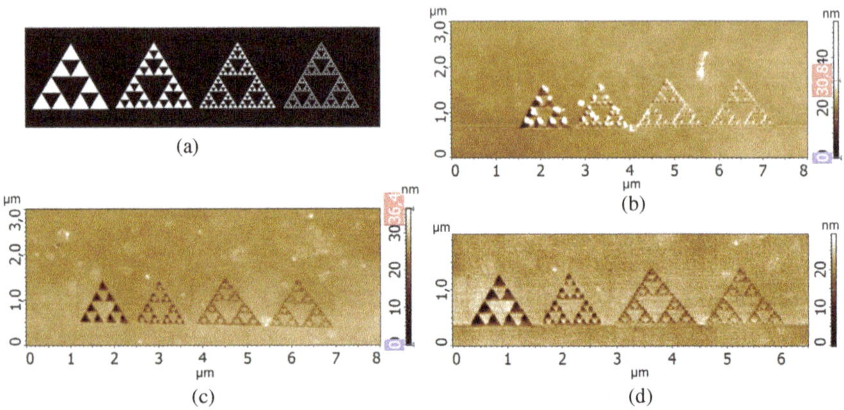

Figure 6.8. Four major steps in creating Sierpinski triangle nanostructures: (a) Greyscale template for AFM scratching, (b) AFM topography imaging of the scratched triangle pattern on PMMA, (c) AFM imaging of the same pattern after wet etching, and (d) AFM imaging of the transferred triangle nanostructures on the underlying silicon nitride layer via plasma etching. (Courtesy of Vincenti et al.,[24] reprinted from open-access data.)

30.8 nm, indicating the height variation. Figure 6.8c shows an AFM image of the same triangular nanostructures after the wet etching process, with the scale bar ranging from 0 nm to 36.4 nm. Figure 6.8d depicts the AFM topography image of the triangular structures transferred onto the Si_xN_y layer after dry etching, where the scale bar ranges from 0 nm to 30 nm. The fabricated nanostructures highlight the potential of the AFM scratching process as a cost-effective, scalable, and versatile method for 3D nanofabrication.

By a similar approach using the greyscale template to guide AFM scratching presented above, Yan et al. applied a diamond tip with an apex radius of 30 nm to create several 3D microstructures with different curved surfaces by using the changing tip force approach.[25] The curved surfaces were designed to have a sine wave profile with a wavelength of 6 μm and a peak valley distance of 180 nm. However, the profile scratched on an Al surface showed more than 20% inaccuracy as compared to its designed value. It was suspected that the inaccuracy was caused by the unreliable relationship used to correlate the depth of the scratched groove with the applied tip force as well as the nonlinearity associated with the scanner or cantilever under large deflection. Moreover, based on a digitized photo, Yan et al. sculpted a human face microstructure in a frame size of $20 \times 20\ \mu m^2$ with a peak valley height of 120 nm.[25] The total time for scratching the micro-sculpture was less than 10 min with an average tip load on the order of 100 N and a scratching speed of 10 m/s. The sculpting experiment demonstrated that the AFM mechanical scratching can be effectively used for fabricating complicated 3D geometries, although the accuracy needs to be improved.

In scratching nanodevices, Tominov et al. reported that AFM scratching was used to fabricate ZnO memristor structures used for resistive random-access memory.[26] The AFM was performed on the photoresist/thinner film using a Solver P47 Pro scanning probe microscope (NT-MDT, Russia). This process resulted in the formation of an array of 9 squared nanostructure grooves. Following this, a thin Ti film was deposited using the BOC Edwards Auto 500 system, and a lift-off process was applied using dimethylformamide.

Kim et al. prepared a three-layered substrate, in which a 100-nm thick mr-I 8010R polymer (mask layer) was spin-coated on the top of a 500-nm thick PMGI SF5 polymer (sacrificial layer), which was on a transparent glass substrate.[27] The authors then applied AFM tips to directly indent the sacrificial layer of the substrate for fabricating plasmonic nanostructures (nanodisks and nanotriangles), with subsequent wet etching and metal deposition. In indentation, a conical diamond-coated AFM tip was used for fabricating nanodisks, while a pyramidal silicon tip was used for creating nanotriangles. The tips penetrated the mask layer and reached the sacrificial layer with an indentation ramp rate of 5 Hz (7 μm/s). The resulting indentations were circular for nanodisks, and triangles for nanotriangles. Their size increased with the indentation force. The size of the nanodisks could be controlled within the range of 60 nm to 120 nm by adjusting the indentation force. After indentation, the sacrificial polymer layer in the indented region was removed through wet etching. Then, a thin layer of titanium (3 nm) was deposited as an adhesion layer, followed by 35 nm of gold or silver deposition using an electron beam evaporator. Finally, the remaining mask and sacrificial layers were removed, leaving the 35 nm of Au or Ag plasmonic nanostructures (nanodisks or nanotriangles) on the substrate.

Li et al. applied a diamond-like-coated AFM tip to scratch nanoripple structures in contact mode on a polycarbonate sheet for creating nanochannel array templates used for fabricating nanofluidic chip for label-free DNA detection.[28] Different normal loads (from 24.3 μN to 36.5 μN) were applied to observe changes in the number of channels and the dimensions of the nanoripple structures. Figure 6.9a shows a 3D AFM image of the nanoripple structures consisting of alternating periodic grooves and bulges dimensions of 25 μm × 25 μm. Figure 6.9b illustrates the profiles of the grooves and bulges of the nanoripple. The figure shows that the bulges are significantly higher than the groove depths, which could lead to channel blockage during bonding if not optimized.

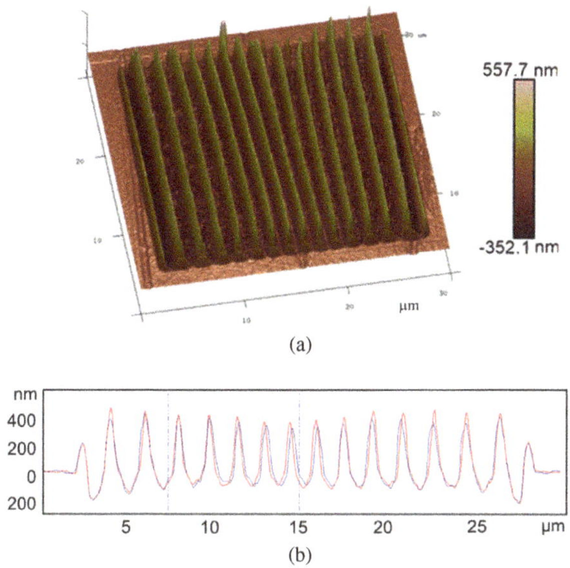

Figure 6.9. Nanoripple structures on polycarbonate sheet: (a) 3D AFM images of nanoripple structure with dimensions of 25 µm × 25 µm, and (b) cross-sections of the nanoripple. (Courtesy of Li et al.,[28] reprinted from open-access data.)

These studies highlight the effectiveness of AFM scratching in achieving precise material removal and fabricating well-defined 3D nanostructures.

6.3.3 *Multiple Probes*

One of the major challenges in the development of AFM nanofabrication is to increase its throughput. Extensive efforts have been made on using multiple probe arrays for parallel-processing nanostructures. Approaches ranging from individual multifunction probes to independently activated array probes have been evaluated.[29,30]

Minne et al. developed a system to operate an array of 50 cantilevered probes for local oxidation patterning in parallel at high speed.[31] IBM has applied the multiple-tip concept, also called Millipede, for data storage applications in which an array of heated tips is used to scratch (write), to image (read), and to melt (erase) nanoscale holes (data) on very thin polymer films (Figure 6.10a). To demonstrate the potential of Millipede for ultrahigh storage density, a 64 × 64 cantilever/tip array for Write/Read/Erase functionality has been fabricated and shown in the photo image in Figure 6.10b.[32,33] The image highlights the central 32 × 32 array section, which measures 3 mm × 3 mm and contains 1,024 cantilevers. Surrounding the array are electrical wiring interconnections that connect the array to the bonding pads located at the chip's periphery. The chip demonstrates the integration of cantilevers with Schottky diodes and multiplexed row/column addressing for efficient operation.

Wang et al. applied a multi-tip probe to scratch or indent nanostructures (nanochannels and nanoindentations) on single-crystal Cu(001) substrates for the use as surface-enhanced Raman scattering substrates.[34] Nanochannels were scratched by

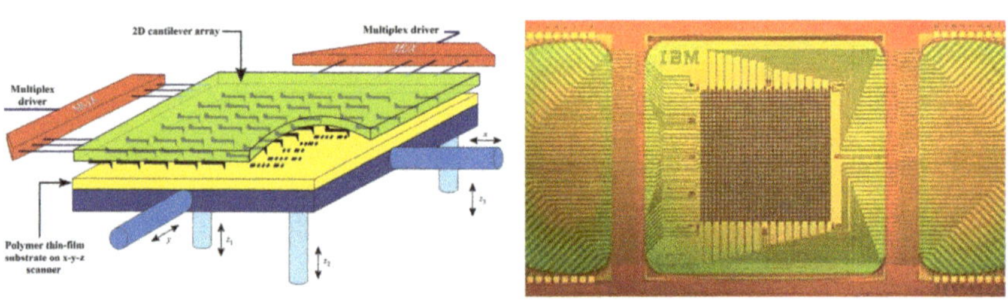

(a) (b)

Figure 6.10. IBM Millipede: (a) Concept to write, read, and to erase nanoscale holes (data) on thin polymer films, and (b) a photo image of fabricated 32 × 32 AFM cantilever array chip (14 mm × 14 mm) located at the center with bond pads distributed on either side. (Courtesy of Vettiger et al.,[32] reprinted from open-access data.)

diamond-coated Si dual-tips with a depth varying from 13 nm to 56 nm and a width varying from 50 nm to 150 nm under the load varying 23 to 57 μN. The distance between two tips was 80 nm with a period from ~229 nm to ~290 nm, depending on the scratched depth. The authors concluded that using multi-tip probes (dual-tip and four-tip) significantly improved the efficiency and quality of fabricating periodic nanostructures.

In the fabrication of multiple probes, not only the height and shape of each tip are important but also the dimensions, which affect the bending and torsion of the cantilever, play crucial roles. These geometric factors have consequences in particular for the variation of the probe properties such as deflection, compliance, and resonant frequency. For example, during scratching, the approach angle for a cantilevered single probe to a planar substrate is not a critical issue. However, in the case of a multiple parallel probe arrangement, the approach angle becomes critical in the plane parallel to the surface. The alignment of the cantilevers and the approach of the array determine which tip comes in contact with the surface first. Often, the two outermost probes are intentionally made longer to serve as adjustment or guide probes.[35]

6.4 Atomic Force Microscopy Scratching Loaded with Dual Sources or Combined Loadings

As indicated in the preceding sections, the cantilevered tip has been mainly loaded with mechanical force or energy to perform AFM scratching. However, the tip sample interaction can also be imposed or influenced by other energy sources, including thermal, electric, and magnetic energies or by the chemicals or contents of the environments, which can create different effects on the nanostructures created by AFM scratching. In this subsection, the effects of the mechanical energy combined with other types of energies on AFM scratching will be examined. Frequently, by combining an additional physical or chemical source, the major limitations on AFM scratching, including the shallowness of scratch depth and tool wear, could be alleviated.

6.4.1 *Vibration-Assisted Atomic Force Microscopy Scratching*

Vibration-assisted scratching (VAS) enhances AFM-based nanomachining by superimposing high-frequency vibrations onto conventional scratching, thereby improving material removal efficiency, machining stability, and surface integrity. This subsection reviews five representative studies that investigate distinct VAS configurations and elucidate their underlying mechanisms across a range of materials.

Wu et al. systematically investigated ultrasonic VAS using a diamond tip to scratch single-crystal 4H-SiC.[36] As schematically illustrated in Figure 6.11, scratching was performed along the X direction, while a sinusoidal ultrasonic vibration was applied along the Y direction. By combining nanoscale experiments with molecular dynamics simulations, the authors analyzed both single asperity and cumulative scratching behaviors. The simulations revealed that ultrasonic vibration significantly increases contact frequency, energy input, and flash heating, while simultaneously reducing dislocation density, lattice distortion, and residual stress. These results demonstrate that VAS is particularly effective for scratching hard and brittle materials at the nanoscale.

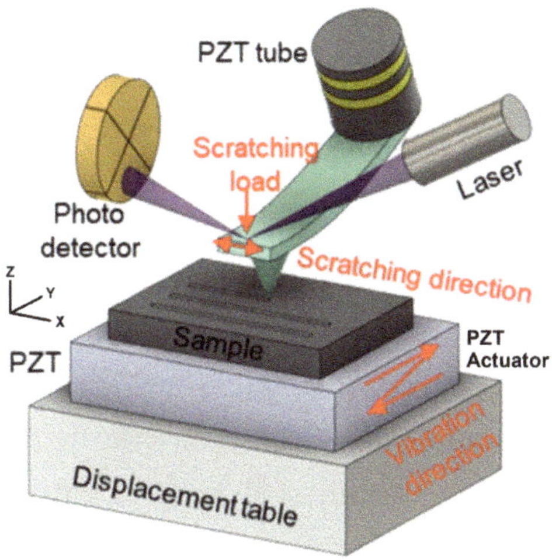

Figure 6.11. Schematic of vibration-assisted scratching (modified figure from Wu et al.,[36] reprinted from open-access data.)

Based on the experimental data, Wu et al.[36] compared scratch depth and material removal volume between conventional no-vibration scratching and ultrasonic VAS (55 kHz, 10 nm amplitude) at an identical scratching speed of 2 μm/s and found that the removal volume during no-vibration scratching remained below 3×10^4 nm^3 for loads up to 48 μN, whereas VAS achieved 1.4×10^6 nm^3 at the same load, approximately 46 times higher. At higher loads, the disparity became even more pronounced: at 64 μN, VAS reached 4.4×10^6 nm^3, while no-vibration scratching achieved only 3.6×10^5 nm^3, more than an order of magnitude lower. Notably, under VAS conditions, increasing the load from 48 μN to 56 μN resulted in a sharp increase in removal volume from 1.4×10^6 nm^3 to 4.4×10^6 nm^3. This abrupt transition suggests that the material removal process approaches a saturation or instability regime at high loads. Such behavior also posed challenges for accurate prediction using molecular dynamics simulations, as checked with the authors' simulation results. These findings imply that ultrasonic vibration may induce operational instability at elevated loads, and that avoiding this high-load regime is advisable for achieving reliable scratching and consistent material removal.

Xu et al. investigated VAS behavior near resonance using a custom AFM system, in which sinusoidal vibration was applied along the Y direction via a PZT actuator (Figure 6.11).[37] Vibration amplitudes ranged from 0 to 0.8 μm, with frequencies between 0 and 10 kHz, while scratching proceeded along the X direction at 3 μm/s under a constant normal load of 16 mN. The scratch force decreased from approximately 13 mN without vibration to 10 mN at 4 kHz, corresponding to a reduction of about 30%. Near the resonance frequency (~4.3 kHz), the average scratch force remained low but exhibited increased fluctuations. The material removal volume peaked at approximately 50 μm^3 near resonance, while the friction coefficient decreased from ~0.85 to ~0.75 under vibration. Groove depth increased substantially, from ~150 nm without vibration to ~600 nm with VAS. Numerical simulations indicated a temperature rise near resonance (up to ~430 K), which contributed to thermal softening and enhanced penetration. Although the von Mises stress generally decreased under vibration, it increased near resonance due to localized heating and deeper cutting. These results underscore the effectiveness of VAS while highlighting the need for careful frequency control to maintain machining stability.

Wang et al. employed a diamond-coated pyramidal AFM tip equipped with a piezoelectric actuator to perform VAS on a polycarbonate substrate for fabricating nanochannel molds.[38] The polycarbonate mold was subsequently used to replicate polydimethylsiloxane nanofluidic chips. The actuator was driven by sinusoidal signals with a 90° phase difference between the X and Y directions, generating a rotary vibration of the AFM tip at frequencies of 100 Hz and 1500 Hz. The resulting vibration amplitude, controlled by driving voltages of 30–150 V, ranged from 350 to 690 nm, with higher voltages and frequencies producing larger amplitudes. The rotary VAS enabled nanochannels with uniform dimensions and reduced leakage during bonding. Fabrication efficiency was significantly improved, achieving feed rates of 1 μm/s at 100 Hz and 15 μm/s at 1500 Hz, with scratching speeds up to 4.10 mm/s. Moreover, the rotary vibration distributed contact stresses more evenly, reducing tip wear, extending tool lifespan, and lowering operational costs.

Tang et al. applied elliptical ultrasonic VAS to hard and brittle materials, including GaAs, Si, and sapphire.[39] The system operated at 26.08 kHz, with vibration amplitudes of 20.5–200.5 nm in the X direction and 15.7–170.8 nm in the Y direction, generated by two PZT transducers driven by sinusoidal voltages with controlled phase differences. Elliptical VAS markedly increased groove dimensions. For GaAs, groove depth increased by approximately 23 times (from 4.95 nm to 115.5 nm) at a load of 46 μN and a peak voltage of 9 V. In silicon, the thickness of the amorphous layer decreased from 26.98 nm to 6.89 nm at 28 μN and 1 V. Notably, elliptical VAS enabled groove formation in sapphire, which could not be scratched under conventional conditions. Overall, elliptical VAS enhanced removal efficiency, reduced subsurface damage, lowered hydrostatic stress, and improved the mechanical and optical quality of the fabricated nanostructures.

Finally, Wang et al. integrated elliptical VAS with laser processing to machine indium phosphide.[40] Elliptical vibrations were applied over a frequency range of 0.5–5 kHz, with amplitudes of 284–340 nm along the X direction and 20–124 nm along the Z direction, generated by sinusoidal inputs with a 90° phase difference. Elliptical VAS alone reduced surface roughness to 2.06 nm, representing a 50% improvement over no-vibration AFM scratching. When combined with laser processing, subsurface damage was further reduced (68 nm compared to 164 nm), strain rate sensitivity was suppressed, and plastic deformation was promoted, thereby minimizing atomic-scale defects. This hybrid approach enabled the fabrication of flatter, high-quality nanosurfaces suitable for high-speed electronic and optoelectronic applications.

Overall, vibration-assisted AFM scratching significantly enhances material removal efficiency, reduces cutting forces, and improves surface and subsurface quality across polymers, semiconductors, and hard–brittle materials. By increasing contact frequency and inducing localized heating, VAS promotes stable plastic deformation while limiting damage when operated within suitable load and frequency ranges. Notably, many reported benefits are closely linked to thermally activated effects such as flash heating and vibration-induced softening. This connection naturally leads to the next subsection on thermal-assisted AFM scratching, where temperature is intentionally employed as a primary parameter to further control nanoscale material removal and machining performance.

6.4.2 Thermal-assisted Atomic Force Microscopy Scratching

An AFM tip can be heated either internally through friction or externally via electrical or optical inputs to reach elevated temperatures, enabling thermally assisted scratching and material removal. Heated AFM tips have been widely used not only to locally melt or soften substrates, but also to supply thermal energy sufficient to activate chemical reactions, break intermolecular bonds, modify molecular structures, or evaporate organic materials. The resulting nanopatterns frequently serve as etching or deposition masks for subsequent lithographic processes, enabling the fabrication of intricate 3D structures and devices.[41,42] In addition, thermally generated nanopatterns can function as guiding templates for directed self-assembly and other bottom-up processes.[43,44]

Thermally heated AFM probes are commonly fabricated from silicon-on-insulator wafers, offering well-controlled resistive heating and facile integration. Platinum heaters are also frequently employed owing to their chemical stability and uniform temperature coefficient of resistance. Chiou et al. demonstrated that a silicon tip heated to 120 °C using an integrated Pt element could scratch an ~800 nm-wide groove in PMMA.[45] Lantz, Wiesmann, and Gotsmann further showed that thermally assisted scratching significantly reduces tip wear, as substrate softening minimizes mechanical abrasion.[46] Consequently, high-resolution features with stable dimensions can be produced while enabling efficient removal of larger material volumes without compromising tip longevity.

Compared with purely mechanical scratching, thermally assisted scratching exhibits greater complexity because the resulting feature geometry depends sensitively on temperature, scanning speed, and loading duration. Okabe et al. investigated diamond tip scratching of silicon from room temperature to elevated temperatures and found that, under identical mechanical conditions, the average groove depth increased from less than 2 nm to more than 4 nm as the substrate temperature rose from 290 to 470 K.[47] Molecular dynamics simulations further revealed an expanded amorphous region surrounding the groove at higher temperatures. Vettiger et al. demonstrated the formation of 15–20 nm holes in a 50 nm-thick PMMA film using heated tips.[33] Because heat transfer occurs through an extremely small contact area (10–40 nm^2), tip temperatures approaching 400 °C were required to initiate polymer softening; once softening began, the enlarged contact area enhanced heat flow and indentation efficiency. At sufficiently high temperatures, polymer evaporation can also occur.

Beyond mechanical softening, nanoscale thermal gradients enable precise control of chemical reactivity, surface energy, solubility, and electrical properties. Szoszkiewicz et al. used a heated silicon tip to write sub-15 nm hydrophilic features into a hydrophobic block copolymer, with reaction outcomes governed by polymer composition, scan speed, interface temperature, and applied load.[48] Such capabilities are particularly attractive for applications in high-density data storage and nanofluidics. Fringes et al. demonstrated thermal decomposition of poly(p-phenylene vinylene) films using a heated cantilever, producing a template for nanoparticle assembly.[49] Similarly, Wang et al. employed a resistively heated tip to thermally convert a sulfonium salt precursor into poly(p-phenylene vinylene) while simultaneously scratching nanogrooves.[50] Using a 100 nm-thick precursor film, grooves were written at 20 μm/s under a 30 nN load, with tip temperatures ranging from 240 to 360 °C. Fluorescence and AFM images revealed enhanced contrast and more pronounced features at higher temperatures, with groove widths of ~70 nm (full width at half maximum) at 240 °C.

The heated tip was used to break down the polymer chains into a volatilized pattern for guiding nanoparticle assembly. Fringes et al. employed a heated AFM tip to locally depolymerize the surface of a thermally sensitive resist material, a 135-nm-thick polyphthalaldehyde (PPA) film.[51] The heat from the tip caused the polymer chains in the polyphthalaldehyde to break down into volatile products, which then evaporated, leaving behind elongated trap patterns approximately 5 μm long and 40 nm deep. These traps were specifically designed to accommodate the high aspect ratio of InAs nanowires and facilitate their precise alignment on predefined electrodes. Subsequently, a nanowire suspension was deposited onto the patterned substrate, and electro-osmotic forces were applied to attract the nanowires into the traps. Once the nanowires, which had a diameter of approximately 35 nm, were stably captured, the polyphthalaldehyde film was thermally removed, allowing the nanowires to link onto the electrodes, thus completing the assembly process. This AFM-assisted process demonstrated precise deposition and stacking of metallic nanowires, enabling enhanced coupling between nanostructures and paving the way for the development of advanced materials like metamaterials.

Thermally assisted AFM scratching was also used to create true 3D nanofabrication through either force-modulated material removal or layer-by-layer patterning. Knoll et al. demonstrated rapid 3D patterning in a self-amplified depolymerization polymer using a heated AFM tip operated at 700 °C with force pulses of 2–14 μs.[41] Under these conditions, rupture of a single polymer bond triggers spontaneous depolymerization of the entire chain, enabling material removal with loading times below 2 μs per pixel. As shown in Figure 6.12a, the measured scratching depth exhibited a linear dependence on applied force, allowing precise depth control. Using this approach, a fractal carpet pattern was fabricated in a 50 nm-thick polyphthalaldehyde layer with uniform 40 × 40 nm^2 pillars and negligible material pile-up as shown in Figure 6.12b. As shown in Figure 6.12c and 6.12e, subsequent reactive ion etching transferred the

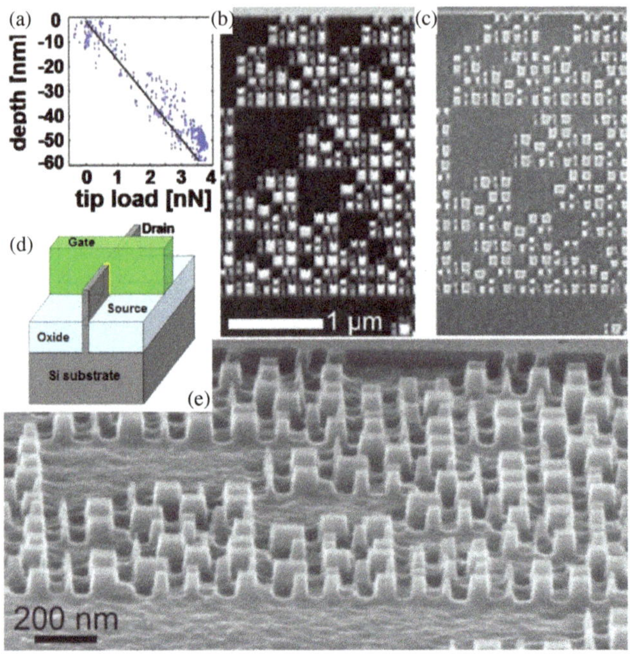

Figure 6.12. Nanoscale fractal-carpet pillar pattern scratched on a 50 nm-thick polyphthalaldehyde layer: (a) scratching depth versus applied force at a 14 μs pulse; (b) AFM image of pillars formed at 700 °C with a 2 μs force pulse; (c) SEM image after reactive ion etching transfer into Si; (d) schematic of a 3D trigate transistor from Intel; and (e) 45 °-tilted SEM view of the etched pillars in (c). (Courtesy of Knoll et al.,[41] reprinted from open-access data.)

pattern into silicon with a vertical amplification factor of ∼6, yielding freestanding silicon pillars ∼30 nm wide and ∼90 nm tall. The resulting steep sidewalls and sharp edges are critical for advanced device architectures such as 3D trigate transistors, where vertical fin-like structures form conduction channels on three sides (Figure 6.12d).

An alternative route to 3D fabrication involves stacking successive patterned layers, treating each as a 2D contour. Pires et al. employed a heated AFM tip (500 ± 30 °C) to thermally desorb a molecular glass resist using pulse durations of 5.5 ms per pixel.[42] The molecular glass, composed of hydrogen-bonded molecules with a glass transition temperature of 126 °C, provides sufficient stability for room temperature imaging while enabling efficient thermal removal at elevated temperatures. Using 120 successive layer-by-layer removal steps, the authors fabricated a nanoscale replica of the Matterhorn mountain with ∼15 nm lateral resolution and a height of 25 nm in a 100 nm-thick film, as shown in the AFM image in Figure 6.13a. Figure 6.13b shows a photograph of the Matterhorn in Switzerland, while Figure 6.13c presents an AFM image of the replica transferred into the underlying silicon substrate using an SF_6/C_4F_8 etchant mixture. This transfer produced a height amplification factor of ∼3, faithfully reproducing fine topographic details at the nanoscale.

Heated AFM scratching has also emerged as a powerful tool for generating guiding templates for directed self-assembly. Gottlieb et al. raised an AFM tip to 700–1000 °C to locally depolymerize a 3.5 nm-thick polyphthalaldehyde resist, well above its decomposition temperature (∼150 °C), producing highly precise nanopatterns.[43] These patterns were subsequently used to direct block copolymer self-assembly, yielding defect-free

Figure 6.13. 3D layer-by-layer patterning using a heated AFM tip: (a) AFM image of a nanoscale Matterhorn replica (25 nm tall) written in phenolic molecular glass using 120 layer-by-layer scratch removing steps; (b) photograph of the Matterhorn in Switzerland; (c) AFM image of the replica transferred into Si with a height amplification factor of ~3 (~75 nm tall). (Courtesy of Pires et al.,[42] reprinted from open-access data.)

sub-10 nm line/space structures. Uranga-Granados et al. extended this approach to fabricate guiding templates for semiconductor quantum devices, transferring the assembled structures into SiO_2 via reactive ion etching to achieve sub-20 nm features.[44]

Related strategies have been demonstrated for device-relevant nanostructures. Pellegrini et al. used a heated AFM probe pulsed to ~880 °C to pattern polyphthalaldehyde films, producing arrays of nanoscale openings that were subsequently converted into TiO_2 nanodisks through lift-off and sputter deposition.[52] Chang et al. showed that heated AFM tips operated at 200–230 °C and extremely low scanning speeds (0.1–1 µm/s) could induce significant polymer softening and thermal expansion, enabling the formation of continuous PMMA nanowires.[53] Cho et al. demonstrated sub-10 nm patterning of polyphthalaldehyde by exploiting self-amplified thermal depolymerization, with the resulting resist patterns exhibiting excellent resistance to reactive ion etching and enabling silicon pattern transfer with nearly tenfold vertical amplification.[54] Lisunova and Brugger further combined heated AFM scratching with deep reactive ion etching to fabricate high-aspect-ratio 3D silicon nanostructures, achieving vertical amplification from ~45 nm-deep polyphthalaldehyde patterns to ~450–500 nm-deep silicon features.[55]

Finally, the versatility of thermally assisted AFM scratching extends beyond conventional lithography. Zheng et al. employed heated-tip-scratched polyphthalaldehyde patterns as guiding templates for the deterministic organization of functionalized DNA origami into nanoarrays.[56] By precisely positioning DNA origami structures decorated with gold nanoparticles and quantum dots, this work highlights the potential of heated AFM scratching for integrating bottom-up nanomaterials with top-down nanofabrication.

6.4.3 *Electrical Bias-Assisted Atomic Force Microscopy Scratching with and without Chemical Contributions*

By equipping an AFM with a conductive tip and applying an electrical potential bias between the tip and a conductive substrate, electrostatic forces are introduced in addition to conventional van der Waals interactions. This principle underlies techniques such as Kelvin probe force microscopy and electric force microscopy, and it can also be exploited to enhance AFM-based scratching. In bias-assisted scratching, the applied electric field improves material removal efficiency, facilitates the formation of deeper and smoother nanostructures, and extends the range of machinable materials. Moreover, when scratching is performed in a chemical solution, the applied bias can drive electrochemical reactions that further enhance material removal. Electrical bias-assisted AFM scratching, with and without chemically mediated effects, is reviewed in this subsection.

Yang and Zhao further demonstrated bias-assisted AFM scratching on copper thin films by combining mechanical loading with a constant electric bias of 4–8 V.[57] Compared with purely mechanical scratching, the addition of an electrical bias produced substantially deeper and smoother features. For instance, repeated scratching under loads of 500 nN and 1,000 nN resulted in maximum groove depths of approximately 10 nm and 30 nm, respectively, whereas the application of an 8 V bias increased the depths to ~40 nm (300% increase) and ~75 nm (150% increase). Importantly, the electrical bias enabled scratching of materials such as platinum, which could not be effectively machined using mechanical loading alone. Overall, the method improved scratching depth, material removal rate, surface smoothness, and material compatibility.

More recently, Wang et al. investigated bias-assisted scratching of single-crystal copper using a customized tungsten tip.[58] Applying a 1 kHz rectangular wave signal with a 30% duty ratio, a normal load of 600 μN, and a cutting speed of 3 μm/s, they observed that the scratch depth increased from ~110 nm to ~160 nm and ~185 nm as the applied bias increased from 0 to 1 V and 2 V, respectively. Concurrently, the surface roughness (R_a) decreased from 4.4 nm to 2.6 nm and 2.1 nm. Their results indicated that increasing voltage, frequency, and duty ratio enhances groove depth and aspect ratio while improving surface quality. The study also examined tip-wear behavior, providing valuable insights for achieving efficient and high-quality scratching of single-crystal copper.

Bias-assisted scratching can also be performed in chemical environments, where concurrent electrochemical reactions further contribute to material removal. This approach is commonly referred to as AFM-based electrochemical nanomachining (AFM-ECM). In AFM-ECM, material removal is conceptually similar to reverse electroplating, in which a current passes between a negatively biased tip (cathode) and a conductive substrate (anode) through an electrolyte. The key distinction is that AFM-ECM operates within a nanoscale electrochemical cell localized at the tip–sample junction. In some cases, material removal is driven predominantly by non-mechanical mechanisms, such as field-emission currents, as demonstrated in carbon nanotube cutting,[59] and electrochemical potentials by Han et al.[60]

Early demonstrations of AFM-ECM were reported by Koinuma and Uosaki,[61] who used a Si_3N_4 tip with a radius of ~40 nm immersed in an electrolyte to induce electrochemical etching at ambient temperature. By applying a tip force of 10 nN and a bias of −0.05 V in a 10 mM H_2SO_4 solution for 30 min, they produced a groove with an ~80 nm-wide opening and ~8 nm depth on a single-crystalline Zn-doped p-GaAs(100) surface. Increasing the potential to +0.05 V enhanced the groove depth to ~18 nm. The results showed that groove depth depended not only on tip force and repeated cycles, but also strongly on the applied voltage bias. AFM-ECM has since been extended to fabricate nanoscale biomolecular patterns,[62] polymer nanostructures,[63] nanoholes and lines on Au and Ti,[64] and nanogrooves on copper thin films.[65]

Lee et al. employed conductive-tip AFM scratching to fabricate nanogrooves on copper surfaces using ultrashort voltage pulses, with pulse durations ranging from 50 to 5000 ns, frequencies from 0.02 to 2 MHz, and amplitudes between 0.5 and 4 V.[65] By immersing the tip–sample system in an electrolyte solution, they demonstrated that higher pulse amplitudes and frequencies improved the precision and uniformity of the fabricated features. Both groove width and depth increased systematically with pulse amplitude and duration; for example, the line width increased from ~100 nm to ~1 μm as the pulse amplitude varied from −0.5 V to −4 V. Notably, the bias-assisted process significantly reduced material pile-up around the grooves, yielding smoother and more well-defined nanostructures.

Electrochemical bias-assisted AFM scratching has also been demonstrated with atomic-scale precision. Han et al. used a $PtIr_5$-coated AFM silicon tip with a curvature radius below 25 nm to electrochemically remove material from highly oriented pyrolytic graphite.[60] A DC voltage of 2.5–8 V was applied between the tip (cathode) and substrate (anode) in the presence of a nanoscale water meniscus formed from a saturated 1 M NaCl solution under high relative humidity (~75%). This configuration enabled anodic dissolution of carbon, converting surface atoms into gaseous CO and CO_2 and allowing atom-by-atom material removal. Three types of nanoscale patterns were produced, including nanoholes with depths ranging from <0.5 nm (single atomic layer) to ~1 nm (three atomic layers), as well as nanolines with widths between 47 and 82 nm, depending on scan speed and tip condition. These results highlight the capability of electrochemical bias-assisted AFM scratching to achieve deterministic, atomic-scale nanofabrication.

Recent advances have expanded AFM-ECM toward non-contact and pulsed-voltage operation. Saxena et al. employed a coated tungsten AFM tip (~50 nm radius) to electrochemically scratch cavities on an Inconel IN718 substrate in non-contact mode with a tip–substrate gap of ~1 μm.[66] The tip was driven by nanosecond voltage pulses with a 50% duty ratio at a frequency of 25 MHz in the presence of a 2% aqueous sodium nitrate electrolyte. Preliminary results showed that cavity size increased from ~2 μm to ~4 μm as the pulse voltage increased from 5 to 10 V, with a lateral resolution of approximately 100 nm. The authors noted that further optimization is required to achieve smaller feature sizes and higher resolution. To ensure electrical isolation and reliable bias delivery, the AFM tip was coated with an inner dielectric layer and an outer gold layer serving as the working electrode.

In this subsection, electrical bias-assisted AFM scratching significantly extends the capabilities of purely mechanical nanomachining by enhancing material removal efficiency, surface quality, and material compatibility. When operated without chemical mediation, the applied bias promotes deeper and smoother grooves through electrostatic and field-assisted effects, while operation in liquid environments enables electrochemical nanomachining with atomic- to nanoscale precision. Together, these approaches bridge top-down mechanical scratching and bottom-up electrochemical reactions, offering versatile pathways for fabricating high-aspect-ratio, chemically selective, and device-relevant nanostructures across a wide range of materials.

6.4.4 *Chemical or Environmental Effects on Scratching*

To mitigate substrate damage during AFM scratching, chemical-aided and environmentally assisted scratching techniques have been studied. Michler et al. demonstrated an etching-assisted approach using an AFM equipped with a diamond tip to scratch 5–7 nm-deep nanopatterns into a 10 nm-thick SiO_2 film, thereby minimizing damage to the underlying silicon substrate.[67] Subsequent hydrofluoric acid etching uniformly removed 4–5 nm of SiO_2 from both scratched and unscratched regions, selectively exposing patterned silicon while preserving a 5–6 nm oxide layer elsewhere. Palladium was then selectively electrodeposited into the scratched regions, illustrating the effectiveness of combining mechanical scratching with controlled chemical etching.

Chemical environments can also directly influence the scratching process when solution etching occurs concurrently with mechanical material removal. Kato et al. investigated aluminum scratching using silicon and diamond tips in four different environments: diluted NaOH, $CuSO_4$ solution, pure water, and a Cu electroless plating solution.[68] The unit scratch depth followed the sequence: diluted NaOH > $CuSO_4$ > pure water > Cu electroless plating solution. The enhanced unit scratch depth in NaOH and $CuSO_4$ was attributed to oxide dissolution, producing $Al(OH)_4^-$ species at pH 9.2 in NaOH and Al^{3+} ions in acidic $CuSO_4$. In contrast, scratching in pure water led to the formation of a thicker, more stable oxide-hydroxide layer, reducing material removal. The smallest unit scratch depth was observed in the Cu electroless plating solution, likely due to Cu nanoparticle deposition that inhibited aluminum dissolution and increased scratching resistance. In addition, the diamond-coated silicon tip exhibited significantly higher wear resistance than the uncoated silicon tip.

Water content and ambient humidity also play critical roles in scratching behavior. Pendergas et al. compared gold scratching in water and under ambient humidity (~60%) using Si_3N_4 and diamond tips.[69] Scratches formed under ambient conditions were slightly shallower than those produced in water, with the difference becoming more pronounced at higher loads. Decomposition of the horizontal force components provided insight into the dominant contributors to scratching friction under different environments.

The influence of ambient humidity was further investigated by Chen et al., who examined scratching of potassium dihydrogen phosphate (KDP (001)) crystals over a humidity range from 17.9% to 78.6% and tip loads between 16.9 and 21.1 μN.[70] A quadrangular pyramid-shaped AFM tip with a radius of ~4 nm was operated at 0.6 Hz to scratch 8 μm-long grooves through repeated back-and-forth motion ten times. At low humidity (17.9%), grooves were relatively shallow with minimal pile-up, whereas high humidity (78.6%) produced deeper grooves with more pronounced pile-up. These effects were attributed to enhanced water meniscus formation and adsorbed surface films at high humidity, which weaken surface integrity through chemical dissolution and surface hydroxylation, thereby facilitating material removal. However, the authors' molecular dynamics simulations suggested that humidity had little direct influence on groove dimensions, indicating that more refined modeling approaches are needed to accurately capture humidity-dependent effects.

Molecular dynamics simulations have also been employed to study the role of aqueous layers in nano-scratching. Cao et al. investigated the effect of water layer thickness on scratching of single-crystal γ-TiAl substrates using a rigid spherical diamond tip with a radius of 3 nm.[71] Their simulations indicated that the presence of an aqueous layer significantly reduced tangential and normal forces, friction coefficient, subsurface damage depth, stress concentration, and substrate temperature, while improving heat dissipation and surface quality. An aqueous layer thickness of ~1 nm was identified as optimal for minimizing damage and enhancing scratching performance. However, given the ambient temperature of 27 °C, the predicted local temperatures in the contact area of the substrate exceeding 1400 °C during scratching appear unrealistically high, raising concerns regarding the accuracy of the simulation assumptions and thermal modeling. More studies on molecular dynamics simulation should be encouraged.

In summary, chemical environments and ambient conditions play critical roles in AFM scratching by modifying surface chemistry, friction, and material removal mechanisms. Controlled solution chemistry, humidity, and interfacial water layers can either enhance or inhibit scratching efficiency, underscoring the importance of coupled mechanical–chemical interactions in nanoscale material removal.

6.5 Material Modification: Atomic Force Microscopy-Induced Local Anodic Oxidation

AFM local anodic oxidation (AFM-LAO) is one of the most widely used techniques in AFM-based nanofabrication. By applying a bias voltage between a conductive AFM tip (typically the cathode) and a substrate (the anode), a highly localized electric field is generated. Under controlled humidity, capillary condensation produces a nanoscale water meniscus that enables electrochemical oxidation, allowing maskless patterning with nanometer precision. A wide range of substrates, including insulating, semiconducting, and even conductive materials, can be oxidized if a conductive pathway or backing layer is present.

During the 1990s, pioneering work by Allee, Quate, and co-workers demonstrated the feasibility of LAO for fabricating silicon-based nanostructures using AFM rather than scanning tunneling microscopy, providing essential insight into oxide growth behavior at the nanoscale.[72,73] Subsequent studies by Gordon, Teuschler, and others further established the electrochemical foundation of the process.[74,75]

The following subsections review AFM-LAO from fundamental principles to recent progress. After introducing the core mechanisms of LAO, three areas of advancement are discussed: precision enhancement, photoresist-free processing, and material removal by LAO. Applications across diverse device structures are also highlighted.

6.5.1 *Fundamentals of Local Anodic Oxidation*

Oxide dots, lines, planes, and cuboids represent the basic building blocks produced in LAO. This subsection introduces the basic principles that govern their formation and the conditions required to generate these elementary structures. In LAO, localized anodic reactions occur within the nanoscale water meniscus formed at the tip–substrate junction. The intense electric field created by the applied bias drives the migration of oxyanions (OH^-, O^-) and other charged species toward the substrate surface, where oxide growth is initiated. Because the reaction volume is confined by the size and stability of the meniscus, precise control of humidity, tip bias, and tip geometry directly determine the resolution and morphology of the resulting oxide features.

Water Meniscus Formation and Oxide Growth: LAO is typically performed in ambient air or in an environment with controlled relative humidity, generally between 40% and 70%. Under this range, a nanoscale water meniscus forms spontaneously due to van der Waals attraction, condensation energy, surface tension, and electrostatic forces.[76] Within this confined water bridge, ionic migration and electrochemical oxidation take place, and the resulting oxide dimensions scale with the radius of the meniscus. By carefully controlling environmental relative humidity, the applied voltage, and tip–sample spacing, sub-10 nm oxide features can be fabricated on semiconductors, metallic thin films, and even certain organic materials.

Oxide Dots and Aspect Ratio Control: Tello and García investigated oxide dot formation in non-contact mode and showed that this mode yields smaller lateral dimensions and greater heights compared with contact mode for equal pulse durations.[77] The resulting dots displayed significantly higher aspect ratios, and growth proceeded at faster rates. For long pulse durations (>1 s), the difference in dot diameter between the two modes can reach 50%. Martinez et al. later applied non-contact LAO to achieve similar improvements in resolution and growth efficiency.[78]

Oxide Lines: Influence of Bias, Speed, and Scanning Mode: The height and width of oxide lines produced by LAO can be tuned by adjusting processing parameters such as applied bias, scanning speed, and environmental humidity. In general, oxide height increases approximately linearly with negative bias and decreases logarithmically with scanning speed.[79] Numerous theoretical and experimental studies have examined additional influences, including scanning mode, path, speed, overlapping pitch, pulse duration, tip geometry, and tip material.[79,80]

As an example, Fontaine, Dubois, and Stievenard compared contact mode and tapping mode LAO on hydrogen-terminated Si(100):H using a Ti/Pt-coated tip.[81] As shown in Figure 6.14a, contact mode oxidation produced thicker oxides (up to 8 nm) at slow scan speeds, with oxide height decreasing logarithmically as speed increased. In tapping mode (Figure 6.14b), oxide height also decreased logarithmically as speed increased but formed thinner oxides (up to 4 nm) with improved spatial resolution and reduced pattern broadening because the tip was not in continuous contact with the surface. Although tapping mode can generate thicker oxides at higher speeds, this typically occurs at

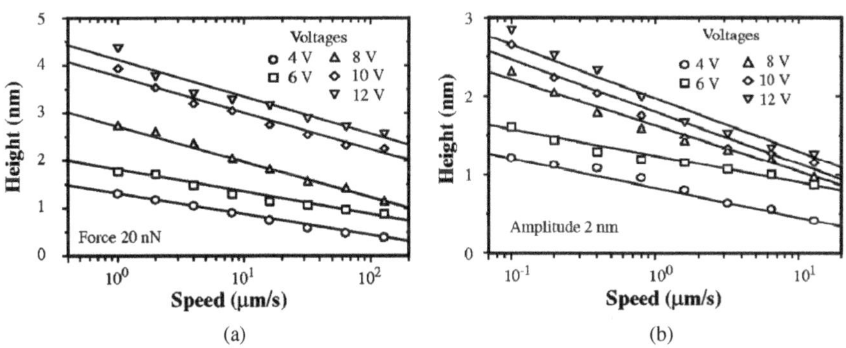

Figure 6.14. Comparison of oxide line heights produced in contact mode and tapping mode AFM-LAO: (a) Oxide height as a function of probe speed for contact mode LAO at applied biases from 4 to 12 V, showing a logarithmic decrease in height with increasing speed and thicker oxides (up to ~8 nm) at low speeds. (b) Corresponding results for tapping mode AFM, yielding thinner oxides (up to ~4 nm) and improved lateral resolution owing to intermittent tip–sample contact. (Courtesy of Fontaine, Dubois, and Stievenard,[81] reprinted from open-access data.)

the expense of reduced resolution due to the intermittent nature of tip–sample interaction.

Environmental Chemical Effects: Seo and Kawamata examined the effect of solution environments on the scratching behavior of anodically oxidized Ta(100).[82] Using a conical diamond tip under a constant normal load of 300 μN, they compared scratching performed in air and in a pH 8.4 borate buffer following anodization at 5.0 V and 50 $\mu A/cm^2$ for 1 h. Scratching in solution produced substantially larger lateral forces than scratching in air, with the force increasing linearly with the logarithm of scratching time, whereas the force remained time-independent in air. Pronounced stick–slip behavior was consistently observed in solution and diminished at higher scratching speeds. AFM imaging revealed protrusions at the ends of grooves formed in solution — features absent in air — which were attributed to the accumulation of anodic oxide at the advancing tip front, contributing to increased friction and lateral force.

Formation of Flat Oxide Planes Through Line Overlap: The controlled assembly of closely spaced oxide lines allows the creation of continuous oxide planes. Tseng et al. used a Si tip at 8 V, 60 nN force, 1 μm/s scan speed, and 55% relative humidity to fabricate LAO line arrays on silicon.[79] They demonstrated that the pitch (p) between adjacent lines determines the uniformity and flatness of the resulting oxide layer. At p = 150 nm ($P = p/w_f = 1.974$), individual lines remain distinct, producing a non-uniform film (Figure 6.15a), where w_f is the full width at half maximum of the oxide lines. Reducing the pitch to 50 nm ($P = 0.658$) resulted in fully overlapped lines and a smooth, continuous oxide plane

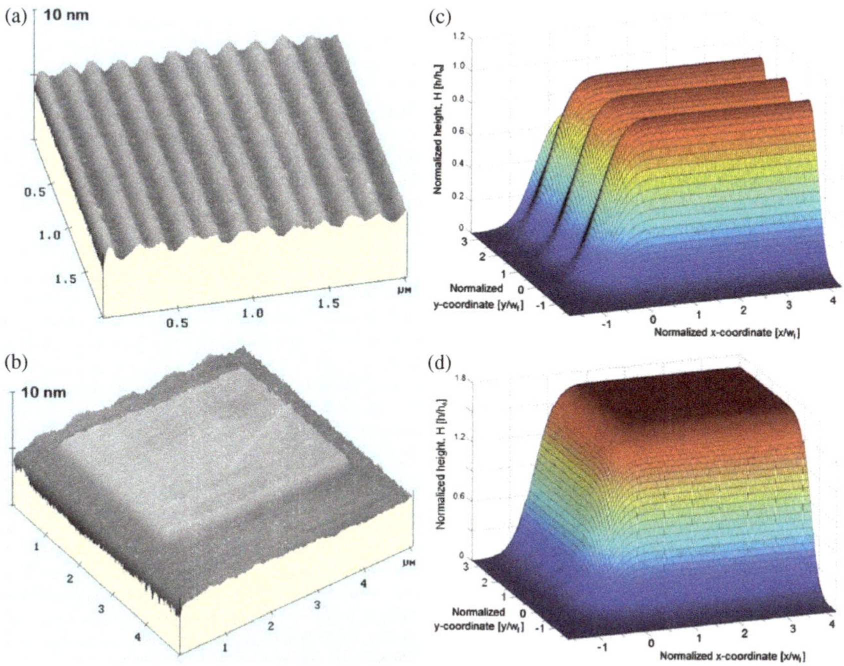

Figure 6.15. Formation of flat oxide planes through controlled line overlap in AFM-LAO: (a) AFM image of oxide lines written at pitch p =150 nm ($P = p/w_f = 1.974$), showing insufficient overlap and a non-uniform oxide layer. (b) AFM image for a reduced pitch of $p = 50$ nm ($P = 0.658$), where extensive overlap produces a continuous, flat oxide plane. (c) Analytical prediction of the normalized line-overlap profile at twice of the threshold value P_t, i.e., $P = 2P_t = 1.274$, indicating persistence of surface unevenness. (d) Predicted profile at the threshold pitch $P_t = 0.637$, showing optimal line merging and uniform layer formation.

(Figure 6.15b). Analytical models (Figure 6.15c, d) showed that a normalized pitch at or below the threshold value $P_t = 0.637$ ensures optimal uniformity, yielding a continuous oxide layer with thickness 1.67 times that of an individual oxide line. This strategy was already applied to construct micro-bowtie structures for nanoscale optical probing.[83]

Oxide Cuboids Nanostructures: With computer-aided design/manufacturing control, Johannes, Cole, and Clark used a Ti-coated Si_3N_4 tip operating in contact mode (5 nN load, 100 nm/s scan speed, 60% relative humidity) to pattern four square oxide cuboids on silicon with cross-sectional areas of ~1 μm² (Figure 6.16).[84] By varying the negative bias from 6 to 9 V, they produced features with heights of 1.5, 2.0, 2.5, and 3.0 nm as shown

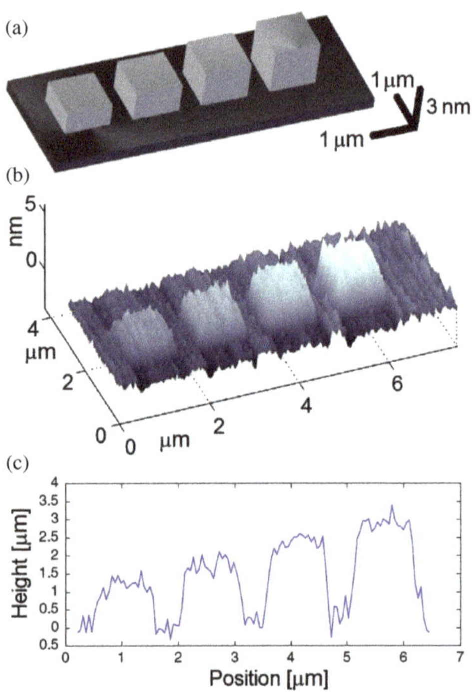

Figure 6.16. Silicon oxide cuboid structures: (a) Computer-aided design model of the desired 3D oxide features (lighter shading). (b) AFM image of the corresponding structures fabricated by AFM-LAO with a Ti-coated Si_3N_4 tip in contact mode. (c) Cross-sectional profiles showing four feature heights (1.5, 2.0, 2.5, and 3.0 nm) obtained by adjusting the applied voltage from 6 to 9 V. Height variations of ~10% arise from local overlapping during oxide accumulation. (Courtesy of Johannes, Cole, and Clark,[84] reprinted from open-access data.)

in Figure 6.16a, b, c, and d, respectively. While the fabricated near-3D nanostructures closely matched their computer-aided designs, the top surfaces exhibited ~10% height variation, likely arising from local overlapping effects during oxide accumulation. This observation suggests that the effective pitch used during rastering may not have been sufficiently small to achieve the optimal overlap threshold discussed earlier in Figure 6.15.[79]

Based on the studies reviewed above, LAO enables controllable formation of nanoscale oxide dots, lines, planes, and cuboids with resolution governed by electric field, humidity, and scanning parameters. These fundamental mechanisms form the basis for advanced developments in precision enhancement, 3D patterning, photoresist-free processing, and material removal discussed in subsequent subsections.

6.5.2 Advancements in High-Precision Local Anodic Oxidation

AFM-LAO enables precise control of nanoscale oxidation processes and is widely used for fabricating high-resolution oxidation patterns for electronic, optoelectronic, and quantum devices. Recent research has focused on pushing the spatial resolution below 10 nm while improving reproducibility, pattern uniformity, and material selectivity. Several representative studies are reviewed below.

Alekseev et al. employed diamond-coated and tungsten carbide-coated conductive AFM tips to perform LAO on epitaxial graphene grown on SiC, producing ~20-nm-wide graphene oxide nanoribbons separated by an unoxidized ~10-nm gap.[85] This narrow unoxidized region formed a graphene nanoconstriction with high geometrical precision. The oxidized and unoxidized regions were distinguished by friction force mapping — graphene oxide exhibits higher friction than pristine graphene. The nanoconstriction width and profile were governed by carefully optimized operating parameters, including relative humidity (~45%), applied voltage (–6 V), probe speed (0.2 μm/s), exposure time (10 ms per point), and a tungsten carbide tip radius of ~20–30 nm. This approach enabled the fabrication of ~10-nm graphene features (resolution) without resist, offering a clean and deterministic method for nanoscale patterning.

Borodin et al. used W_2C-coated Si probes to oxidize $MoSe_2$ flakes deposited on Au/Si substrates, forming ~2-nm-thick oxide layers.[86] They identified two oxidation regimes controlled by relative humidity: isotropic oxidation at 60–65% relative humidity and anisotropic oxidation at 40–50% relative humidity. Oxygen atoms bonded with Mo and Se to form Mo oxide and Se oxide species, allowing precise modulation of oxidation pathways. Under optimal conditions (relative humidity ~60%, 15 V bias, 0.1 ms exposure), spatial resolutions of ~10 nm with ~2 nm average error were achieved. Their results confirm that AFM-LAO can finely tune the surface chemistry and morphology of transition metal dichalcogenides, enabling 10-nm-scale device patterning.

Komonov et al. used tungsten carbide-coated probes to create oxidized nanohills (nanodots) on amorphous VO_x films.[87] The nanohills, composed of V_2O_5, were water-soluble; dissolving them revealed nanoholes with depths below 0.3 nm and lateral dimensions under 50 nm. The oxidation kinetics followed a logarithmic time dependence consistent with the Cabrera–Mott model, with a critical oxide thickness of 0.45 nm, a growth rate of 0.64 nm/ms, and an activation energy of 1.9 eV. The threshold oxidation voltage was ~5 V at 15% relative humidity. By adjusting voltage, humidity, and film thickness, nanostructure geometry could be controlled with sub-nanometer vertical precision (0.3–0.45 nm). Subsequent annealing enabled the transformation of these nanoholes into ordered VO_2 nanocrystal arrays suitable for photonic and nanoelectronic applications.

Using AFM datacube mode, Huang et al. systematically investigated the influence of mechanical load on oxide morphology and current-voltage characteristics during LAO.[88] They showed that adhesive forces enhance vertical oxide growth while minimizing lateral spreading. This control enabled oxide bumps with heights as small as 0.37 nm and standard deviations below 0.1 nm. Their results demonstrate that mechanical contributions, often overlooked compared to voltage and humidity, play a critical role in achieving near-atomic-level control over oxide dimensions.

Based on the four studies reviewed above, AFM-LAO has demonstrated remarkable progress in achieving high-resolution, chemically selective, and geometrically precise oxide nanostructures. Optimization of tip coatings, humidity, voltage, and mechanical loading has enabled reliable fabrication of sub-10 nm features and even sub-nanometer vertical control. These advancements further establish AFM-LAO as a clean, resist-free, and highly versatile nanofabrication technique for next-generation electronic, photonic, and quantum device applications.

6.5.3 Advancements in Photoresist-Free Process

Unlike traditional lithography methods, AFM-LAO eliminates the need for photoresist, avoiding issues like residual contamination and simplifying the fabrication process. The AFM-LAO is a low-cost technology that has no photoresist residue and can oxidize materials to be used for masks to slow down or stop the subsequent etching processes. The advancements in making LAO masks for replacing the conventional photomasks using photoresists are reviewed in this subsection.

Chen et al. applied a conductive tip for making grayscale Si oxide masks by controlling the variation of LAO thickness or height on a Si(100) substrate for the subsequent reactive ion etching for making Si microlenses.[89] The oxidation voltage, typically ranging from 3 V to 10 V, directly influences the oxide height, with higher voltages producing thicker layers. A preprogrammed voltage matrix (512 × 512 in size) in the AFM controller allowed precise spatial control of oxide thickness. The scan speed of 4 μm/s and ambient humidity of 60% were maintained during patterning. Figure 6.17 showcases the fabrication of Si microlenses using AFM-based grayscale oxidation and reactive ion etching. Figure 6.17a displays convex microlenses with diameters of 2 μm, 3 μm, and 4 μm, with focal lengths of 4 μm, 9 μm, and 16 μm, respectively. Figure 6.17b shows a concave microlens (4.5 μm diameter). Figure 6.17c presents a convex microlens (4.5 μm diameter) for comparison. Figure 6.17d illustrates AFM profiles of microlenses before and after reactive ion etching, highlighting smooth surfaces with roughness below a few nanometers, demonstrating the precision of this fabrication method. A concave microlens is challenging to create with conventional methods but feasible using AFM oxidation and reactive ion etching, since the pixel size and pitch could be controlled within the order of tens of nanometers with desired grayscale levels.

Martinez et al. utilized a n+-type Si tip operated in non-contact mode to perform LAO on the Si surface of a silicon-on-insulator

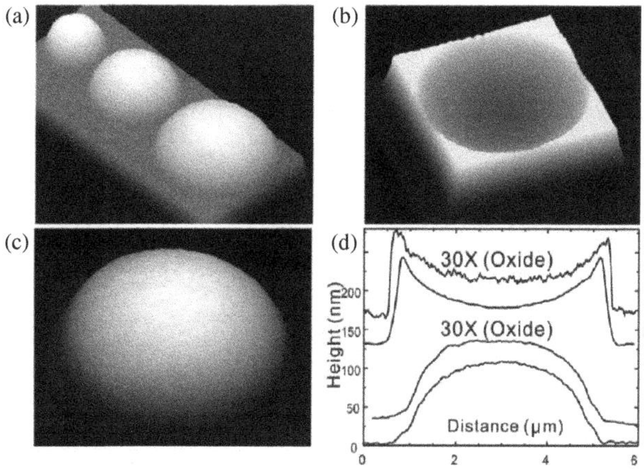

Figure 6.17. AFM images of Si microlenses fabricated by AFM grayscale oxidation with reactive ion dry etching: (a) Microlenses with diameters of 2, 3, and 4 μm, (b) Concave microlens with a diameter of 4.5 μm, (c) Convex microlens with a diameter of 4.5 μm, (d) AFM-measured profiles of microlenses shown in (b) and (c) before and after reactive ion etching. (Courtesy of Chen et al.,[89] reprinted from open-access data.)

substrate.[78] The process involved applying voltage pulses of 36 V with a pulse duration of 100 μs to pattern two perpendicular silicon nanowires (SiNWs) for field-effect transistors (FETs). The LAO pattern served as an etching mask, while the unmasked silicon was selectively removed using wet etching with a 20% KOH solution at 50 °C. Finally, the oxide mask was removed using hydrofluoric acid to complete the fabrication process. The AFM image of the cross-oxide pattern for a four-arm SiNW-FET is depicted in Figure 6.18a. The cross-oxide lines were obtained by controlling the size of the individual oxide dots to be larger than the lateral distance between pulses. Figure 6.18b shows an AFM image of two perpendicular SiNWs (A-C and B-D) obtained after KOH etching, while Figure 6.18c shows an optical image

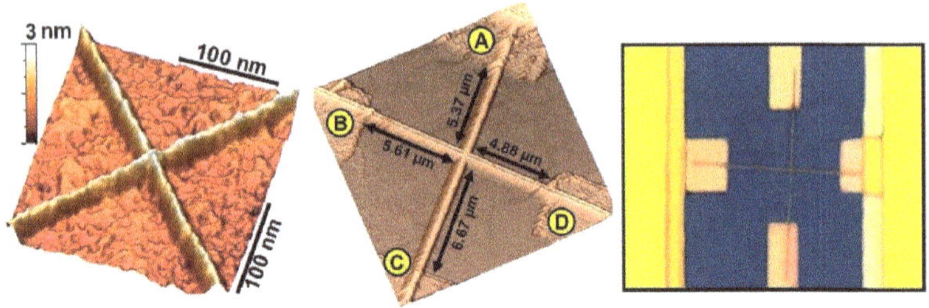

(a) (b) (c)

Figure 6.18. Images of a multiple-level SiNW transistor, which was formed by two perpendicular SiNWs: (a) AFM image of oxide mask fabricated by AFM-LAO, (b) AFM image of two perpendicular SiNWs, marked as A-C and B-D, obtained after etching with KOH, and (c) optical image of final device. (Courtesy of Martinez et al.,[78] reprinted from open-access data.)

of the final circuit. As indicated by the authors, the SiNW could be made as narrow as 4 nm wide.[78]

Alias et al. utilized a Cr/Pt-coated conductive AFM tip with a bias voltage of 9 V and a writing speed of 0.3 µm/s in contact mode to perform LAO for patterning a wire-shaped SiNW array on a p-type silicon-on-insulator substrate with a (100) crystallographic orientation.[90] The resulting silicon dioxide array pattern served as a mask for a selective etching process using a mixture of tetramethylammonium hydroxide, isopropyl alcohol, and Triton X-100. As shown in a field emission SEM (FESEM) image of Figure 6.19a, the lateral dimensions of the oxide nanowires in the SiNW array (before etching) are approximately 300–400 nm while the lateral dimensions of the Si nanowires (after first etching) are 412 nm as shown in Figure 6.19b, where the lateral dimensions after etching being larger than those before etching is due to slight isotropic etching beyond the oxide mask boundary. The SiNW array underwent

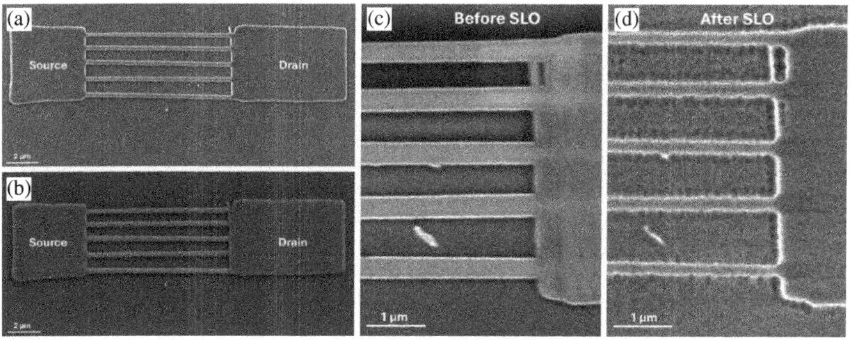

Figure 6.19. Morphological characterization of the fabricated SiNW array: (a) FESEM image of SiNW structure patterned via AFM-LAO before etching, (b) FESEM image after etching, (c) Magnified FESEM image of nanowires prior to SLO treatment, and (d) Corresponding image after three SLO cycles at 1000 °C, indicating dimensional reduction. (Courtesy of Alias et al.,[90] reprinted from open-access data.)

further refinement through a self-limiting oxidation (SLO) process at 1,000 °C, where a silicon dioxide layer (through thermal oxidation) formed on the nanowire surface, consuming approximately 46% of the original silicon while 54% forms an external oxide layer. This oxide was removed using hydrofluoric acid, leading to significant dimensional reduction. At 1000 °C and after three SLO cycles, vertical height dropped from 118 nm to 37 nm (68.6% reduction), while lateral dimensions decreased from 412 nm to 220 nm (46.6% reduction), as shown in the FESEM images before (Figure 6.19c) and after (Figure 6.19d). The present process (AFM-LAO + SLO) not only effectively reduced the lateral and vertical dimensions of SiNW, but also enhanced uniformity and surface quality, making it highly suitable for SiNW-FETs.

6.5.4 *Advancements in Local Anodic Oxidation for Controlled Material Removal*

AFM-LAO has emerged as a powerful approach for nanoscale material removal via electrically induced oxidation and gasification, enabling high-resolution patterning for nanodevice fabrication. This subsection highlights three representative studies that demonstrate recent advances in precision, material scope, and achievable feature size.

Cohen et al. employed a Pt/Ir-coated AFM probe driven by an 18 V peak-to-peak excitation at 150 kHz to oxidize exfoliated ~3 nm-thick graphite flakes on SiO_2/Si substrates.[91] The high-frequency bias promoted the conversion of graphite into gaseous and amorphous oxidation byproducts, likely carbon oxides, enabling clean sub-nanoscale material removal. Using this approach, they fabricated sub-100 nm device features, including X-shaped quantum point contact gates, hole arrays for quantum dot superlattices, and edge-state interferometers for Fabry-Pérot cavities. The minimum linewidths achieved were ~60–70 nm, increasing to ~100 nm after ~150 μm of cutting due to tip wear. Owing to the weak van der Waals bonding between graphene layers, AFM-LAO yielded smooth, residue-free interfaces essential for high-quality mesoscopic device performance.

Tang et al. investigated AFM-LAO on hydrogen-terminated diamond surfaces using a platinum-coated silicon tip under high DC bias voltages (≤–5 V).[92] The anodic oxidation replaced hydrogen termination with oxygen, thereby reducing surface conductivity and increasing local thermal resistance. The resulting heat accumulation weakened the interfacial bonding between the oxidized layer and the underlying diamond, promoting graphitization or amorphization and facilitating material detachment under the mechanical load of the AFM tip. An approximately constant removal depth of ~0.5 nm was observed across a range of bias voltages, accompanied by a pronounced reduction in surface hardness. These effects significantly enhance diamond machinability and enable controlled layer-by-layer removal for nanoscale fabrication and surface planarization.

Wu et al. demonstrated high-resolution electrochemical etching of graphene using a platinum-coated AFM tip driven by a high-frequency AC bias to induce a nanoscale water meniscus.[93] The confined electrochemical reactions oxidized and removed carbon atoms from the graphene lattice. By optimizing the AC voltage amplitude, tip velocity, and applied normal force, trenches as narrow as ~15 nm were achieved, substantially surpassing the previously reported limits of 60–100 nm. This residue-free process enables the fabrication of graphene nanostructures suitable for quantum devices and electronic band structure engineering. Notably, the effective removal depth reached approximately 0.5 nm, exceeding the nominal tip size and highlighting the non-local nature of LAO-driven electrochemical material removal.

Overall, recent advancements in AFM-based LAO have enabled highly controlled, residue-free material removal across layered carbon and wide bandgap materials. By coupling electrical bias, interfacial chemistry, and mechanical action, AFM-LAO provides a versatile and scalable pathway for fabricating high-resolution nanostructures essential to next-generation electronic and quantum devices.

6.6 Atomic Force Microscopy-based Material Addition: Fundamentals

This section outlines the fundamental mechanisms that govern AFM-assisted additive nanofabrication. In these approaches, the AFM tip serves as a nanoscale delivery tool—functioning as a dip pen, fountain pen, or electrochemical activator—transporting material to the substrate through pathways such as meniscus-mediated diffusion, thermally induced melting and flow, force-driven de-passivation for electrochemical growth, or hollow-tip polymer-nanoparticle dispensing. Ink transfer is dictated by coupled thermal, chemical, mechanical, and environmental factors, including water-meniscus formation and stability which, as shown by recent theoretical analysis, play a central role in controlling molecular transport efficiency and pattern resolution.

6.6.1 *Atomic Force Microscopy Tip-Activated Chemical or Electrochemical Deposition*

One of the earliest demonstrations of AFM-assisted additive nanofabrication using the tip as an electrochemical activator was reported by LaGraff and Gewirth.[94] Using a conductive AFM tip in a dilute $Cu(ClO_4)_2$ electrolyte (10^{-4} M), they achieved highly localized epitaxial Cu deposition on a Cu(110) substrate by mechanically depassivating an otherwise inert surface. A modest tip bias of −70 mV combined with an increased normal force of ~25 nN exposed reactive sites beneath surface passivation layers, enabling Cu nucleation exclusively within tip-defined regions, while the surrounding passivated surface remained electrochemically inactive. In this configuration, the Cu substrate served as the working electrode, providing precise spatial control over metal growth.

Later, Obermair et al. further studied this using AFM tip for de-passivation and activation as "mechano-electrochemical tip".[95] They made this selective electrochemical deposition fully reversible for writing, reading, deleting, and rewriting copper nanostructures on a polycrystalline gold surface. Figure 6.20a shows an AFM image of the digit "6" pattern formed in the writing step, in which a copper structure of the digit "6" was selectively deposited on a gold surface by scanning the AFM tip along the desired path while applying a deposition potential of −60 mV relative to a Cu/Cu^{2+} reference electrode for 5 seconds. Figure 6.20b shows the AFM image that the copper structure was completely dissolved by applying a dissolution potential of +0.2 V for 180 seconds, leaving no surface defects. In Figure 6.20c, a new copper structure in the shape of the digit "9" was successfully rewritten at the same location. The copper nanopatterns shown in Figure 6.20a and 6.20c have a typical vertical dimension

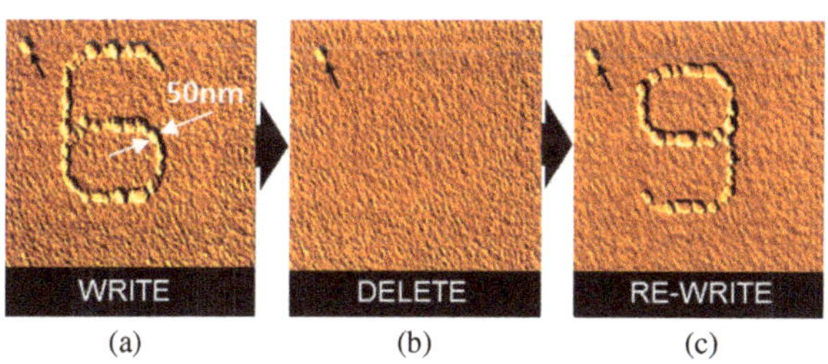

Figure 6.20. Reversible writing, deleting, and rewriting of copper nanostructures using an AFM tip: (a) A "6"-shaped copper structure was deposited on a gold surface with a −60 mV potential; (b) The structure was dissolved using a +0.2 V potential; (c) A "9"-shaped structure was rewritten at the same location without surface defects. The height and line width of the "6" and "9" patterns are 15 nm and 50 nm, respectively, with a scan area imaged by an AFM being 1.6 μm × 1.6 μm. (Courtesy of Obermair et al.,[95] reprinted from open-access data.)

(height) of 15 nm and a lateral dimension (line width) of 50 nm. The scan area shown is 1.6 μm × 1.6 μm, representing the dimensions imaged by the AFM.

By directly activating the deposition (eliminating the first material removing step), Tanabe et al. employed an Rh-coated Si AFM tip with a curvature radius of ~10 nm to electrochemically deposit Ag and Au nanoparticles onto conductive Nb-doped TiO_2 substrates.[96] Metal ions from aqueous $AgNO_3$ or $H[AuCl_4]$ solutions were first adsorbed onto the substrate surface by immersion. The thin adsorbed water layer acted as an electrolyte, enabling localized electrochemical deposition when a bias voltage was applied between the tip and substrate. By applying −6.0 V for 0.5 s, silver nanorods with heights of ~30 nm were deposited, while nanoparticle arrays exhibited lateral diameters of ~70 ± 30 nm and heights of ~20 ± 10 nm with ~5 nm spacing. Gold nanoparticles showed slightly smaller lateral dimensions (~50 ± 30 nm) and heights of ~30 ± 10 nm with uniform spacing. Reliable deposition required a relative humidity of at least 90%, as lower humidity limited ion transport and reduced growth. The nanoparticle size increased systematically with more negative bias voltages and longer deposition times. Compared with contact mode operation, tapping mode AFM, using a vibration amplitude of ~70 nm at ~120 Hz, provided improved control over particle size, shape, and spacing.

Similarly, García et al. demonstrated AFM tip-activated electrochemical deposition driven by extremely high electric fields.[97] In their experiments, a conductive AFM tip was positioned 2–5 nm above a silicon substrate, and voltage pulses of 10–40 V were applied, generating electric fields on the order of

1–30 V/nm. These intense local fields polarized CO_2 molecules and lowered the activation energy for molecular dissociation, leading to the deposition of solid carbonaceous material on the substrate. As shown in Figure 6.21a, carbonaceous dots deposited at a constant voltage of 21 V exhibited heights monotonically increasing from ~1.5 nm to ~11 nm as pulse duration increased from 0.1 ms to 2 s. Results also show that the dot size increased with the applied voltage (Figure 6.21b). The dot height increased linearly from ~2 to ~9.5 nm as the voltage changes from 20 to 36 V pulses with a pulse duration of 0.1 ms. These results demonstrate that using AFM to create carbonaceous nanopatterns by field-induced splitting of CO_2 is effective in producing sub-25 nm structures and enables precise control over dot size.

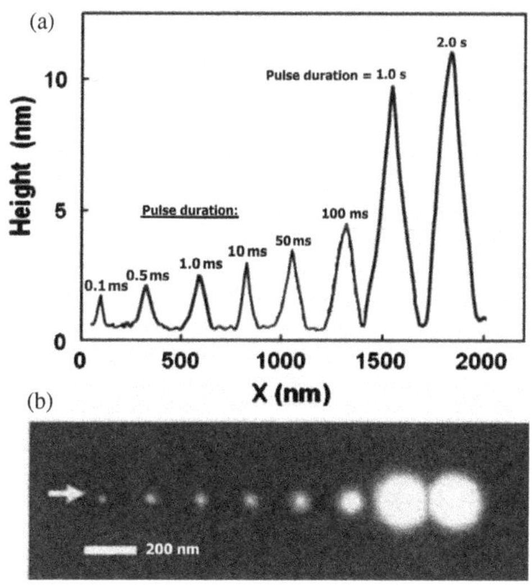

Figure 6.21. Sequence of carbonaceous dots deposited by applying a constant voltage of 21 V for pulse durations from left to right at 0.1, 0.5, 1, 10, 50, and 100 ms, 1 s and 2 s: (a) AFM profile measurements, and (b) AFM dot images. (Courtesy of Garcia et al.,[97] reprinted from open-access data.)

Beyond electrochemical activation, AFM tips can also initiate tribochemical reactions for additive nanopatterning. Khare et al. introduced a nanotribological printing approach using a diamond-like carbon-coated Si tip with a radius of ~7 nm to deposit zinc dialkyl dithiophosphate (ZDDP) molecules onto polished silicon or stainless steel substrates.[98] Experiments were conducted in a fluid chamber filled with polyalphaolefin oil containing dissolved ZDDP molecules. Under applied normal loads, the sharp 7-nm tip generated high normal and shear stresses at the contact interface, kinetically activating tribochemical reactions by lowering the energy barrier for bond breaking and formation. This process transformed ZDDP molecules into solid, surface-bound films. Single-line patterns were produced by scanning parallel (0°) or perpendicular (90°) to the cantilever's longitudinal axis. The pattern height increased nearly linearly from 1.8 to 8 nm as the applied load rose from 376 to 1,004 nN, while the full width at half maximum remained largely independent of load. A scan speed of 11 μm/s and 200 reciprocation cycles enabled controlled pattern growth. The resulting films exhibited robust mechanical properties, with hardness and elastic modulus comparable to macroscale ZDDP-derived coatings and surface roughness below 1 nm.

Collectively, the above studies demonstrate that an AFM tip can function as a localized chemical or electrochemical activator, enabling site-selective deposition through de-passivation, high-field electrochemistry, or tribochemical reactions. Precise control of tip force, electrical bias, and environment allows deterministic growth of metals and molecular films, establishing a fundamental basis for AFM-based additive nanofabrication.

6.6.2 Atomic Force Microscopy-Based Dip-Tip Deposition

In their pioneering work on using dip-pen for depositing live DNA, Demers et al. utilized a surface-modified silicon nitride AFM tip for direct patterning of DNA (hexanethiol-modified oligonucleotides) inks on gold (Au).[99] The AFM tip was first coated with 3-aminopropyltrimethoxysilane to enhance adhesion of the DNA ink to the tip surface. Dot and line DNA patterns were formed, with heights of 2 to 5 nm and lateral dimensions from 50 nm to several μs. Note that the hexanethiol groups on the oligonucleotides formed covalent bonds with gold due to their strong chemical affinity, while a nanoscale water meniscus facilitated ink transfer. Figure 6.22a shows an AFM image of the patterned thiol-modified DNA dots on the Au surface, with varying contact times at 45% relative humidity, while the plot below illustrates the linear relationship between dot diameter and the square root of contact time. Figure 6.22b depicts the DNA dots on the Au substrate with a fixed contact time of 10 seconds at varying relative humidities, where the plot below shows how dot diameter increases with higher relative humidity. These figures highlight the role of the contact times and humidity in precise control over feature size during deposition.

Demers et al. also patterned the second DNA type, 5′-acrylamide-modified oligonucleotides, on silicon oxide (SiO_x), producing DNA spots with diameters as small as 50 nm.[99] Arrays with 100,000 spots were created within a 100 μm × 100 μm area. The SiO_x surface was pre-treated with 3-aminopropyltrimethoxysilane to introduce thiol groups, which reacted with acrylamide groups via Michael addition under controlled humidity, forming

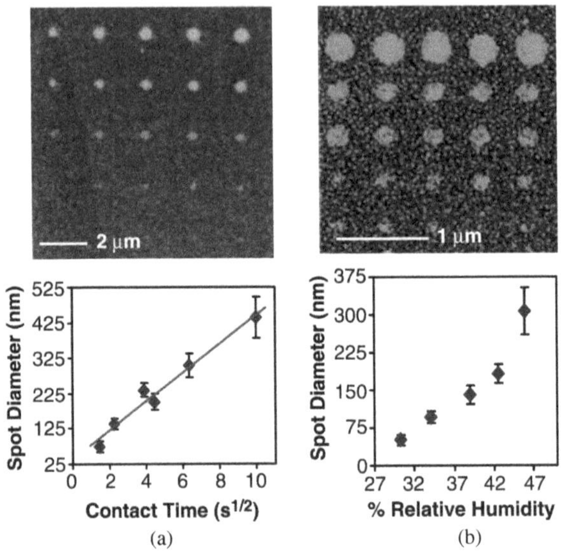

Figure 6.22. AFM tip patterning DNA dots on Au: (a) AFM image of dots on Au for various contact times at 45% humidity (top) and plot of dot diameter versus square root of contact time (bottom). (b) AFM image of DNA dots on Au with 10 s contact time per dot at varying humidity levels (top) and plot of dot diameter versus humidity (bottom). The error bars were calculated from the standard deviation of at least five dot measurements. (Courtesy of Demers et al.,[99] reprinted from open-access data.)

covalent bonds that securely anchor the oligonucleotides to the substrate. Importantly, the patterned oligonucleotides on the SiO_x surface were verified as remaining biologically active, showing sequence-specific binding to complementary fluorophore-labeled DNA.

Noy et al. utilized AFM tips to contact inked paper membranes, applying four different inks (one at a time) to successfully create distinct nanopatterns on glass and gold surfaces.[100] The nanopatterns created using these inks include: 1) *luminescent nanopatterns* on glass with rhodamine 6G dye, 2) *protein patterns* on glass with human chorionic gonadotropin antibody tagged with tetramethylrhodamine dye, 3) *polymer nanowires* on glass with MEH-PPV polymer ink, and 4) *alkanethiol lines* on gold substrates with mercaptohexadecanoic acid (MHA) ink. These patterns were unique for their ability to achieve single-molecule deposition with precise control over size, shape, and position. Optical detection enabled visualization of nanoscale features beyond the resolution of traditional AFM imaging. The method minimized non-specific adsorption in protein patterning, creating high-density, well-defined arrays. It also demonstrated versatility by fabricating luminescent patterns, protein arrays, and polymer nanowires, paving the way for advancements in biochips, molecular studies, and optoelectronic devices.

In their work on creating self-assembled nanodots and nanolines, Chung et al. employed a heated, MHA-coated AFM tip operating as a *dip pen* to pattern MHA to form self-assembled monolayers on gold surfaces under 46% relative humidity.[101] For dot fabrication, the tip–surface contact time was fixed at 3 s, and increasing the tip temperature from 25 °C to 42 °C enlarged the dot diameter from ~250 nm to ~1900 nm (Figure 6.23a-b). For line writing, the MHA-coated tip was heated between 25 °C and 50 °C, below the MHA melting point of 65–69 °C, and scanned at 0.3 μm/s under 36–46% relative humidity, producing 0.9-μm long lines with a 3-s contact duration. As shown in Figure 6.23b, line width increased systematically with tip temperature at this fixed writing speed. Both dot radius and line width exhibited a clear logarithmic dependence on tip temperature (inset in Figure 6.23b) and conformed to an Arrhenius relationship corresponding to an activation energy of 165 kJ/mol. This study clearly highlights the dominant role of tip temperature in controlling

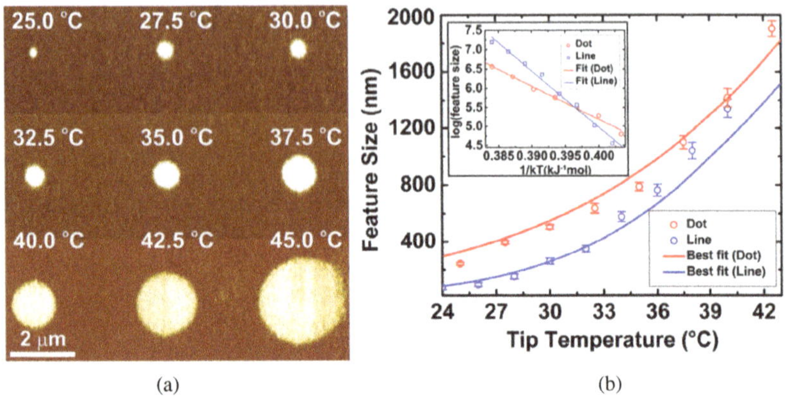

Figure 6.23. Temperature-dependent feature sizes of MHA dots and lines produced by AFM-assisted thermal deposition: (a) AFM images of MHA dot patterns generated on a gold substrate at various tip temperatures, using a 3-s contact time and 46% relative humidity. (b) Measured diameters of MHA dots (red circles) and widths of MHA lines (blue circles), including error bars, plotted as functions of tip temperature. The inset shows the logarithmic dependence of dot radius (red circles) and line width (blue squares) on the cantilever temperature. (Courtesy of Chung et al.,[101] reprinted from open-access data.)

feature size during thermally assisted AFM dip-pen deposition of MHA nanostructures.

Sheehan et al. used a Si tip coated with two monolayers of octadecylphosphonic acid to deposit 500-nm square plateaus on smooth mica.[102] The plateaus were arranged in a grid with 1-μm spacing. Each square was raster-scanned four times at 2 Hz with 128 lines per scan, giving a total writing time of 256 s per feature. The tip was heated in short pulses up to ~700 °C, with heating durations of 1–20 s and cooling periods of 1–50 s, temperatures far exceeding octadecylphosphonic acid's melting point (99 °C). At a tip temperature of 98 °C, only light deposition occurred, producing an approximate 1.1-nm film (less than half a monolayer), whereas a temperature of 122 °C yielded robust transfer, forming a full monolayer with a height of 2.5 nm.

A notable distinction between the two studies lies in the required tip temperature relative to the ink's melting point. In Chung et al.,[101] effective MHA deposition occurred with tip temperatures (25–50 °C) well below the MHA melting point, whereas Sheehan et al. required heating the octadecylphosphonic acid-coated tip above,[102] and in some cases far above, its melting temperature to enable material transfer. This discrepancy likely reflects fundamental differences in both ink transport mechanisms and material properties. In the MHA–gold system, deposition is facilitated by water meniscus-mediated diffusion and strong thiol–gold chemisorption, which promote molecular transfer without the need for melting. In contrast, octadecylphosphonic acid on mica forms a more crystalline, less mobile coating that must be thermally softened or melted to permit viscous flow and effective wetting of the substrate. Additional contributing factors may include differences in ink viscosity, coating morphology, substrate thermal conductivity, and the extreme pulsed heating used by Sheehan et al.,[102] which can suppress meniscus formation and shift the dominant transfer pathway from diffusion to molten-fluid flow. Further systematic studies are warranted to clarify how ink molecular structure, thermal transport, and environmental conditions determine whether AFM-mediated additive deposition requires heating above or below the ink's melting point.

Recently, Chen et al. applied theoretical and computational methods to investigate the fundamental mechanisms of AFM-assisted deposition using a molecular ink-coated AFM tip to transfer material onto substrates such as mica, glass, and gold through diffusion or liquid flow.[103] Their analysis focused on how operating parameters influence the formation and behavior of the condensed water meniscus. They found that high relative humidity, low ambient temperature, small surface contact angle, large tip radius, and a small tip–substrate gap all enlarge the meniscus volume, improving molecular transport efficiency of these ink molecules studied, but compromising pattern resolution. Relative humidity values above 80% were shown to induce meniscus instability, whereas lower ambient temperature and smaller surface contact angle promoted stable, larger menisci conducive to effective deposition. Also, larger tip–substrate gap reduces meniscus size, and beyond a certain gap, menisci fail to condense, which is beneficial for AFM non-contact mode stability. Overall, the study provides theoretical insight into optimizing meniscus characteristics to balance deposition efficiency, pattern fidelity, and operational stability.

6.6.3 Atomic Force Microscopy-Based Hollow Tip Deposition

AFM-based hollow tip deposition employs contained liquid inks to generate a wide range of ink-based nanopatterns with high spatial precision. Figure 6.24 schematically illustrates this process using an AFM hollow tip functioning as a nanoscale fountain pen. As shown in Figure 6.24a, liquid ink is stored in an on-chip reservoir and transported through enclosed microchannels to annularly apertured writing tips via capillary action; the inset presents an SEM image of an apertured tip with a scale bar of 2 μm, also known as volcano hollow tips. Figure 6.24b depicts direct-write nanopatterning, where the tip contacts the substrate

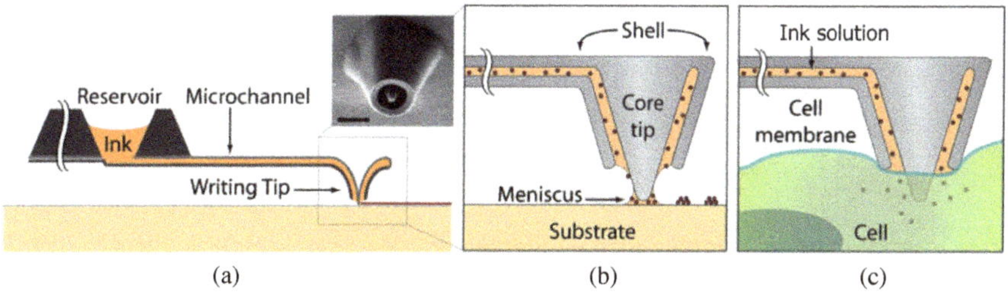

Figure 6.24. Schematic illustration of AFM-based hollow-tip deposition: (a) liquid ink stored in an on-chip reservoir is transported through microchannels to apertured tips by capillary action (inset: SEM image of an apertured tip; scale bar: 2 μm); (b) direct-write nanopatterning via meniscus-mediated ink transfer when the tip contacts the substrate; (c) cellular injection or deposition, where ink is delivered into the pre-coated cell membrane under controlled force. (Courtesy of Loh et al.,[104] reprinted from open-access data.)

and an ink meniscus is formed to enable controlled material transfer. Figure 6.24c shows cellular deposition, in which the hollow tip penetrates the pre-coated cell membrane under a prescribed insertion force. Together, these schematics highlight the dual functionality of hollow AFM tips for both nanoscale patterning and single cell injection, underscoring their versatility in nanomanufacturing and biological applications. The fundamental principles underlying these capabilities are reviewed in this subsection.

For direct-write nanopatterning, Loh et al. utilized an AFM-based hollow tip filled with functionalized nanodiamonds suspended in liquid ink.[104] Upon contact with a glass coverslip, an ink meniscus formed between the tip and substrate (Figure 6.24a, b). By precisely controlling the dwell time, defined amounts of drug-coated nanodiamond ink were dispensed, yielding highly accurate arrays of nanodiamond dots. The ink was prepared by physisorbing the chemotherapeutic drug doxorubicin hydrochloride onto nanodiamonds. The resulting circular dot features were densely packed, with diameters ranging from 93 nm to 351 nm and heights of approximately 5–20 nm, both strongly dependent on dwell time. Importantly, TUNEL assay results confirmed that the patterned doxorubicin hydrochloride retained its biological activity, demonstrating the suitability of this approach for spatially resolved drug delivery with minimal collateral damage.

The same study also demonstrated direct cellular injection using hollow AFM tips (Figure 6.24c).[104] Functionalized nanodiamonds suspended in water or dimethyl sulfoxide were delivered into cells by penetrating the preloaded membrane under controlled force, thereby minimizing mechanical damage. Experiments conducted on MCF-7 human breast adenocarcinoma cells and RKO human colorectal carcinoma cells showed localized fluorescence following injection, indicating successful intracellular delivery while maintaining cell viability. These results highlight the potential of hollow tip AFM for targeted single-cell drug delivery and mechanistic biological studies.

In polymer deposition, Meister et al. modified a commercial silicon nitride AFM probe by fabricating a ~200 nm aperture at the tip apex using focused ion beam milling.[105] This hollow probe enabled the deposition of low-volatility inks, such as glycerol, onto polished silicon substrates. Ink filled the hollow tip and aperture via capillary action and was dispensed upon tip–surface contact. Glycerol inks containing 20 nm polystyrene nanoparticles were deposited as arrays of nanodroplets on hydrophobic substrates. After solvent evaporation, the resulting features consisted of non-volatile nanoparticle residues. The smallest dots measured approximately 20 nm in diameter, with spacings of 410 nm on hydrophobic surfaces and 800 nm on hydrophilic substrates.

For polymer–nanoparticle composite deposition, Lee et al. employed a heated hollow AFM probe to dispense molten composite inks.[106] Various substrates, including gold, mica, and silicon oxide, were patterned using polymer matrices such as PMMA (with Fe_3O_4 magnetic or Au nanoparticles), P(VDF-TrFE) (with metal–organic molecules), polyethylene (with CdSe–ZnS core–shell quantum dots), and PDDT (with Au nanoparticles). The deposited nanocomposites could be used directly, preserving the polymer–nanoparticle architecture, or selectively processed by oxygen plasma to remove the polymer phase, yielding either continuous nanoparticle lines or discrete arrays depending on particle size and surface chemistry. PMMA–Fe_3O_4 composites produced 400 nm wide lines that shrank to ~10 nm after plasma etching, forming rows of single nanoparticles. PDDT–Au patterns yielded 250 nm wide lines after etching. Hexamethyldisilazane-functionalized Fe_3O_4 nanoparticles formed rows that narrowed from 439 nm to 187 nm, while untreated Fe_3O_4 nanoparticles focused into ~10 nm rows. The smallest achieved line widths were ~10 nm for nanoparticle rows and 78 nm for nanoparticle–polymer composites.

Hollow tip AFM has also been applied to metal nanowire fabrication. Fan et al. developed a nanotube-based hollow tip using a single copper-filled carbon nanotube produced by thermal chemical vapor deposition and connected to an ink reservoir.[107] Continuous Cu nanowires were deposited onto conductive Au-coated substrates under an applied tip–substrate bias. The nanotube aperture diameter ranged from 11 to 19 nm, which yielded the deposited nanowires with diameters as small as 11 nm. While the initially filled Cu mass was estimated at 1.9 fg, the deposited mass reached up to 29.0 fg due to continuous replenishment from the reservoir. Complex 3D patterns, including alphanumeric characters, were fabricated by controlled tip motion, with feature dimensions governed by the applied electrical bias. Larger-diameter nanotube tips required higher electrical energy to overcome transport barriers and initiate metal flow.

Finally, electrode-based hollow tip deposition was demonstrated by Hu and Yu, who developed a micropipette functioning as a hollow electrode for localized electrodeposition of metals.[108] Using a 0.05 M $CuSO_4$ aqueous solution for copper and chloroplatinic acid for platinum, metal structures were deposited by applying an electrical potential between the electrolyte-filled micropipette (cathode) and the substrate (anode). Straight vertical metal wires with diameters as small as 100 nm and lengths up to 80 μm were achieved at an applied potential of 0.2 V relative to an Au-coated substrate, demonstrating the feasibility of hollow tip AFM for high-aspect-ratio metallic micro- and nanostructures.

AFM-based hollow tip deposition enables direct, volumetric delivery of diverse inks, including polymers, nanoparticles, metals, and biomolecules, extending AFM material addition beyond diffusion-limited dip-pen approaches. While highly versatile, these methods remain predominantly serial and single tip in nature. To address scalability, throughput, and patterning complexity, the next section explores material addition strategies based on multiple steps, repeated ink loading, and multi-tip systems.

6.7 Material Addition by External Loadings, Multiple Tips, and Multiple Steps

In material addition, an AFM tip can function as an electrochemical activator, a dip pen, or a hollow pen, to deliver materials onto a substrate, either directly or with assistance from thermal, mechanical, chemical, or electrical means. These additive processes enable direct and indirect nanopatterning of metals, polymers, DNA, proteins, peptides, lipids, viruses, and bacteria on various substrates for diverse applications. This section highlights representative additive nanoscale processes, focusing on depositions using multiple external sources and multiple tips, as well as multi-step approaches for nanodevice fabrication.

6.7.1 *Tip-Induced Material Deposition by Additional Electrochemical Contributions*

AFM-induced material addition can be enhanced not only by thermal or mechanical effects but also by chemical, biological, magnetic, and electrical contributions. These additional mechanisms significantly broaden the range of depositable materials and the achievable functionalities of nanoscale structures. This subsection reviews techniques utilizing additional electrochemical loadings, while contributions from other energy sources, including self-assembly, are discussed in the following subsection.

Tip-induced material deposition with electrochemical effects mainly includes two categories: electrochemical deposition and hollow tip deposition. Complementary approaches such as AFM-assisted chemical vapor deposition, AFM-assisted electron beam-induced deposition, and hybrid techniques exploit multiple physical or chemical sources. In electrochemical deposition, the AFM tip typically acts as a nanoscale dip tip, transferring atoms or nanoparticles from tip to substrate under a voltage bias. In hollow tip deposition, the tip functions as a nanoscale fountain pen, delivering liquid-phase materials directly.

In small-scale electrochemical deposition, atoms or nanoparticles are transported from the AFM tip to the substrate by applying a voltage bias. For example, Calleja et al. used an oscillating Au-coated AFM tip to fabricate gold nanowires on SiO_2 surfaces.[109] A pulsed negative voltage of 10–15 V (0.5–5 ms) applied at a tip–surface separation of ~3 nm enabled sequential deposition to form continuous nanowires with extended tip lifetime. Electrical transport measurements confirmed the metallic behavior of the fabricated nanowires.

In large-scale electrochemical deposition, the AFM tip and substrate act as electrodes forming a nanoscale electrochemical cell, where chemical reactions dominate deposition. Croitoru et al. demonstrated electrically induced chemical vapor deposition using a Pt-coated tip in a chamber filled with dimethylgold(III) trifluoroacetylacetonate.[110] A bias voltage between the tip and substrate electrodes triggered precursor decomposition, yielding conductive deposits on Pd/Cr- and Au/Pd-coated Si substrates.

Gaitas advanced AFM-assisted chemical vapor deposition using a heated Si_3N_4-coated tip to deposit Cu and CuO line patterns on SiO_2-coated Si substrates.[111] The tip, resistively heated to ~298.8 °C, scanned over the surface while $Cu(acac)_2$ vapor at 130 °C condensed and decomposed at the tip–substrate interface. O_2 and Ar flows (200 sccm and 30 sccm) maintained the chemical vapor deposition environment, and patterning over 70 μm × 70 μm areas was achieved using a piezoelectric XYZ stage with 0.2 nm resolution. Pausing the tip every 5 μm for 10 s along X and advancing 5 μm along Y produced high spatial selectivity, demonstrating precise, site-specific deposition.

Electrochemical deposition has also been adapted for low-energy electron beam-induced deposition. Holz et al. used electrons (<50 eV) emitted from a diamond-coated AFM tip to deposit Pt nanostructures from $MeCpPtMe_3$ on boron-doped Si under high vacuum.[112] Dot arrays with full width at half maximum ~80 nm and heights up to 38 nm were achieved, while line deposition at tip biases 20–80 V yielded structures with high aspect ratios. Such low-energy deposition is promising for single-electron devices, medical devices, and catalysis.

Because electrochemical deposition is limited in material diversity, hollow tip deposition using hollow AFM tips allows direct delivery of liquid-phase materials. Kim et al. and Lovsky et al. demonstrated hollow tip deposition for proteins, achieving higher deposition rates than conventional electrochemical deposition, especially with voltage-driven hollow probes.[113–114] Lovsky et al. deposited bovine serum albumin on protein-spotting glass using low voltages localized at the tapered probe tip, enhancing protein concentration in the deposited volumes.[114]

Rodolfa et al. developed a double-barreled hollow probe for depositing two different biological molecules from a single tip.[115] DNA molecules migrated toward the positive electrode, enabling selective deposition from each barrel. As a demonstration, a two-color image of *Degas Dancers* (75 × 61 pixels) was printed on a streptavidin-coated glass surface using rhodamine green- and Alexa-647-labeled DNA consecutively, reproducing

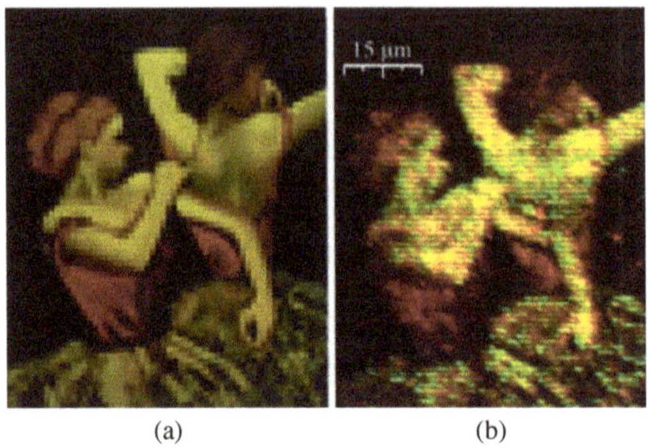

Figure 6.25. Micropatterning by double-barrel hollow probe on streptavidin-coated glass: (a) optical image of *Degas Dancers* (75 × 61 pixels); (b) fluorescence image printed with rhodamine green- and Alexa 647-labeled DNA consecutively from a double-barreled pipette, 1 μm pixels, covering 75 × 61 μm². (Courtesy of Rodolfa et al.,[115] reprinted from open-access data).

the image with high fidelity despite minor substrate non-uniformities (Figure 6.25).

Hollow tip deposition has also been extended to metal deposition. Ito et al. used a tapered hollow probe (~200 nm aperture) filled with $CuSO_4$, with a Cu wire as anode and Au substrate as cathode.[116] Applying a 1-V DC bias produced Cu dots with heights ranging from ~118 nm to 530 nm and full width at half maximum from ~240 nm to 1.2 μm, demonstrating controllable 3D metal deposition.

Hollow tip deposition and related techniques are also known as dip-pen nanolithography (DPN), where ink molecules diffuse from a coated cantilever tip to the substrate. Collectively, tip-induced electrochemical deposition techniques, including electrochemical deposition, AFM-assisted chemical vapor deposition and electron beam-induced deposition, and hollow tip deposition, enable highly localized material addition. Electrochemical deposition offers superior spatial precision, while hollow tip deposition and related approaches expand material diversity and deposition throughput, providing versatile platforms for nanoscale fabrication.

6.7.2 Atomic Force Microscopy-Assisted Assembly and Deposition with Additional External Sources

This subsection reviews AFM-assisted assembly and deposition processes driven by electrochemical and biochemical energy sources, in which the AFM tip functions as a patterning tool, specifically as an active guide for directing nanoscale self-assembly. Biomolecules and nanoparticles typically undergo spontaneous self-assembly through non-covalent interactions, often resulting in random surface attachment. In contrast, AFM tip guidance, frequently combined with applied electrical bias, chemical functionalization, or environmental control, enables deterministic assembly by locally modifying surface chemistry, charge distribution, or energy landscapes. The studies discussed below illustrate nanopatterning strategies that integrate multiple energy sources to achieve spatially controlled assembly and deposition.

An early example of chemically guided self-assembly was reported by Liu, Maoz, and Sagiv, who demonstrated AFM-assisted fabrication of predefined nanostructures composed of colloidal gold nanoparticles on Si substrates.[117] An AFM equipped with a W_2C-coated conductive tip was used to inscribe carboxylic acid (–COOH) patterns within an n-octadecyltrichlorosilane self-assembled monolayer on Si. Voltage pulses locally oxidized terminal $-CH_3$ groups into chemically active –COOH functionalities, which subsequently served as templates for the self-assembly of citrate-stabilized gold nanoparticles (2–6 nm in diameter). Following pattern inscription, the –COOH regions were selectively reacted with a nonadecenyltrichlorosilane overlayer and further modified *in situ* to generate amino (–NH_2) terminal groups. The resulting positively charged bilayer selectively attracted negatively charged gold nanoparticles via electrostatic interactions. Figure 6.26a shows an AFM image reproducing Picasso's *World Without Weapons* (1962), while Figure 6.26b presents the corresponding digital input image.[117] The pattern was written using 800 × 800 raster-scanned points with a dwell time of 3.3 ms per point at a tip bias of –8.5 V. The assembled nanoparticles exhibited heights of 2–6 nm and apparent lateral dimensions of ~25 nm, with lateral resolution primarily limited by tip convolution. The close correspondence between the assembled pattern and the predefined template confirms the effectiveness of AFM-guided chemical self-assembly.

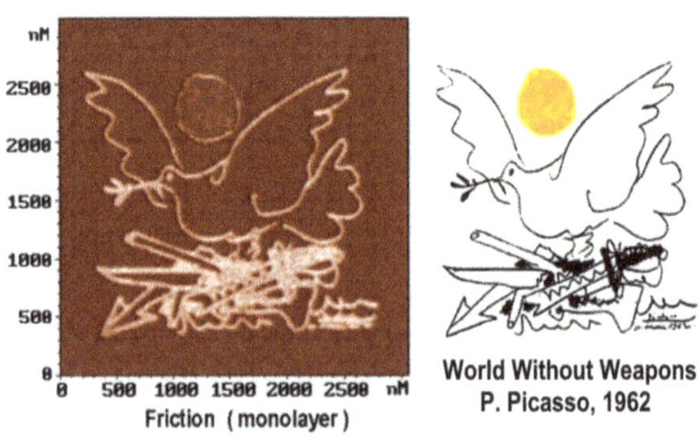

(a) (b)

Figure 6.26. Patterning the "*World Without Weapons*" poster by Pablo Picasso in 1962: (a) AFM image of gold particle-assembled pattern by AFM-assisted assembly. (b) digital image of the original poster. (Courtesy of Liu et al.,[117] reprinted from open-access data.)

AFM tip-guided assembly has also been extended to semiconductor nanoparticles. Xing et al. employed a Pt-coated conductive AFM tip operating in contact mode under high bias voltages (–40 to –90 V) to write charge patterns on a CYTOP substrate under controlled relative humidity (20–40%) (Figure 6.27a).[118] The applied bias induced localized surface potentials (–0.15 to –1.5 V) and nanoscale defluorination, increasing local surface energy and creating preferential adsorption sites (Figure 6.27b). Subsequent deposition resulted in selective nanoparticle assembly, including dual-color patterns formed by sequential assembly of two quantum dot species without cross-contamination. Figure 6.30c shows a hollow "N" pattern assembled from 12 nm $CsPbBr_3$ quantum dots, while Figure 6.30d presents the

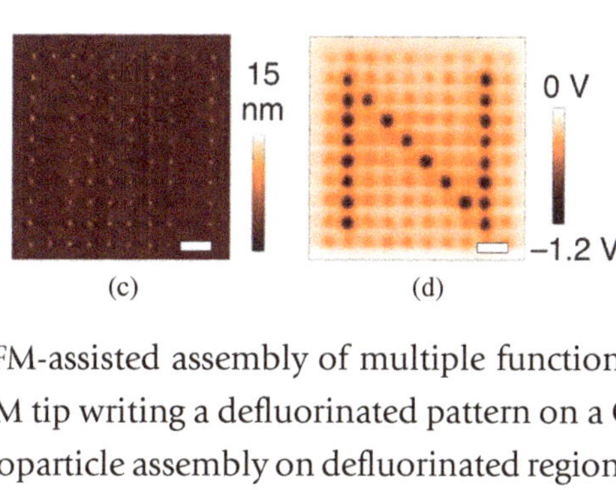

Figure 6.27. AFM-assisted assembly of multiple functional nanoparticles: (a) high-bias AFM tip writing a defluorinated pattern on a CYTOP substrate; (b) selective nanoparticle assembly on defluorinated regions; (c) AFM height image of a hollow "N" pattern assembled from 12-nm $CsPbI_3$ quantum dots; (d) electrical potential image after writing a second charge pattern for $CsPbI_3$ quantum dot assembly (scale bars: 1 μm). (Courtesy of Xing et al.,[118] reprinted from open-access data.)

surface potential image after writing a second pattern for $CsPbI_3$ quantum dot assembly, demonstrating high spatial alignment and material selectivity.

AFM-assisted assembly has also been widely applied to biomolecular deposition. Agarwal et al. demonstrated using molecule-coated AFM tips to deposit histidine-tagged peptides (MH2), proteins, and porphyrins onto ionized nickel substrates under applied biases of −2 to −3 V.[119] The AFM was operated in tapping mode with elevated drive amplitudes and slow scan rates. Distinct self-assembled morphologies were observed, reflecting intrinsic molecular organization: MH2 formed upright features (1.0–1.7 nm), TlpA assembled into filamentous structures, and TlpA-8 formed globular features (2–5 nm). GFP-R5 and porphyrins retained fluorescence after rinsing, confirming stable immobilization and preserved functionality.

DNA represents a particularly versatile self-assembly material due to its predictable base-pairing interactions and linear structure. However, precise manipulation of individual DNA molecules remains challenging. AFM-based techniques offer a solution by enabling controlled single-molecule handling. Kim et al. demonstrated that an AFM tip could reliably pick up a single DNA molecule with a success rate of ~75% and subsequently extend it via hybridization and ligation, forming longer conjugated DNA constructs.[120] This work highlights the potential of AFM-assisted single-molecule engineering.

AFM-assisted assembly has also been applied to living cells. Hsiao et al. demonstrated live cell patterning using a DNA-functionalized AFM tip.[121] A short DNA strand was immobilized on a Si_3N_4 tip to capture individual Jurkat cells under a normal force of ~400 pN. The captured cells were then transferred to

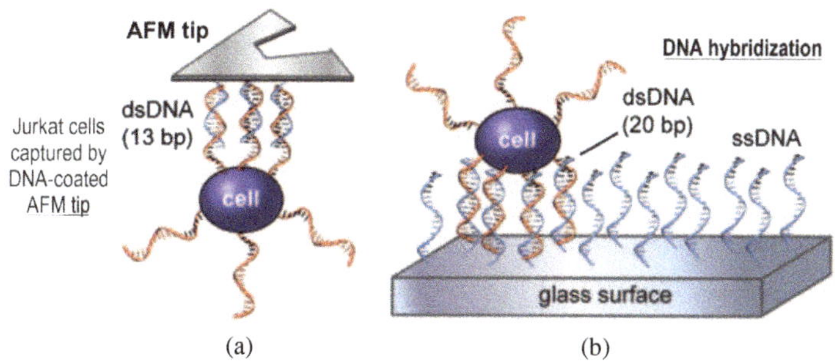

Figure 6.28. Live cell patterning using a DNA-coated AFM tip: (a) capture of living Jurkat cells by a DNA-functionalized AFM tip; (b) transfer and printing of the Jurkat cell onto a predefined location on a glass substrate functionalized with complementary DNA strands via DNA hybridization. (Courtesy of Hsiao et al.,[121] from open-access data.)

predefined substrate locations functionalized with complementary DNA strands, enabling precise placement via DNA hybridization (Figure 6.28). This biomolecule-mediated process minimized mechanical damage and preserved cell viability.

Beyond DNA, a broad range of biomolecules including ligands, peptides, proteins, and viruses have been exploited for AFM-assisted self-assembly. Chung et al. developed a multi-phase platform combining nanografted self-assembled monolayers (SAMs), biomolecular recognition, and nanoscale liquid meniscus confinement to achieve hierarchical organization of viruses, oligonucleotides, and metallic nanocrystals, culminating in Au nanoparticle nanowires.[122] This work demonstrates a general strategy for integrating chemical patterning, biological specificity, and physical confinement.

More recently, Ibrahim et al. reported an AFM-assisted molecular self-assembly process for nanoprinting molecular-scale devices using hollow AFM probes to deliver thiol- and silver-containing solutions onto SAM-functionalized Au substrates.[123] Spontaneous assembly led to hierarchical carbon nanosheets mediated by metallic silver atoms, although quantitative morphological characterization was not reported.[124]

Overall, AFM-assisted assembly and deposition processes that integrate electrochemical and biochemical energy sources enable deterministic control over self-assembly across molecular, nanoparticle, and cellular length scales. By combining tip-induced surface modification, programmable biomolecular interactions, and bottom-up assembly, these approaches provide powerful routes for constructing complex, multifunctional nanoarchitectures with high spatial precision.

6.7.3 Atomic Force Microscopy-Assisted Deposition with Multiple Tips

As emphasized by Mirkin,[125] the central advantage of massively parallel coated tip deposition, known as dip-pen nanolithography, lies in its ability to dramatically increase patterning throughput while preserving nanoscale resolution. Recent advances in multi-tip AFM systems have transformed AFM-assisted deposition from a serial, single-tip process into a scalable, parallel nanofabrication platform. This subsection reviews key developments employing multiple dip-tips and hollow tips (fountain pen) for high-throughput nanoscale deposition.

A landmark demonstration of massively parallel AFM-based patterning was reported by Salaita et al., who used a large-scale multi-tip system to fabricate ∼55,000 nanoscale replicas of a five-cent coin.[126] The system employed a 55,000-tip array with low-aspect-ratio pyramidal tips and curved cantilevers, achieving sub-100-nm resolution via gravity-driven transport

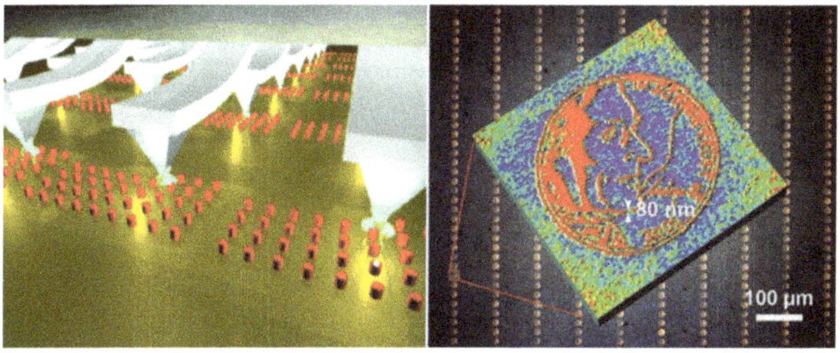

(a) (b)

Figure 6.29. Large-scale AFM-assisted deposition using multiple tips: (a) Schematic of parallel molecular deposition with 55,000 curved cantilevers and low-aspect-ratio pyramidal tips achieving sub-100-nm resolution; (b) AFM image of a nanoscale five-cent coin replica composed of 8,773 dots fabricated by 1-octadecanethiol deposition and chemical etching, overlaid on an optical micrograph of a region containing ~55,000 replicas. (Courtesy of Salaita et al.,[126] reprinted from open-access data.)

(Figure 6.29a). Each tip, coated with 1-octadecanethiol, deposited molecular patterns onto a gold-on-SiO_x substrate that served as an etch mask to form raised gold reliefs. Spanning a 1 cm^2 area, the array produced coin features with lateral dimensions of ~80 nm. Individual tips contacted the surface for 0.08 s under their own weight (~20 nN) as the array translated at 60 μm/s. Figure 6.29b shows an AFM image of a representative replica overlaid on an optical micrograph, demonstrating that ~55,000 replicas were generated in under 30 min without active feedback. This work highlights the efficiency and scalability of massively parallel AFM deposition.

Moldovan, Kim, and Espinosa developed a multi-ink array probe consisting of 12 microfluidic cantilevers connected to two on-chip ink reservoirs, enabling simultaneous use of two inks for combinatorial biochemical patterning.[127] As shown in

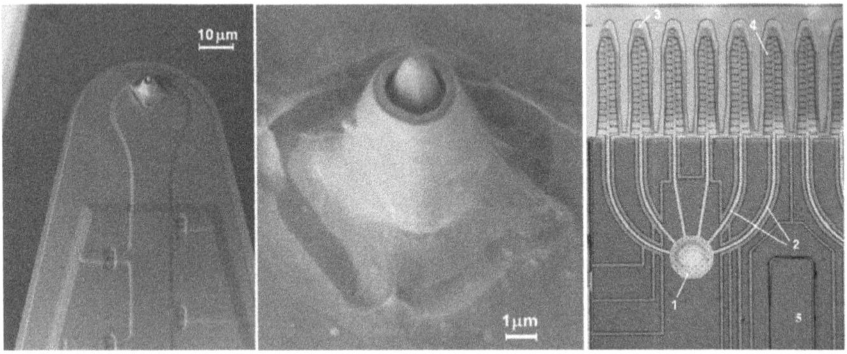

(a) (b) (c)

Figure 6.30. Volcano hollow tip array: (a) SEM image of a single volcano tip; (b) enlarged view of the hollow orifice; (c) optical image of part of the 12-tip array showing 1: ink reservoir, 2: microchannels, 3: volcano tip, 4: stabilization beam, and 5: central spacing post. (Courtesy of Moldovan et al.,[127] reprinted from open-access data.)

Figure 6.30a-b, the volcano-shaped hollow tips (7–8.5 µm tall, radii from tens of nanometers to 300 nm) deliver ink through a small orifice, analogous to a rapidograph nib (Figure 6.24a). This design minimizes evaporation and cross-contamination while preventing clogging and enabling smooth writing of large molecules. Line patterns with minimum widths of ~78 nm were produced on Au substrates using MHA ink, and dot arrays of fluorescein isothiocyanate-labeled DNA were deposited at 70 ± 5% relative humidity.

Okbi et al. and Yadav et al. further demonstrated multi-tip utility using a twelve-tip array to deposit polymer–ligand ink clusters on gold electrodes for meta-chemical surface sensors.[128,129] PMMA-based inks containing diaminonaphthalene, D-penicillamine, or no ligand were patterned with a dwell time of 1 s at 18–19 °C and 40–42% relative humidity. The resulting clusters (100–300 nm high) were UV-cured to stabilize the structures. Ink-1 and Ink-2 functioned as active sensing electrodes for Pb(II), Cu(II), and Cd(II), while Ink-3 served as a control. The multi-tip approach ensured uniform morphology and enhanced throughput.

Jo et al. employed both a linear 12-tip array and a 2D 55,000-tip array to fabricate positive- and negative-tone nanopatterns on Au-coated Si substrates.[130] For positive-tone patterning, MHA-SAMs served as etch resists during metal-assisted chemical etching, producing recessed Si wells with diameters down to ~104 nm (Figure 6.31a, b). For negative-tone patterning, polyethylene glycol inks yielded protruding Si nanowires via metal-assisted chemical etching with diameters of 570 ± 69 nm and lengths of 4.1 ± 0.24 μm (Figure 6.31d, e). In both cases, feature dimensions scaled linearly with the square root of dwell time

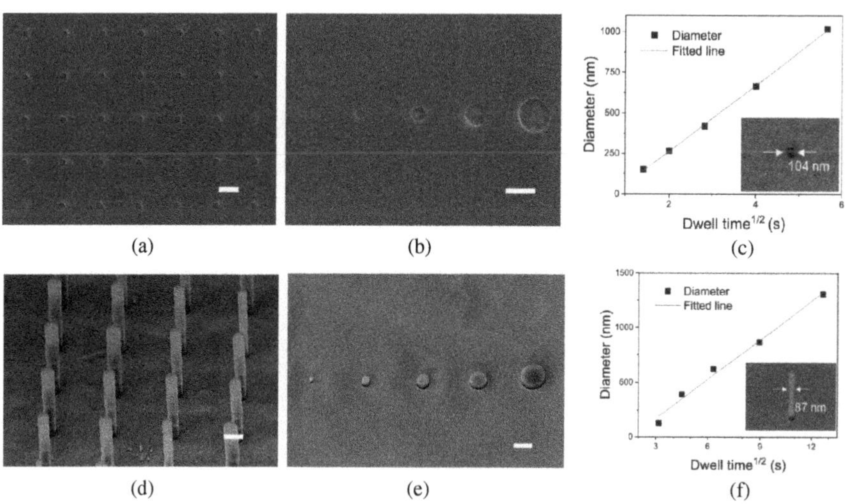

Figure 6.31. Multi-tip AFM-assisted deposition: (a,b) SEM images of recessed Si wells formed via MHA-SAM deposition and metal-assisted chemical etching; (c) well diameter versus square root of dwell time; (d,e) SEM images of protruding Si nanowires; (f) nanowire diameter versus square root of dwell time. All scale bars: 1 μm. (Courtesy of Jo et al.,[130] reprinted from open-access data.)

(Figure 6.31c, f), demonstrating predictable and tunable fabrication using multi-tip deposition.

Extending probe parallelization to electrochemical deposition, Oh et al. introduced a parallel hollow tip array comprising over 10,000 probes operating in a three-electrode electrochemical cell.[131] The system enabled high-throughput deposition of Ni (up to 900 nm in height), Ag, Pt (166 ± 32 nm in height), and Ni–Co alloy (~80 nm in height) structures with tunable composition and geometry, demonstrating its potential for catalysis, electronics, and materials discovery.

Complementing these approaches, Gaitas and French developed a seven-tip micromachined cantilever array integrating on-chip heating and piezoresistive deflection sensing for thermo-chemical patterning, nano-chemical vapor deposition, and dip-tip deposition.[132] The system eliminated optical detection and achieved nanometer-scale imaging and patterning with high sensitivity and minimal probe-to-probe variation, underscoring the potential of multifunctional probe arrays.

Dip-tip arrays can also be photoactuated. Huang et al. developed a photoactuated probe array containing 250,000 polydimethylsiloxane-carbon nanotube (PDMS–CNT) tips coated with MHA to deposit SAM patterns on gold substrates.[133] Local light-induced expansion on PDMS–CNT tips drove individual tips into contact with the substrate, enabling reversible, localized actuation (Figure 6.32a, b). SEM images showed triangular arrays of MHA-SAM dots ~2 nm thick with lateral dimensions from 500 nm to 2 μm (Figure 6.32c).

Overall, AFM-assisted deposition using multi-tip or tip-array configurations has evolved into a powerful and scalable nanofabrication strategy. These systems enable simultaneous patterning

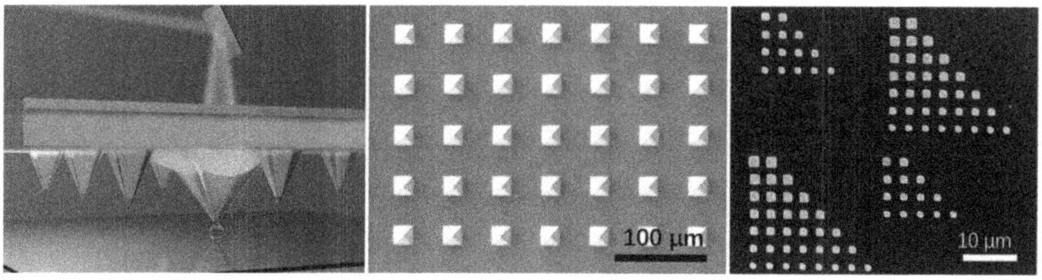

Figure 6.32. Photoactuated tip array and SAM patterns: (a) schematic of light-driven tip actuation; (b) SEM image of a portion of the PDMS–CNT tip array; (c) SEM image of deposited triangular MHA-SAM dot arrays (~2 nm thick). (Courtesy of Huang et al.,[133] reprinted from open-access data.)

of molecular, polymeric, and metallic materials over large areas while maintaining nanoscale precision, tunable feature geometry, and high reproducibility. Whether employing dip-tips or hollow tips, multi-tip architectures dramatically enhance throughput and broaden the scope of AFM-based deposition for applications in electronics, sensing, catalysis, and advanced materials manufacturing.

6.7.4 Atomic Force Microscopy-Assisted Multi-Step Deposition for Fabricating Nanodevices

AFM-assisted deposition techniques described in preceding sections have been extended to enable the fabrication of more sophisticated nanostructures for functional nanodevices and components. These approaches typically integrate AFM patterning with pre- or post-processing steps, where the AFM-generated nanopattern serves as an etch mask or alignment guide for subsequent pattern transfer. This subsection presents six representative studies illustrating how such multi-step methodologies are applied to fabricate nanodevices.

In 2005, Chung et al. employed an AFM Si_3N_4 tip coated with thiol-modified DNA to selectively deposit DNA on gold electrode surfaces, particularly within nanogaps, under controlled humidity (45–50% relative humidity, 23 °C).[134] Gold nanoparticles were subsequently assembled via DNA hybridization, forming tunnel junction circuits for biosensing applications (Figure 6.33). The electrodes were initially fabricated by photolithography and electron beam lithography to define 20–100 nm gaps, followed by O_2 plasma cleaning.

In 2009, Jang et al. combined AFM deposition with photolithography to fabricate Cr photomasks for microelectrode arrays.[135] A polyethylene glycol-coated AFM tip deposited negative patterns (100–240 nm thick) on hexamethyldisilazane-treated quartz, followed by Cr evaporation and polyethylene glycol lift-off to form the photomask (Figure 6.34a). This mask was used to fabricate interdigitated Au electrodes (40 nm thick), which were subsequently connected to pre-defined leads (Figure 6.34b).

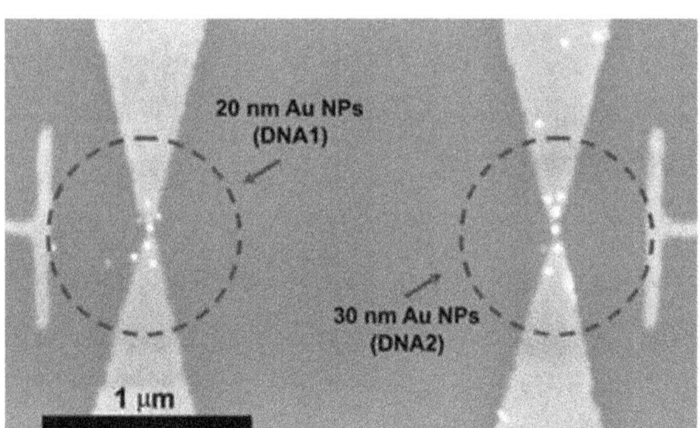

Figure 6.33. FESEM image of 20- and 30-nm Au nanoparticles bridging adjacent nanoelectrode junctions. (Courtesy of Chung et al.,[134] reprinted from open-access data.)

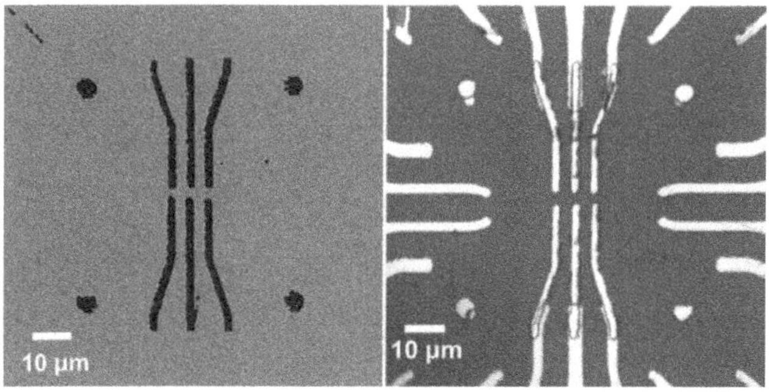

Figure 6.34. (a) SEM image of a critical region of the Cr photomask created by AFM deposition; (b) optical image of interdigitated Au electrodes fabricated using the AFM-defined photomask, after connecting to electrode leads previously defined by photolithography with a conventional mask. (Courtesy of Jang et al.,[135] reprinted from open-access data.)

The achieved resolution and patterning speed were comparable to laser- and e-beam-based techniques.

In 2016, Lee et al. directly patterned aligned single-walled CNT bundles on SiO_2 substrates using an AFM tip coated with single-walled CNT ink for FET applications.[136] Excess polyethylenimine was removed by methanol vapor, yielding single-walled CNT lines as narrow as 8 nm at a writing speed of 0.1 μm/s and a tip force of 50 nN. The patterned single-walled CNT line bundle (~1.5 μm wide) was incorporated into p-type FETs with conventional photolithographically defined 100 nm-thick Au source and drain electrodes.

Hu et al. used a heated AFM tip (260 °C) to deposit the dipped molten polystyrene nanowires on silicon-on-insulator substrates (Figure 6.35a, b).[137] By scanning at 150 nm/s, polystyrene nanowires were deposited to bridge pre-patterned Al micropads and served as high-resolution etch masks for Bosch silicon etching (Figure 6.35c). Subsequent removal of polystyrene and

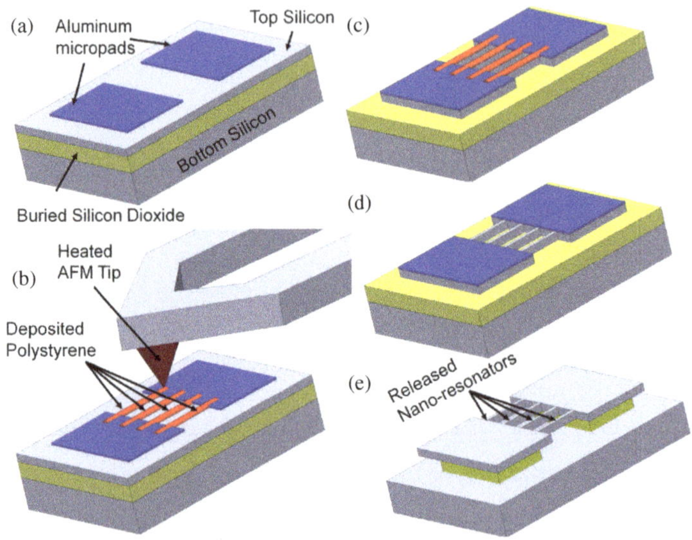

Figure 6.35. Fabrication sequence for Si nanomechanical resonators: (a) Formation of aluminum micropads via conventional lithography; (b) using dipped tip to deposit polystyrene nanowires; (c) transfer of the polystyrene nanowire pattern (as etch mask) into the silicon-on-insulator device layer by Bosch Si etching; (d) removal of polystyrene nanowires using acetone and oxygen plasma; (e) buffered HF etching of the buried SiO_2 layer to release suspended silicon resonators. (Courtesy of Hu et al.,[137] reprinted from open-access data.)

selective etching of the buried oxide layer released suspended silicon nano-resonators (Figure 6.35d, e), demonstrating precise alignment and compatibility with standard microfabrication.

Furthermore, Hu et al. applied their AFM-assisted deposition technique to fabricate Si and PDMS molds for nanofluidic channels used in biomolecule analysis.[138] The process began with conventional photolithography to define two curvilinear SiO_2 features (70 nm thick, separated by a 12-μm gap) on a Si wafer. A polystyrene-dipped AFM tip (heated to 260 °C) was then operated in contact mode at a tip speed of 150 nm/s to deposit polystyrene nanowires along predefined paths within the gap, thereby bridging the two SiO_2 features. The combined SiO_2 patterns and deposited polystyrene nanowires served as etch masks, producing a silicon mold with well-defined nanofluidic channels (Figure 6.36a). This Si mold was subsequently used to fabricate a PDMS mold via replica molding, enabling the formation of final nanofluidic channels (Figure 6.36b–d). More complex multi-channel and curved channel geometries were also demonstrated. Such nanofluidic channels are essential for the manipulation and analysis of biomolecules, including DNA and proteins, and are key components in nanoscale sensing, separation, and lab-on-a-chip applications.

Chen et al. utilized AFM-assisted deposition to pattern a narrow PMMA etch mask for fabricating monolayer MoS_2 nanoribbon FETs.[139] After chemical vapor deposition growth of

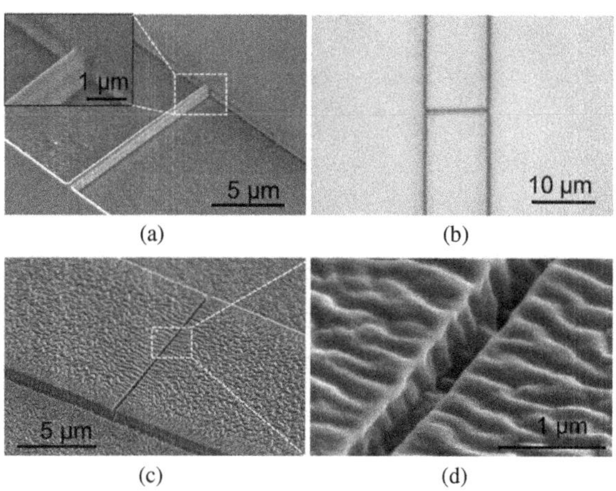

Figure 6.36. (a) SEM image of a silicon single linear nanochannel mold; inset shows a magnified section of the channel. (b) Optical micrograph of the PDMS nanofluidic channel replicated from the Si mold in (a). (c) SEM image of the replicated PDMS channel. (d) Enlarged view of a section of the PDMS nanochannel. (Courtesy of Hu et al.,[138] reprinted from open-access data.)

monolayer MoS_2 and lithographic definition of source–drain electrodes, a heated AFM tip dipped with PMMA ink to deposit a PMMA channel mask (Figure 6.37a, b). XeF_2 etching formed the MoS_2 nanoribbon, which was subsequently encapsulated with a 10 nm Al_2O_3 layer by atomic layer deposition to enhance device performance (Figure 6.37c, d). Here, MoS_2 is a 2D material and the monolayer (a single atomic layer) was used as the channel material, in which charge carriers flow between the source and drain terminals.

Overall, these studies demonstrate the versatility of AFM-assisted multi-step deposition for fabricating complex nanodevices by integrating nanoscale AFM patterning with conventional lithographic processes. Using AFM-defined features as precise

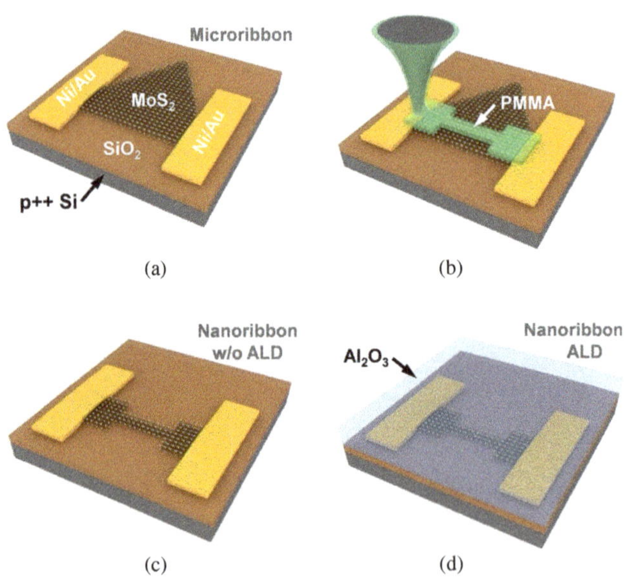

Figure 6.37. Process flow for fabricating monolayer MoS_2 nanoribbon transistors: (a) pattern contact electrodes; (b) deposit PMMA; (c) etch unmasked MoS_2 by XeF_2; (d) encapsulate device. (Courtesy of Chen et al.,[139] reprinted from open-access data.)

templates or etch masks enables the transfer of nanoscale fidelity into functional devices such as silicon resonators, graphene nanoribbons, and 2D material transistors.

6.8 Summary, Challenges, Recommendations, and Prospects

AFM-based nanofabrication encompasses a family of enabling nanofabrication techniques that employ a nanometer-scale tip to create a wide range of nanoscale structures. AFM was invented in 1985 and formally introduced in March 1986 by Binnig, Quate, and Gerber.[140] Over the past four decades, AFM has evolved from a high-resolution imaging instrument into a versatile platform for nanomanipulation and nanofabrication, playing an increasingly important role across many areas of science and technology. This review has surveyed both the fundamental principles and major advancements of AFM-enabled nanofabrication techniques. Beyond highlighting major technological advances, this final section also identifies persistent scientific and technical challenges and offers recommendations to guide future research and development, particularly as AFM-based nanofabrication becomes increasingly integrated with complementary nanofabrication processes.

6.8.1 *Challenges and Recommendations*

Despite remarkable progress over the past three decades, AFM-based nanofabrication continues to face fundamental challenges that limit its broader adoption and technological impact. These challenges span precision, reliability, throughput, 3D capability, application scope, and accessibility. This section critically examines these limitations and offers targeted recommendations to guide future research and development. Addressing these challenges will require coordinated advances in instrumentation, data-driven modeling, automation, and hybrid integration with complementary nanofabrication techniques. Continued innovation in these areas is essential for transitioning AFM-based nanofabrication from a powerful laboratory tool into a scalable, reliable, and versatile manufacturing platform for next-generation nanotechnology.

6.8.1.1 *Machine Learning-Assisted Precision Control for High Resolution and Reliability*

Achieving reliable, reproducible sub-10-nm resolution is the central performance objective of AFM-based nanofabrication, yet it remains difficult to attain in practice despite the technique's intrinsic nanoscale precision. The primary obstacle is the complex and strongly coupled relationship between fabrication outcomes — such as groove geometry in scratching, oxide growth in LAO, or deposition rate in additive processes — and operating parameters including tip force, scan trajectory, tip geometry, material properties, and environmental conditions. As feature sizes approach the fundamental limits of AFM, even minor fluctuations in these parameters can lead to significant variability in resolution and pattern fidelity.

Establishing reliable parameter-resolution relationships typically requires extensive experimental effort supported by theoretical models. For example, AFM-LAO has been shown to follow logarithmic growth behavior consistent with the Cabrera–Mott model, in which oxide thickness increases as the electric field driving ion migration weakens during oxide growth.[87] While such models provide valuable physical insight, they are often insufficient to fully capture system-level variability arising from tip wear, piezo creep, thermal drift, environmental noise, and instrument dynamics. These factors collectively limit resolution consistency and undermine reproducibility, particularly when pushing toward sub-10-nm patterning.

Machine learning offers a practical and data-driven pathway to address these limitations by directly linking high-dimensional process parameters to achievable resolution and fabrication fidelity. When trained on sufficiently large and well-characterized datasets, machine learning models can identify hidden correlations, predict fabrication outcomes, and map optimal operating windows that are difficult to derive using conventional approaches. As demonstrated in recent studies and summarized in the review presented in this monograph,[141] machine learning-based modeling has already improved understanding and control of a few AFM-based nanofabrication processes, leading to enhanced resolution and reduced experimental trial and error.

Beyond process prediction, machine learning enables precision control through real-time feedback and adaptive automation. By integrating machine learning algorithms with AFM sensing and control systems, it becomes possible to compensate dynamically for drift, noise, and tip degradation, thereby stabilizing resolution over extended fabrication times. Machine learning-assisted closed-loop control and autonomous parameter tuning are therefore strongly recommended for next-generation AFM-based nanofabrication platforms. Such capabilities are essential for transforming AFM-based nanofabrication from a precision-limited laboratory technique into a reliable, high-resolution nanofabrication technology suitable for complex and scalable applications.

6.8.1.2 *Bridging the Throughput Gap*

A key challenge for AFM-based nanofabrication is scaling to industrial mass production, as its inherently serial operation limits both throughput and large-area coverage. Parallelized AFM systems have demonstrated sub-10 nm nanostructure fabrication, yet they remain confined to laboratory prototyping. The bottleneck extends beyond tip-array density to the integration of high-speed sensing and control systems required for uniform tip-to-surface performance across large wafers. By contrast, multi-beam electron beam lithography has successfully scaled from single-beam setups to industrial production. Companies such as Mapper Lithography and NuFlare Technology have harnessed advanced electron optics to achieve high-throughput photomask fabrication, showing that massive parallelism can rival conventional photolithography.[142,143] For multi-tip AFM to follow this trajectory, development must shift from single-probe precision toward MEMS-integrated actuator arrays with automated tip-wear compensation. With substantial industrial R&D investment, AFM could evolve from a specialized nanofabrication tool into a robust platform for high-volume nanomanufacturing.

6.8.1.3 *Ability in Creating 3D Nanostructures*

The fabrication of high-fidelity 3D nanostructures remains a major challenge in AFM-based nanofabrication. Current approaches often combine AFM with conventional etching or deposition, but such hybrid workflows are inefficient and difficult to automate. Existing AFM-based removal, modification, and additive techniques lack integrated strategies for deterministic 3D construction.

Future progress requires automated AFM platforms with artificial intelligence-based computer-aided design/manufacturing tools to coordinate multiple tips for precise 3D nanofabrication. Hollow or ink-coated tips, effective for 2D patterning, face limitations in vertical stacking, necessitating improvements in ink chemistry, interfacial control, and instrument design. Hybrid approaches that integrate complementary AFM-based nanofabrication methods, rather than conventional macro-scale processes, will further expand capabilities.

Advancing these areas is critical for enabling efficient fabrication of functional 3D nanodevices and realizing the full potential of AFM-based nanofabrication in sophisticated nanomanufacturing applications.

6.8.1.4 *Expanding Application Domains*

Although AFM-based nanofabrication has demonstrated impact across electronics, optics, biology, and materials science, its application space remains far from fully explored. Future efforts should focus on leveraging AFM-based nanofabrication for rapid prototyping of nanoscale components in emerging fields such as quantum devices, neuromorphic systems, advanced sensors, biomedical interfaces, and chemical nanoreactors. Continued exploration of extreme miniaturization effects will likely reveal new physical phenomena and functional advantages unique to AFM-based nanofabrication. Developing application-specific AFM-based nanofabrication methodologies, rather than adapting generic processes, will be essential for translating nanoscale precision into system-level performance gains.

6.8.1.5 *Cost, Accessibility, and Standardization*

The high cost and complexity of AFM instrumentation and large-scale tip-array fabrication present significant barriers to widespread adoption. Developing cost-effective, scalable manufacturing methods for probes, actuators, and control electronics is therefore critical. Standardized platforms with modular designs could reduce system costs while improving maintainability and reproducibility. In parallel, open-source software, shared datasets, and standardized benchmarking protocols would lower entry barriers for new users and accelerate innovation. Improving accessibility without compromising resolution or throughput is essential for expanding the AFM-based nanofabrication user community and fostering broader academic and industrial engagement.

6.8.2 Concluding Remarks: Prospects

This review has comprehensively examined the fundamentals, advances, challenges, and future prospects of AFM-based nanofabrication. Across material removal, modification, and additive processes, AFM-based nanofabrication has demonstrated unparalleled flexibility and nanoscale precision, enabling the deterministic fabrication of 2D and 3D nanostructures across a wide range of materials. Despite these achievements, the central challenge remains the realization of reliable, reproducible sub-10-nm resolution with sufficient throughput and robustness for practical applications.

As feature sizes approach fundamental limits, system complexity, parameter coupling, and instrument instability increasingly constrain resolution and reproducibility. Addressing these limitations will require a paradigm shift from manual, empirically tuned operation toward data-driven, automated control. Machine learning has emerged as a key enabling technology for linking fabrication outcomes to multidimensional process parameters, compensating for drift and noise, and enabling adaptive, closed-loop optimization. When combined with advances in instrumentation, multi-tip architectures, and hybrid fabrication strategies, machine learning-assisted AFM-based nanofabrication offers a credible pathway toward enhanced resolution, reliability, and scalability.

Looking forward, the integration of AFM-based nanofabrication with artificial intelligence, advanced CAD/CAM frameworks, and complementary nanofabrication technologies is expected to significantly expand its application space. With continued interdisciplinary research and industrial investment, AFM-based nanofabrication is poised to evolve from a versatile laboratory technique into a powerful platform for next-generation nanomanufacturing and device fabrication.

Acknowledgements

The author gratefully acknowledges financial support from the National Science Council of the Republic of China (Grant No. NSC99-2811-E-007-014) and the Czech Ministry of Education, Youth and Sports (Project No. HEATEAM — CZ.1.07/2.3.00/20.0188), which endowed Chair Professorships at National Tsing Hua University (Taiwan) and Brno University of Technology (Czech Republic), respectively, during preparation of the first manuscript. He also thanks the late Professor Wen-Hwa Chen, Professor Hung Hocheng, and Professor Chien-Chung Fu of National Tsing Hua University for their generous hospitality and encouragement.

The author used ChatGPT, an artificial intelligence language model by OpenAI, to assist with literature review questions and language refinement. All artificial intelligence-generated content was critically evaluated and edited by the author, who takes full responsibility for the accuracy, originality, and integrity of the final manuscript.

References

1. Tseng, Ampere A. "Advancements and challenges in development of atomic force microscopy for nanofabrication." *Nano Today* 6, no. 5 (2011): 493–509.
2. Tseng, Ampere A. "A comparison study of scratch and wear properties using atomic force microscopy." *Applied Surface Science* 256, no. 13 (2010): 4246–4252.
3. Tseng, Ampere A., Jun-ichi Shirakashi, Shyankay Jou, Jen-Ching Huang, and T. P. Chen. "Scratch properties of nickel thin films using atomic force microscopy." *Journal of Vacuum Science & Technology B* 28, no. 1 (2010): 202–210.
4. Kawasegi, Noritaka, Noboru Takano, Daisuke Oka, Noboru Morita, Shigeru Yamada, Kazutaka Kanda, Shigeto Takano, Tsutomu Obata, and Kiwamu Ashida. "Nanomachining of silicon surface using atomic force microscope with diamond tip." *ASME Journal of Manufacturing Science and Engineering* 128 (2006): 723–729.
5. Kawasegi, Noritaka, Noboru Morita, Shigeru Yamada, Noboru Takano, Tatsuo Oyama, Kiwamu Ashida, and Hitoshi Ofune. "Nanomachining Zr-based metallic glass surfaces using an atomic force microscope." *International Journal of Machining and Machinability of Materials* 2, no. 1 (2007): 3–16.
6. Bourne, Keith, Shiv G. Kapoor, and Richard E. DeVor. "Study of a high performance AFM probe-based microscribing process." *Journal of Manufacturing Science and Engineering* 132 (2010): 030906.
7. Mills, B., A. H. Redford, *Machinability of Engineering Materials*. Applied Science, London, 1983.
8. Filho, H. D. F., M. H. P. Mauricio, C. R. Ponciano, and R. Prioli. "Metal layer mask patterning by force microscopy lithography." *Materials Science and Engineering: B* 112, no. 2–3 (2004): 194–199.
9. Tseng, Ampere A., Jun-ichi Shirakashi, Shinya Nishimura, Kazuya Miyashita, and Zhuang Li. "Nanomachining of permalloy for fabricating nanoscale ferromagnetic structures using atomic force microscopy." *Journal of Nanoscience and Nanotechnology* 10, no. 1 (2010): 456–466.

10. Tseng, Ampere A., Jun-ichi Shirakashi, Shinya Nishimura, Kazuya Miyashita, and Andrea Notargiacomo. "Scratching properties of nickel-iron thin film and silicon using atomic force microscopy." *Journal of Applied Physics* 106, no. 4 (2009): 044314.

11. Yamamoto, Shin-ichi Yamamoto Shin-ichi, Hirofumi Yamada Hirofumi Yamada, and Hiroshi Tokumoto Hiroshi Tokumoto. "Nanometer modifications of non-conductive materials using resist-films by atomic force microscopy." *Japanese Journal of Applied Physics* 34, no. 6S (1995): 3396–3399.

12. Fang, Te-Hua, Cheng-I. Weng, and Jee-Gong Chang. "Machining characterization of the nano-lithography process using atomic force microscopy." *Nanotechnology* 11, no. 3 (2000): 181–187.

13. Tsuchitani, Shigeki, Reizo Kaneko, and Shigeru Hirono. "Effects of humidity on nanometer scale wear of a carbon film." *Tribology International* 40, no. 2 (2007): 306–312.

14. Ogino, Takumi, Shinya Nishimura, and Jun-ichi Shirakashi. "Scratch nanolithography on Si surface using scanning probe microscopy: influence of scanning parameters on groove size." *Japanese Journal of Applied Physics* 47, no. 1S (2008): 712–714.

15. Gnecco, E., R. Bennewitz, and E. Meyer. "Abrasive wear on the atomic scale." *Physical Review Letters* 88, no. 21 (2002): 215501.

16. Prioli, R., M. Chhowalla, and F. L. Freire. "Friction and wear at nanometer scale: a comparative study of hard carbon films." *Diamond and Related Materials* 12, no. 12 (2003): 2195–2202.

17. Tseng, Ampere A. "Nanoscale scratching with single and dual sources using atomic force microscopes." In *Tip-Based Nanofabrication: Fundamentals and Applications,* edited by Ampere A. Tseng, pp. 1–64. Springer, 2011.

18. Feng, Junyuan, Xiaofang Huang, Shu Yang, Jianbo Qing, Peng Tang, and Zhenping Wan. "Speed effect on the material behavior in high-speed scratching of BK7 glass." *Ceramics International* 47, no. 14 (2021): 19978–19988.

19. Ren, Jiaqi, Jinsheng Zhao, Zeguang Dong, and Pinkuan Liu. "Molecular dynamics study on the mechanism of AFM-based nanoscratching process with water-layer lubrication." *Applied Surface Science* 346 (2015): 84–98.

20. Sugihara, Hideki, Atsushi Takahara, and Tisato Kajiyama. "Mechanical nanofabrication of lignoceric acid monolayer with atomic force microscopy." *Journal of Vacuum Science & Technology B* 19, no. 2 (2001): 593–595.

21. Zhang, Y., Eugeniu Balaur, and P. Schmuki. "Nanopatterning of an organic monolayer covered Si (1 1 1) surfaces by atomic force microscope scratching." *Electrochimica Acta* 51, no. 18 (2006): 3674–3679.

22. Firtel, Max, Grant Henderson, and Igor Sokolov. "Nanosurgery: observation of peptidoglycan strands in Lactobacillus helveticus cell walls." *Ultramicroscopy* 101, no. 2–4 (2004): 105–109.

23. Muir, Benjamin W., Andrew Fairbrother, Thomas R. Gengenbach, Florian Rovere, Michael A. Abdo, Keith M. McLean, and Patrick G. Hartley. "Scanning probe nanolithography and protein patterning of low-fouling plasma polymer multilayer films." *Advanced Materials* 18, no. 23 (2006): 3079–3082.

24. Vincenti, Lorenzo, Paolo Pellegrino, Isabella Farella, Mariafrancesca Cascione, Valeria De Matteis, Fabio Quaranta, and Rosaria Rinaldi. "Raster-Atomic force nanolithography: New insights towards the fabrication of 3D nanostructures on PMMA and Silicon Nitride." *Nano Trends* 9 (2025): 100088.

25. Yan, Yongda, Zhenjiang Hu, Xueshen Zhao, Tao Sun, Shen Dong, and Xiaodong Li. "Top-down nanomechanical machining of three-dimensional nanostructures by atomic force microscopy." *Small* 6, no. 6 (2010): 724–728.

26. Tominov, R. V., V. A. Smirnov, V. I. Avilov, A. A. Fedotov, V. S. Klimin, and N. E. Chernenko. "Formation of ZnO memristor structures by scratching probe nanolithography." In *IOP Conference Series: Materials Science and Engineering*, vol. 443, no. 1, p. 012036. IOP Publishing, 2018.

27. Kim, Jongwoo, Jeong Seop Lee, Ji-woong Kim, Peter De Wolf, Seunghyun Moon, Dong Hwan Kim, Joo-Hyun Song et al. "Fabrication of plasmonic arrays of nanodisks and nanotriangles by nanotip indentation lithography and their optical properties." *Nanoscale* 13, no. 8 (2021): 4475–4484.

28. Li, Zihan, Yongda Yan, Wenhan Zhu, Jiqiang Wang, and Yanquan Geng. "Nanofluidic chip with ripple nanostructures fabricated by atomic force microscopy tip-based nanoscratching for label-free deoxyribonucleic acid detection." *ACS Applied Nano Materials* 8, no. 9 (2025): 4814–4823.

29. Takami, Kazuhiro, Megumi Akai-Kasaya, Akira Saito, Masakazu Aono, and Yuji Kuwahara. "Construction of independently driven double-tip scanning tunneling microscope." *Japanese Journal of Applied Physics* 44, no. 1L (2005): L120.

30. Wang, Xuefeng, and Chang Liu. "Multifunctional probe array for nano patterning and imaging." *Nano Letters* 5, no. 10 (2005): 1867–1872.

31. Minne, S. C., J. D. Adams, G. Yaralioglu, S. R. Manalis, Abdullah Atalar, and C. F. Quate. "Centimeter scale atomic force microscope imaging and lithography." *Applied Physics Letters* 73, no. 12 (1998): 1742–1744.

32. Vettiger, P., M. Despont, U. Drechsler, U. Durig, W. HaÈberle, M. I. Lutwyche, H. E. Rothuizen, R. Stutz, R. Widmer, G. K. Binnig. "The "millipede" — more than one thousand tips for future AFM data storage." *IBM Journal of Research and Development* 44, no. 3 (2000): 323–340.

33. Vettiger, P., G. Cross, M. Despont, U. Drechsler, U. Durig, B. Gotsmann, W. Haberle, M. A. Lantz, H. E. Rothuizen, R. Stutz, G. K. Binnig, "The "millipede"-nanotechnology entering data storage." *IEEE Transactions in Nanotechnology* 1 (2002): 39–54.

34. Wang, Jiqiang, Qiang Zhang, Yongda Yan, Yu Liu, and Yanquan Geng. "Fabrication of periodic nanostructures for SERS substrates using multi-tip probe-based nanomachining approach." *Applied Surface Science* 576 (2022): 151790.

35. Santschi, C., J. Polesel-Maris, J. Brugger, and H. Heinzelmann. "Scanning probe arrays for nanoscale imaging, sensing and modification." In *Nanofabrication: Fundamentals and Applications*, edited by A. A. Tseng, pp. 65–126, World Scientific, Singapore, 2008.

36. Wu, Hanqiang, Shibo Zhang, Ximin Ye, Jian Guo, Linhe Sun, Chen Xiao, and Yongbo Wu. "Unveiling the underlying mechanism of ultrasonic vibration assisted machining: Tip-based single asperity nanoscratching experiments and insights from molecular dynamics simulations." *Journal of Manufacturing Processes* 140 (2025): 78–90.

37. Xu, Rui, Yanhui Zhang, Yi Liu, Zhimu Yang, Yifan Li, Wei Yu, and Jianli Wang. "Investigation of machining characteristics near resonance based on a custom-built vibration platform for atomic force microscopy." *Precision Engineering* 96 (2025): 444–454.

38. Wang, Jiqiang, Yongda Yan, Yanquan Geng, Yang Gan, and Zhuo Fang. "Fabrication of polydimethylsiloxane nanofluidic chips under AFM tip-based nanomilling process." *Nanoscale Research Letters* 14, no. 1 (2019): 136.

39. Tang, Jinyan, Zhong-Hao Cao, Zhongwei Li, and Yuan-Liu Chen. "Ultrasonic elliptical vibration-assisted AFM tip-based nanofabrication on brittle materials." *Precision Engineering* 96 (2025): 787–802.

40. Wang, Jiqiang, Yue Liu, Yongda Yan, Shaoqin Liu, Chen Li, and Yanquan Geng. "The mechanisms of material removal and atomic-scale damage in InP during elliptical vibration-assisted nanoscratching integrated with

laser processing." *Journal of Materials Processing Technology* 345 (2025): 119069.

41. Knoll, Armin W., David Pires, Olivier Coulembier, Philippe Dubois, James L. Hedrick, Jane Frommer, and Urs Duerig. "Probe-based 3-D nanolithography using self-amplified depolymerization polymers." *Advanced Materials* 22, no. 31 (2010): 3361–3365.
42. Pires, David, James L. Hedrick, Anuja De Silva, Jane Frommer, Bernd Gotsmann, Heiko Wolf, Michel Despont, Urs Duerig, and Armin W. Knoll. "Nanoscale three-dimensional patterning of molecular resists by scanning probes." *Science* 328, no. 5979 (2010): 732–735.
43. Gottlieb, Steven, Matteo Lorenzoni, Laura Evangelio, Marta Fernández-Regúlez, Yu Kyoung Ryu, C. Rawlings, Martin Spieser, Armin W. Knoll, and Francesc Perez-Murano. "Thermal scanning probe lithography for the directed self-assembly of block copolymers." *Nanotechnology* 28, no. 17 (2017): 175301.
44. Uranga-Granados, Iker, Berke Erbas, Pol Torres-Vila, Arnaud Bertsch, Jordi Llobet, Joan Bausells, Francesc Pérez-Murano, Juergen Brugger, and Marta Fernández-Regúlez. "Combination of thermal scanning probe lithography and directed self-assembly of block copolymers." In *Novel Patterning Technologies 2025*, vol. 13427, pp. 27–34. SPIE, 2025.
45. Chiou, Chi-Han, Shang-Jen Chang, Gwo-Bin Lee, and Huei-Huang Lee. "New fabrication process for monolithic probes with integrated heaters for nanothermal machining." *Japanese Journal of Applied Physics* 45, no. 1R (2006): 208.
46. Lantz, Mark A., Dorothea Wiesmann, and Bernd Gotsmann. "Dynamic superlubricity and the elimination of wear on the nanoscale." *Nature Nanotechnology* 4, no. 9 (2009): 586–591.
47. Okabe, H., T. Tsumura, Jun Shimizu, Li Bo Zhou, and Hiroshi Eda. "Experimental and simulation research on influence of temperature on nano-scratching process of silicon wafer." *Key Engineering Materials* 329 (2007): 379–384.
48. Szoszkiewicz, Robert, Takashi Okada, Simon C. Jones, Tai-De Li, William P. King, Seth R. Marder, and Elisa Riedo. "High-speed, sub-15 nm feature size thermochemical nanolithography." *Nano Letters* 7, no. 4 (2007): 1064–1069.
49. Fringes, Stefan, Christian Schwemmer, Colin D. Rawlings, and Armin W. Knoll. "Deterministic deposition of nanoparticles with sub-10 nm resolution." *Nano Letters* 19, no. 12 (2019): 8855–8861.

50. Wang, Debin, Suenne Kim, William D. Underwood, Anthony J. Giordano, Clifford L. Henderson, Zhenting Dai, William P. King, Seth R. Marder, and Elisa Riedo. "Direct writing and characterization of poly (p-phenylene vinylene) nanostructures." *Applied Physics Letters* 95, no. 23 (2009): 233108.

51. Fringes, Stefan, Christian Schwemmer, Colin D. Rawlings, and Armin W. Knoll. "Deterministic deposition of nanoparticles with sub-10 nm resolution." *Nano Letters* 19, no. 12 (2019): 8855–8861.

52. Pellegrini, Paloma E. S., Luana De-Moraes, Freddy Jara Poma, Silvia Vaz Guerra Nista, Frederico H. Cioldin, Hugo Enrique Hernández-Figueroa, and Stanislav Moshkalev. "Nanofabrication of all-dielectric metasurfaces through pulsed thermal scanning probe lithography." *Scientific Reports* 15, no. 1 (2025): 23469.

53. Chang, Shunyu, Yongda Yan, Bo Li, and Yanquan Geng. "Nanoscale manipulation of materials patterning through thermomechanical nanolithography using atomic force microscopy." *Materials & Design* 202 (2021): 109547.

54. Cho, Ryu, Yu Kyoung, Colin D. Rawlings, Heiko Wolf, Martin Spieser, Samuel Bisig, Steffen Reidt, Marilyne Sousa, Subarna R. Khanal, Tevis DB Jacobs, and Armin W. Knoll. "Sub-10 nanometer feature size in silicon using thermal scanning probe lithography." *ACS Nano* 11, no. 12 (2017): 11890–11897.

55. Lisunova, Y., and J. Brugger. "Combination of thermal scanning probe lithography and ion etching to fabricate 3D silicon nanopatterns with extremely smooth surface." *Microelectronic Engineering* 193 (2018): 23–27.

56. Zheng, Tingting, Caoimhe O'Neill, John F. Marshall, Thomas Iskratsch, and Matteo Palma. "Selective placement of functionalised DNA origami via thermal scanning probe lithography patterning." *Materials Advances* 5, no. 23 (2024): 9376–9382.

57. Yang, Ye, and Wansheng Zhao. "Fabrication of nanoscale to microscale 2.5 D square patterns on metallic films by the coupling AFM lithography." *Journal of Manufacturing Processes* 46 (2019): 129–138.

58. Wang, Tao, Yanling Tian, Zhilai Lu, Weijie Wang, Zhao Zhang, Guangwei Zhu, Hui Tang, and Dawei Zhang. "Experimental study of electrical-assisted nanomachining of monocrystalline copper using customized tungsten tip." *Nanomanufacturing and Metrology* 7, no. 1 (2024): 12.

59. Kim, Dal-Hyun, Ja-Yong Koo, and Jong-Jean Kim. "Cutting of multiwalled carbon nanotubes by a negative voltage tip of an atomic force microscope: A possible mechanism." *Physical Review B* 68, no. 11 (2003): 113406.

60. Han, Wei, Paven Thomas Mathew, Srikanth Kolagatla, Brian J. Rodriguez, and Fengzhou Fang. "Toward single-atomic-layer lithography on highly oriented pyrolytic graphite surfaces using AFM-based electrochemical etching." *Nanomanufacturing and Metrology* 5, no. 1 (2022): 32–38.

61. Koinuma, Michio, and Kohei Uosaki. "AFM tip induced selective electrochemical etching of and metal deposition on p-GaAs (100) surface." *Surface Science* 357 (1996): 565–570.

62. Piner, Richard D., Jin Zhu, Feng Xu, Seunghun Hong, and Chad A. Mirkin. ""Dip-pen" nanolithography." *Science* 283, no. 5402 (1999): 661–66363.

63. Li, Guangming, and Larry W. Burggraf. "Controlled patterning of polymer films using an AFM tip as a nano-hammer." *Nanotechnology* 18, no. 24 (2007): 245302.

64. Hu, Xiaodong, Jing Yu, Jinping Chen, and Xiaotang Hu. "Analysis of electric field-induced fabrication on Au and Ti with an STM in air." *Applied Surface Science* 187, no. 3–4 (2002): 173–178.

65. Lee, Gyudo, Huihun Jung, Jongsang Son, Kihwan Nam, Taeyun Kwon, Geunbae Lim, Young Ho Kim, Jongbum Seo, Sang Woo Lee, and Dae Sung Yoon. "Experimental and numerical study of electrochemical nanomachining using an AFM cantilever tip." *Nanotechnology* 21, no. 18 (2010): 185301.

66. Saxena, Krishna Kumar, Muhammad Hazak Arshad, and Dominiek Reynaerts. "AFM-ECM: Electrochemical micro/nano machining on an AFM platform." *Procedia CIRP* 137 (2025): 288–295.

67. Michler, J., R. Gassilloud, Ph Gasser, L. Santinacci, and P. Schmuki. "Defect-free AFM scratching at the Si/SiO_2 interface used for selective electrodeposition of nanowires." *Electrochemical and Solid-State Letters* 7, no. 3 (2004): A41.

68. Kato, Z., M. Sakairi, and H. Takahashi. "Nanopatterning on aluminum surfaces with AFM probe." *Surface and Coatings Technology* 169 (2003): 195–198.

69. Pendergast, Megan, Alex A. Volinsky, Xiaolu Pang, Robert Shields. "Moisture effects on gold nanowear." In *Nanoscale Tribology- Impact for Materials and Devices,* 1085-T05-10, *MRS Proceedings*, Vol. 1085, Cambridge University Press, 2008.

70. Chen, Guang, Jian Cheng, Linjie Zhao, Mingjun Chen, Wenyu Ding, Hongqin Lei, and Qi Liu. "Influences of the ambient humidity and temperature on the material removal and wear behaviors on the soft-brittle KDP optics lithographed by the AFM probes." *Wear* 571 (2025): 205826.

71. Cao, Hui, Qianqian Huang, Hanzong Xu, Baocheng Zhou, Wenke Chen, Mianlai Yu, and Ruicheng Feng. "Effect of aqueous layer thickness on nano-scratching of single-crystal γ-TiAl alloys." *Molecular Simulation* 50, no. 14 (2024): 1076–1091.

72. Day, H. C., and D. R. Allee. "Selective area oxidation of silicon with a scanning force microscope." *Applied Physics Letters* 62, no. 21 (1993): 2691–2693.

73. Minne, S. C., H. T. Soh, Ph Flueckiger, and C. F. Quate. "Fabrication of 0.1 μm metal oxide semiconductor field-effect transistors with the atomic force microscope." *Applied Physics Letters* 66, no. 6 (1995): 703–705.

74. Gordon, A. E., R. T. Fayfield, D. D. Litfin, and T. K. Higman. "Mechanisms of surface anodization produced by scanning probe microscopes." *Journal of Vacuum Science & Technology B* 13, no. 6 (1995): 2805–2808.

75. Teuschler, T., K. Mahr, S. Miyazaki, M. Hundhausen, and L. Ley. "Nanometer-scale field-induced oxidation of Si (111): H by a conducting-probe scanning force microscope: doping dependence and kinetics." *Applied Physics Letters* 67, no. 21 (1995): 3144–3146.

76. Garcia-Martin, Antonio, and Ricardo Garcia. "Formation of nanoscale liquid menisci in electric fields." *Applied Physics Letters* 88, no. 12 (2006): 123115.

77. Tello, Marta, and Ricardo García. "Nano-oxidation of silicon surfaces: Comparison of noncontact and contact atomic-force microscopy methods." *Applied Physics Letters* 79, no. 3 (2001): 424–426.

78. Martinez, Javier, Ramses Valentin Martinez, and Ricardo Garcia. "Silicon nanowire transistors with a channel width of 4 nm fabricated by atomic force microscope nanolithography." *Nano Letters* 8, no. 11 (2008): 3636–3639.

79. Tseng, Ampere A., Taewoo Lee, Andrea Notargiacomo, and T. P. Chen. "Formation of uniform nanoscale oxide layers assembled by overlapping oxide lines using atomic force microscopy." *Journal of Micro/Nanolithography, MEMS and MOEMS* 8, no. 4 (2009): 043050.

80. Yu, Jing, Abdulsalam Aji Suleiman, Jing-Wen Shi, Ruey Jinq Ong, Francis Chi-Chung Ling, and Weiwen Zhang. "Recent advances in atomic force microscopy-based local anodic oxidation nanolithography of 2D materials." *Advanced Materials Interfaces* 12 (2025): 2500137.

81. Fontaine, P. A., Emmanuel Dubois, and D. Stievenard. "Characterization of scanning tunneling microscopy and atomic force microscopy-based techniques for nanolithography on hydrogen-passivated silicon." *Journal of Applied Physics* 84, no. 4 (1998): 1776–1781.

82. Seo, Masahiro, and Daisuke Kawamata. "Nano-scratching in solution to the single-crystal Ta (100) subjected to anodic oxidation." *Journal of Physics D: Applied Physics* 39, no. 15 (2006): 3150.

83. Tseng, Ampere A., Chii D. Chen, Cen-Shawn Wu, Rodolfo E. Diaz, and Michael E. Watts. "Electron beam lithography of microbowtie structures for next-generation optical probe." *Journal of Micro/Nanolithography, MEMS and MOEMS* 1, no. 2 (2002): 123–135.

84. Johannes, Matthew S., Daniel G. Cole, and Robert L. Clark. "Three-dimensional design and replication of silicon oxide nanostructures using an atomic force microscope." *Nanotechnology* 18, no. 34 (2007): 345304.

85. Alekseev, Prokhor Anatol'evich, Bogdan Romanovich Borodin, Mikhail Sergeevich Dunaevskii, Aleksandr Nikolaevich Smirnov, V. Yu Davydov, Sergei Pavlovich Lebedev, and Alexander Alexandrovich Lebedev. "Local anodic oxidation of graphene layers on SiC." *Technical Physics Letters* 44, no. 5 (2018): 381–383.

86. Borodin, Bogdan R., Fedor A. Benimetskiy, and Prokhor A. Alekseev. "Study of local anodic oxidation regimes in $MoSe_2$." *Nanotechnology* 32, no. 15 (2021): 155304.

87. Komonov, A. I., N. D. Mantsurov, B. V. Voloshin, V. A. Seleznev, and S. V. Mutilin. "Nanopatterning of thin amorphous vanadium oxide films by oxidation scanning probe lithography." *Applied Surface Science* 658 (2024): 159869.

88. Huang, Ning, Ping Zhou, and Fengzhou Fang. "Effect of mechanical load on atomic-scale fabrication based on local anodic oxidation." *Nanomanufacturing and Metrology* 8, no. 1 (2025): 11.

89. Chen, C-F., S-D. Tzeng, H-Y. Chen, and S. Gwo. "Silicon microlens structures fabricated by scanning-probe gray-scale oxidation." *Optics Letters* 30, no. 6 (2005): 652–654.

90. Alias, Nurain Najihah, Muhammad Aniq Shazni Mohammad Haniff, and Khatijah Aisha Yaacob. "Fabrication and dimensional scaling of SiNW-FET array via AFM-LAO lithography and self-limiting oxidation." *Scientific Reports* 15, no. 1 (2025): 32902.

91. Cohen, Liam A., Noah L. Samuelson, Taige Wang, Kai Klocke, Cian C. Reeves, Takashi Taniguchi, Kenji Watanabe, Sagar Vijay, Michael P. Zaletel, and Andrea F. Young. "Nanoscale electrostatic control in ultraclean van der Waals heterostructures by local anodic oxidation of graphite gates." *Nature Physics* 19, no. 10 (2023): 1502–1508.

92. Tang, Jinyan, Zhong-Hao Cao, Zhongwei Li, and Yuan-Liu Chen. "Material removal on hydrogen-terminated diamond surface via AFM tip-based local anodic oxidation." *Micromachines* 16, no. 9 (2025): 981.

93. Wu, Zhenghan, Xianliang Zhou, Kunqi Xu, Zhichun Zhang, Yufeng Xie, Kenji Watanabe, Takashi Taniguchi, and Zhiwen Shi. "High-precision AFM cutting of graphene via improved electrode-free local anodic oxidation for electronic band engineering." *Nanoscale* 17, no. 44 (2025): 25657–25663.

94. LaGraff, John R., and Andrew A. Gewirth. "Enhanced electrochemical deposition with an atomic force microscope." *The Journal of Physical Chemistry* 98, no. 44 (1994): 11246–11250.

95. Obermair, Christian, Marina Kress, Andreas Wagner, and Thomas Schimmel. "Reversible mechano-electrochemical writing of metallic nanostructures with the tip of an atomic force microscope." *Beilstein Journal of Nanotechnology* 3, no. 1 (2012): 824–830.

96. Tanabe, Ichiro, and Tetsu Tatsuma. "Size-and shape-controlled electrochemical deposition of metal nanoparticles by tapping mode atomic force microscopy." *The Journal of Physical Chemistry C* 116, no. 6 (2012): 3995–3999.

97. Garcia, R., N. S. Losilla, J. Martínez, R. V. Martinez, F. J. Palomares, Y. Huttel, Matteo Calvaresi, and Francesco Zerbetto. "Nanopatterning of carbonaceous structures by field-induced carbon dioxide splitting with a force microscope." *Applied Physics Letters* 96, no. 14 (2010): 143110.

98. Khare, H. S., N. N. Gosvami, Imène Lahouij, Z. B. Milne, J. B. McClimon, and Robert W. Carpick. "Nanotribological printing: a nanoscale additive manufacturing method." *Nano Letters* 18, no. 11 (2018): 6756–6763.

99. Demers, L. M., D. S. Ginger, S-J. Park, Z. Li, S-W. Chung, and C. A. Mirkin. "Direct patterning of modified oligonucleotides on metals and insulators by dip-pen nanolithography." *Science* 296, no. 5574 (2002): 1836–1838.

100. Noy, Aleksandr, Abigail E. Miller, Jennifer E. Klare, Brandon L. Weeks, Bruce W. Woods, and James J. DeYoreo. "Fabrication of luminescent nanostructures and polymer nanowires using dip-pen nanolithography." *Nano Letters* 2, no. 2 (2002): 109–112.

101. Chung, Sungwook, Jonathan R. Felts, Debin Wang, William P. King, and James J. De Yoreo. "Temperature-dependence of ink transport during thermal dip-pen nanolithography." *Applied Physics Letters* 99, no. 19 (2011): 193101.

102. Sheehan, P. E., L. J. Whitman, William P. King, and Brent A. Nelson. "Nanoscale deposition of solid inks via thermal dip pen nanolithography." *Applied Physics Letters* 85, no. 9 (2004): 1589–1591.

103. Chen, Guang, Zhaoyang Yin, Linjie Zhao, Jian Cheng, Mingjun Chen, Hongqin Lei, Zishuai Zhang, and Jixiang Chen. "Toward condensing stable water meniscus for dip-pen nanolithography considering ambient condition, surface wettability, and SPM probe structure." *Surfaces and Interfaces* 65 (2025): 106500.

104. Loh, Owen, Robert Lam, Mark Chen, Nicolaie Moldovan, Houjin Huang, Dean Ho, and Horacio D. Espinosa. "Nanofountain-probe-based high-resolution patterning and single-cell injection of functionalized nanodiamonds." *Small* 5, no. 14 (2009): 1667–1674.

105. Meister, A., M. Liley, J. Brugger, R. Pugin, and H. Heinzelmann. "Nanodispenser for attoliter volume deposition using atomic force microscopy probes modified by focused-ion-beam milling." *Applied Physics Letters* 85, no. 25 (2004): 6260–6262.

106. Lee, Woo Kyung, Zhenting Dai, William P. King, and Paul E. Sheehan. "Maskless nanoscale writing of nanoparticle–polymer composites and nanoparticle assemblies using thermal nanoprobes." *Nano Letters* 10, no. 1 (2010): 129–133.

107. Fan, Zheng, Xinyong Tao, Xudong Fan, Xiaobin Zhang, and Lixin Dong. "Nanotube fountain pen: Towards 3D manufacturing of metallic nanostructures." *Carbon* 86 (2015): 280–287.

108. Hu, Jie, and Min-Feng Yu. "Meniscus-confined three-dimensional electrodeposition for direct writing of wire bonds." *Science* 329, no. 5989 (2010): 313–316.

109. Calleja, Montserrat, M. Tello, J. Anguita, F. Garcıa, and R. Garcıa. "Fabrication of gold nanowires on insulating substrates by field-induced mass transport." *Applied Physics Letters* 79, no. 15 (2001): 2471–2473.

110. Croitoru, M. D., A. Höchst, G. Bertsche, S. Krauß, S. Roth, and D. P. Kern. "Single-step nanopatterning with a non-contact scanning force microscope by electrically induced local chemical vapour deposition." *Microelectronic Engineering* 67 (2003): 696–701.

111. Gaitas, Angelo. "Tip-based chemical vapor deposition with a scanning nano-heater." *Applied Physics Letters* 102, no. 13 (2013): 133104.

112. Holz, Mathias, Frances I. Allen, Christoph Reuter, Ahmad Ahmad, Martin Hofmann, Alexander Reum, Tzvetan Ivanov, and Ivo W. Rangelow.

"Tip-based electron beam induced deposition using active cantilevers." *Journal of Vacuum Science & Technology B* 37, no. 6 (2019): 061812.

113. Kim, Keun-Ho, Nicolaie Moldovan, and Horacio D. Espinosa. "A nanofountain probe with sub-100 nm molecular writing resolution." *Small* 1, no. 6 (2005): 632–635.

114. Lovsky, Yulia, Aaron Lewis, Chaim Sukenik, and Eli Grushka. "Atomic-force-controlled capillary electrophoretic nanoprinting of proteins." *Analytical and Bioanalytical Chemistry* 396, no. 1 (2010): 133–138.

115. Rodolfa, Kit T., Andreas Bruckbauer, Dejian Zhou, Yuri E. Korchev, and David Klenerman. "Two-component graded deposition of biomolecules with a double-barreled nanopipette." *Angewandte Chemie* 117, no. 42 (2005): 7014–7019.

116. Ito, So, Takuya Keino, and Futoshi Iwata. "Volume control of metal-plating deposition using a nanopipette probe by controlling electric charge." *Japanese Journal of Applied Physics* 49, no. 8S3 (2010): 08LB16.

117. Liu, Shantang, Rivka Maoz, and Jacob Sagiv. "Planned nanostructures of colloidal gold via self-assembly on hierarchically assembled organic bilayer template patterns with in-situ generated terminal amino functionality." *Nano Letters* 4, no. 5 (2004): 845–851.

118. Xing, Xing, Zaiqin Man, Jie Bian, Yadong Yin, Weihua Zhang, and Zhenda Lu. "High-resolution combinatorial patterning of functional nanoparticles." *Nature Communications* 11, no. 1 (2020): 6002.

119. Agarwal, Gunjan, Rajesh R. Naik, and Morley O. Stone. "Immobilization of histidine-tagged proteins on nickel by electrochemical dip pen nanolithography." *Journal of the American Chemical Society* 125, no. 24 (2003): 7408–7412.

120. Kim, Duckhoe, Nak-Kwan Chung, Jung Sook Kim, and Joon Won Park. "Immobilizing a single DNA molecule at the apex of AFM tips through picking and ligation." *Soft Matter* 6, no. 16 (2010): 3979–3984.

121. Hsiao, Sonny C., Ailey K. Crow, Wilbur A. Lam, Carolyn R. Bertozzi, Daniel A. Fletcher, and Matthew B. Francis. "DNA-coated AFM cantilevers for the investigation of cell adhesion and the patterning of live cells." *Angewandte Chemie* 120, no. 44 (2008): 8601–8605.

122. Chung, Sung-Wook, Andrew D. Presley, Selim Elhadj, Saphon Hok, Sang Soo Hah, Alex A. Chernov, Matthew B. Francis, Bruce E. Eaton, Daniel L. Feldheim, and James J. Deyoreo. "Scanning probe-based fabrication of 3D nanostructures via affinity templates, functional RNA, and

meniscus-mediated surface remodeling." *Scanning: The Journal of Scanning Microscopies* 30, no. 2 (2008): 159–171.

123. Ibrahim, M. M., H. Hamoudi, F. Xia and K. Youcef-Toumi, "A molecular self-assembly AFM nanoprinting system towards the fabrication of functional technorganic machines," In *2024 IEEE 17th International Conference on Nano/Molecular Medicine & Engineering (NANOMED)*, pp. 142–147, IEEE, 2024, doi: 10.1109/NANOMED64244.2024.10946037.
124. Ibrahim, Malek M. *Molecular Self-Assembly of Carbon Nanosheets via AFM Nanoprinting*. M.S. Thesis, Massachusetts Institute of Technology, 2024.
125. Mirkin, Chad A. "The power of the pen: development of massively parallel dip-pen nanolithography." *ACS Nano* 1, no. 2 (2007): 79–83.
126. Salaita, Khalid, Yuhuang Wang, Joseph Fragala, Rafael A. Vega, Chang Liu, and Chad A. Mirkin. "Massively Parallel Dip–Pen Nanolithography with 55 000-Pen Two-Dimensional Arrays." *Angewandte Chemie International Edition* 45, no. 43 (2006): 7220–7223.
127. Moldovan, N., K. H. Kim, and H. D. Espinosa. "A multi-ink linear array of nanofountain probes." *Journal of Micromechanics and Microengineering* 16, no. 10 (2006): 1935–1942.
128. Okbi, Rahma, Mohammed Alkrenawi, Krishna Kumar Yadav, Dror Shamir, Haya Kornweitz, Yael Peled, Moshe Zohar, and Ariela Burg. "Dip-pen nanolithography-based fabrication of meta-chemical surface for heavy metal detection: Role of poly-methyl methacrylate in sensor sensitivity." *Small Science* 5, no. 2 (2025): 2400459.
129. Yadav, Krishna K., Dror Shamir, Haya Kornweitz, Yael Peled, Moshe Zohar, and Ariela Burg. "Development of meta-chemical surface by dip-pen nanolithography for precise electrochemical lead sensing." *Small Methods* 8, no. 3 (2024): 2301118.
130. Jo, Jeong-Sik, Jihoon Choi, Seung-Hoon Lee, Changhoon Song, Heeso Noh, and Jae-Won Jang. "Mass fabrication of 3D silicon nano-/microstructures by fab-free process using tip-based lithography." *Small* 17, no. 4 (2021): 2005036.
131. Oh, EunBi, Rustin Golnabi, David A. Walker, and Chad A. Mirkin. "Electrochemical polymer pen lithography." *Small* 17, no. 28 (2021): 2100662.
132. Gaitas, Angelo, and Paddy French. "Piezo-thermal probe array for high throughput applications." *Sensors and Actuators A: Physical* 186 (2012): 125–129.
133. Huang, Zhongjie, Le Li, Xu A. Zhang, Nourin Alsharif, Xiaojian Wu, Zhiwei Peng, Xiyuan Cheng, Peng Wang, Keith A. Brown, and YuHuang

Wang. "Photoactuated pens for molecular printing." *Advanced Materials* 30, no. 8 (2018): 1705303.

134. Chung, Sung-Wook, David S. Ginger, Mark W. Morales, Zhengfan Zhang, Venkat Chandrasekhar, Mark A. Ratner, and Chad A. Mirkin. "Top-down meets bottom-up: dip-pen nanolithography and DNA-directed assembly of nanoscale electrical circuits." *Small* 1, no. 1 (2005): 64–69.

135. Jang, Jae Won, Raymond G. Sanedrin, Andrew J. Senesi, Zijian Zheng, Xiaodong Chen, Seongpil Hwang, Ling Huang, and Chad A. Mirkin. "Generation of metal photomasks by dip-pen nanolithography." *Small* 5, no. 16 (2009): 1850–1853.

136. Lee, Jae-Hyeok, Choolakadavil Khalid Najeeb, Gwang-Hyeon Nam, Yonghun Shin, Jung-Hyurk Lim, and Jae-Ho Kim. "Large-scale direct patterning of aligned single-walled carbon nanotube arrays using dip-pen nanolithography." *Chemistry of Materials* 28, no. 18 (2016): 6471–6476.

137. Hu, Huan, Hanna Cho, Suhas Somnath, Alexander F. Vakakis, and William P. King. "Silicon nano-mechanical resonators fabricated by using tip-based nanofabrication." *Nanotechnology* 25, no. 27 (2014): 275301.

138. Hu, Huan, Yue Zhuo, Muhammed E. Oruc, Brian T. Cunningham, and William P. King. "Nanofluidic channels of arbitrary shapes fabricated by tip-based nanofabrication." *Nanotechnology* 25, no. 45 (2014): 455301.

139. Chen, Sihan, Sun Phil Kim, Weibing Chen, Jiangtan Yuan, Rashid Bashir, Jun Lou, Arend M. Van Der Zande, and William P. King. "Monolayer MoS_2 nanoribbon transistors fabricated by scanning probe lithography." *Nano Letters* 19, no. 3 (2019): 2092–2098.

140. Binnig, Gerd, Calvin F. Quate, and Ch Gerber. "Atomic force microscope." *Physical Review Letters* 56, no. 9 (1986): 930–933.

141. Tseng, Ampere A., Miroslav Raudensky, and Jun-ichi Shirakashi. "Machine learning in emerging nanofabrication processes: from imprinting and tip-based to ion-beam, laser, and capillary approaches." In *Nanofabrication in the Era of Machine Learning: Advances and Emerging Frontiers*, edited by Ampere A. Tseng, Chapter 4, pp. 149-208. Singapore: World Scientific, 2026.

142. de Boer, Guido. "Multi e-beam direct write: what are the options for the future?." *Journal of Micro/Nanopatterning, Materials, and Metrology* 22, no. 4 (2023): 041407–041407.

143. Matsumoto, Hiroshi, Haruyuki Nomura, Jumpei Yasuda, Shunsuke Isaji, Kenichi Yasui, Hayato Kimura, and Yoshinori Kojima. "Multi-beam mask writer MBM-4000 for high-NA EUV era." In *Novel Patterning Technologies 2025*, vol. 13427, p. 1342706. SPIE, 2025.

https://doi.org/10.1142/9789819830763_0007

CHAPTER 7

Nanoscale Additive Manufacturing: Enabling Precision-Engineered Mechanical Metamaterials

Muhammad Rizwan ul Haq[1], Muhammad Hammad[1], Kashif Azher[2], Aamer Nazir[2,3*]

[1]School of Mechanical and Manufacturing Engineering (SMME), National University of Sciences and Technology (NUST), Islamabad 44000, Pakistan
[2]Department of Mechanical Engineering, King Fahd University of Petroleum & Minerals, Dhahran, 31261, Saudi Arabia
[3]Interdisciplinary Research Center for Advanced Materials, King Fahd University of Petroleum & Minerals, Dhahran, 31261, Saudi Arabia
*Corresponding author: aamer.nazir@kfupm.edu.sa

Abstract

Nanoscale additive manufacturing is a revolutionary step forward in the fabrication of architected materials and mechanical metamaterials, enabling new levels of design freedom and spatial resolution. This chapter examines its evolving domains, concentrating on its applications in the design and fabrication of mechanical metamaterials. These engineered structures have better adjustable mechanical properties, such as modifiable stiffness, stimulus responsiveness, and energy absorption. Nanoscale control over geometry, material composition, structural arrangement, and hierarchy makes these developments possible.

This chapter reviews nanoscale additive manufacturing routes, namely two-photon polymerization, focused electron/ion beam-induced deposition, nanoscale jetting, and projection micro-stereolithography, through the lens of process, structure, and property for mechanical metamaterials. For each method, we discuss controllable parameters (e.g., voxel aspect ratio, hatch spacing, and precursor chemistry) to unit cell geometry, flaw distributions, and fatigue behavior. We then outline material systems (shape-memory polymers, nanocomposites, and ceramics/metals) that enable graded stiffness, negative thermal expansion, and energy absorption in architected lattices. The chapter closes with design strategies, topology optimization, and data-driven inverse design, along with pragmatic limits (resolution/throughput trade-offs and repeatability) and remedies (hybrid processing and artificial intelligence-guided defect mitigation). This framing aims to help practitioners select a nanoscale additive manufacturing route based on a target mechanical response rather than process familiarity.

Keywords: nanoscale additive manufacturing, mechanical metamaterials, architected materials, two-photon polymerization, projection micro-stereolithography, topology optimization

7.1 Introduction

Mechanical metamaterials are defined by their architecture along with the materials; for instance, the allocation of the type of structure or material at different places within the structure creates distinct effects.[1-3] Nanoscale additive manufacturing (nano-AM) makes architectural decisions more critical for applications because fabrication parameters such as scan path, feature size, and aspect ratio directly determine the individual shapes of the design, including struts, buckling, and defect propagation, which in turn affect stiffness, hysteresis, and failure modes. This chapter focuses on that coupling. We compare two-photon polymerization (2PP), focused electron/ion beam-induced deposition (FEBID/FIBID), nanoscale jetting, and projection micro-stereolithography (PμSL) on what they give you mechanically, then survey materials that preserve those gains, and finally show how topology optimization and machine learning inverse design can close the loop from property target to printing parameter.

Nano-AM is best understood as the tightly coupled relationship among resolution, chemistry, and architecture, enabling structures whose properties are governed far more by geometry than by bulk composition. Within this model, a small set of enabling processes supports most demonstrations: 2PP routinely writes sub-micrometer features for free-form lattices and thin shells;[4] PμSL replaces serial voxel scanning with layer-wise optical projection, delivering sub-micron precision at significantly higher throughput;[5] FEBID/FIBID draws metals and functional compounds in 3D with tens-of-nanometers placement, excelling at site-specific reinforcement, contacts, and nanodevices;[6] and nanoscale material jetting, notably electrohydrodynamic jet printing, patterns inks and colloids at sub-micrometer line widths to add conductive, catalytic, or bioactive phases.[7] At these dimensions, surface forces, voxel overlap, exposure kinetics, resin diffusion, cure shrinkage, and post-cure densification become deliberate design variables: voxel aspect ratio sets strut ellipticity; hatch spacing tunes inter-voxel necking; dose and scan trajectory govern conversion and cross-link density; and thermal or pyrolytic post-processing determines final density, residual stress, and flaw distribution.[5,8] Since the same parameters that determine geometry also ensure mechanical fidelity, nano-AM benefits from a design that is specifically aware of the process.

Hybrid processing closes the performance gap between polymers and engineering materials. Pyrolysis converts 2PP/PμSL polymer scaffolds into glassy carbon or polymer-derived ceramics with dramatic strength gains. At the same time, atomic layer deposition (ALD) adds nanometric, conformal coatings that enhance stiffness, fatigue resistance, and environmental stability without occluding pores.[9,10] These routes have yielded ultralight, ultrastiff lattices and plate-based nanolattices with strengths approaching theoretical limits, benchmarks unlikely without nano-AM.[11] When architectural mechanics are exploited, auxetic and other programmable lattices provide lateral expansion under tension, spatially graded stiffness, tailored dispersion for elastic waves, and on-demand nonlinearity (via reconfigurable, snap-through, or multi-stable motifs) for energy absorption and impact mitigation.[12,13] Performance can be pushed further by converting polymer templates to glassy carbon or polymer-derived ceramics, or by applying ALD coatings, thereby boosting stiffness and durability while preserving open porosity for mass transport.[10,11,14,15]

The same architectural freedom supports meta-biomaterials: unit cell dimensions set pore size and interconnectivity for nutrient flow; strut orientation and shell thickness tune apparent modulus to match tissue; and patterned surface chemistries or deposited bioactive phases guide cell attachment and differentiation.[5,16] Multi-material nano-AM also supports sacrificial channels or hydrogel phases for microvascular networks, and degradable scaffolds whose mechanics evolve with tissue regeneration.[16]

Finally, scaling from demonstrators to devices requires rigorous process-structure-property-performance mapping: pre-distortion to compensate for shrinkage, exposure strategies that limit heat and voxel overlap, and statistical process control with *in situ* sensing and inverse modeling to suppress defect modes (e.g., missing struts, under-cured nodes, microvoids). These process-level requirements are especially important because mechanical metamaterials derive their response also from geometry, making them highly sensitive to fabrication imperfections.[2,12] In summary, nano-AM provides the fabrication freedom needed to realize architected matter with prescribed deformation pathways, wave responses, and mass transport and, when coupled with hybrid post-processing and rigorous metrology, delivers mechanical metamaterials that are not only extraordinary in theory but also reproducible, scalable, and deployable in aerospace, protective structures, soft microrobotics, and regenerative medicine.

7.1.1 *Two-Photon Polymerization*

2PP is a nonlinear femtosecond-laser lithography in which two near-infrared photons are absorbed simultaneously within a tightly focused voxel, initiating polymerization only at the beam waist and enabling true 3D free-form fabrication in photosensitive resins.[8,17] Modern direct laser writing implementations routinely achieve sub-micrometer features and complex free-form architectures; strategies that push beyond the diffraction limit (e.g., stimulated emission depletion-inspired inhibition, dip-in objectives) and process windows for reliable feature control are comprehensively reviewed.[18] Print accuracy emerges from the full exposure-chemistry-transport stack: photoinitiator efficiency and resin formulation set the polymerization threshold, while laser power, scan speed, hatch spacing, and voxel overlap govern aspect ratio, necking, and cross-link density relationships quantified in processing-accuracy studies.[19] Resolution can be driven further below the far-field limit using peripheral photoinhibition/ stimulated emission depletion-like schemes and other inhibition optics, demonstrated at practical scan speeds.[18,20,21] Because serial voxel writing constrains raw throughput, parallelization via spatial-light-modulator holography and multi-beamlet optics enables multi-focus 2PP and large-area stitching, substantially accelerating build rates while preserving fine features.[22,23] For mechanical metamaterials, 2PP's geometric freedom underpins stretch-dominated lattices, re-entrant motifs, and shell-truss hybrids; hybrid post-processing then elevates performance: pyrolysis of polymer templates yields glassy carbon nanolattices with specific strengths approaching theoretical limits, and conformal ALD skins add angstrom-level control over coating thickness and composition, hence hybrid post-processing routes such as pyrolysis may help to enhance the mechanical performance such as stiffness, fatigue, and reduced porosity.[10,24] In biomedicine, the same voxel-level control enables scaffolds with graded stiffness, prescribed pore size/interconnectivity, and engineered surface cues; recent reviews survey 2PP-fabricated micro-/nano-scaffolds for cell guidance, organoid culture, and tissue regeneration.[25]

7.1.2 *Focused Electron/Ion Beam-Induced Deposition*

FEBID/FIBID is a maskless, direct-write nanolithography method in which a focused electron or ion beam dissociates adsorbed precursor molecules, "writing" material voxel-by-voxel with tens-of-nanometers placement precision and full 3D capability inside the microscope.[26,27] Literature establishes the growth physics (adsorption, surface diffusion, electron/ion dissociation, secondary electron yield) and typical deposit chemistries (granular metal-carbon composites) that govern rate, resolution, and purity.[26] The recent 3D nanoprinting roadmap demonstrates free-form nanowires, overhangs, and meshes, and summarizes key limits in throughput, precursor chemistry, and mechanical/electrical property optimization.[27] Material quality can be improved by *in situ*/post-growth purification (electron/thermal treatments and reactive gases) and electron beam curing, which densify deposits and increase conductivity; field overviews and perspectives document the progress and remaining bottlenecks.[28,29] To mitigate serial writing speed limits, cryo-FEBID/FIBID condenses precursor layers and then patterns with dramatically lower doses, achieving orders-of-magnitude faster growth for selected materials while preserving nanoscale features.[30] In the context of mechanical metamaterials, FEBID/FIBID are valuable as site-specific tools to stitch or reinforce lattice junctions, add conductive interconnects, and locally tune mechanics via hybrid overcoat capabilities, highlighted across recent reviews and perspectives.[27-29]

7.1.3 *Nanoscale Inkjet Printing — Micro/Nano Material Jetting*

Material jetting spans piezoelectric drop-on-demand inkjet and electrohydrodynamic jet printing; the former is governed by fluid property/printing number constraints (Ohnesorge, Weber, Reynolds), which set droplet stability and feature size limits,[31] while electrohydrodynamic jetting employs strong electric fields to produce sub-micrometer to sub-100 nm jets/droplets, patterning metals, polymers, and bioinks with high placement accuracy.[32] Electrohydrodynamic jetting has demonstrated nanoscale, 3D free-standing architectures via electrostatic autofocusing of nanodroplets, offering vertical buildup and fine lateral control well beyond conventional inkjet.[33] Recent 3D electrohydrodynamic "rapid nanodripping" further shows high-aspect-ratio metallic (e.g., Au) features with low resistivity approaching bulk values, underscoring its potential for post-functionalizing micro-architected lattices with conductive, catalytic, or sensing pathways.[34] Together, inkjet (for broader materials and gentle processing) and electrohydrodynamic jetting (for ultimate resolution) provide complementary routes for multi-material patterning on or within nano-architected metamaterials.[31-34] Nanoimprint lithography is not a jetting process; it is often paired with jetting or photopolymer routes as a high-throughput replication step. A high-throughput lithography technique with 25-nanometer resolution and smooth vertical sidewalls is proposed and validated. This method enables the manufacture of sub-10-nanometer structures and contributes to global sustainability, resource efficiency, and product life cycles.[35]

7.1.4 *Projection Micro-Stereolithography — High-Throughput Micro 3D Printing*

PμSL is an area-projection photopolymerization process that employs a digital micromirror device as a dynamic mask to expose an entire sliced layer per shot, enabling parallel fabrication of complex 3D microstructures with millisecond-scale exposures.[36] In PμSL, binary (or grayscale) images of the computer-aided design slices are relayed through reduction optics into a UV-curable resin bath; scanning-free projection yields high throughput, and bottom-up configurations with rapid recoating/peel strategies further accelerate layer turnover for fast builds.[37] State-of-the-art systems report sub-micrometer in-plane resolution down to ~0.6 μm and sub-micrometer layer thicknesses, while supporting a broad palette of photocurable materials (including hydrogels and shape-memory formulations) and multi-material implementations.[5] Critically for mechanical metamaterials, PμSL can be combined with nanoscale coatings or conversion steps to realize ultralight, ultrastiff lattices at exceptionally low relative densities.[11] Moreover, multi-material PμSL has enabled architected lattices with negative thermal expansion by integrating constituents with contrasting coefficients of thermal expansion into a single printed topology, demonstrating programmable thermomechanical responses in 3D.[38]

7.1.5 *Atomic Layer Deposition and Other Hybrid Techniques*

ALD deposits angstrom-level, conformal films via sequential, self-limiting surface reactions, uniformly coating ultra-high-aspect-ratio and porous architectures with precise thickness and composition control; typical growth-per-cycle values are on the order of ~0.1–0.2 nm per cycle, making ALD ideal for strengthening fragile micro/nanolattices with reduced porosity.[24] When applied to polymer-templated lattices, ALD enables hollow-tube ceramic shells with wall thicknesses of only a few to tens of nanometers and densities down to ≈6 kg m^{-3}; such structures exhibit high strength, energy absorption, and recoverability under large compressive strains.[39] Complementary hybridization pyrolysis of polymer scaffolds to glassy carbon pushes specific strength toward theoretical limits while preserving topology, providing a route to robust all-carbon nanolattices.[10] Beyond stiffness and strength, ALD coatings may help to tune surface energy, fatigue resistance, and environmental stability, which is valuable for integrating mechanical metamaterials into devices.[10,24,39] In parallel, polymer-derived ceramics produced by stereolithography followed by pyrolysis offer a direct path to ceramic lattices with relatively uniform shrinkage and high reliability.[14]

7.2 Innovations in Material Systems for Nanoscale Additive Manufacturing

7.2.1 *Nanocomposites*

Nanoscale reinforcements of sizes generally under 100 nm are added to the standard materials to create the nanocomposites. Various properties of the matrix are improved by these reinforcements, including the functional, thermal, physical, and mechanical properties. There are three major categories of nanocomposites:

1. Polymer matrix nanocomposites
2. Ceramic matrix nanocomposites
3. Metal matrix nanocomposites

7.2.1.1 *Polymer Matrix Nanocomposites*

Polymers are the most suitable materials for most of the additive manufacturing techniques. That is why polymer matrix nanocomposites are the most widely studied nanocomposites to date. Polymer matrices can be further divided into thermoplastics and thermosets. There are several properties of the materials that can be enhanced, including the functional properties, thermal stability, flame retardancy, and mechanical strength. There are some issues in the uniform dispersion and robust interfacial adhesion of the nanofibers and the polymer matrix, which cause difficulty in the formation of these composites and affect the overall performance.

7.2.1.2 *Metal Matrix Nanocomposites*

Metals like aluminum, titanium, or nickel can be strengthened by ceramics or other nanoparticles. The additive manufacturing techniques that use high temperatures, like selective laser melting and directed energy deposition, are used to fabricate the metal matrix nanocomposites.

7.2.1.3 *Ceramic Matrix Nanocomposites*

The applicability of ceramic matrices is constrained by their inherent properties of brittleness and fracture toughness. Here, the nanocomposite techniques play their role and reduce the intensity of these constraints by adding reinforcements that redirect the cracks and increase the energy absorption capability of the ceramic matrices.

7.2.1.4 *Hybrid Nanocomposites*

Nanofillers from other categories of materials can also be added to have a synergetic effect on the parent material. One example is the amalgamation of carbon nanotubes with graphene or other ceramic nanoparticles. It can give multifunctional properties, including mechanical and electrical reinforcement.[40,41]

7.2.2 *Multi-Functional Photoresins*

Multi-functional photoresins have emerged as a key class of materials in advanced additive manufacturing, particularly in 4D printing, where light-induced polymerization enables the fabrication of dynamic structures with programmable shapes, properties, and functionalities. These photoresin systems are typically composed of photocurable monomers or oligomers, photoinitiators, and functional additives that allow precise spatial and temporal control over network formation during processes such as stereolithography and digital light processing, as shown in Figure 7.1.

By incorporating stimuli-responsive components into photocurable formulations, multi-functional photoresins can be engineered to respond to external stimuli such as light, electric fields, temperature, magnetic fields, and moisture. This capability has significantly increased the interest in photoresin-based 4D printing across various research communities, as it enables

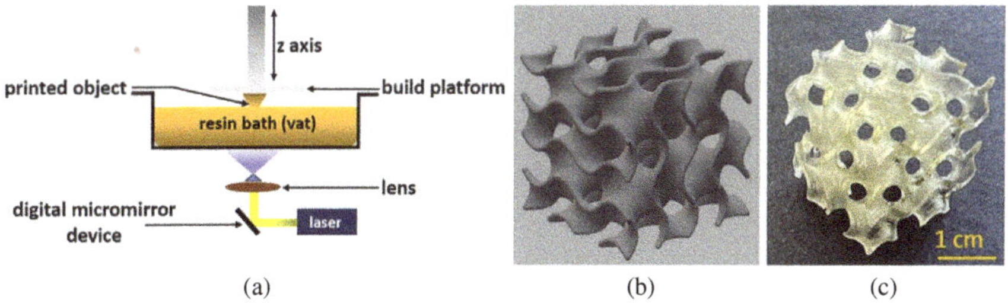

Figure 7.1. Schematic illustration of a digital light processing 3D printer and photographs of the printed objects using different resins: (a) A typical digital light processing 3D printer involving different components; (b) representative 3D computer-aided design model of a gyroid cube; and (c) corresponding 4D-printed gyroid using acrylated polyol.[42]

the production of structures that can undergo controlled and reversible transformations after fabrication.

In particular, these smart photoresin systems have demonstrated substantial potential in biomedical applications. Light-curable, stimuli-responsive polymers have been utilized in soft robotics, flexible bioelectronics, actuators, smart biomedical devices, drug delivery systems, and tissue engineering scaffolds. The ability to fabricate complex geometries with high resolution, combined with programmable functionality, makes photoresin-based systems especially suitable for applications requiring precision and adaptability.

Among these, temperature-responsive photoresins—often based on shape-memory polymer networks—are of particular interest. These materials can be programmed into a temporary shape and subsequently recover their original configuration when exposed to temperatures above their transition temperature. This behavior is enabled by the tailored crosslinked network formed during photopolymerization. Such properties make them promising candidates for minimally invasive biomedical applications, including enophthalmic invagination correction, cardiac occlusion devices, endovascular embolization, maintenance of respiratory tract patency, and vascular grafting.[42]

7.2.3 Stimuli-Responsive Polymers

Stimuli-responsive materials are programmable substances that can alter their geometries from transient configurations back to their original forms in response to environmental stimuli such as moisture, light, heat, pH, electric fields, and magnetic fields, following the shape memory cycle depicted in Figure 7.2. These materials demonstrate a range of characteristics, including self-folding, self-assembly, and self-adaptability. Moreover, self-healing attributes can be readily instilled in soft, responsive materials.

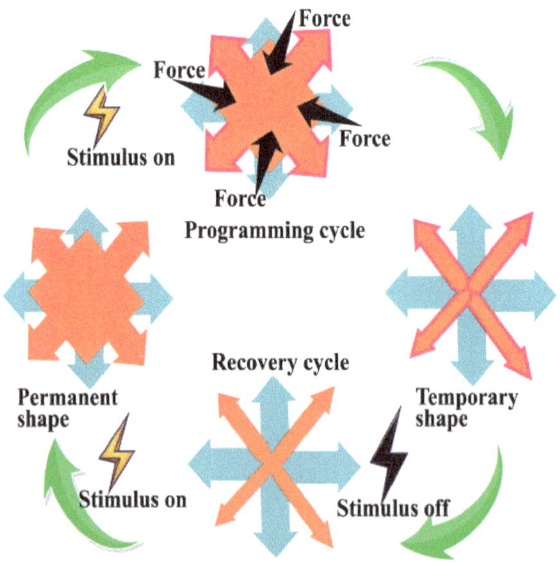

Figure 7.2. Schematic illustration of the shape-memory cycle for stimuli-responsive polymers.[43]

Stimuli-responsive materials can be engineered to undergo preprogrammed changes in shape and to alter color or other functional properties on demand. Because they can tune attributes such as thermal conductivity, stiffness, and optical appearance under external cues, these materials are drawing strong interest for tissue engineering, energy-absorbing structures, optical devices, and biosensing.

Within this class, shape-memory polymers (SMPs) stand out for their easy processing, low cost, large elastic stretch, tunability, reversible deformation, and robust shape-memory behavior. They also offer desirable traits including biodegradability, biocompatibility, and versatility that make them suitable for complex tasks such as scaffold fabrication. Functionally, the response arises from stimulus-driven changes in polymer chain conformation mediated by shifts in polymer–solvent or polymer–polymer interactions. Despite rapid progress, 3D printing of new SMP-based composites and systematic enhancement of their performance via targeted stimulation strategies remain early-stage topics that warrant deeper study.[43]

Stimuli-responsive polymers more broadly require an external input to trigger changes in structure, properties, or color. Depending on the stimulus, these flexible polymers can twist, bend, spiral, or fold.[43] Figure 7.3 summarizes representative stimuli and the macroscopic responses observed in stimuli-responsive polymer composites; the appropriate trigger should be chosen to match the intended operating environment.

Recent years have seen extensive investigation into the stimulus responses of various polymers (Figure 7.4), while Figure 7.5 illustrates the classification of stimulus-responsive polymers according to the type of external stimulation.

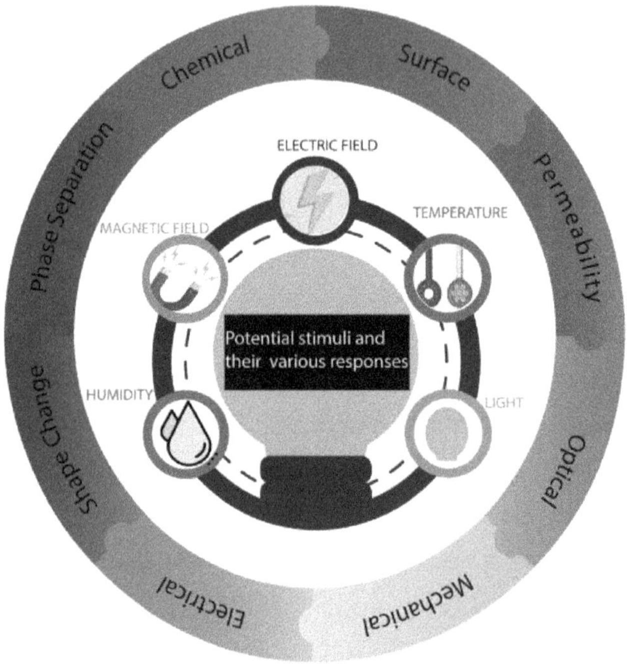

Figure 7.3. Potential stimuli and response of polymers.[43]

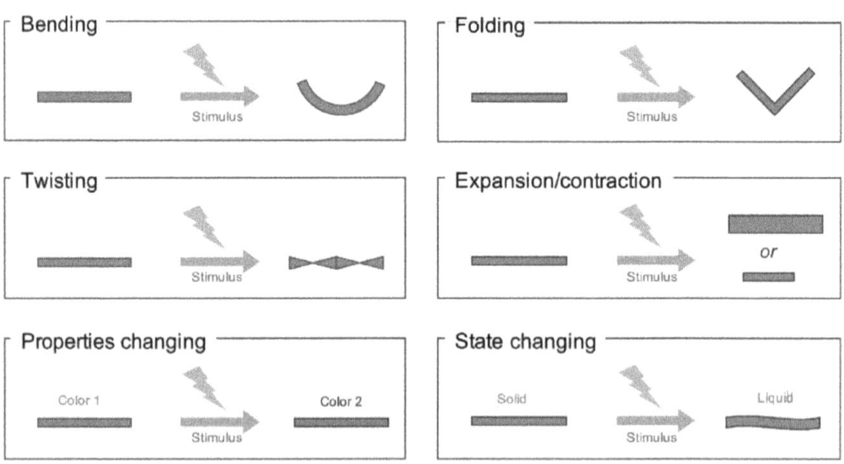

Figure 7.4. Stimuli-responsive polymers show a wide range of behaviors when triggered by external cues. Adapted with permission.[43,44]

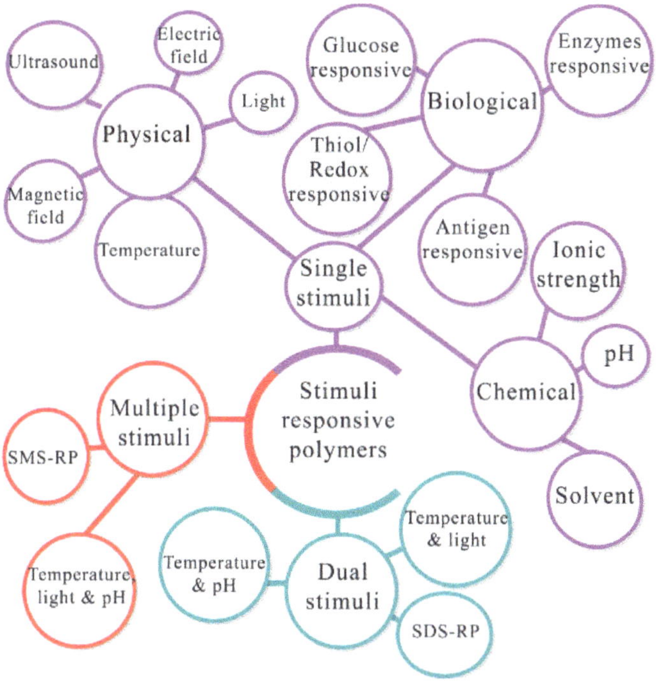

Figure 7.5. Classification of stimuli-responsive macromolecules.[43]

This section includes a comprehensive analysis of several stimuli-responsive polymers, encompassing temperature, light, humidity, electro, magneto, pH, and multi-responsive types.

7.2.3.1 *Temperature-Responsive Polymers*

Thermo-responsive polymers are soft, smart materials that switch behavior with temperature, often undergoing sol–gel transitions. This temperature-triggered response makes them useful in tissue engineering, controlled drug release, and gene delivery.[43]

7.2.3.2 *Photoreactive Polymers*

Photo-responsive polymers alter their functional properties when exposed to light. They usually print well, work well with other materials, and are easy to work with. Near-infrared, a clean, remotely deliverable, and directionally controllable energy source, is widely used to drive light-actuated structures with minimal damage to materials or operators. Actuation is commonly achieved by embedding photoactive moieties (e.g., azobenzene derivatives) into the polymer matrix.[43]

7.2.3.3 *Magnetoactive Polymers*

Magneto-active polymers are advanced materials that can modify their structure and mechanical characteristics when subjected to a magnetic field stimulus. These polymers consist of two fundamental components: a non-magnetic matrix and magnetic particles. Magnetic nanoparticles are incorporated into polymer matrices to achieve magneto-responsive properties. The responsive behavior of magneto-active polymers in a non-magnetic matrix is ascribed to hard or soft magnetic fillers.[43]

7.2.4 *Shape-Memory Polymers*

SMPs are a type of polymer that revert to their original shape as an external stimulus is applied. Based on the activation source or mechanism, they can be divided into two categories:

1. Thermally activated SMPs
2. Non-thermally activated SMPs

SMPs offer more strain recovery capability, enhanced flexibility, and reduced density as compared to conventional metal materials. Nevertheless, there are still some areas of advancement or improvement like recovery speed, cycle flexibility, and durability.

Intricate SMPs can be produced by advanced additive manufacturing techniques with precision that facilitates customization and functionalization. The properties of SMPs, like the shape-memory effect and mechanical properties, can be investigated using additive manufacturing techniques. With the advancement of additive manufacturing technologies, the applications of SMPs will also expand to a large extent, as can be seen in Figure 7.6.[44]

An important benefit of additive manufacturing, encompassing 4D printing, is the capability to simultaneously process many materials. The production of SMPs allows for the attainment of more intricate shape-memory behaviors and performance by using polymers with varying properties inside the same component.[44] For instance, intelligent and functional components can be produced by employing polymers with enhanced shape recovery rates in essential regions and polymers with superior mechanical qualities in alternative places. Multi-material 4D printing technology is pioneering a novel trend in drug delivery and biomedical applications.[44]

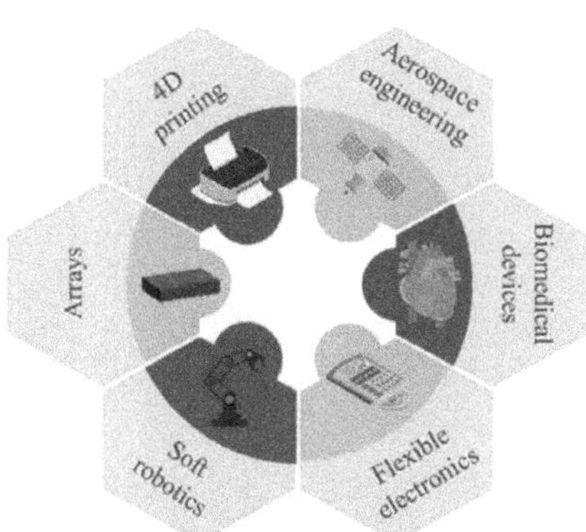

Figure 7.6. Utilization of SMPs and their composite materials in different applications.[44]

7.2.5 Functional Ceramics and Metals via Nanoscale Additive Manufacturing

Additive manufacturing of ceramics is fundamentally different from that of polymers and metals. Parts are built in two stages: first, an additively made green body is printed from a feedstock of ceramic powders mixed with polymer binders/additives; then the polymer is debonded, and the part is sintered to develop ceramic properties. Because this green body is essentially a polymer-rich composite containing dispersed ceramic particles, many polymer additive manufacturing processes can be adapted to produce ceramic components.[45] Multi-material design of a ceramic forming tool is shown in Figure 7.7.

Ceramic additive manufacturing routes can be organized by raw material form, printing method, and shaping mechanism. Under ASTM/ISO 52900:2021, additive manufacturing processes fall into seven families: material extrusion, material jetting, vat photopolymerization, powder bed fusion, sheet lamination, directed energy deposition, and binder jetting. For ceramics, these

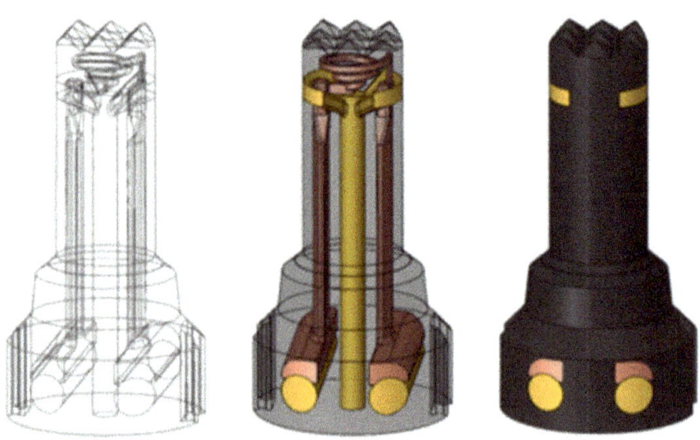

Figure 7.7. Multi-material design of the ceramic-forming tool. Yellow indicates support material.[45]

map cleanly to four mechanism-focused groups: extrusion-based, photopolymerization-based, powder fusion-based, and powder bonding-based methods.

7.3 Design Strategies for Nano-Architected Metamaterials

7.3.1 *Topology Optimization and Computational Design*

Additive manufacturing enables intricate geometries and material efficiency, which in turn has accelerated advanced design tools such as topology optimization, lattice structures, and lattice-infill topology optimization aimed at cutting weight while preserving, or improving, structural performance. These methods are especially valuable in automotive, aerospace, and biomedical applications where strength-to-weight and integrity are critical. Recent research on multi-objective multi-patch iso-geometric topology optimization (MP-ITO) enhances solid, porous, and hybrid infills in region-partitioned designs, yielding global structural enhancements while maintaining practicality for engineering applications. MP-ITO has also been used to build non-conforming meshes for periodic and graded cellular architectures, with performance validated in numerical studies.[46]

The structural integrity, mechanical behavior, and manufacturability must also be taken into account. To ensure the load-bearing capacity of the material, the anisotropy and the variability induced by the additive manufacturing techniques are monitored. For this purpose, a Lagrangian–Eulerian particle flow has been introduced,[47] in which the authors optimized the topology based on the iso-geometric material point method, which improved the structural performance. Some other hybrid interval random models have been employed by related studies, which guide creation of robust designs. It is essential to model the real world's performance of these materials before bringing them into practice.

Strong energy absorption and durability have been exhibited by lattice designs. Furthermore, multi-lattice topologies enable the designers to create more designs by mixing the unit cell types and maintaining the manufacturability and connectivity.[46]

There is another technique, known as lattice-infill topology optimization, which combines lattice structures with topology optimization. This enables the designing of load-adaptive geometries and lightweight cellular layouts. This approach is mainly used in biomedical contexts, where porous architectures support bone ingrowth and match tissue mechanics. It has been observed that the hybrid-strut Ti-6Al-4V lattices have outperformed solid-strut counterparts, showing +32% strength, +15% stiffness, and +65% toughness, underscoring their value in lightweight components. In aerospace, Ti-6Al-4V lattices reduce mass while maintaining integrity. In biomedicine, porous Ti-6Al-4V implants promote osseointegration and reduce stress shielding. Its strong energy absorption capacity supports automotive and defense crash protection, and its thermal properties benefit heat exchange applications. The alloy's ductility also facilitates intricate lattice geometries and advanced topology optimization, reinforcing its practical importance.[46] The attributes of this material are further enhanced by optimization methods, leading to substantial advancements in performance and safety for demanding applications. However, while topology optimization is effective under specified loads and can be tolerant of some print defects, it does not directly resolve residual stresses generated during additive manufacturing, which can threaten long-term integrity.[46]

Significant benefits have been observed in the lattice structures for mass reduction, although they exhibit reduced safety margins under operational stresses in high-performance applications like aeronautical technology. Lattices excel at mass reduction, but in high-demand environments (e.g., aerospace), they may carry narrower safety margins under operational loads. Lattice-infill topology optimization addresses such issues by co-optimizing the macro-layout and the micro-lattice to tune the strength–weight trade-off. Meeting these targets still requires explicit consideration of safety factors, process limits, and service performance.

Despite significant examination of face-centered cubic (FCC), body-centered cubic (BCC), and their combination structures, research is scarce on fully integrated lattice structures that connect the central nodes of both FCC and BCC configurations, particularly in complex geometries and on load-adaptive unit cells that vary density and strut thickness inside a single geometry. To address this gap, a study proposed FBCore, a hybrid metamaterial unit cell that fuses FCC and BCC through a core-centric layout. The architecture is geometrically complex and load-adaptive, with spatially graded strut thicknesses to improve local load distribution.[46]

7.3.2 *Nano-topology Optimization*

A growing objective in this field is to advance Feynman's atom-by-atom vision through a genuinely *de novo*, top-down design framework that specifies nanoscale materials with atomic-level precision. To distinguish this atomistic route from conventional finite element method formulations, we refer to the approach as nano-topology optimization (nano-TO). In a recent study, nano-TO was employed to design nanostructured materials with targeted elastic responses. The optimized layouts not only outperformed well-known references such as gyroids and other triply periodic minimal surface architectures, but by deliberately leveraging nanoscale surface effects, they also exceeded the classical Hashin–Shtrikman upper bound for bulk modulus. Put differently, by shaping and tuning the energetics of surfaces at the nanometer scale, nano-TO can achieve an apparent stiffness that outstrips continuum-level theoretical limits. The broader significance lies in providing a computational platform that allows computers to propose and realize novel materials, atom by atom, without any pre-imposed motif, thereby enabling the development of diverse families of high-performance nanomaterials and nanomachines.[48]

In the design process of the nanostructured solids, the surface phenomenon is exploited by the nano-TO techniques. Atomistic coordination affects the bonding strength of the materials, especially at the surfaces, edges, and corners, as shown in Figure 7.8. But this effect can only be captured by the atomistic simulations. Thus, the traditional finite element method-based

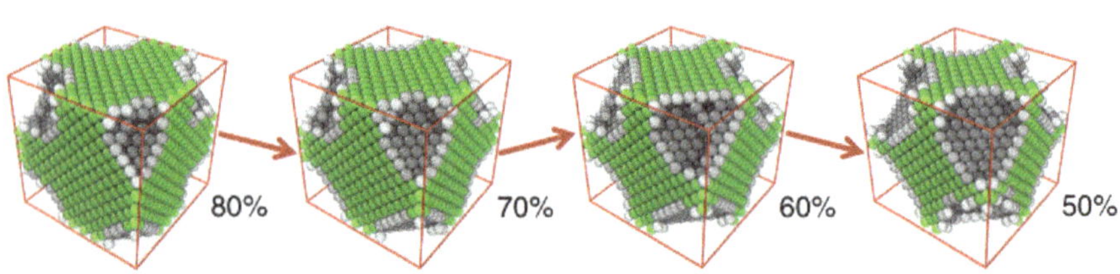

Figure 7.8. The nano-TO designs aim to optimize the bulk modulus with volume fractions ranging from 50% to 80%. The bulk atoms are depicted in green, whereas the surface atoms are represented in gray.[48]

topology optimization cannot be reproduced by the nano-TO designs. The framework adopted in the study[48] was used to enhance the bulk modulus and elastic constant. They used the weighted sum strategy to get the multi-objective nano-TO results. The sensitivity values were computed for each atom with respect to each objective and then combined into a weighted sensitivity using designer-selected weights that encode the relative priority of those objectives. This procedure yielded families of nano-structures with tunable trade-offs, providing practical handles for tailoring performance at the atomic scale.[48]

By construction, nano-TO supports multi-objective design, allowing it to search for architectures that strike a balanced trade-off among competing targets rather than optimizing only a single metric. Although most of the examples focus on tuning the elastic moduli, the same framework can be extended to engineer other nanoscale functionalities as well. In metallic systems, many key mechanical behaviors are tied, directly or indirectly, to the elastic modulus. Measures of elastic anisotropy, for example, can be used to flag alloy chemistries and microstructures associated with super-elasticity, exceptionally high strength, and unusual ductility, traits characteristic of so-called gum metals. If the quantitative links between a material's elastic modulus and its other mechanical responses (e.g., failure strain, yield/ultimate strength, toughness) are mapped with sufficient clarity, then the same nano-TO machinery used to tailor elastic properties can be repurposed to steer those other properties. In that scenario, per-atom sensitivities and multi-objective weighting would target not only stiffness but also damage tolerance and deformability, enabling systematic, atom-level design of nanomaterials with co-optimized mechanical performance.[48]

7.3.3 *Artificial Intelligence and Machine Learning in Metamaterial Design*

The literature frames artificial intelligence in metamaterial design around three linked activities: data preparation/analysis, algorithm and model development, and model evaluation. Before any artificial intelligence tool is deployed, the underlying datasets, typically drawn from experiments or numerical simulations, must be cleaned and organized since raw measurements are often incomplete, inconsistent, or noisy. Preparation steps reorganize relevant portions of the data, remove artifacts, and establish a reliable structure for learning; machine learning can also help denoise signals and impute missing entries to recover usable information. On the modeling side, approaches span supervised, unsupervised, and hybrid learning and are often discussed in two broad families: generative models and optimization methods.

Generative models such as generative adversarial networks and variational autoencoders explore latent design spaces to propose new metamaterial candidates that would be difficult to discover manually. On the other hand, efficiency and convergence are the main focus of topology optimization that moves the design towards the target specifications. The generative modes tend to excel in novelty and ideation, while the optimization models try to refine the performance.[49]

Artificial intelligence can take advantage of artificial metamaterials, as they can provide the physical mechanisms for faster computation and improved signal handling within machine learning pipelines. These materials can accelerate core neural network operations, particularly matrix multiplications and convolutions, that use electromagnetic physical neural networks to exploit diffraction through engineered media with programmable amplitude and phase.[50]

Typical digital computing systems can be improved, and parallel processing at the speed of light can be achieved by modifying the characteristics of the metamaterials. Efficient data management and computation in neural networks is achieved by establishing an innovative framework. Also, the energy loss is reduced, and data transmission is optimized by the deployment of artificial metamaterials.[50]

7.3.4 *Tandem Neural Networks*

The tandem architecture is a baseline setup for generative inverse designs. It couples a decoder (a forward surrogate model) with an encoder (the inverse mapper). In the forward model, feed in the design variables and train the decoder to predict the physical response (e.g., a reflectance spectrum). While locking the decoder's weights, attach an encoder in front and supply the target response (e.g., the desired spectrum) to the encoder. The encoder proposes a candidate design, which is then passed through the frozen decoder to produce a predicted response. Compare the predicted response with the target; the difference defines the training loss. Backpropagation carries gradients through the frozen decoder to the encoder, and repeated updates drive the encoder to recover the essential structural parameters that yield the target behavior.[50–52]

7.3.5 *Variational Autoencoders*

Variational autoencoders address the one-to-many nature of inverse problems by learning a structured latent space in which multiple valid designs map to different latent vectors. After training converges, inverse design amounts to steering or sampling the latent code and decoding it into candidate geometries. The strength of variational autoencoders is their stable training and systematic exploration of latent space, which tends to reveal diverse, viable metamaterial designs rather than a single solution.[50,53–55]

7.3.6 *Generative Adversarial Networks*

Generative adversarial networks offer a complementary generative route. Whereas variational autoencoders emphasize reconstruction and latent regularization, often yielding smoother, slightly blurred outputs, generative adversarial networks pit a generator against a discriminator in an adversarial game, pushing the generator to produce high-fidelity, detail-rich designs. Although they can suffer from mode collapse, the adversarial objective steadily pressures the generator toward increasingly realistic candidates. This makes generative adversarial networks well-suited to metamaterials, where the design space is highly nonlinear, and the outputs can be intricate in both geometry and property distribution.[50,56–59] The four basic artificial intelligence network architectures are represented in Figure 7.9.

7.4 Current Challenges and Potential Solutions in Nanoscale Additive Manufacturing

7.4.1 *Resolution vs. Throughput Trade-offs*

In its broadest definition, manufacturing is the process of transforming information into ordered matter. Three interconnected subprocesses make up this process:

1. Information processing and memory retrieval;
2. Hardware functionality; and
3. Material reaction.

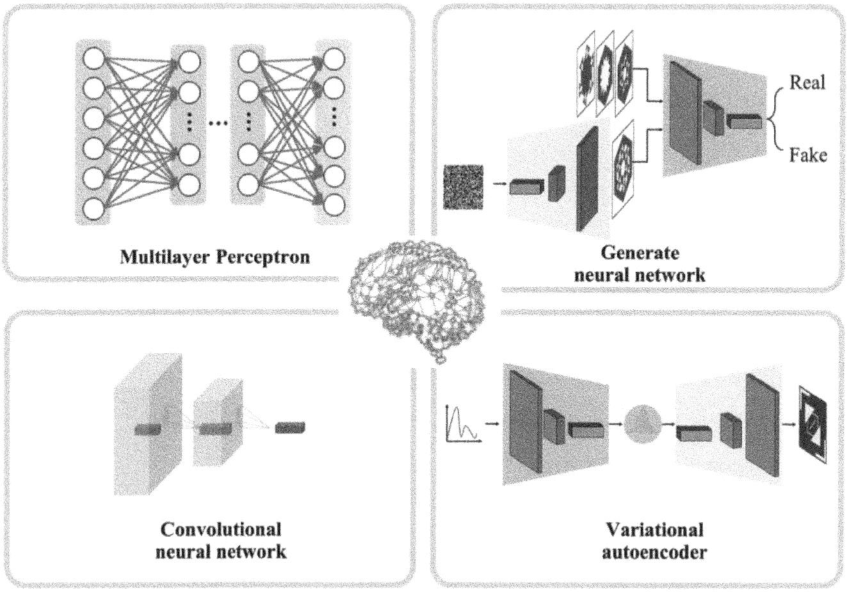

Figure 7.9. Graphical representation of the four basic artificial intelligence network architectures.[50]

The slowest subprocess serves as the limiting function (or "bottleneck") for the entire manufacturing process's pace, and each subprocess has a maximum rate or throughput that can be established.

Thanks to several advances in resolution, material generality, and throughput, assembly-based 3D nanofabrication has gained more traction in nanotechnology in recent decades. This approach is anticipated to have a higher impact on improving structural control, heterogeneous fabrication techniques, fabrication efficiency, and environmental friendliness in scalable applications, despite its enormous unrealized potential.

Because fully printed electronics techniques can be directly printed on a wide range of flexible and stretchable substrates, including paper, thermoplastic polyurethane, polyethylene terephthalate, polyimide, and polydimethylsiloxane, they are more affordable, effective, and scalable than traditional MEMS/NEMS processes. Nevertheless, these methods are also constrained by product yields and throughputs, which make it exceedingly challenging to produce hybrid devices in a way that would enable their widespread use. Aerosol jet printing, for example, has shown extremely high print resolutions and control over material heights and microarchitectures; yet it is inevitably a low-speed manufacturing alternative due to the additive deposition of several nano-thickness layers.

7.4.2 Manufacturing Repeatability and Scalability

Even though both additive manufacturing technology and nanomaterial development have advanced significantly on their own, methodologies for incorporating nanomaterials into additive manufacturing processes have mainly stayed the same and lack the intelligence and flexibility needed for real-time process optimization. With their limited ability to adapt dynamically to the unpredictability present in additive manufacturing environments, such as variations in temperature, material flow, and laser power, existing techniques mostly concentrate on pre-processing parameters and post-processing activities. The creation of a cohesive, intelligent system that combines real-time data collection, predictive control, and real-time decision-making to govern nanomaterial behavior during production is thus severely lacking. This lack of flexibility restricts scalability across many applications and jeopardizes the uniformity and quality of printed components. Technology's dependability, functionality, and industrial viability are all threatened by this gap's continued existence.

7.4.3 *Nanoscale Mechanical Reliability and Metrology*

Many difficulties are observed in metrology due to the difference in material behavior at the nanoscale versus their bulk equivalents. Accurate measurement of characteristics including elasticity, plasticity, viscosity, and fracture toughness at the nanometer is crucial from microelectronics to biomedical devices. Important difficulties in the fabrication and use of these technologies include:

- **Sensitivity to Scale:** Some factors, like surface interactions and quantized deformation at the nanoscale, become equivalent to bulk forces and necessitate the use of an extremely sensitive instrument.
- **Tool Limitations:** Extremely precise operation is required for certain techniques, such as electron backscatter diffraction, confocal Raman microscopy, nanoindentation, and atomic force microscopy.
- **Material Behavior:** Some phenomena like quantized plasticity, compliant grain boundaries in nanocrystalline materials, and higher elastic moduli in nanowires need new measurement methods and standards.
- **Environmental Effects:** Adhesion and contact mechanics are greatly impacted by variables like humidity, which makes measurements more difficult.
- **Device Reliability:** A thorough understanding of stress distributions and fracture mechanics at the nanoscale is crucial for the design of robust nanoelectromechanical systems.

7.4.4 *Material Restrictions and Process Harmony*

To enable the usability of additive manufacturing as a TL4-level technology and facilitate its transition from research labs to actual companies, a number of obstacles must be fully addressed (Figure 7.10). The compatibility of raw materials with additive manufacturing techniques is one of the primary problems. Another issue that additive manufacturing materials encounter is the limitations imposed by the manufactured parts' surface smoothness, dimensional accuracy, anisotropic behavior, etc. The absence of testing facilities for certain parts and materials made via additive manufacturing is yet another difficulty. The emergence of multiple faults in items manufactured using additive manufacturing techniques is one of the biggest challenges.

Materials are essential to additive manufacturing operations. Currently, additive manufacturing raw materials such as polymers, ceramics, composites, metals, alloys as well as functionally graded, smart, and hybrid materials are widely used. The raw

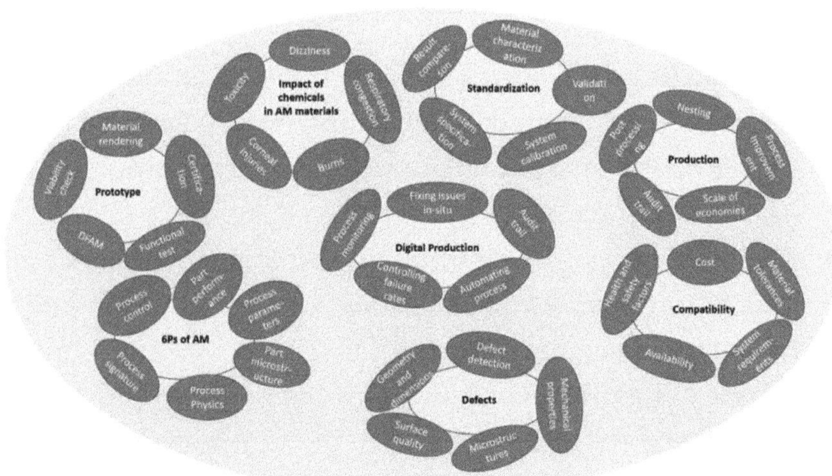

Figure 7.10. Challenges of additive manufacturing materials.[60]

materials and the additive manufacturing machines should work together. Rigidity in raw material selection is a major obstacle for the majority of novel additive manufacturing applications. Hybrid and functionally graded materials provide some relief from these problems.

A few factors about currently available raw materials that limit the full exploitation of additive manufacturing technology include poor mechanical performance, high cost, lack of appropriate machine availability, health risks associated with various materials, restrictions on testing, as well as standardization and material characterization techniques. To comprehend the kinetics of the process, one must first comprehend the physics underlying the binding mechanisms connected to the materials used in various additive manufacturing procedures.

Important research areas related to additive manufacturing materials include methods to enhance mechanical and microstructural properties, ways to create specially engineered materials, recycling of additive manufacturing materials, the importance of material particle size and distribution, investigating thermal issues, synthesizing newer materials specifically for customized in-house applications, creating a material database, etc. Particularly when producing parts for important industries like biomedical, construction, aerospace, automotive, etc., these concerns require careful study.[60]

The uniform dispersion of reinforcement particles inside the matrix is a crucial prerequisite for the synthesis of feedstock materials and has a big impact on the final properties of the composite. Additionally, the build-up of filler particles may have an impact on the vat photopolymerization process's uncured photopolymer and the material extrusion process's clogged nozzle. Although nanoparticles have a significant impact on enhancing mechanical properties, their aggregation and poor interfacial adhesion might result in decreased ductility or failure strain in nanocomposites. The aggregated nanoparticles deviate from the anticipated material characteristics by acting as stress-concentrating flaws. For nanocomposites to function at their best, homogeneous dispersion of nanoparticles is essential. To guarantee proper dispersion, a high shear mixing method such as a twin-screw extruder should be used.

Achieving better electrical or thermal conductivity in the direction of oriented nanoparticles may be advantageous, depending on the functional application of nanocomposites. The mechanical characteristics of 3D-printed nanocomposites are significantly impacted by printing orientation. There are numerous reports of anisotropic effects brought on by printing direction. Things manufactured vertically may have up to five times lower mechanical qualities than things printed in the XY direction.[61] In this regard, a polymer blend can be used to provide customized mechanical properties with reduced mechanical anisotropy. The successful additive manufacturing process in the molten state is driven by the rheological properties of the polymer material, which are changed by nanoparticle loading.

The mobility of the polymer chains is limited when nanoparticles are added to the polymer matrix; additionally, agglomeration influences the rheological behavior, which impacts the material's printability. The feedstock filament diameter is a critical factor in the quality of the final product in the fused filament fabrication process. As a result, during the filament extrusion process, a constant filament diameter must be reached. Because the particle profile might not be appropriate to offer adequate interfacial interaction between the two phases, the size and shape of the reinforcement are crucial. The quality of the end product in the fused filament fabrication process is also greatly influenced by several internal and external parameters that Popescu et al.[62] reported.

The homogeneity of the filler material within the composite matrix is also influenced by its nature; most polymers are hydrophobic. Combining a polymer solution with a hydrophilic filler is challenging, if not impossible. For instance, graphene can be employed in cellulose composites by dispersing it using an aqueous suspension of nano-fibrillated cellulose. In mechanical, thermal, and electrical applications, graphene has shown itself to be a potential filler material. Therefore, in order to properly utilize it, fresh ways and processes are needed. Grafting polymers is a successful way to increase dispersion. The efficient distribution of particles inside the polymer matrices nevertheless presents certain difficulties. The resolution of the additive manufacturing process is crucial to the quality of the finished product in terms of mechanical strength, dimensional tolerance, surface finish, and better resolution. Because it is easier to manage the light source or powder particle size in vat photopolymerization and powder bed fusion processes, respectively, they are often better at controlling product surface morphology than material extrusion techniques. Higher resolutions allow for more sophisticated process repeatability, which is advantageous especially for crucial technical applications like synthetic bone models.

However, lengthier printing times are typically associated with higher resolution. In order to properly take into account the requirements for the finished product, intended cost, lead time, and initial polymer composite properties, the best process windows (3D printing speed, temperature) and parameters must be created. The qualities of the printed part are greatly influenced by process variables such as printing temperature, ambient temperature, infill pattern, infill density, and printing speed. Porosity and voids would decrease as a result of high-temperature processes. For consistent printing, the material's printability is crucial. Simple polymers are easily printed, but additive manufacturing of composites is difficult for several reasons, including temperature gradients and the flow properties of the composite materials discussed in earlier sections.

The mechanical and adhesive characteristics of the composites are correlated with the interaction between the polymer and reinforcement, which may also result in structural flaws. Modified morphology, the ideal aspect ratio, surface roughness, or the addition of functional groups to reinforcements can all improve this interaction. Better mechanical properties will come from increased interfacial adhesion between the two phases. 3D printed composites' interlayer adhesion has been enhanced by the use of annealing.

To determine the ideal temperatures for various polymers or create innovative post-printing procedures, thorough research is still necessary. The physical, mechanical, or thermal qualities of the finished product are directly impacted by the reinforcement content or material composition. Finding the ideal polymer ratio to strengthen a particular particle's size, shape, and polymer type is crucial. The existence of structural flaws, which continue to be a major barrier to high-performance applications, shapes the dependability of final products. In general, conducting the procedure under vacuum circumstances helps prevent structural flaws. For low-cost applications, this condition might not be the best option. Additionally, end-of-life products and waste material recycling need to be taken into account at the design and production stages.[63]

7.5 Applications of Nanoscale Additively Manufactured Mechanical Metamaterials

7.5.1 *Tissue Engineering*

When the area of biomaterials was still in its infancy in the early 1960s, the idea of tissue engineering was conceived. It became necessary to incorporate cells and physiologically active chemicals into these systems in order for them to function as seamless tissues, as biomaterials research broadened to include all biocompatible materials created especially for use in the body. Scaffolds must be created by weaving biomaterials, cells, and physiologically active chemicals in the matrix to attain this level of complexity and replicate native tissue characteristics. This will encourage cell adhesion, growth, differentiation, and even migration as required.

By offering structural support to promote cell adhesion and new cellular growth, and preserving cell physiology and intercellular molecular signaling, tissue engineering aims to create bio-composites that can repair damaged tissues.

Restoring sizable tissue segments may also result in the preservation of an entire organ. Scaffolds must offer a milieu that is conducive to cell adhesion and protein adsorption. The physicochemical properties of scaffold microstructures must be tailored to the target tissue to support:

a) the adsorption of adhesive proteins released by surrounding cells before their attachment, and
b) the sustained release of a variety of biomolecules (such as growth factors) incorporated into the matrix to promote cell growth and differentiation; for instance, osteoblast attachment, followed by cellular growth and the formation of new bone structures.[64]

In tissue engineering, nanoparticles have been employed for several purposes, including molecular detection, biosensing, gene delivery, DNA transfection, viral transduction, patterning of cells, and improving biological, electrical, and mechanical characteristics. Using the proper kind of nanoparticles can greatly improve the scaffolds' qualities. Depending on the application, they can also perform a variety of other tasks, as detailed below.

7.5.1.1 *Cardiovascular Tissue Engineering*

Nanomaterials play a pivotal role in enhancing scaffold performance for cardiac tissue regeneration:[65]

- **Gold Nanotubes/Nanowires in Polyurethane Scaffolds:** These improve electrical conductivity and cell communication. When cardiomyocytes were seeded on gold-integrated scaffolds and electrically stimulated, gene expression related to cardiac differentiation (Nkx2.5, ANF, NPPB) significantly increased.
- **Chitosan-Gold Nanoparticle Hydrogels:** These thermosensitive hydrogels showed enhanced electrical conductivity and supported mesenchymal stem cell differentiation into cardiac lineages. The porous structure facilitated cell migration and proliferation.
- **Fullerenol/Alginate Hydrogels:** Designed to combat oxidative stress, these hydrogels activated ERK and p38 pathways while inhibiting JNK, promoting cell survival and angiogenesis in myocardial infarction models.
- **Carbon Nanotube-Enhanced Electrospun Vascular Scaffolds:** Carbon nanotubes increased tensile strength and

cell alignment. Endothelial cells seeded on these scaffolds showed improved viability and elongation.

7.5.1.2 *Neural Tissue Engineering*

The following are the main applications of neural tissue engineering:[65]

- **Carbon Nanotubes:** Carbon nanotubes are widely used for their conductivity and biocompatibility. Modified carbon nanotubes can be used to reduce cytotoxicity and support neural stem cell adhesion.
- **Carbon Nanotube-Based Hydrogels:** These hydrogels increase the porosity and conductivity, which in turn help in enhancing synaptogenesis.
- **Graphene Derivatives:** Reduced graphene oxide incorporated into poly(3,4-ethylenedioxythiophene), polycaprolactone, and silk fibroin scaffolds can improve electrical properties and neural differentiation.
- **Gold Nanoparticles and Nanorods:** If they are integrated into chitosan or silk fibroin scaffolds, they can enhance Schwann cell viability and neurite extension.
- **Electroactive Polymers:** Polypyrrole-based scaffolds support the PC12 cell adhesion and neurite growth, with tunable conductivity and biocompatibility.

7.5.1.3 *Bone Tissue Engineering*

Nanomaterials enhance osteogenesis, scaffold strength, and anti-bacterial properties. Graphene oxide and hydroxyapatite form porous, conductive hydrogels that improve mesenchymal stem cell adhesion and elongation. Silver nanoparticles incorporate into collagen scaffolds to induce osteoblast differentiation and accelerate fracture healing. Zirconium oxide nanocomposites support MG-63 cell proliferation and provide some antibacterial effects, while copper and zinc nanoparticles improve scaffold roughness and porosity, facilitating fibroblast adhesion and migration.[65]

7.5.2 *Energy Harvesting and Dynamic Mechanical Systems*

For small, low-power devices, tapping ambient energy is a practical alternative to relying solely on batteries. Using the harvested power near the source often reduces or eliminates the need for long transmission runs or large storage systems. Recent micro/nano transducers, triboelectric, electromagnetic, piezoelectric, and electrostatic devices have demonstrated electrical outputs from tens to hundreds of μW.[66,67]

Common ambient inputs include wind, sunlight, ocean waves, human motion, rainfall, and environmental vibrations. These can be converted through several mechanisms: electromagnetic, piezoelectric, triboelectric, thermoelectric, pyroelectric, or hybrid schemes that combine more than one pathway. Additive manufacturing is now widely used to build the structural elements of these devices, e.g., solar "trees," turbine blades, housings, piezoelectric substrates, and micro/nanotextures for triboelectric nanogenerators (TENGs), enabling customized form factors and rapid iteration. Nano-mechanical energy harvesting (NMEH) underpins many applications: wearables, wireless sensor networks, navigation beacons, wastewater treatment, phone charging, biomedical monitoring, LED lighting, weather stations, structural health monitoring, and cloud-connected sensors.

The raw output typically passes through power management and rectification circuits before use, and surplus energy is buffered in batteries or supercapacitors.[66,67]

An overview of the predominant methods used in 3D-printed NMEH and an illustrative prototype are discussed, and Figure 7.11 shows that most 3D-printed NMEH systems use electromagnetic,

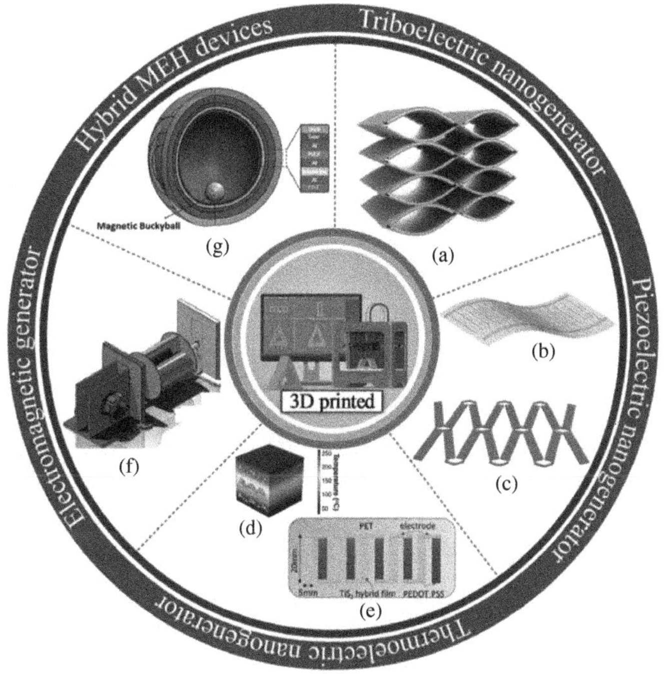

Figure 7.11. The classification of 3D-printed energy harvesting systems is based on their operating principles. (a) Flexible TENG used as a self-powered electro-Fenton degrader; (b) 3D-printed piezoelectric device for autonomous sensing; (c) flexible piezoelectric nanogenerator with a non-protruding kirigami layout; (d) 3D-printed BiSbTe thermoelectric generator; (e) p-/n-type organic superlattice thermoelectric device on flexible polyethylene terephthalate; (f) miniature airflow-driven electromagnetic harvester; (g) hybrid inertial sensor combining EMG, TENG, and PENG.[66,67]

piezoelectric, triboelectric, or thermoelectric methods, or a mix of these, to convert energy. One example is a flexible 3D-printed TENG that can be used for self-powered electro-Fenton degradation and blue energy harvesting in wastewater treatment. The prototype delivered a maximum power density of 6 W m^{-2}, a short-circuit current of 2 mA, and an open-circuit voltage of 610 V, achieving ~98% degradation of methylene blue in about one hour.[66,67]

7.5.2.1 *Electromagnetic Energy Harvesters*

NMEH devices that are based on electromagnetic generators (EMGs) operate on the basis of Faraday's law of electromagnetic induction, which establishes a connection between the time rate at which the magnetic flux changes through a coil and the induced electromotive force across it. A current is induced in the external circuit by the relative velocity of the rotor coils and the nearby magnets. Both stationary and moving coils and magnets are possible.

7.5.2.2 *Nanogenerators that Use Triboelectric Power*

TENGs convert mechanical motion into electricity through contact electrification coupled with electrostatic induction. They are promising for nanoscale energy harvesting because they combine high conversion efficiency, low fabrication cost, and useful power density, and they work across many everyday sources: wind, water waves, human motion, and ambient vibrations. To boost output, researchers increase the effective frictional contact area between the opposing triboelectric layers by engineering micro/nanostructures on their surfaces.

Common fabrication routes include electrospinning, surface etching, and spin coating, each of which increases roughness or patterning to intensify charge transfer during contact–separation cycles.[67]

7.5.2.3 *Nanogenerators that Use Piezoelectric Power*

Piezoelectricity describes the two-way coupling between mechanical and electrical states in certain materials. Under the direct effect, applying mechanical stress generates an electric potential across the device terminals. Under the inverse effect, an applied electric field produces a mechanical strain in the material. Devices that exploit these effects (often called PENGs, piezoelectric nanogenerators) harvest vibration or motion as electrical energy or actuate when driven electrically.[67]

7.5.2.4 *Mechanisms of Hybrid 3D-Printed Nano-mechanical Energy Harvesting*

To maximize energy capture from diverse ambient inputs, systems often combine multiple transduction mechanisms in a single platform. At the nano/microscale, PENGs and TENGs are especially flexible in form and structure, which makes them natural partners for EMGs. As a result, the two most common hybrids are PENG–EMG and TENG–EMG. Additive manufacturing supports these hybrids by supplying custom structures and components, including substrates, blades, frames, shells, casings, and other geometries, tailored to the operating environment and frequency band. This manufacturing freedom accelerates prototyping and integration of hybrid NMEH architectures for self-powered sensing and similar applications.[67,68]

7.5.3 *Nanorobotics and Micro-robotics*

Targeted drug delivery for tumors is one of the many therapeutic applications for which bio-carrier micro/nanorobots (MNRs) hold tremendous promise. These MNRs' remarkable maneuverability and autonomous actuation capabilities, which enable precise drug delivery to otherwise inaccessible lesions, are made possible by the special autonomous biological functions of bio-carriers. The efficiency and safety of these MNRs are also markedly enhanced by the bio-derived structure. However, more research and development are still required to determine the stability and viability of bio-hybrid MNRs in complicated human contexts.

7.5.3.1 *Immune Cell Carriers*

Macrophages, neutrophils, natural killer cells, dendritic cells, and monocytes are used for their phagocytic, chemotactic, and immune-penetrating abilities. These cells can be loaded with nanoparticles or drugs and guided via external fields (magnetic, optical, acoustic) to deliver therapies, especially in cancer treatment. Applications include targeted chemotherapy and immunotherapy, with promising animal study results; however, clinical adoption needs further work on toxicity and immune responses.

7.5.3.2 *Non-Immune Cells*

Stem cells (e.g., mesenchymal stem cells) are used for tissue repair and targeted regenerative therapies, manipulated by MNRs for precise delivery and differentiation. Red blood cells serve as drug vehicles due to their biocompatibility and long circulation, but have limits in carrying capacity and membrane fragility. Fibroblasts and other cells are used for wound healing and tissue engineering.

7.5.3.3 *Sperm, Microbe, and Drug Carriers*

Sperm cells offer autonomous propulsion and chemotaxis, used for reproductive and precision medicine drug delivery. Protective exoskeletons and magnetic steering enhance their use. Microbial carriers (bacteria, algae) provide unique movement properties and can deliver therapeutics or enable bio-detection. Engineered bacteria can penetrate tumors and stimulate immune response, while light-driven algae provide biocompatible, biodegradable delivery systems.[69]

Thrombotic diseases arise from abnormal blood clot formation in arteries and veins, causing vascular occlusion, stroke, myocardial infarction, and pulmonary embolism. Current antithrombotic drugs are limited by poor targeting, short half-lives, risk of bleeding, and narrow therapeutic windows. Nanotechnology offers precise targeting, controlled drug release, reduced dosage, and lower hemorrhagic risk, making it promising for thrombotic disease management. MNRs combine robotics and nanotech to remotely execute therapeutic or diagnostic tasks at the microscale, accessing hard-to-reach sites within the body.

Nano-drug delivery systems (inorganic, organic, and biological nanoparticles) enhance targeting via surface modifications with ligands like peptides targeting thrombus molecules. Stimuli-responsive systems release drugs in the thrombus microenvironment (pH, shear stress, reactive oxygen species). Nanoparticles also enable high-resolution thrombus imaging.

Nanorobots provide self-propulsion and higher drug retention in thrombi compared to passive nanoparticles. Magnetic field-driven nanorobots with heparin polymer brushes demonstrated enhanced thrombus penetration and dissolution. Platelet membrane-coated silica/platinum nanomotors achieved sequential thrombolytic and anticoagulant drug release under near-infrared light, optimizing treatment stages. DNA nanorobots enabled precise, thrombin-triggered clot-targeting and drug release.[70]

7.5.3.4 *External Field Actuation*

There are four kinds of field actuation that can be achieved.

Magnetic Field Actuation: External magnetic fields, generated by electromagnetic coils or magnetic resonance, are used to remotely control and propel nanorobots. Precise navigation, deep tissue penetration, and low invasiveness are offered by magnetic actuation, which makes it highly suitable for targeted drug delivery, microsurgery, and biomedical imaging.

Electric Field Actuation: Nanorobots are manipulated through electrophoresis or di-electrophoresis under the effect of electric fields. Precise nanomotor control can be achieved without complex mechanical parts in the devices powered by electric coils and microelectrode-based propulsion.

Light Field Actuation: Light-based energy, such as photocatalytic reactions and photothermal effects, is reutilized to induce movement or control functionalities in nanorobots. High accuracy and non-contact control used for drug release and theragnostic can be achieved through light actuation.

Acoustic Field Actuation: This technology uses ultrasound waves or focused ultrasound to create pressure differences or vaporize droplets, driving nanorobot movement. It has found a lot of applications in drug delivery and microsurgery, as it can propel microrobots at high speeds.

7.5.3.5 *Self-Actuation*

Self-actuation can be achieved by chemical or biological processes. In chemical self-actuation, chemical reactions are used to perform certain tasks such as catalytic breakdown of hydrogen peroxide on platinum surfaces, creating bubbles or concentration gradients that propel nanorobots autonomously.

Biological self-actuation uses biological materials like bacterial flagella, sperm cells, or magnetotactic bacteria for propulsion. These biohybrid nanorobots combine living components with synthetic elements to enhance mobility and targeting.[71]

7.6 Outlook and Conclusion

Nano-AM enables architected materials whose properties are governed by geometry as well as by composition, creating a practical route to mechanical metamaterials with auxetic, high-specific-stiffness, and well-defined properties. This chapter summarizes how the process would affect the structure and property. It compares mainly four enabling fabrication routes: 2PP for free-form, sub-micrometer lattices and shells; PμSL for high-throughput, layer-wise projection; FEBID/FIBID for site-specific metallic/ceramic features; and nanoscale jetting (electrohydrodynamic/inkjet) for patterning functional inks, showing how exposure, voxel overlap, precursor/ink chemistry, and scan/peel strategies map to unit cell geometry, defect modes, and fatigue behavior.

The innovations with materials for nano-AM have emphasized how advanced photoresists, nanocomposites, shape-memory and stimuli-responsive polymers, polymer-derived ceramics, and functional metallic and ceramic coatings expand what these processes can achieve, enabling structures that combine low density with high stiffness, tailored damping, and environmental robustness. The chapter then presented hybrid routes, notably pyrolysis (polymer to glassy carbon/ceramics) and ALD coatings, to bridge polymer printability with engineering-grade stiffness, strength, and environmental durability, while preserving open porosity. On the design side, it outlined topology optimization and data-driven inverse design (including artificial intelligence/machine learning surrogates) that close the loop from target property to print parameters. Finally, it addressed key bottlenecks (resolution–throughput trade-offs, repeatability, nanoscale reliability) and practical remedies (process windows, *in situ*/artificial intelligence-guided monitoring, hybrid post-processing) and highlighted applications across tissue engineering, energy harvesting, and micro/nanorobotics. This approach enables selecting the nano-AM route based on the mechanical response required, not on process familiarity.

References

1. Bertoldi, Katia, Vincenzo Vitelli, Johan Christensen, Martin Van Hecke. "Flexible mechanical metamaterials." *Nature Reviews Materials* 2, no. 11 (2017): 1–11.

2. Zadpoor, Amir A. "Mechanical meta-materials." *Materials Horizons* 3, no. 5 (2016): 371–381.

3. Zhao, Zeang, Jiangtao Wu, Xiaoming Mu, Haosen Chen, H. Jerry Qi, Daining Fang. "Origami by frontal photopolymerization." *Science Advances* 3, no. 4. (2017): e1602326.

4. Harinarayana, V., Y. C. Shin. "Two-photon lithography for three-dimensional fabrication in micro/nanoscale regime: A comprehensive review." *Optics & Laser Technology* 142, (2021): 107180.

5. Ge, Qi, Zhiqin Li, Zhaolong Wang, Kavin Kowsari, Wang Zhang, Xiangnan He, Jianlin Zhou, Nicholas X. Fang. "Projection micro stereolithography based 3D printing and its applications." *International Journal of Extreme Manufacturing* 2, no. 2 (2020): 022004.

6. Huth, M., F. Porrati, O. V. Dobrovolskiy. "Focused electron beam induced deposition meets materials science." *Microelectronic Engineering* 185–186 (2018): 9–28.

7. Huang, Yongan, Ningbin Bu, Yongqing Duan, Yanqiao Pan, Huimin Liu, Zhouping Yin, Youlun Xiong. "Electrohydrodynamic direct-writing." *Nanoscale* 5, no. 24 (2013): 12007–12017.

8. Sun, Hong Bo, Satoshi Kawata. "Two-Photon Photopolymerization and 3D Lithographic Microfabrication." *Advances in Polymer Science* 170 (2004): 169–273.

9. Lacelle, Thomas, Kathleen L. Sampson, Hamidreza Yazdani Sarvestani, Amir Rahimizadeh, Julieta Barroeta Robles, Mohammad Mirkhalaf, Mohammad Rafiee, Michael B. Jakubinek, Chantal Paquet, Behnam Ashrafi. "Additive manufacturing of polymer derived ceramics: Materials, methods, and applications." *APL Materials* 11, no. 7 (2023): 70602.

10. Bauer, J., A. Schroer, R. Schwaiger, O. Kraft. "Approaching theoretical strength in glassy carbon nanolattices." *Nature Materials* 15, no. 4 (2016): 438–443.

11. Zheng, Xiaoyu, Howon Lee, Todd H. Weisgraber, Maxim Shusteff, Joshua DeOtte, Eric B. Duoss, Joshua D. Kuntz, Monika M. Biener, Qi Ge, Julie A. Jackson, Sergei O. Kucheyev, Nicholas X. Fang, Christopher M. Spadaccini. "Ultralight, ultrastiff mechanical metamaterials." *Science* 344, no. 6190 (2014): 1373–1377.

12. Overvelde, Johannes T.B., Twan A. De Jong, Yanina Shevchenko, Sergio A. Becerra, George M. Whitesides, James C. Weaver, Chuck Hoberman, Katia Bertoldi. "A three-dimensional actuated origami-inspired transformable metamaterial with multiple degrees of freedom." *Nature Communications* 7, no. 1 (2016): 1–8.

13. Lakes, Roderic. "Foam Structures with a Negative Poisson's Ratio." *Science* 235, no. 4792 (1987): 1038–1040.

14. Eckel, Zak C., Chaoyin Zhou, John H. Martin, Alan J. Jacobsen, William B. Carter, Tobias A. Schaedler. "Additive manufacturing of polymer-derived ceramics." *Science* 351, no. 6268 (2016): 58–62.

15. Meza, Lucas R., Alex J. Zelhofer, Nigel Clarke, Arturo J. Mateos, Dennis M. Kochmann, Julia R. Greer. "Resilient 3D hierarchical architected metamaterials." *Proceedings of the National Academy of Sciences of the United States of America* 112, no. 37 (2015): 11502–11507.

16. Zadpoor, Amir A. "Meta-biomaterials." *Biomaterials Science* 8, no. 1 (2019): 18–38.

17. Maruo, Shoji, Satoshi Kawata, Osamu Nakamura. "Three-dimensional microfabrication with two-photon-absorbed photopolymerization." *Optics Letters* 22, no. 2 (1997): 132–134.

18. Fischer, Joachim, Martin Wegener. "Three-dimensional optical laser lithography beyond the diffraction limit." *Laser & Photonics Reviews* 7, no. 1 (2013): 22–44.

19. Zhou, Xiaoqin, Yihong Hou, Jieqiong Lin. "A review on the processing accuracy of two-photon polymerization." *AIP Advances* 5, no. 3 (2015): 30701.

20. Liu, Xi, Qiulan Liu, Qiulan Liu, Mengdi Luo, Liang Xu, Cuifang Kuang, Cuifang Kuang, Cuifang Kuang, Xu Liu, Xu Liu. "Super-resolution direct laser writing via multiphoton and peripheral photoinhibition photolithography." *Optics Letters* 50, no. 5 (2025): 1675–1678.

21. Zhu, Dazhao, Liang Xu, Chenliang Ding, … Xu Liu. "Direct laser writing breaking diffraction barrier based on two-focus parallel peripheral-photoinhibition lithography." *Advanced Photonics* 4, no. 6 (2022): 066002.

22. Obata, Kotaro, Jürgen Koch, Ulf Hinze, Boris N Chichkov, D Du, X Liu, G Korn, J Squier, G Mourou. "Multi-focus two-photon polymerization technique based on individually controlled phase modulation." *Optics Express* 18, no. 16 (2010): 17193–17200.

23. Maibohm, Christian, Oscar F. Silvestre, Jérôme Borme, Maina Sinou, Kevin Heggarty, Jana B. Nieder. "Multi-beam two-photon polymerization for fast large area 3D periodic structure fabrication for bioapplications." *Scientific Reports* 10, no. 1 (2020): 1–10.

24. George, Steven M. "Atomic layer deposition: An overview." *Chemical Reviews* 110, no. 1 (2010): 111–131.

25. Jing, Xian, Hongxun Fu, Baojun Yu, Meiyan Sun, Liye Wang. "Two-photon polymerization for 3D biomedical scaffolds: Overview and updates." *Frontiers in Bioengineering and Biotechnology* 10, (2022): 994355.

26. Utke, Ivo, Patrik Hoffmann, John Melngailis. "Gas-assisted focused electron beam and ion beam processing and fabrication." *Journal of Vacuum Science & Technology B: Microelectronics and Nanometer Structures Processing, Measurement, and Phenomena* 26, no. 4 (2008): 1197–1276.

27. Winkler, R., J. D. Fowlkes, P. D. Rack, H. Plank. "3D nanoprinting via focused electron beams." *Journal of Applied Physics* 125, no. 21 (2019): 210901.

28. Huth, Michael, Fabrizio Porrati, Sven Barth. "Living up to its potential — Direct-write nanofabrication with focused electron beams." *Journal of Applied Physics* 130, no. 17 (2021): 170901.

29. Huth, M., F. Porrati, O. V. Dobrovolskiy. "Focused electron beam induced deposition meets materials science." *Microelectronic Engineering* 185–186 (2018): 9–28.

30. De Teresa, José María, Pablo Orús, Rosa Córdoba, Patrick Philipp. "Comparison between Focused Electron/Ion Beam-Induced Deposition at Room Temperature and under Cryogenic Conditions." *Micromachines* 10, no. 12 (2019): 799.

31. Derby, Brian. "Inkjet printing of functional and structural materials: Fluid property requirements, feature stability, and resolution." *Annual Review of Materials Research* 40, (2010): 395–414.

32. Huang, Yongan, Ningbin Bu, Yongqing Duan, Yanqiao Pan, Huimin Liu, Zhouping Yin, Youlun Xiong. "Electrohydrodynamic direct-writing." *Nanoscale* 5, no. 24 (2013): 12007–12017.

33. Galliker, P., J. Schneider, H. Eghlidi, S. Kress, V. Sandoghdar, D. Poulikakos. "Direct printing of nanostructures by electrostatic autofocussing of ink nanodroplets." *Nature Communications* 3, no. 1 (2012): 1–9.

34. Rohner, Patrik, Alain Reiser, Freddy T. Rabouw, Alla S. Sologubenko, David J. Norris, Ralph Spolenak, Dimos Poulikakos. "3D electrohydrodynamic printing and characterisation of highly conductive gold nanowalls." *Nanoscale* 12, no. 39 (2020): 20158–20164.

35. Chou, Stephen Y., Peter R. Krauss, Preston J. Renstrom. "Imprint Lithography with 25-Nanometer Resolution." *Science* 272, no. 5258 (1996): 85–87.

36. Sun, C., N. Fang, D. M. Wu, X. Zhang. "Projection micro-stereolithography using digital micro-mirror dynamic mask." *Sensors and Actuators A: Physical* 121, no. 1 (2005): 113–120.

37. Pan, Yayue, Chi Zhou, Yong Chen. "A fast mask projection stereolithography process for fabricating digital models in minutes." *Journal of Manufacturing Science and Engineering* 134, no. 5 (2012): 1–9.

38. Wang, Qiming, Julie A. Jackson, Qi Ge, Jonathan B. Hopkins, Christopher M. Spadaccini, Nicholas X. Fang. "Lightweight Mechanical Metamaterials with Tunable Negative Thermal Expansion." *Physical Review Letters* 117, no. 17 (2016): 175901.

39. Meza, Lucas R., Satyajit Das, Julia R. Greer. "Strong, lightweight, and recoverable three-dimensional ceramic nanolattices." *Science* 345, no. 6202 (2014): 1322–1326.

40. Yaqoob, Saba, Zulfiqar Ali, Alberto D'Amore, Alessandro Lo Schiavo, Antonio Petraglia, Mauro Rubino. "Enhanced Mechanical and Electrical Performance of Epoxy Nanocomposites Through Hybrid Reinforcement of Carbon Nanotubes and Graphene Nanoplatelets: A Synergistic Route to Balanced Strength, Stiffness, and Dispersion." *Journal of Composites Science* 9, no. 7 (2025): 374.

41. Liu, Bo, Jialin Sun, Jun Zhao, Xialun Yun. "Hybrid graphene and carbon nanotube–reinforced composites: polymer, metal, and ceramic matrices." *Advanced Composites and Hybrid Materials* 8, no. 1 (2025): 1.

42. Ghosal, Krishanu, Nagham Rashed, Bassma Khamaisi, Shady Farah. "Precision 4D Printing of Multifunctional Olive Oil-Based Acrylate Photo-resin for Biomedical Applications." *Advanced Functional Materials* 35, no. 13 (2025): 2414129.

43. Tariq, Ali, Zia Ullah Arif, Muhammad Yasir Khalid, Mokarram Hossain, Peerzada Ifham Rasool, Rehan Umer, Seeram Ramakrishna. "Recent Advances in the Additive Manufacturing of Stimuli-Responsive Soft Polymers." *Advanced Engineering Materials* 25, no. 21 (2023): 2301074.

44. Dimassi, Saoussen, Frédéric Demoly, Christophe Cruz, H. Jerry Qi, Kyoung-Yun Kim, Jean-Claude André, Samuel Gomes. "An ontology-based framework to formalize and represent 4D printing knowledge in design." *Computers in Industry* 126 (2021): 103374.

45. Scheithauer, Uwe, Robert Johne, Lisa Gottlieb, Steven Weingarten, Hajo Wiemer, Stefan Holtzhausen. "Recent developments in the field of additive manufacturing of ceramic-based multi-material components open the door to highly functionalized components." *Progress in Additive Manufacturing* 10, no. 6 (2025): 4051–4059.

46. Efa, Dame Alemayehu, Dejene Alemayehu Ifa, Naol Dessalegn Dejene. "Computational design of hybrid metamaterials: Optimized topology and lattice infill for complex multi-geometric structures." *Next Research* 2, no. 3 (2025): 100449.

47. Lin, Daji, Liang Gao, Jie Gao. "The Lagrangian-Eulerian described Particle Flow Topology Optimization (PFTO) approach with isogeometric material point method." *Computer Methods in Applied Mechanics and Engineering* 440 (2025): 117892.

48. Chen, Chun-Teh, Daryl C. Chrzan, Grace X. Gu. "Nano-topology optimization for materials design with atom-by-atom control." *Nature Communications* 11, no. 1 (2020): 3745.

49. Tezsezen, Ece, Defne Yigci, Abdollah Ahmadpour, Savas Tasoglu. "AI-Based Metamaterial Design." *ACS Applied Materials and Interfaces* 16 (2024): 29547–29569.

50. Si, Liming, Rong Niu, Chenyang Dang, Xiue Bao, Yaqiang Zhuang, Weiren Zhu. "Advances in artificial intelligence for artificial metamaterials." *APL Materials* 12, no. 12 (2024): 120602.

51. Rossi, Marco, Lei Zhang, Xiao Qing Chen, Che Liu, Giuseppe Castaldi, Tie Jun Cui, Vincenzo Galdi. "Machine-Learning-Enabled Multi-Frequency Synthesis of Space-Time-Coding Digital Metasurfaces." *Advanced Functional Materials* 34, no. 40 (2024): 2403577.

52. He, Xiangxu, Xiaohan Cui, C. T. Chan. "Constrained tandem neural network assisted inverse design of metasurfaces for microwave absorption." *Optics Express* 31, no. 24 (2023): 40969–40979.

53. Neto, L. Shelling, J. Dickmann, S. Kroker. "Deep learning assisted design of high reflectivity metamirrors." *Optics Express* 30, no. 2 (2022): 986–994.

54. Xi, Jiawei, Tsz Kit Yung, Hong Liang, Tan Li, Wing Yim Tam, Jensen Li. "Coincidence imaging for Jones matrix with a deep-learning approach." *npj Nanophotonics* 1, no. 1 (2024): 1–8.

55. Liu, Zhaocheng, Lakshmi Raju, Dayu Zhu, Wenshan Cai. "A hybrid strategy for the discovery and design of photonic structures." *IEEE Journal on Emerging and Selected Topics in Circuits and Systems* 10, no. 1 (2020): 126–135.

56. Tranter, A. D., H. J. Slatyer, M. R. Hush, A. C. Leung, J. L. Everett, K. V. Paul, P. Vernaz-Gris, P. K. Lam, B. C. Buchler, G. T. Campbell. "Multiparameter optimisation of a magneto-optical trap using deep learning." *Nature Communications* 9, no. 1 (2018): 1–8.

57. Tian, Jie, Keke Tang, Xianyan Chen, Xianqiao Wang. "Machine learning-based prediction and inverse design of 2D metamaterial structures with tunable deformation-dependent Poisson's ratio." *Nanoscale* 14, no. 35 (2022): 12677–12691.

58. Jiang, Jiaqi, Jonathan A. Fan. "Global Optimization of Dielectric Metasurfaces Using a Physics-Driven Neural Network." *Nano Letters* 19, no. 8 (2019): 5366–5372.

59. Liu, Peiqin, Liushifeng Chen, Zhi Ning Chen. "Prior-Knowledge-Guided Deep-Learning-Enabled Synthesis for Broadband and Large Phase Shift Range Metacells in Metalens Antenna." *IEEE Transactions on Antennas and Propagation* 70, no. 7 (2022): 5024–5034.

60. Srivastava, Manu, Sandeep Rathee, Vivek Patel, Atul Kumar, Praveennath G. Koppad. "A review of various materials for additive manufacturing: Recent trends and processing issues." *Journal of Materials Research and Technology* 21 (2022): 2612–2641.

61. Torrado Perez, A. R., D. A. Roberson, R. B. Wicker. "Characterizing the effect of additives to ABS on the mechanical property anisotropy of specimens fabricated by material extrusion 3D printing." *Additive Manufacturing* 6 (2015): 16–29.

62. Popescu, Diana, Aurelian Zapciu, Catalin Amza, Florin Baciu, Rodica Marinescu. "FDM process parameters influence over the mechanical properties of polymer specimens: A review." *Polymer Testing* 69 (2018): 157–166.

63. Al Rashid, A., S. A. Khan, S. G. Al-Ghamdi, M. Koç. "Additive manufacturing of polymer nanocomposites: Needs and challenges in materials, processes, and applications." *Journal of Materials Research and Technology* 15 (2021): 2838–2869.

64. Kandhola, Gurshagan, Sunho Park, Jae Woon Lim, Cody Chivers, Young Hye Song, Jong Hoon Chung, Jangho Kim, Jin Woo Kim. "Nanomaterial-Based Scaffolds for Tissue Engineering Applications: A Review on Graphene, Carbon Nanotubes and Nanocellulose." *Tissue Engineering and Regenerative Medicine* 20 (2023): 411–433.

65. Habibzadeh, Faezeh, Seyed Mahdi Sadraei, Roghayeh Mansoori, Narendra Pal Singh Chauhan, Ghasem Sargazi. "Nanomaterials supported by polymers for tissue engineering applications: A review." *Heliyon* 8, no. 12 (2022): 12193.

66. Calautit, Katrina, Diana S.N.M. Nasir, Ben Richard Hughes. "Low power energy harvesting systems: State of the art and future challenges." *Renewable and Sustainable Energy Reviews* 147 (2021): 111230.

67. Ahmed, Ammar, Ali Azam, Yanen Wang, Zutao Zhang, Ning Li, Changyuan Jia, Ray Tahir Mushtaq, Mudassar Rehman, Thierno Gueye, Muhammad Bilal Shahid, Basit Ali Wajid. "Additively manufactured nano-mechanical

energy harvesting systems: advancements, potential applications, challenges and future perspectives." *Nano Convergence* 8, no. 1 (2021): 37.

68. Bairagi, Satyaranjan, Shahid-ul-Islam, Mohammad Shahadat, Daniel M. Mulvihill, Wazed Ali. "Mechanical energy harvesting and self-powered electronic applications of textile-based piezoelectric nanogenerators: A systematic review." *Nano Energy* 111 (2023): 108414.

69. Lv, Yu, Ruochen Pu, Yining Tao, Xiyu Yang, Haoran Mu, Hongsheng Wang, Wei Sun. "Applications and Future Prospects of Micro/Nanorobots Utilizing Diverse Biological Carriers." *Micromachines*. Epub ahead of print May 1, 2023. DOI: 10.3390/mi14050983.

70. Lv, Siqi, Liang V. Tang, Yu Hu. "Application of nanotechnology and micro/nanorobots in thrombotic diseases." *EngMedicine* 2, no. 2 (2025): 100061.

71. Weerarathna, Induni Nayodhara, Praveen Kumar, Hellen Yayra Dzoagbe, Lydia Kiwanuka. "Advancements in Micro/Nanorobots in Medicine: Design, Actuation, and Transformative Application." *ACS Omega* 10 (2025): 5214–5250.

https://doi.org/10.1142/9789819830763_0008

CHAPTER 8

Advancing Nanoparticle Synthesis with Machine Learning: Fundamentals and Insights from an Umbrella Review

Ampere A. Tseng
Manufacturing Institute, Arizona State University, Tempe, Arizona 85287-6106, USA
ampere.tseng@asu.edu

Abstract

This umbrella review synthesizes findings from 16 review articles on machine learning-assisted nanoparticle synthesis, covering various synthesis processes, nanoparticle types, and applications. Nanoparticle synthesis is critical for applications in optoelectronics, catalysis, biomedicine, and photovoltaics, yet conventional trial-and-error methods are time- and resource-intensive. Machine learning offers revolutionary potential by predicting synthesis outcomes, optimizing experimental conditions, and accelerating discovery. Key algorithms, including artificial neural networks, random forests, gradient boosting, and Bayesian optimization, have demonstrated high accuracy in predicting nanoparticle properties such as size, shape, and quantum yield. Machine learning-assisted approaches have enabled precise control over synthesis parameters, reduced experimental iterations, and enhanced scalability. Challenges include data scarcity, limited model interpretability, and scalability issues which hinder broader adoption. Future directions emphasize integrating machine learning with autonomous systems, expanding datasets, and improving model explainability. By consolidating evidence across nanoparticle types and synthesis methods, this umbrella review provides a comprehensive overview of machine learning's role in advancing nanoparticle synthesis and guiding the development of next-generation nanofabrication.

Keywords: carbon dots, green synthesis, multi-principal element nanoparticles, lipid nanoparticles, machine learning, nanoparticle synthesis, quantum dots

8.1 Introduction

The International Union of Pure and Applied Chemistry defines a nanoparticle as a particle with one or more characteristic dimensions between approximately 1 and 100 nm.[1] Nanoparticles can take various forms, such as spheres, nanorods, nanowires, nanosheets, and quantum dots, which are produced through diverse synthesis processes. Their properties, such as optical, electrical, and magnetic behaviors, vary dramatically with size and shape, even when composition is identical. Precise control of these parameters enables broad applications in optoelectronics, medical diagnostics, drug delivery, catalysis, and photovoltaics.

Conventional nanoparticle synthesis typically relies on trial-and-error experimentation, which is time-consuming, labor-intensive, and resource-intensive. More predictive and efficient approaches are needed to design nanoparticles with application-specific properties. Machine learning (ML), a branch of artificial intelligence, addresses this challenge by developing algorithms that learn patterns from data to make predictions and guide optimization without explicit programming.[2] Unlike rule-based systems, ML models improve as they process more data and have shown strong potential to accelerate nanoparticle synthesis by uncovering hidden patterns in high-dimensional datasets.[3,4]

The rapid growth of publications in recent years underscores the importance of this topic, reflecting the convergence of nanotechnology, materials science, and artificial intelligence as priority research areas. Although several systematic and narrative reviews exist, no prior work has synthesized their findings at a higher level. This umbrella review therefore fills an important gap by integrating evidence across review types and findings, offering both a comprehensive overview of how ML is transforming nanoparticle synthesis and a higher-level perspective on this emerging field.[5]

An umbrella review goes beyond a conventional review by systematically evaluating and synthesizing evidence across multiple review types — meta-analyses, systematic reviews, state-of-the-art surveys, and narrative reviews.[6] This approach not only consolidates current knowledge but also highlights consensus, divergence, and gaps across the literature, giving readers a clearer map of the field. Whereas individual reviews often emphasize specific techniques or applications, an umbrella review provides a bird's eye perspective, enabling researchers to identify the most robust findings, avoid duplication of effort, and prioritize directions for future investigation. For readers, this means greater clarity on which ML strategies are broadly validated and where open questions remain.

Database searches (Google Scholar, Engineering Village, Web of Science) initially yielded more than a thousand publications on ML-assisted nanoparticle synthesis in the past five years. The number was narrowed to about 50 when only review-type papers were included. Ultimately, 16 reviews were selected based on their direct focus on synthesis processes rather than design- or application-centered studies. Selection was further guided by scholarly judgment, considering factors such as journal impact factor, citation count, review type, and recency of publication. The final selection comprises two meta-analyses, two systematic reviews, two state-of-the art reviews, and ten narrative reviews.

This chapter is organized as follows. After background information on ML and nanoparticle synthesis, the selected reviews are discussed in three sections: the first covers four 0D nanoparticles (quantum dots), the second focuses on three 1D or 2D nanostructures (nanotubes, graphene, atomic monolayers), and the third includes nine 3D nanoparticles (average diameter of 10–100 nm). Each review is presented chronologically. The concluding section synthesizes findings across reviews, highlights the most effective ML approaches, and identifies key challenges and future directions for ML-assisted synthesis. By consolidating current evidence, this umbrella review aims to guide the development of next-generation ML tools for more efficient and precise nanoparticle synthesis.

Moreover, in reviewing each article, the subject and goals of the study are first outlined, followed by descriptions of the ML algorithms and models applied to nanoparticle synthesis. The role and performance of these models are then assessed, including metrics of predictive accuracy. Finally, the original authors' recommendations and perspectives are noted.

8.2 Background

This section outlines the foundations of applying ML to nanoparticle synthesis. It introduces key ML algorithms and models, describes the ten-step workflow for model development, and reviews nanoparticle classifications and applications across diverse fields. Finally, it examines synthesis methods, highlighting current challenges and the potential of ML to accelerate discovery and optimization.

8.2.1 *Machine Learning Algorithms/Models*

Unlike rule-based programming, ML algorithms learn patterns directly from data and improve predictive accuracy as more information becomes available. In nanoparticle synthesis, where outcomes are influenced by numerous interdependent variables, ML provides a framework to optimize processes and uncover correlations that are difficult to detect with conventional methods. Broadly, ML approaches can be categorized into: (1) supervised learning, (2) unsupervised learning, (3) reinforcement learning, and (4) deep learning.

Supervised learning requires labeled input data to train models capable of predicting outcomes for new, unseen data.[7] It operates with a feedback mechanism and is typically divided into regression and classification tasks, which are widely applied to predict nanoparticle size, shape, and yield from synthesis parameters. Common algorithms include decision trees (DTs), k-nearest neighbor (k-NN), random forest (RF), support vector machines (SVMs), and ensemble methods such as extreme gradient boosting (XGBoost). These algorithms excel at capturing nonlinear relationships between synthesis conditions and nanoparticle properties. They are referenced frequently throughout the subsequent reviews.

Unsupervised learning analyzes unlabeled data to uncover hidden structures, clusters, or associations without human-provided feedback. It is used for clustering, association rule mining, and dimensionality reduction. For nanoparticle research, techniques such as principal component analysis (PCA) and t-distributed stochastic neighbor embedding help reveal natural groupings, detect anomalies, or identify key synthesis parameters. For example, Yan et al. applied PCA to carbon dot synthesis, enabling dimensionality reduction, visualization of relationships, and identification of factors affecting quantum yield and synthesis optimization.[8]

Reinforcement learning trains software to make sequential decisions by mimicking human trial-and-error strategies. The software acts as an autonomous agent, interacting with an environment (problem space), receives rewards or penalties for its actions, and refines its strategy to maximize long-term gains. While still less common in nanoparticle synthesis, reinforcement learning shows promise in adaptive experimental design and real-time decision-making. In a review article, Park et al. reported its ability to optimize synthesis conditions in constrained parameter spaces,[9] while Loke et al. highlighted its potential for iterative, feedback-driven optimization.[10]

Deep learning uses multilayered neural networks to automatically extract features and learn complex patterns from large datasets.[11] Architectures may range from a few layers to hundreds, enabling powerful pattern recognition in high-dimensional data. In nanoparticle synthesis, deep learning has been applied for classification, regression, and representation learning. Convolutional neural networks (CNNs) and recurrent neural networks have been particularly effective for analyzing microscopy images, spectral signatures, and time-series synthesis data. Deep generative models such as generative adversarial networks (GANs) and autoencoders are increasingly used to design nanoparticles with targeted properties. For example, Al Zoubi et al. reported that CNNs achieved automated characterization of multi-principal element nanoparticles (MPENs), while ResNet50 exceeded 90% accuracy in crystallographic phase identification from electron backscatter diffraction patterns.[12] Also, GANs were further employed to generate synthetic datasets, and autoencoders supported dimensionality reduction and feature extraction in MPEN analysis.

8.2.2 Development of a Machine Learning Model for Nanoparticle Synthesis

The development of a robust ML model for nanoparticle synthesis requires a systematic and well-structured workflow. Figure 8.1 illustrates a ten-step framework for model development and deployment.

1) *Experiment Design or/and Simulation Planning*: Data generation begins with carefully planned experiments or simulations. Experiments employ synthesis methods such as chemical vapor deposition (CVD) or phytosynthesis, depending on the desired material and nanoparticle structure. Simulations can be based on physical theory (e.g., density functional theory), probabilistic methods (e.g., kinetic Monte Carlo), or atomic-level modeling (e.g., molecular dynamics), which have been used to simulate CVD synthesis of 2D monolayers.[13] Data can also be sourced from public repositories, such as the approximately 390 publications reporting CdSe nanoparticle synthesis, which provide valuable datasets for ML training.[3,14] Critical process parameters — such as

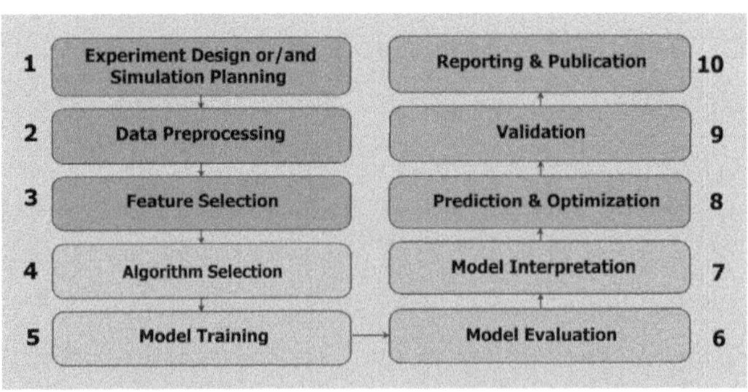

Figure 8.1. Ten-step workflow for developing a machine learning model for nanoparticle synthesis.

temperature, concentration, pH, and reaction time — must be systematically varied and reliably recorded to establish a solid foundation for model development.

2) *Data Preprocessing*: Raw data often contains inconsistencies, missing values, or errors. Preprocessing involves cleaning the dataset, normalizing features to comparable scales, and removing outliers that may bias predictions. For sparse datasets, augmentation techniques such as interpolation or synthetic data generation can enhance robustness. Effective data preprocessing is critical for enhancing model performance.[15]
3) *Feature Selection*: Selecting relevant input variables improves both model accuracy and interpretability. Researchers identify synthesis parameters that most significantly affect nanoparticle properties. Dimensionality reduction methods such as PCA help minimize redundancy and highlight the most informative features.[8,10]
4) *Algorithm Selection*: Choosing an appropriate ML algorithm depends on data characteristics and task complexity. For example, RFs perform well with nonlinear relationships and moderate-sized datasets, while neural networks are suited for high-dimensional and complex tasks.[3,16] Selection must balance performance, interpretability, and computational cost. Often, multiple algorithms are tested to determine the best fit.
5) *Model Training*: The model is exposed to historical data to learn patterns and dependencies. Data are typically divided into training, validation, and test sets to ensure generalization.[17] Training is performed iteratively, with parameter tuning to minimize errors and improve predictive accuracy.

6) *Model Evaluation*: By comparing with either experiment or simulation results, the performance of model outcomes is assessed using metrics such as the coefficient of determination (R^2), mean absolute error, and root mean square error (RMSE). Cross-validation ensures robustness by testing the model across multiple subsets of the dataset.
7) *Model Interpretation*: Understanding why a model produces specific predictions is essential for building scientific trust and gaining deeper insights. Some algorithms, such as linear regression and tree-based models, are inherently more interpretable than others. Linear regression offers direct relationships between input variables and outputs, while tree-based models such as RF rely on decision paths that can be visualized and understood by humans. Tao et al. noted that gradient-boosted regression trees can generate and test hypotheses in computational experiments, thereby providing a pathway to extract meaningful scientific insights.[3]

   Although explainability techniques have been applied only sparingly in nanoparticle synthesis, model-agnostic tools are particularly valuable for interpreting complex models. SHAP (SHapley Additive exPlanations) quantifies each feature's contribution to predictions using a game-theoretic framework, while LIME (Local Interpretable Model-agnostic Explanations) approximates complex models with simpler surrogates such as linear regressions to provide local explanations.[18] Together, these methods move beyond raw predictive accuracy, helping researchers uncover the underlying scientific drivers of model behavior.
8) *Prediction and Optimization*: A validated model can predict nanoparticle properties under new synthesis conditions.

Optimization algorithms are then used to identify favorable parameter sets for achieving desired outcomes such as size, morphology, or stability.

9) *Validation*: Model predictions must be confirmed through independent experiments or simulations. If predictions fall short, further hyperparameter tuning, additional data collection, or iterative refinement of both model and experimental design may be necessary.
10) *Reporting and Publication*: The final step is transparent dissemination. Reports should document methods, model performance, and insights to ensure reproducibility and advance ML-assisted nanoparticle synthesis.

This ten-step framework ensures that ML models developed for nanoparticle synthesis are accurate, reliable, and scalable.

8.2.3 *Nanoparticle Classifications and Applications*

According to the International Union of Pure and Applied Chemistry, a nanoparticle is defined as a particle with one or more characteristic dimensions between ~1 and 100 nm.[1] Nanoparticles can take diverse forms, such as spheres, nanorods, nanowires, nanosheets, and quantum dots, and each can be produced through a variety of synthesis methods.

For this review, nanoparticles are categorized by dimensionality. 1) 0D: Examples include quantum dots (typically <10 nm in diameter), whose properties are dominated by quantum confinement effects. 2) 1D: Nanowires and nanotubes exhibit anisotropic mechanical, electronic, and optical properties. 3) 2D: Materials such as graphene and atomic monolayers offer high surface areas and unique electronic structures, making them promising for catalysis, sensing, and energy storage. 4) 3D: Nanoparticles with average diameters of ~10–100 nm, exhibiting a wide range of aspect ratios. Challenges in this category include defect engineering and structural uniformity.

The unique physicochemical properties of nanoparticles enable applications across multiple fields, including:[19] 1) Biomedicine: Nanoparticles support targeted drug delivery, biosensing, cancer therapies, and advanced imaging modalities such as magnetic resonance imaging. 2) Electronics: Nanowires, quantum dots, and related nanostructures improve conductivity and enable novel device architectures in next-generation electronics. 3) Catalysis: High surface-to-volume ratios make nanoparticles highly efficient catalysts, accelerating industrial chemical reactions. 4) Environmental remediation: Nanoparticles can degrade dyes, adsorb pollutants, and neutralize contaminants in water and soil.

Each dimensional class and application domain presents unique opportunities but also distinct challenges for controlled synthesis, optimization, and integration into real-world systems.

8.2.4 *Synthesis of Nanoparticles*

Nanoparticle synthesis refers to the processes by which nanoscale materials with controlled size, shape, and composition are produced. Tailoring these parameters is essential, as they directly determine the physical, chemical, and biological properties of the resulting nanoparticles. Like other nanofabrication processes, synthesis methods are categorized into top-down and bottom-up approaches.[20]

Top-down methods break down bulk materials into nanoscale structures using physical or mechanical techniques such as lithography, milling, etching, and laser ablation. While capable of producing nanoparticles with well-defined dimensions, these approaches often suffer from high cost, low throughput, and difficulties in maintaining uniformity at scale.

Bottom-up methods assemble nanoparticles atom by atom or molecule by molecule. Representative techniques include CVD, sol–gel processing, hydrothermal/solvothermal synthesis, micro-emulsion or colloidal methods, and biological/green synthesis. Bottom-up approaches generally provide better control over composition and morphology, making them highly attractive for designing application-specific nanoparticles. However, challenges remain in achieving reproducibility and scalability.

Different synthesis methods yield nanoparticles with distinct structures and properties, making precise control over morphology and functionality critical. Current technologies still rely heavily on trial-and-error experimentation, which is both time- and resource-intensive. Outcomes are often highly sensitive to variations in parameters such as temperature, pH, precursor concentration, or reaction time. The complexity of this multidimensional parameter space makes optimization difficult.

Here, ML provides a powerful complement. By learning from past experimental or simulation data, ML models can predict optimal synthesis conditions, reduce failed trials, and accelerate the discovery of novel nanoparticle structures with tailored properties.

8.3 Machine Learning-Assisted Synthesis of Quantum Dots

This section provides an overview of four review articles focused on the synthesis of quantum dots (0D nanoparticles), which are nanoparticles less than 10 nm in diameter that display pronounced quantum effects. The key data and major findings of each review are consolidated in Table 8.1, where they are organized chronologically for the ease of comparison and reference.

8.3.1 *Machine Learning-Assisted Synthesis of Carbon Dots for Photoelectrochemical Sensor Applications*

Mohammadzadeh Kakhki and Mohammadpoor reviewed recent advances in ML-assisted synthesis of carbon dots (CDs) and their applications in photoelectrochemical sensors for chemical and biological detection.[21] The article examined ML algorithms used in CD synthesis, studied their benefits, and discussed the integration of CDs with other materials in photoelectrochemical sensors. The review highlighted how ML optimizes CD synthesis to enhance fluorescence, stability, and quantum yield while improving the sensitivity, selectivity, and performance of photoelectrochemical sensors.

More than ten ML algorithms/models were surveyed for different tasks. XGBoost predicted quantum yield, optimized synthesis parameters such as temperature and precursor type, and identified impactful features like ethylenediamine. Artificial neural networks (ANNs) predicted emission wavelengths and particle sizes, while RF accurately predicted photoluminescence peaks. Deep convolutional neural networks (DCNNs) predicted optical properties such as fluorescence color. Genetic algorithms (GAs) optimized ANN architectures for particle size prediction. PCA aided feature extraction, and SVMs improved process-related properties. Advanced methods such as multivariate curve resolution–alternating least squares and non-negative matrix

Table 8.1. ML-assisted synthesis of quantum dots.*

Section. Author (Year)[Ref.] Review Type	Subject or Process	ML Models, Algorithms, and Platforms	High-Performing ML Models	Key Findings for Model Performance
8.3.1. Mohammadzadeh et al. (2024),[21] Narrative	CD synthesis for photoelectrochemical sensor applications	ANNs, DCNNs, GAs, MCR-ALS, NMF-ARD-SO, PCA, RF, SVMs, XGBoost	ANNs achieved $R^2 \approx 1$ in size prediction, AC = 94% in CD color prediction; DCNNs achieved AC = 92% for predicting optical emission property	CDs exhibited unique emission properties for photoelectrochemical sensing and bioimaging
8.3.2. Duman et al. (2024),[22] Narrative	CD synthesis and optimization for synthesis parameters, CD properties, and chemical space exploration	AdaBoost, ANNs, CNNs, DTs, GAs, GBDTs, k-NN, LDA, LiR, LoR, MCR, MLR, MLP, NMF, PCA, PR, RF, SVM, XGBoost	ANNs (AC = 94%) in color prediction; k-NNs and SVMs (AC = 100%) for protein classification; PR ($R^2 = 0.986$) for predicting optical properties; XGBoost ($R^2 = 0.840$) for predicting photoluminescence quantum yield	ML predicted photoluminescence quantum yield, emission wavelengths and fluorescence intensity of CDs and enhanced material discovery
8.3.3 El-Azazy et al. (2024),[23] State-of-the-art	Synthesizing carbon quantum dots for water remediation and electrochemical applications	Basic neural networks, deep neural networks, GANs, GBDTR, Gaussian process regression, k-RR, MVLR, RFDTs	GBDTR ($R^2 > 0.9$) for predicting quantum yield; XGBoost ($R^2 > 0.96$) in predicting optical properties; ANNs (MAE = 25.8 nm) for predicting emission wavelengths	CDs/carbon quantum dots were effective in water remediation, heavy metal sensing, and electrocatalytic hydrogen evolution reactions

(*Continued*)

Table 8.1. (*Continued*)

Section. Author (Year)[Ref.] Review Type	Subject or Process	ML Models, Algorithms, and Platforms	High-Performing ML Models	Key Findings for Model Performance
8.3.4. Yan et al. (2025),[8] State-of-the-art	Synthesis and analysis of CDs for enhancing detection and optimizing optical properties	ANNs, DCNNs, ETs, GBT, k-NN, LDA, MLP, PCA, PLSR, RF, SA, SNs, SVMs, TPOT, XGBoost, YOLO-v3	GBT (AC = 100%) for protein detection; SVMs (RSD = 0.76–2.02%) for antibiotic detection; YOLO-v3 (AC = 97.3%) for glutathione detection	Enabled detecting inorganic ions, bacteria, protein, antibiotics and glutathione with high precision

*The representing words or phrases for the acronyms or abbreviations used in the above table are defined below in alphabetical order: 1) AC = Accuracy, 2) ANNs = Artificial neural networks, CD = Carbon dot, 3) CNNs = Convolutional neural networks, 4) DCNNs = Deep convolutional neural networks, 5) DTs = Decision trees, 6) ETs = Extra trees, 7) GAs = Genetic algorithms, 8) GBDTs = Gradient-boosting decision trees, 9) GBDTR = Gradient-boosting decision tree regression, 10) GBT = Gradient-boosted tree, 11) k-NN = k-nearest neighbor, 12) k-RR = Kernel ridge regression, 13) LDA = Linear discriminant analysis, 14) LiR = Linear regression, 15) LoR = Logistic regression, 16) MAE = Mean absolute error, 17) MCR = Multivariate curve resolution, 18) MCR-ALS = Multivariate curve resolution-alternating least square, 19) MLR = Multiple linear regression, 20) MLP = Multilayer perceptron, 21) MVLR = multivariate linear regression, 22) NMF = Non-negative matrix factorization, 23) NMF-ARD-SO = Non-negative matrix factorization with auto relevance determination and soft orthogonal constraint, 24) PCA = Principal component analysis, 25) PLSR = Partial least squares regression, 26) PR = Polynomial regression, 27) R^2 = Coefficient of determination, 28) RF = Random forest, 29) RFDTs = Random forest decision trees, 30) RSD = Relative standard deviation, 31) SA = Simulated annealing, 32) SNs = Siamese networks, 33) SVMs = Support vector machines, 34) TPOT = Tree-based pipeline optimization technique, and 35) YOLO-v3 = You only look once version 3.

factorization with auto relevance determination and soft orthogonal constraint (NMF-ARD-SO) analyzed photoluminescence datasets to determine optimal excitation wavelengths.

The review also reported model performance metrics. Hybrid ANNs achieved 94% accuracy in CD color prediction, while ANNs with GA optimization reached near-perfect accuracy ($R^2 \approx 1$) for size prediction. NMF-ARD-SO was recommended for analyzing complex photoluminescence data. RF excelled in photoluminescence peak prediction, while DCNNs achieved 92% accuracy for predicting excitation-dependent emission and excitation-independent emission properties. These results demonstrate the high accuracy and efficiency of ML in optimizing CD synthesis.

The authors also reviewed ML's broader roles in: (1) ML-assisted synthesis of CDs, (2) integration of CDs into photoelectrochemical sensors, and (3) applications of CDs in photoelectrochemical sensors. These methods accelerated material discovery, improved quantum yield, and enabled doping and surface passivation for precise control of CD properties in sensing, bioimaging, and energy conversion. ML-driven synthesis also enhances efficiency, reliability, and scalability.

Key challenges include controlling CD size, achieving uniformity, and fine-tuning surface properties. Aggregation and functional group tuning for higher quantum yields are persistent hurdles. While bottom-up synthesis methods are cost-effective and eco-friendly, they lack precise size control. Achieving accurate parameter control and scalable production remain critical challenges.

Looking forward, the authors recommend combining ML with theoretical and experimental approaches to develop stable

CD-based catalysts for applications in green chemistry and clean energy. ML advancements will further improve synthesis, support innovative sensing and sustainable applications, and enable scalable production to meet industrial demand. Among the reviewed approaches, XGBoost, ANNs, and DCNNs emerged as the most impactful algorithms, consistently delivering high accuracy and broad applicability across CD synthesis and property optimization.

With 57 cited references, this article is a narrative review, synthesizing insights rather than systematically selecting and analyzing studies. The key data and major findings of this review paper are summarized in Table 8.1.

8.3.2 *Machine Learning-Assisted Synthesis and Optimization of Carbon Dots for Various Applications*

Duman et al. reviewed the applications of ML in the synthesis and optimization of CDs, emphasizing ML's ability for accelerating exploration of chemical space, comprehending synthesis parameters, and enhancing CD properties for the applications including medical imaging and diagnostics, optoelectronics, chemical sensing, drug delivery, and photocatalysis.[22]

The authors discussed 19 key ML algorithms/models, categorized according to their primary roles. Six algorithms were applied to chemical space exploration by identifying relationships between synthesis inputs and CD properties: (1) PCA for dimensionality reduction and feature selection; (2) Multivariate curve resolution for analyzing pH-dependent photoluminescence spectra; (3) Non-negative matrix factorization for identifying photoluminescence mechanisms; (4) Linear regression for exploring synthesis–property correlations; (5) SVMs for classification and prediction in multidimensional datasets; and (6) k-NNs for pattern recognition in classification and regression tasks.

Eight models were highlighted for optimizing synthesis parameters to achieve target CD properties: (1) Multiple linear regression to evaluate synthesis–performance relationships; (2) Polynomial regression for capturing nonlinear synthesis–property dependencies; (3) RF for identifying influential synthesis parameters and predicting their effects; (4) Gradient-boosting decision trees (GBDTs) for optimizing fluorescence quantum yield; (5) XGBoost for predicting optimal conditions for high photoluminescence quantum yield and fluorescence intensity; (6) GA for optimizing CD-based corrosion inhibitors; (7) DTs for predicting synthesis outcomes; and (8) AdaBoost for evaluating precursor combinations.

Five additional models were used to predict and enhance CD properties for specific applications: (1) ANNs for emission color, wavelength, and amino acid classification; (2) CNNs for fluorescence color prediction and ethanol content estimation in beverages; (3) Multilayer perceptrons for photoluminescence quantum yield prediction and sustainability-focused optimization; (4) Logistic regression for protein classification using CD fluorescence sensor arrays; and (5) Linear discriminant analysis for distinguishing between tetracyclines. This categorization illustrates how ML models are tailored to different challenges, ranging from exploring chemical space to optimizing synthesis and enhancing functionality.

Model performance varied across tasks. For chemical space exploration, PCA, multivariate curve resolution, and non-negative matrix factorization effectively simplified datasets, while linear regression achieved an R^2 of 0.6751 for predicting optical properties. SVMs and k-NN excelled in classification, reaching 100% accuracy in protein classification tasks. For synthesis optimization, polynomial regression achieved the highest accuracy ($R^2 = 0.9860$) for predicting the optical properties of BN-doped graphene quantum dots. XGBoost attained an R^2 of 0.8402 for photoluminescence quantum yield prediction, RF outperformed others in emission wavelength and photoluminescence quantum yield prediction, and GBDTs were particularly effective in biochar-based CD fluorescence prediction. For property enhancement, ANNs achieved 94% training accuracy for emission color prediction and significantly reduced wavelength prediction errors, while CNNs excelled in ethanol content prediction. Logistic regression achieved 100% accuracy in protein classification tasks.

Overall, ML significantly improved the synthesis and application of CDs by predicting material properties, uncovering complex parameter relationships, and enhancing performance outcomes. It reduced costs and experimental trial-and-error, accelerated material discovery, and enabled the tailored design of CDs for advanced applications such as imaging, sensing, and optoelectronics. With 125 references cited, this narrative review synthesizes knowledge and provides insights rather than systematically evaluating all available evidence.

8.3.3 *Machine Learning-Assisted Synthesis of Carbon Dots for Water Remediation and Electrochemical Applications*

El-Azazy et al. reviewed recent advancements in carbon quantum dots (CQDs), exploring the role of ML in optimizing CQD synthesis, tailoring their properties, and expanding applications in water remediation and electrochemical applications.[23]

The authors analyzed eight ML algorithms/models, each contributing uniquely to improve CQD synthesis and properties. GBDT regression was applied to predict the fluorescence quantum yield of CQDs from biochar, achieving high accuracy. Multivariate linear regression offered simplicity and interpretability but was limited by overfitting and weak performance with complex data. Kernel ridge regression enabled nonlinear modeling but faced challenges in hyperparameter tuning. Gaussian process regression provided uncertainty-aware predictions but was computationally intensive. RFDTs used ensemble learning for robust predictions with intrinsic error estimation. Basic neural networks demonstrated adaptability but required careful optimization due to numerous hyperparameters. Deep neural networks (DNNs) excelled at dimensionality reduction and feature extraction, particularly in advanced architectures such as GANs. Finally, XGBoost combined ensemble learning with logistic regression, delivering high-accuracy predictions to guide CQD synthesis optimization.

Performance metrics highlighted the strengths of several models. GBDT regression achieved an R^2 greater than 0.9, an RMSE below 0.02, and a mean absolute percentage error under 3%, demonstrating strong predictive capability for CQD quantum yield. XGBoost achieved an R^2 exceeding 0.96 in predicting optical properties, underscoring its accuracy in optimizing synthesis conditions. An ANN-based hybrid classification–regression model predicted CQD photoluminescence emission wavelengths with a mean absolute error of 25.8 nm. While models such as multivariate linear regression, kernel ridge regression, Gaussian process regression, RFDTs, basic neural networks, and DNNs were discussed, detailed performance metrics were not provided.

The review further emphasized CQDs' intrinsic advantages: tunable fluorescence, high surface area, and low toxicity, making them highly effective in heavy metal sensing, adsorption, and electrocatalytic hydrogen evolution reactions. In water remediation, CQDs enabled femtomolar-level heavy metal detection and enhanced removal of pollutants through adsorption. In hydrogen evolution reactions, they improved catalytic kinetics and stability, supporting efficient hydrogen production for clean energy applications.

ML played a central role in advancing CQDs by improving synthesis precision, enabling predictive modeling, and supporting scalable design strategies. Key challenges remain, however, including synthesis control, long-term stability, and large-scale production. Future directions call for innovative synthesis approaches, multifunctional CQD-based materials, and interdisciplinary collaboration to unlock their potential in sustainable water treatment, energy generation, and environmental monitoring.

Overall, this article qualifies as a state-of-the-art review, offering a comprehensive and critical assessment of the latest developments in ML-assisted CQD research, while mapping out future opportunities and challenges for the field.

8.3.4 *Machine Learning-Assisted Synthesis and Analysis of Carbon Dots*

Yan et al. reviewed the integration of ML in the synthesis and analysis of CDs, focusing on optimizing their optical and functional properties, enhancing detection capabilities, and combining ML with density functional theory (DFT) for functional property optimization.[8]

The authors examined 16 ML algorithms/models, each contributing distinctively to CD synthesis and analysis. PCA was applied for dimensionality reduction and identifying key parameters affecting quantum yields and optical performance. SVMs classified multidimensional datasets, while k-NNs facilitated correlation analysis. RFs predicted optical properties and optimized features for multicolor CDs and pesticide detection. XGBoost refined synthesis conditions, quantum yields, and catalytic activity. ANNs predicted emission wavelengths and antibiotic concentrations, while DCNNs captured complex features for emission predictions. Gradient-boosted trees and partial least squares regression supported protein detection and ion quantification, respectively. Simulated annealing optimized synthesis sensitivity, and multilayer perceptrons generalized classification tasks. Extra trees and Siamese networks improved antibiotic detection, while the tree-based pipeline optimization technique automated sensing model refinement. Linear discriminant analysis enabled data classification, and YOLO-v3 provided real-time ion and biomolecule detection. Collectively, these models advanced both the synthesis and analytical performance of CDs.

Performance metrics highlighted the strengths of different models. PCA identified critical variables influencing quantum yields via eigenvalue and variance analysis. SVMs achieved high precision, with relative standard deviations between 0.76% and 2.02% for antibiotic detection. k-NN improved dataset correlation accuracy, while RF yielded low RMSE and mean absolute error in corrosion inhibition efficiency. XGBoost achieved

98% accuracy in cancer cell classification under tenfold cross-validation, demonstrating strong generalization. ANNs minimized variance in emission wavelength predictions, and DCNNs provided reliable frequency prediction. Gradient-boosted trees reached 100% accuracy in protein detection, while linear discriminant analysis and partial least squares regression classified and quantified chromium ions with high precision. YOLO-v3 achieved 97.3% accuracy in glutathione detection, and the tree-based pipeline optimization technique refined Cu^{2+} detection models through cross-validation. These results underscore the effectiveness of ML models in optimizing CD synthesis and functional performance.

The review concluded that ML models such as PCA, SVM, RF, XGBoost, and YOLO-v3 not only enhance predictive accuracy but also streamline experimental workflows and improve CDs' optical properties, including fluorescence and quantum yields. Furthermore, SVM, gradient-boosted trees, and YOLO-v3 have demonstrated high precision in detecting antibiotics, proteins, and glutathione. Integrating ML with DFT further advanced understanding of CDs' reaction mechanisms and functional properties. The authors recommended establishing unified data standards, expanding accessible and diverse datasets, and broadening ML applications to include synthesis, detection, and functional optimization. They also highlighted the potential of ML to impart novel properties to CDs for industrial and consumer applications.

Overall, this article is a state-of-the-art review as it critically evaluates recent progress in ML-assisted CD research. By categorizing findings into synthesis, applications, and analysis, it provides insights into emerging trends, current challenges, and future directions while emphasizing innovative methodologies and technologies shaping the field rather than adhering to traditional systematic protocols.

8.4 Machine Learning-Assisted Synthesis of One- and Two-Dimensional Nanoparticles

In this section, three recent review articles on ML-assisted syntheses of 1D and 2D nanoparticles, which include nanotubes, graphene, and atomic monolayers, are studied. Each of the following subsections deals with one review article according to chronological order. Table 8.2 maps the key data and major findings of the respective reviews.

8.4.1 *Machine Learning-Assisted Chemical Vapor Deposition for Graphene and Monolayer Growth*

Bhowmik and Rajan reviewed the application of ML to CVD for synthesizing 2D materials, including graphene, hexagonal boron nitride, and transition metal dichalcogenides, as well as emerging materials such as silicene, phosphorene, and borophene.[13] The review emphasized the integration of ML with modeling and simulation to better understand nucleation and growth mechanisms, while optimizing reaction conditions for scalable synthesis. For context, a single layer of graphene or hexagonal boron nitride is ~0.33–0.34 nm thick; monolayer transition metal dichalcogenides such as MoS_2 or WS_2 are ~0.6–0.7 nm, reflecting the metal layer sandwiched between two chalcogen layers; silicene, phosphorene, and borophene exhibit thicknesses of ~0.4–0.5 nm, 0.8–1.0 nm, and 0.5–0.6 nm, respectively.

Table 8.2. ML-assisted synthesis of 1D and 2D nanoparticles.*

Section. Author (Year)[Ref.] Review Type	Subject or Process	ML Models, Algorithms, and Platforms and	High-Performing ML Models	Key Findings for Model Performance
8.4.1. Bhowmik & Rajan (2022),[13] Systematic	CVD for graphene and monolayer growth	ANNs, DNNs, FNNs, NNs, PA, RF, XGBoost	XGBoost (AUC = 0.96) for predicting CVD growth condition; PA for identifying gas flow rate and reaction temperature as critical parameters	ML-assisted simulations using DFT, KMC and MD are effectively for CVD of 2D monolayers
8.4.2. Chen & Tang (2024),[24] Narrative	Optimizing CNT synthesis	ANNs, BO, CNNs, DTR, ETR, GAs, GBR, LaR, LiR, MLP, RF, SVMs, XGBoost	GBR ($R^2 = 0.96$) for strength prediction; DTR (AC = 98.6%) for mass density prediction; ETR (AC = 97.5%) for hardness prediction; CNNs (AC = 98%) for CNT classification	In CNT synthesis, BO enhances experimentation speed fivefold; RF and XGBoost are effective for synthesis optimization

8.4.3. Loke et al. (2025)[10] Narrative	Pyrolysis optimization for CNT synthesis	ANFIS, ANNs, CaB, DNNs, DTR, GBDTs, GPs, k-NN, LiR, RFR, RL, SVMs, XGBR	ANNs ($R^2 > 0.99$) for predicting thermogravimetric curves and CNT yields; ANFIS and DNNs ($R^2 = 0.9999$) for adsorption and pyrolysis yield prediction; SVMs (AC = 99%) for yield predictions	Identifying key factors for CNT synthesis, including feedstock type, catalyst properties, and operating conditions

*The representing words or phrases for the acronyms or abbreviations used in the above table are defined below in alphabetical order: 1) AC = Accuracy, 2) ANFIS = Adaptive neuro-fuzzy inference systems, 3) ANNs = Artificial neural networks, 4) AUC = Area under the receiver operating characteristic curve, 5) BO = Bayesian optimization, 6) CaB = Categorical boosting, 7) CNNs = Convolutional neural networks, 8) CNTs = Carbon nanotubes, 9) CVD = Chemical vapor deposition, 10) DFT = Density functional theory, 11) DNNs = Deep neural networks, 12) DTR = Decision tree regressor, 13) ETR = Extra trees regressor, 14) FNNs = Feedforward neural networks, 15) GAs = Genetic algorithms, 16) GBDTs = Gradient-boosting decision trees, 17) GBR = Gradient-boosting regressor, 18) GPs = Gaussian processes, 19) KMC = Kinetic Monte Carlo, 20) k-NN = k-nearest neighbor, 21) LaR = Lasso regression, 22) LiR = Linear regression, 23) MD = Molecular dynamics, 24) MLP = Multilayer perceptron, 25) NNs = Neural networks, 26) PA = Progressive adaptive, 27) R^2 = Coefficient of determination, 28) RF = Random forest, 29) RFR = Random forest regression, 32) RL = Reinforcement learning, 33) SVMs = Support vector machines, 34) XGBoost = eXtreme gradient boosting, and 35) XGBR = eXtreme gradient boosting regression.

The authors examined several ML algorithms/models applied to CVD growth. DNNs analyzed microscopy images of graphene, enabling defect identification and tracking of atomic transformations. Another neural network developed force fields for molecular dynamics simulations, accelerating predictions and validating results against DFT. ANNs, trained on DFT data, simulated transition metal dichalcogenide heterostructures to evaluate stability at reduced computational cost. RF models identified key features in MoS_2 growth, highlighting precursor temperature as the dominant parameter. The progressive adaptive model iteratively refined synthesis outcomes using feedback loops. Gradient-boosted classifiers (XGBoost) predicted successful growth conditions for MoS_2, distinguishing between growth and non-growth scenarios. Feedforward neural networks analyzed reactive molecular dynamics data, revealing critical reaction pathways relevant to experimental growth. Collectively, these models optimized reaction conditions, predicted growth outcomes, and accelerated insights into CVD mechanisms.

Performance results underscored the potential of ML in CVD. XGBoost achieved an area under the receiver operating characteristic curve (AUC) of 0.96 in predicting MoS_2 growth, demonstrating high discriminative power between growth and non-growth regimes. The progressive adaptive model identified gas flow rate and reaction temperature as critical synthesis parameters. While other models (DNNs, ANNs, RF, feedforward neural networks) were validated against experimental or simulation data, detailed numerical metrics were not reported. Nevertheless, they proved instrumental in detecting defects, improving computational efficiency, and elucidating reaction mechanisms.

The review also highlighted the value of ML-assisted simulations — including DFT, kinetic Monte Carlo, and reactive molecular dynamics — in uncovering nucleation pathways, growth dynamics, and defect formation in materials such as graphene, hexagonal boron nitride, and transition metal dichalcogenides. Models such as XGBoost and progressive adaptive demonstrated effectiveness in predicting growth outcomes, accelerating synthesis, and identifying key process parameters. Despite these advances, significant knowledge gaps remain, particularly in understanding grain boundary formation, nanopore evolution, and substrate effects. The authors emphasized the need for further computational and ML-driven research to improve reproducibility, scalability, and control in CVD growth of both established and emerging 2D materials, including silicene, phosphorene, and borophene.

This article qualifies as a systematic review, as it organizes and synthesizes current research on ML applications in CVD growth of 2D materials. By addressing nucleation, growth, substrate effects, and knowledge gaps, it provides a structured and comprehensive analysis while proposing future directions for advancing the field.

8.4.2 *Machine Learning-Assisted Carbon Nanotube Synthesis, Characterization, and Applications*

Chen and Tang reviewed how ML advances carbon nanotube (CNT) research by optimizing synthesis, automating characterization, and accelerating application development.[24] They emphasized ML's ability to foster innovation, reduce trial and error, and enable precise property control. While focusing primarily on

CNT synthesis, the review also briefly covered CNT-reinforced AlSi10Mg nanocomposites, ABS composites, and functionally graded composites in a subsection of CNT applications in composite materials.

Thirteen ML algorithms were discussed. Linear regression established basic relationships between inputs and outputs for CNT synthesis and sensor calibration. SVMs handled nonlinear classification and prediction, aiding synthesis optimization and property prediction. RFs improved accuracy through multiple DTs, supporting high-throughput screening, property prediction, and DNA sequence identification for functionalization. XGBoost optimized synthesis trade-offs and growth conditions. Neural models, including multilayer perceptrons and ANNs, supported nonlinear modeling for synthesis optimization and DNA sequence prediction. Bayesian optimization (BO) incorporated real-time feedback in autonomous synthesis. CNNs automated chirality determination from transmission electron microscopy images and identified CNT structures via Raman spectroscopy. Decision tree regressor, gradient-boosting regressor, and extra trees regressor predicted composite properties, while lasso regression reduced overfitting. GAs analyzed nanoparticle morphologies from transmission electron microscopy images.

Collectively, linear regression, SVMs, RF, XGBoost, multilayer perceptrons, BO, and ANNs advanced CNT synthesis optimization; CNNs, ANNs, and GAs improved characterization; and RF, decision tree regressor, gradient-boosting regressor, extra trees regressor, and lasso regression enabled property prediction. RF, multilayer perceptrons, and CNNs further supported functionalization and applications. Together, these models enhance efficiency and accuracy in addressing synthesis, characterization, and application challenges.

The review reported notable performance metrics. BO improved experimentation speed fivefold in autonomous CNT synthesis. RF and XGBoost models showed high prediction accuracy in CNT synthesis optimization though details were limited. Gradient-boosting regressor reached an R^2 of 0.96 for predicting compressive strength of CNT-reinforced ABS composites. For AlSi10Mg nanocomposites, decision tree regressor predicted relative mass density with 1.42% error, and extra trees regressor predicted hardness with 2.47% error. CNNs achieved 90% accuracy in classifying the type of CNTs from Raman spectra at low signal-to-noise ratio, 98% at high signal-to-noise ratio, and 90.5% in chirality determination from simulated transmission electron microscopy images. Other models such as SVMs, multilayer perceptrons, and lasso regression were mentioned but lacked detailed benchmarks.

The authors concluded that ML has transformed CNT research by optimizing multivariable systems, reducing reliance on trial and error, and enabling autonomous synthesis. This has enhanced crystallinity, growth efficiency, and overall quality. In characterization, ML has automated chirality determination and spectral analysis, significantly improving speed and accuracy. For applications, ML has facilitated design and optimization of CNT-based electronics, sensors, composites, and biomedical constructs through accurate property prediction, efficient fabrication, and targeted functionalization. Remaining challenges include the need for well-structured datasets and integration with broader artificial intelligence systems.

As a narrative review, the article synthesizes current knowledge, outlines methodologies, and discusses applications while offering perspectives on future directions. Rather than systematically analyzing data or conducting meta-analysis, it provides a cohesive overview of how ML is driving CNT innovation and expanding possibilities in nano-synthesis.

8.4.3 *Machine Learning-Assisted Pyrolysis Optimization for Carbon Nanotube Synthesis*

Loke et al. reviewed the integration of ML techniques to enhance pyrolysis for CNT synthesis, with emphasis on parameter optimization, yield prediction, and CNT production from plastic waste.[10] The article, which appears to be narrative in nature, also discussed challenges and future prospects.

Thirteen key ML models were reviewed. Among supervised learning models, ANNs, SVMs, decision tree regression, RF regression, and GBDTs received detailed coverage. ANNs were highlighted for handling complex nonlinear relationships, ideal for predicting CNT yields and analyzing thermogravimetric curves. DNNs effectively processed large datasets with high accuracy. SVMs performed well on small datasets, predicting liquid, gas, and char yields. Decision tree regressor and RF regression were useful for regression tasks, with RF regression particularly accurate in thermogravimetric and pyrolysis yield predictions. GBDTs optimized bio-oil yields and synthesis conditions.

Other models included Gaussian processes for uncertainty quantification, categorical boosting for co-pyrolysis yield prediction, linear regression for optimizing CNT-based sensor conditions, k-NN for CNT-reinforced concrete strength, XGBoost regression for oil yield prediction, and adaptive neuro-fuzzy inference systems for adsorption studies. Reinforcement learning was briefly noted for potential real-time decision-making, while DNNs were mentioned on handling large datasets. Collectively, these models enhanced process efficiency, optimized parameters, and improved CNT yield and quality.

Performance outcomes were extensively reported. ANNs consistently achieved high accuracy ($R^2 > 0.99$) for predicting thermogravimetric curves, CNT yields, and product distributions, and $R^2 = 0.885$ for concrete strength, outperforming linear regression and k-NN. Adaptive neuro-fuzzy inference systems reached $R^2 = 0.9999$ in CNT adsorption studies, while DNNs achieved similar accuracy for pyrolysis gas and oil yields. SVMs delivered $R^2 \sim 0.85$ for yield predictions and matched experimental values up to 99% in some cases. RF regression models achieved $R^2 = 0.9696$ for thermogravimetric curve predictions, and GBDTs reached $R^2 = 0.90$ with RMSE = 5.04 for bio-oil optimization. Categorical boosting was noted for co-pyrolysis predictions, XGBoost regression reported $R^2 = 0.83$ for oil yield, and Gaussian processes models showed reliable predictions with RMSE < 6%.

The review emphasized environmental benefits of converting plastic waste into CNTs, contributing to United Nations Sustainable Development Goals 11 and 12. Pyrolysis was underscored as an efficient thermochemical recycling pathway for CNT precursors. Key factors influencing CNT production include feedstock type, catalyst properties, and operating conditions. Supervised models such as ANNs, adaptive neuro-fuzzy inference systems, and DNNs consistently demonstrated superior predictive performance ($R^2 > 0.999$), while unsupervised methods like PCA provided critical insights into parameter sensitivity.

Challenges include the need for large datasets, high computational demands, and complexity in modeling catalyst–gas–CNT interactions. The review concluded that ML integration enhances pyrolysis efficiency, economic feasibility, and environmental sustainability, supporting circular economy initiatives. In total, 355 studies on "pyrolysis using ML" published since 2021 were reviewed.

8.5 Machine Learning-Assisted Synthesis of Three-Dimensional Nanoparticles

This section reviews nine articles on the synthesis of 3D nanoparticles (with a diameter between 10 to 100 nm). The key data and major findings of each review are consolidated in Table 8.3, which are listed chronologically to facilitate comparison and reference.

8.5.1 *Machine Learning-Assisted Nanoparticle Synthesis: Prediction, Optimization, and Discovery*

Tao et al. reviewed how ML can advance nanoparticle synthesis by predicting outcomes, optimizing experimental conditions, and accelerating the discovery of new nanoparticles.[3] The review highlighted ML's potential to improve efficiency, reduce resource consumption, and enhance scalability while also identifying challenges in precision, scalability, data quality, and automation. Applications include semiconductor, metallic, carbon-based, and polymeric nanoparticles.

The authors identified three primary ML tasks aligned with the goals of synthesis: prediction, optimization, and discovery. In prediction, supervised learning algorithms — including neural

Table 8.3. Summary of ML-assisted synthesis of 3D nanoparticles.*

Section. Author (Year)[Ref.] Review Type	Subject or Process	ML Models, Algorithms, and Platforms	High-Performing ML Models	Key Findings for Model Performance
8.5.1. Tao (2021)[3] Systematic	Prediction and optimization in NP synthesis	BO, CMA-ES, CuD, GAs, GBRTs, GPs, IM, k-NN, NNs, RF, SNOBFIT, SVMs, ULAs	NNs ($R > 0.9$) in quantum yield prediction; RF ($R^2 = 0.75–0.77$) in NP size prediction; SVMs (RMSE ~ 0.45 nm) in predicting nanoplatelet thickness	ML improved prediction and optimization in NP synthesis; SVMs outperformed linear and quadratic models
8.5.2. Liu et al. (2021)[26] Meta-analysis	Cytotoxicity in Ag NP phytosynthesis	DT1 and DT2, RF1 and RF2	For classification of cell viability outcomes, RF (AC = 79.2–82.5%, AUC = 0.866–0.904) and DT (AC = 73.1–75.8%, AUC = 0.815–0.839)	Cytotoxicity of Ag NPs was dictated by dose, time, cell type, particle size and plant family; RF outperformed DT
8.5.3. Lv and Chen (2022)[27] Narrative	Microfluidic systems for NP synthesis	ANNs, BO, DNNs, DTs, GAs, Kriging, RF, SVMs	ANNs ($R = 0.978$) for size prediction; SVM (RMSE = 0.7556 nm) for predicting nanosheet thickness; BO and GAs effective for optimization	ML optimized NP synthesis in microfluidic systems for desired functionalities
8.5.4. Park et al. (2023)[9] Narrative	Closed-loop optimization for NP synthesis	AL, BO, CNNs, GAs, RA, RL, RNNs, TM, TSEMO	Highlighted advantages of BO and reinforcement learning for optimization and GAs for experiment planning (performance metrics unreported)	Automation enhanced reliability and reproducibility in synthesizing homogeneous NPs

(*Continued*)

Table 8.3. (*Continued*)

Section. Author (Year)[Ref.] Review Type	Subject or Process	ML Models, Algorithms, and Platforms and	High-Performing ML Models	Key Findings for Model Performance
8.5.5. Karakus (2024)[28] Meta-analysis	Optimizing magnetite NP synthesis	ANNs, CARTs, RF, RTs	ANNs (AC = 75%) for magnetic property prediction; RF is robust in predictions	Organic modifiers and $FeSO_4$-based precursors highly enhanced magnetization
8.5.6. Wang et al. (2025)[16] Narrative	NP design and synthesis by bibliometric analysis with applications	ANNs, BO, DL, DNNs, DTs, GAs, GPs, k-NN, LiR, RF, SVMs, SX model	ANNs (AC = 98.1%) in cancer diagnosis; SX model (AC = 96.5%) for detecting ions; DL (AC = 94%) for pattern recognition	ML enabled precise property predictions, optimized synthesis, and enhanced applied performance
8.5.7. Al Zoubi et al. (2025)[4] Narrative	Synthesizing high-entropy MPENs	BNR, CNNs, EPI, GAs, GPC, RNNs, SISSO, TL	BNR ($R^2 = 0.997$) for energy prediction; GAs (AC > 90%) for predicting crystal structures	Mixing entropy and enthalpy dictated MPEN properties and optimization
8.5.8. Al Zoubi et al. (2025)[12] Narrative	Strategies, material properties, and ML prediction in synthesizing MPENs	Autoencoders, ANNs, BR, CCA, CNNs, DSs, EMs, GAs, GANs, GB, GBDTs, GNNs, GPC, k-NN, LiR, LRR, MLE, NNs, PR, RF, ResNet50, SVMs	ANNs (R = 0.982) for composition prediction; BR and LRR ($R^2 > 0.95$) for strength prediction; CNNs, ResNet50 (AC > 90%) and GPC (AUC = 0.97) for phase prediction; RF (AC = 100%) for labeled data prediction and (AC = 92.3–97.5%) for strength prediction	Interplay of Gibbs free energy equation with enthalpy and entropy was critical in MPEN formation; ML was effective for prediction and optimization in MPEN synthesis

8.5.9. Li et al. (2025)[29] Narrative	Design and synthesis of lipid NPs for gene therapy	GNNs, GT, LiR, MLP, RF, SVMs, XGBoost	Highlights GNNs, GT, and RF as highly accurate for certain tasks (performance metrics unreported)	Microfluidic mixing was the best method for lipid NP synthesis

*The representing words or phrases for the acronyms or abbreviations used in the above table are defined below in alphabetical order: 1) AC = Accuracy, 2) AL = Active learning, 3) ANNs = Artificial neural networks, 4) AUC = Area under the curve, 5) BNR = Bayesian normalized regression, 6) BO = Bayesian optimization, 7) BR = Bayesian regression, 8) CARTs = Classification and regression trees, 9) CCA = Canonical-correlation analysis, 10) CMA-ES = Covariance matrix adaptation-evolution strategy, 11) CNNs = Convolutional neural networks, 12) CuD = Curiosity-driven, 13) DL = Deep learning, 14) DNNs = Deep neural networks, 15) DSs = Deep sets, 16) DTs = Decision trees, 17) EMs = Ensemble methods, 18) EPI = Effective pair interaction, 19) GA = Generative algorithm, 20) GANs = Generative adversarial networks, 21) GAs = Genetic algorithms, 22) GB = Gradient boosting, 23) GBDTs = Gradient-boosted decision trees, 24) GBRTs = Gradient-boosted regression trees, 25) GNNs = Graph neural networks, 26) GPC = Gaussian process classification, 27) GPs = Gaussian processes, 28) GT = Graph transformer, 29) IM = Intrinsic motivation, 30) k-NN = k-nearest neighbor, 31) LiR = Linear regression, 32) LRR = Linear ridge regression, ML = Machine learning, 33) MLE = Machine learning evolution, 34) MPENs = Multi-principal element nanoparticles, 35) NNs = Neural networks, 36) NPs = Nanoparticles, 37) PR = Polynomial regression, 38) R = Correlation factor, 39) R^2 = Coefficient of determination, 40) RA = Regression algorithm, 41) RF = Random forest, 42) RL = Reinforcement learning, 43) RMSE = Root mean square error, 44) RNNs = Recurrent neural networks, 45) RTs = Regression trees, 46) SISSO = Sure independence screening and sparsifying operator, 47) SNOBFIT = Stable noisy optimization by branch and fit, 48) SVMs = Support vector machines, 49) TL = Transfer learning, 50) TM = Text mining, 51) TSEMO = Thompson sampling efficient multi-objective optimization, and 52) ULAs = Unsupervised learning algorithms.

networks, k-NN, RF, Gaussian processes, and SVMs were used to predict nanoparticle properties (e.g., size, shape, yield) from experimental conditions. Algorithm selection depends on predictive power, interpretability, and computational cost. For example, neural networks achieved high accuracy with Pearson correlation $R > 0.9$ for predicting quantum yield, wavelength, and reaction yield while RF reached $R^2 = 0.75–0.77$ for predicting nanoparticle size. k-NN was effective with small datasets, and Gaussian processes worked well with datasets <1,000 samples though at higher computational cost. Also, for small to medium datasets, SVMs achieved an RMSE of 0.7556 atomic layers for predicting the thickness of $CsPbBr_3$ nanoplatelets, outperforming linear (1.0266 layers) and quadratic models (0.8835 layers). Since a single atomic layer of $CsPbBr_3$ nanoplatelets is typically around 0.6 nm, RMSE = 0.7556 layers corresponds to approximately 0.45 nm, which implies that SVMs offered very accurate predictions.

In optimization, experimental optimization algorithms including GAs and BO were used to plan new experiments that optimize nanoparticle properties with fewer trials. GAs were specifically effective for high-dimensional optimization and parallel experiments but typically required more experimental trials than BO, which was more efficient for experiment planning. Other algorithms such as covariance matrix adaptation evolution strategy were used for optimization of synthesis yield in high-dimensional spaces while stable noisy optimization by branch and fit was for optimizing quantum dot synthesis in low dimensional spaces with ~100 experiments.

In discovery of new nanoparticles, ML algorithms like intrinsic motivation and curiosity-driven algorithms were useful for exploring unknown chemical spaces and discovering nanoparticles with novel properties. Gradient-boosted regression trees could generate and test hypotheses for computational experiments to provide scientific insights. These capabilities were crucial for understanding the relationships between experimental conditions and nanoparticle properties, a step toward improving interpretability. Also, unsupervised learning algorithms were able to suggest alternative experimental conditions and identify new classes of nanoparticles with targeted functionalities.

Datasets for ML training and testing were typically assembled from published literature or experimental data, including reaction conditions and nanoparticle properties. Tao et al. further advocated integrating ML with autonomous platforms, such as robotic or microfluidic systems, to enable high-throughput automated synthesis and characterization.

Looking ahead, several challenges were emphasized. A key limitation is the interpretability of ML models, which affects scientific insight and trust in predictions. Inconsistent and non-standardized datasets hinder model training, as data often vary in format and quality. Automation also faces bottlenecks, particularly in multistep purification and *in situ* characterization. While real-time characterization techniques are promising, their precision remains limited. Moreover, translating results from robotic or microfluidic systems to conventional synthesis remains difficult due to differences in reaction environments.[25]

8.5.2 *Machine Learning-Assisted Meta-Analysis for Predicting Cytotoxicity in Phytosynthesized Silver Nanoparticles*

Liu et al. conducted an ML-assisted meta-analysis of cytotoxicity data for phytosynthesized silver nanoparticles (AgNPs) using

ML algorithms for analyzing and predicting their toxic effects based on nanoparticle properties, biosynthesis parameters, and experimental conditions.[26] The goal was to identify key variables influencing AgNP-mediated cytotoxicity and provide guidance for optimizing green AgNPs for biomedical and therapeutic applications. In this study, phytosynthesis is a green process for synthesizing materials (such as AgNPs) using plant-based extracts to avoid the use of harmful chemicals typically involved in nanoparticle synthesis.

This meta-analysis employed two ML algorithms, DTs and RF, to classify and predict the cytotoxicity of phytosynthesized AgNPs. DTs and RF classify the cytotoxicity of AgNPs into binary categories: high cell viability (>50%) or low cell viability (<50%). The DTs were constructed using a tree structure with two variations: a) DT1, which utilized seven features — particle size, extraction solvent, exposure dose, zeta potential, normal/cancer cells, animal/human cells, and exposure time. b) DT2, which included the same seven features as DT1, with the addition of plant family as an eighth feature. Similarly, RF models were developed using a tree-based approach. RF1 was built using the same seven features as DT1, while RF2 incorporated plant family as an additional feature, mirroring the approach of DT2.

Including plant family as a feature significantly improved classification accuracy, with RF2 achieving the highest performance at 82.5% accuracy (correctly classified 82.5% of the data points) and AUC of 0.904, where AUC = 0.5 and 1.0 indicate random guessing and perfect classification, respectively. The performances of RF1, DT1, and DT2 were 79.2%, 73.1%, and 75.8% in accuracy, respectively, and 0.866, 0.815, and 0.839 for AUC. These models identify key predictors, such as exposure regime and cell type, enabling researchers to understand the factors influencing cytotoxicity and classify outcomes effectively across diverse experimental conditions.

In prediction, the ML models predict the cytotoxicity of AgNPs by analyzing features like exposure dose, time, cell type, and biosynthesis parameters. RF models excel in prediction accuracy, with RF2 achieving the highest performance (AUC = 0.904). They rank variable importance, identifying exposure regime, cell type, and plant family as the most significant predictors. By integrating wide-ranging data, the models forecast cell viability outcomes, helping researchers anticipate cytotoxic effects under specific conditions. This predictive framework supports the design of safer AgNPs by guiding experimental setups and optimizing biosynthesis parameters. Additionally, the models enable the virtual design of "safe-by-design" nanoparticles tailored for therapeutic applications, minimizing risks while maximizing benefits. Their predictive capabilities are crucial for advancing the development of green AgNPs in biomedical and healthcare fields.

The models were trained and tested using a tenfold cross-validation method. The dataset was divided into 10 independent subsets, where nine subsets were used for training and the remaining subset was used for testing. This process was repeated 10 times, ensuring each subset was used for testing once. The model's performance was evaluated using metrics such as precision, recall, accuracy, F-measure, and AUC. A total of 690 data points were used for training and testing of ML models. They were extracted from 81 articles through a literature search, covering peer-reviewed articles published between 2003 and 2020. The data features include particle size, zeta potential, cell type, exposure dose, exposure time, plant family, and extraction solvent. Units for particle size (nm), zeta potential (mV), and exposure dose (mg/L) were harmonized to ensure consistency.

Overall, the meta-analysis highlights that cytotoxicity of phytosynthesized AgNPs is influenced by nanoparticle properties, exposure regime, cell type, and plant family. ML models, especially RF, demonstrated high predictive accuracy. Incorporating plant family improved performance, offering valuable guidance for designing safer, tailored green AgNPs for biomedical and therapeutic applications.

The authors indicate several challenges in evaluating the cytotoxicity of AgNPs. A major issue is the lack of comprehensive and standardized data on nanoparticle properties, such as zeta potential and colloidal stability, which limits their utility in predictive modeling. Variations in experimental conditions, cell assays, and characterization methods further complicate comparisons across studies. Another challenge is the unclear chemical composition of plant extracts, which makes it difficult to identify the key ingredients influencing nanoparticle properties and toxicity.

8.5.3 *Machine Learning-Assisted Nanoparticle Synthesis Via Microfluidic Systems*

Lv and Chen reviewed the applications of ML techniques to assist in nanoparticle synthesis using microfluidic systems.[27] The microfluidic systems are employed due to their advantages, such as efficient mixing, reduced reagent loss, uniform heat transfer, and precise control over reaction conditions. Microreactors within microfluidic systems enable the synthesis of nanoparticles with controlled size, shape, and surface chemistry, which are critical for applications in electronics, optics, sensing, and medical fields.

Many ML algorithms were reviewed for the intelligent control and optimization of nanoparticle synthesis in microfluidic systems. ML algorithms such as ANNs, SVMs, RF and DTs are used to predict the size, shape, and surface chemistry of nanoparticles based on reaction conditions (e.g., temperature, time, reagent concentrations). ANNs, SVMs, GAs, BO, and Kriging can be used to optimize reaction parameters (e.g., temperature, flow rates, precursor ratios) to achieve desired properties, reducing trial-and-error experimentation. ANNs, SVMs, DTs, BO, and DNNs are useful for analyzing large datasets to uncover complex relationships between reaction conditions and nanoparticle characteristics, accelerating synthesis processes. Autonomous ML-assisted platforms using ANNs, BO, and DNNs are effective for exploring reaction spaces and optimizing synthesis pathways, saving time and costs. Techniques like BO and CNNs improve prediction accuracy and reduce errors in nanoparticle characterization. ANNs, BO, DNNs, GAs, and Kriging also facilitate the design of novel nanoparticles with unique properties by exploring uncharted reaction spaces. Overall, these ML algorithms transform nanoparticle synthesis, making it more efficient, scalable, and precise for applications in electronics, optics, sensing, and medicine.

The performance of these algorithms in nanoparticle synthesis was also reported. ANNs demonstrated high accuracy in predicting nanoparticle size, achieving a correlation factor (R) of 0.978 for Au nanoparticles and $R = 0.937$ for CdSe quantum dots, making it highly effective for nonlinear relationships and optical property predictions. SVM showed good predictive accuracy with an RMSE of 0.7556 for perovskite nanosheets. DT models were prone to overfitting, while RF improved generalization for

CdSe quantum dots. GAs optimized Ag nanoparticle synthesis, achieving better size distribution and lower polydispersity. BO proved efficient for small datasets, minimizing dissatisfaction factors in quantum dot synthesis with fewer iterations. Kriging was reliable for spatial modeling of CdSe quantum dots. Overall, ANNs were the most robust, while BO and GAs excelled in experimental optimization. Each algorithm was selected based on problem complexity and dataset size.

The authors highlight challenges in ML-assisted nanoparticle synthesis, including difficulties in achieving precise control over nanoparticle size and shape due to errors in characterization methods like online spectrometers. Emission spectra with similar shapes may correspond to nanoparticles of varying sizes, introducing bias in predictions. Additionally, the time and effort required for training ML models for different nanoparticle types remain significant. Addressing these issues requires improved characterization techniques and efficient model training strategies.

Future directions include integrating advanced characterization tools like Fourier transform spectrometers and dynamic light scattering for accurate nanoparticle size measurement. CNNs could process nanoparticle images for precise size determination. Transfer learning may reduce time costs by reusing trained model parameters for new nanoparticle types, accelerating research and improving efficiency. These advancements promise breakthroughs in ML-assisted nanoparticle synthesis.

8.5.4 *Machine Learning-assisted Closed-Loop Optimization for Nanoparticle Synthesis*

Park et al. reviewed the development of ML-assisted closed-loop optimization system for enhancing the efficiency and reproducibility of nanoparticle synthesis.[9] This system is a self-sustaining process that integrates ML-assisted optimization with robotic synthesis, automated materials characterization, and computational prediction to continuously accelerate the discovery and production of structurally homogeneous nanoparticles with desired properties. The system operates in a feedback loop, where data from each step informs the next, allowing for efficient exploration of synthesis parameters and optimization of nanoparticle properties to reduce reliance on manual trial-and-error processes. Also, a continuously enriched database links synthesis parameters, nanoparticle properties, and structural information, enabling iterative improvement.

Here, ML-assisted optimization used ML algorithms to analyze the data collected from synthesis and characterization to predict optimal synthesis conditions or suggest the next experiments, while robotic synthesis used automated systems for performing nanoparticle synthesis by controlling experimental parameters such as precursors, solvents, temperature, and reaction time. Also, automated materials characterization used automated tools like spectroscopy, X-ray diffraction, and transmission electron microscopy to measure physical, chemical, and structural properties of synthesized nanoparticles, and computational prediction was using computational methods like DFT or molecular dynamics simulations to provide additional insights into nanoparticle properties. Together, these components created a self-driving laboratory, paving the way for scalable and efficient synthesis tailored to desired properties and applications.

Various ML models to enhance nanoparticle synthesis were employed in the closed-loop system. Regression models predicted nanoparticle properties based on synthesis conditions, establishing synthesis-property relationships. Generative models enabled inverse design by suggesting experimental conditions for desired properties. BO planned sequential experiments, optimizing parameters efficiently for expensive data generation. GAs facilitated parallel batch experiments, accelerating optimization. Active learning minimized the number of experiments by selecting the most informative data points for training. Reinforcement learning optimized conditions in limited parameter spaces by maximizing cumulative rewards. Text mining built initial datasets from literature, while Thompson sampling efficient multi-objective optimization optimized costly scale-up parameters. CNNs and recurrent neural networks encoded image and process data, respectively.

While these algorithms/models are promising, their performance depends on data quality, experimental setup, and application. However, no specific metrics were provided for these ML models. Thus, the efficiency and accuracy of the ML-assisted closed-loop optimization system reviewed by the authors cannot be quantitatively assessed.

Key challenges in nanoparticle synthesis include managing complex parameters, ensuring structural homogeneity, scaling up processes, and automating characterization techniques such as transmission electron microscopy and X-ray diffraction. Current robotic systems face limitations in handling intricate tasks and responding to unexpected issues. Recommendations include advancing robotic manipulators for delicate operations, integrating modular synthesis systems for parallel experiments, and developing algorithms for fully automated characterization. ML models like BO and GAs should be leveraged for efficient parameter exploration. Additionally, creating comprehensive databases with both positive and negative experimental results can improve ML accuracy.

8.5.5 *Machine Learning-Assisted Meta-Analysis for Optimizing Magnetite Nanoparticle Synthesis*

Karakus conducted an ML-assisted meta-analysis to systematically compile and analyze data from 123 independent studies (255 data points) on the synthesis of magnetite nanoparticles.[28] The study aimed to aggregate existing knowledge, identify key synthesis parameters, and optimize synthesis conditions.

Several ML algorithms were employed, including classification and regression trees (CARTs), regression trees, RF, and ANNs. CART and regression trees were used to partition the data space and identify critical synthesis parameters influencing the magnetic properties of magnetite nanoparticles. Regression tree models were pruned using cross-validation to prevent overfitting. RF models evaluated the relative importance of synthesis parameters by quantifying their contributions to magnetic properties, with modifier type identified as the most influential factor. ANN models, built using a feed-forward network with two hidden layers, were applied to predict magnetic properties based on synthesis conditions. Collectively, these models provided a comprehensive assessment of the relationships between synthesis variables and nanoparticle properties, enabling data-driven optimization.

The dataset (255 data points) was divided into a training set (75%) and a test set (25%) for model development and evaluation. CART models were pruned via tenfold cross-validation while ANN models, trained using resilient backpropagation and optimized by adjusting the number of hidden layers and nodes, achieved the highest performance. In this meta-analysis, CARTs and ANNs were able to identify modifier type, precursor type, reaction temperature, and environment as critical synthesis parameters. ANN models achieved better prediction accuracy (75%) for magnetic properties compared to CARTs, which was the best-performing tree in explaining 62% of the variation in magnetic properties.

In contrast to traditional trial-and-error methods, ML approaches effectively captured the complex interplay among synthesis parameters, reducing both experimental time and cost. Overall, the integration of ML transformed raw data into actionable insights, accelerating the design of nanoparticles with targeted properties.

The author emphasized the advantages of ML, particularly ANN, in optimizing magnetite nanoparticle synthesis. Expanding datasets to include more diverse synthesis methods and parameters would further improve model accuracy. Future research should also incorporate additional material characteristics and synthesis strategies to enable precise, cost-effective development of tailored nanomaterials for biomedical and industrial applications, in line with the quality-by-design framework.

8.5.6 *Machine Learning-Enhanced Nanoparticle Design and Synthesis for Cancer Drug Delivery*

Wang et al. reviewed the transformative role of ML in advancing nanoparticle synthesis and design for precision cancer drug delivery.[16] The review emphasized three critical areas where ML contributed: optimization of nanoparticle synthesis processes, prediction of nano–bio interactions, and enhancement of therapeutic efficacy.

In optimizing nanoparticle synthesis, ML facilitated systematic navigation of complex design parameters such as size, shape, surface chemistry, and drug-loading efficiency. Algorithms such as linear regression, RF, Gaussian processes, DNNs, and BO were highlighted. Linear regression was used to predict drug release kinetics and encapsulation efficiency, while RF helped identify drug-loading efficiency and encapsulation properties. DNNs simulated thin-film nanocomposite reverse osmosis membranes and predicted their performance, while Gaussian processes modeled drug loading in solid lipid nanoparticles. BO explored parameter spaces and refined synthesis conditions based on DNN predictions.

The second area was predicting nano–bio interactions, where ML models elucidated nanoparticle interactions with biological systems, including protein corona formation, blood circulation, tumor penetration, and cellular uptake. SVMs, RF, and SX model had proven effective tools. SVMs modeled antibiotic removal from water and excelled at predicting protein adsorption and pharmacokinetics, RF identified determinants of nanoparticle–protein interactions and immune effects, while the SX model applied a stepwise prediction strategy to optimize the detection process for metal nanoparticles using a multi-emission fluorescent sensor array. These models offered valuable insights for optimizing nanoparticles to improve stability, targeting efficiency, and compatibility with the tumor microenvironment.

In improving therapeutic efficacy, ML aided in therapeutic outcome prediction and bioactivity assessment, ensuring selective targeting of cancer cells while minimizing side effects. DTs and k-NN predicted *in vivo* efficacy, GAs optimized nanoparticle properties for cytotoxicity, and ANNs modeled cellular internalization patterns and therapeutic outcomes as well as predicted adsorption behavior and optimized nanomaterial properties.

The review reported performance metrics for certain ML models applied in nanomaterial design and related fields. ANNs demonstrated an impressive accuracy rate of 98.1% in cancer diagnosis. Additionally, a multi-emission fluorescent sensor array, combined with the SX model, achieved a high accuracy rate of 95.6% in detecting heavy metal ions like Cr^{6+}, Fe^{2+}, and Cu^{2+}. Furthermore, a MXene flexible pressure sensor which applied deep learning algorithms demonstrated an average accuracy of 94% in sound (pattern) classification tasks, including recognizing 26 letters and polite expressions. These metrics highlight the effectiveness of ML in advancing nanoparticle applications.

Despite promising results, several challenges remain. Limited dataset size and quality constrain the training of robust models, particularly deep learning architectures. Data bias, such as underrepresentation of rare tumor types, undermines generalizability. The absence of standardized nanoparticle representation methods hampers data integration, and current descriptors often fail to capture biologically relevant interactions. Furthermore, the interpretability of complex ML models remains a barrier to clinical translation. The authors recommended using tree-based models like RF, which are relatively more interpretable. The authors also suggested integrating interpretability frameworks with tools from network science to analyze complex dependencies between nanoparticle properties and biological outcomes.

Furthermore, the authors called for interdisciplinary collaboration to overcome these barriers. Large-scale, high-quality datasets generated through high-throughput screening, microfluidics, and automated synthesis platforms are needed. Standardized experimental protocols and biologically informed descriptors would further enhance model reliability. The integration of ML with multi-omics, imaging, and automated experimentation is envisioned to accelerate the transition of nanomedicine into clinical practice, enabling more precise and effective cancer therapies.

8.5.7 *Machine Learning-Assisted Synthesis of High-Entropy Multi-Principal Element Nanoparticles*

Al Zoubi et al. reviewed recent progress in ML-assisted synthesis of high-entropy multi-principal element nanoparticles (HE-MPENs), emphasizing ML's role in accelerating design and discovery.[4] HE-MPENs, composed of five or more metallic elements homogeneously mixed to form stable single-phase solid structures, exhibit distinctive features such as high configurational entropy, multi-element synergy, and enhanced stability. These properties enable superior catalytic activity, efficient energy conversion, biomedical applications, innovative material designs, and durable industrial alloys for advanced technologies.

The review highlights how ML can address challenges in HE-MPEN synthesis by predicting relationships among lattice structures, material properties, and phase formations. Several ML algorithms and models are discussed. Bayesian normalized regression is used to predict energy changes and quantify uncertainties, offering precise and stable predictions of structural and electronic properties. The effective pair interaction algorithm accelerates DFT calculations, enabling faster molecular dynamics simulations and improving dataset representativeness. Gaussian process classification predicts promising solutes in ductile Mg alloys and reveals correlations between physical mechanisms and descriptors. GAs generate models and descriptors to predict HE-MPEN crystal structures for phase classification. Transfer learning enhances performance with limited datasets, while SISSO (Sure Independence Screening and Sparsifying Operator) identifies optimal descriptors for phase classification, creating interpretable phase diagrams. In addition, CNNs and recurrent neural networks demonstrate versatility in analyzing image data and generating captions.

The authors reported favorable predictive accuracy across these models. Bayesian normalized regression achieved exceptional precision, with an R^2 of 0.997 and an RMSE of 0.43 meV when compared with DFT-calculated energy values, while also effectively quantifying uncertainties in effective pair interaction predictions. The effective pair interaction model improved short-range order parameter estimations and accelerated DFT-based simulations. Gaussian process classification reliably identified ductile Mg alloys, and GAs combined with 70 descriptors reached over 90% accuracy in predicting MPEN crystal structures. SISSO successfully classified phases with only two descriptors, producing interpretable 2D phase diagrams, while cross-validation strategies reduced overfitting. Transfer learning showed promise but limited accuracy with small datasets. Collectively, these ML tools accelerate HE-MPEN discovery by predicting bulk properties, optimizing synthesis parameters, and reducing experimental trial-and-error, while ensuring consistency between computational and experimental results.

The review also underscores persistent challenges in HE-MPEN synthesis, including difficulties in achieving uniform mixing due to compositional complexity, phase segregation, and the scarcity of experimental data for ML training. The authors recommend developing tunable synthesis methods to better control HE-MPEN characteristics, integrating experimental data with ML for multi-objective optimization, and improving the scale and quality of data collection. Addressing miscibility issues and refining ML-driven predictions are considered essential for advancing applications in catalysis, materials science, and beyond.

8.5.8 *Machine Learning-Assisted Synthesis of Multi-Principal Element Nanoparticles*

Al Zoubi et al. reviewed synthesis strategies, material properties, and ML-assisted predictions for MPENs, with emphasis on catalytic performance, stability, and the role of ML in synthesis.[12] This article is closely related to another review by the same group discussed in the preceding subsection.[4] While that work focused specifically on HE-MPENs, which are a single-phase subclass and their ML-predicted properties, the present review provides a broader overview of synthesis strategies and ML applications across diverse MPEN systems. Given its wider scope, it is included in this umbrella review.

The authors examined a wide range of ML algorithms and models applied to MPEN synthesis and optimization. Key models include linear ridge regression, polynomial regression, Bayesian regression, ANNs, CNNs, GANs, GBDTs, Gaussian process classification, RF, SVMs, and gradient boosting. These models accelerate MPEN discovery by uncovering hidden correlations between elemental composition and material properties. For instance, linear ridge regression, Bayesian regression, polynomial regression, and GBDTs have been used to predict hardness, yield strength, and other mechanical properties using datasets of varying sizes. ANNs and Gaussian process classification predicted solid-solution phases and lattice structures, while GANs generated synthetic data to compensate for limited experiments. CNNs enabled automated characterization of MPENs from microscopy images, and ensemble methods such as RF and gradient boosting identified key descriptors, including enthalpy and entropy, that govern phase stability. While autoencoders were used for feature extraction and dimensionality reduction in MPEN datasets, ResNet50 further advanced crystallographic phase identification from electron backscatter diffraction patterns. Collectively, these tools support inverse materials design, guiding MPEN synthesis toward optimal compositions and processing conditions while reducing trial-and-error experimentation.

The review summarized diverse performance metrics. RF achieved 100% accuracy in predicting labeled data (134 samples) and a yield-strength prediction error of 2.5–7.7%. ANNs showed 80% accuracy for phase selection (118 samples) and a Pearson's R of 0.983 for predicting binary to quaternary systems (321 samples). Gradient boosting achieved 61% accuracy for phase prediction, with Young's modulus errors ranging from 0.1% to 42% (329 samples). Gaussian process classification achieved an AUC of 0.97 for phase prediction (1,252 samples) and 93% accuracy for solid-solution prediction, with AUC = 1 considered ideal classification. GBDTs reached 96.41% accuracy for single-phase solid-solution prediction (1,807 samples). SVMs and polynomial regression both yielded an RMSE of ~50 Vickers Hardness, while Bayesian regression and linear ridge regression achieved $R^2 > 0.95$ for yield strength. CNNs and ResNet50 exceeded 90% accuracy for crystallographic phase identification. GANs and autoencoders contributed by generating synthetic datasets and reducing dimensionality. Together, these results highlight the effectiveness of ML in predicting MPEN properties and phase behavior, underscoring its potential to accelerate synthesis and design.

The review highlights the synthesis strategies and thermodynamic principles governing MPENs, emphasizing that the Gibbs free energy equation plays a pivotal role in predicting the stability and phase formation of MPENs. Synthesis methods including wet chemistry, carbothermal shock, and microwave heating are explored to achieve uniform mixing and control nanoparticle size and morphology. ML models are increasingly used to predict phase formation, lattice structures, and physical properties, enabling efficient material design and optimization. Despite these advances, challenges remain. Major limitations include scarcity of high-quality datasets, risk of overfitting with small datasets, and incomplete reporting of experimental conditions, which reduces prediction reliability. The vast compositional space of MPENs complicates data collection, while the absence of negative (failed) results limits model training. The authors recommend broader data sharing — including unsuccessful experiments — and closer integration of experimental and computational data. They also note that for small datasets, simpler models such as SVMs may outperform more complex ones by avoiding overfitting. With 119 cited references, the article adopts a narrative rather than systematic review format.

8.5.9 *Machine Learning-Assisted Design and Synthesis of Lipid Nanoparticles for Gene Therapy and Drug Delivery*

Li, Zhao, and Xu reviewed the application of ML techniques in the design and synthesis of lipid nanoparticles (LNPs), with a particular emphasis on ionizable lipids for nucleic acid-based therapeutics.[29] The article introduced ML workflows, algorithms, and tools that can accelerate LNP development for gene therapy and drug delivery by enhancing safety, efficacy, and targeting. LNPs are nanoscale, lipid-based carriers designed to deliver nucleic acids such as mRNA or siRNA into cells, where they protect the cargo, promote uptake, and facilitate endosomal escape, making them central to modern gene therapy and vaccines.

The authors examined seven representative ML algorithms/models, each serving distinct roles in LNP design, synthesis, and optimization. Linear regression identified linear correlations between LNP properties and target attributes, offering interpretability but limited predictive power. SVMs classified complex, nonlinear data by defining optimal hyperplanes in high-dimensional spaces. RFs employed ensembles of DTs to reduce overfitting, handle mixed data types, and improve robustness. XGBoost constructed sequential trees to minimize errors, achieving high accuracy and tolerance for missing data. Multilayer perceptrons captured nonlinear relationships through supervised learning. Graph neural networks modeled molecular structures as graphs, uncovering atom–bond interactions, while graph transformers such as Grover extended this capability to capture long-range dependencies with self-supervised training.

Although the authors did not provide numerical performance metrics (e.g., accuracy, precision, mean squared error), they discussed the relative strengths and limitations of each model. For example, RFs offered strong interpretability and accuracy in smaller trees (fewer decision points), whereas graph neural networks and graph transformers achieved superior predictive power but required higher computational resources and specialized interpretability methods.

The review also outlined a four-step workflow for ML-driven LNP development: (1) synthesis of ionizable lipids and LNPs, (2) conversion of experimental data into machine-readable formats, (3) ML model selection and training, and (4) model evaluation and refinement. Among synthesis methods, microfluidic mixing was highlighted as the preferred technique for LNP assembly due to its precision, reproducibility, scalability, and ability to preserve sensitive biomolecules such as mRNA. The Ugi and Passerini reactions were noted as high-throughput strategies for generating diverse ionizable lipid libraries, thereby enabling large datasets suitable for ML modeling.

The authors further addressed key challenges, including data scarcity, overfitting, and limited model generalization. They suggested hybrid modeling, simpler algorithms, and public data sharing as strategies to overcome these limitations. Emphasizing interdisciplinary collaboration and the development of accessible ML tools, the review advocates for democratizing ML approaches to accelerate LNP research and advance therapeutic delivery systems.

Overall, this narrative review provides a comprehensive overview of how ML can be leveraged to optimize LNP design and synthesis, underscoring its potential to transform nucleic acid–based therapeutics.

8.6 Concluding Remarks

This umbrella review synthesizes findings from 16 review articles on ML-assisted nanoparticle synthesis, spanning 0D, 1D, 2D, and 3D nanoparticles. The reviewed synthesis approaches include CVD, pyrolysis, microfluidic platforms, closed-loop optimization, green methods, and many others. Across these processes, ML has shown remarkable potential to predict nanoparticle properties, optimize synthesis conditions, and accelerate discovery, offering transformative solutions to the inefficiencies of traditional trial-and-error approaches.

As indicated in Tables 8.1, 8.2, and 8.3, more than 60 different ML algorithms/models were discussed and the key findings are summarized as follows. Supervised learning algorithms such as RF, XGBoost, ANNs, and GBDTs consistently achieved high accuracy in predicting nanoparticle size, shape, yield, and optical properties. Deep learning models including CNNs and DNNs excelled in analyzing microscopy images, spectral data, and other complex datasets, often exceeding 90% accuracy. Optimization methods such as BO and GAs reduced experimental workloads by refining synthesis parameters. Meanwhile, emerging approaches like graph neural networks, transfer learning, and GANs showed promise for handling small datasets and generating synthetic data. Hybrid models that combine simpler algorithms such as SVMs and RF with advanced techniques like GANs and transfer learning can balance accuracy and interpretability, especially for small datasets. These advancements have enabled progress in biomedicine, catalysis, optoelectronics, environmental remediation, and energy storage, with notable examples including CDs for sensors, LNPs for gene therapy, and HE-MPENs for catalysis.

Despite these successes, challenges remain. Chief among them is the scarcity of standardized, high-quality datasets, particularly those including negative or failed experiments, which limits model generalizability. The interpretability of complex ML models — especially deep learning architectures — also remains a barrier, as their "black box" nature constrains scientific insight and trust. Scalability is another concern: results obtained from automated or microfluidic systems do not always translate to conventional synthesis environments. Furthermore, resource-intensive models such as GANs pose accessibility challenges for smaller research groups.

Looking forward, integrating ML with autonomous systems, including robotic synthesis, automated characterization, and computational prediction, offers a pathway toward self-driving workshops or laboratories for scalable and reproducible nanoparticle synthesis. Interdisciplinary collaboration among materials scientists, manufacturing engineers, and artificial intelligence researchers will be crucial for advancing data collection, model development, and experimental validation. Establishing unified standards for nanoparticle synthesis and characterization will also strengthen model reliability and reproducibility. Future progress will depend on generating diverse, high-quality datasets, including negative results; developing hybrid models that balance accuracy with interpretability; leveraging transfer learning to reduce training costs; and enhancing explainability through tools such as SHAP or through interpretability frameworks from network science. Expanding access to ML platforms and computational resources will further democratize applications across research groups.

In conclusion, ML-assisted nanoparticle synthesis marks a paradigm shift in nanofabrication by enabling unprecedented efficiency, precision, and scalability. Addressing current challenges through interdisciplinary collaboration, data standardization, and explainable ML will be key to unlocking its full potential in accelerating innovation across nanofabrication and its applications.

Acknowledgements

The author gratefully acknowledges financial support from Pacific Technology, LLC (Phoenix, USA), which enabled the author to spend time to prepare substantial portions of this manuscript. In addition, the author used ChatGPT, an artificial intelligence language model developed by OpenAI, to assist in addressing technical questions arising from the review of related literature and to help refine the language of the manuscript. All artificial intelligence-assisted content was critically evaluated, verified, and edited by the author, who takes full responsibility for the accuracy, originality, and integrity of the final manuscript.

References

1. Vert, Michel, Yoshiharu Doi, Karl-Heinz Hellwich, Michael Hess, Philip Hodge, Przemyslaw Kubisa, Marguerite Rinaudo, and François Schué. "Terminology for biorelated polymers and applications (IUPAC Recommendations 2012)." *Pure and Applied Chemistry* 84, no. 2 (2012): 377–410.
2. Koza, John R., Forrest H. Bennett III, David Andre, and Martin A. Keane. "Automated design of both the topology and sizing of analog electrical circuits using genetic programming." In *Artificial Intelligence in Design'96*, edited by John S. Gero & Fay Sudweeks, pp. 151–170. Springer, 1996.

3. Tao, Huachen, Tianyi Wu, Matteo Aldeghi, Tony C. Wu, Alan Aspuru-Guzik, and Eugenia Kumacheva. "Nanoparticle synthesis assisted by machine learning." *Nature Reviews Materials* 6, no. 8 (2021): 701–716.
4. Al Zoubi, Wail, Yujun Sheng, Iftikhar Hussain, Heo Seongjun, Mohammad R. Thalji, and Nokeun Park. "Synthesis and machine learning prediction of high entropy multi-principal element nanoparticles." *Small* 21, no. 22 (2025): 2501444.
5. Fernandez, Ritin Santiago, Amir Masoud Sharifnia, and Hanan Khalil. "Umbrella reviews: A methodological guide." *European Journal of Cardiovascular Nursing* 24, no. 6 (2025): 996–1002.
6. Grant, Maria J., and Andrew Booth. "A typology of reviews: An analysis of 14 review types and associated methodologies." *Health Information & Libraries Journal* 26, no. 2 (2009): 91–108.
7. Mohri, Mehryar, Afshin Rostamizadeh, and Ameet Talwalkar. *Foundations of Machine Learning*. MIT Press, 2018.
8. Yan, Fanyong, Ruixue Bai, Juanru Huang, Xihui Bian, and Yang Fu. "Machine learning-assisted carbon dots synthesis and analysis: State of the art and future directions." *TrAC Trends in Analytical Chemistry* 184 (2025): 118141.
9. Park, Jungwon, Young Min Kim, Seonghun Hong, Byungchan Han, Ki Tae Nam, and Yousung Jung. "Closed-loop optimization of nanoparticle synthesis enabled by robotics and machine learning." *Matter* 6, no. 3 (2023): 677–690.
10. Loke, Kah Yee, Xiu Xian Lim, Mohd Azam Osman, Siew Chun Low, and Wen-Da Oh. "Enhancing plastic pyrolysis for carbon nanotubes synthesis through machine learning integration: A review." *Journal of Analytical and Applied Pyrolysis* 187 (2025): 106989.
11. LeCun, Yann, Yoshua Bengio, and Geoffrey Hinton. "Deep learning." *Nature* 521, no. 7553 (2015): 436–444.
12. Al Zoubi, Wail, Yujun Sheng, Iftikhar Hussain, Heo Seongjun, and Nokeun Park. "Multi-principal element nanoparticles: Synthesis strategies and machine learning prediction." *Coordination Chemistry Reviews* 535 (2025): 216656.
13. Bhowmik, Sayan, and Ananth Govind Rajan. "Chemical vapor deposition of 2D materials: A review of modeling, simulation, and machine learning studies." *iScience* 25, no. 3 (2022): 103832.

14. Chan, Emory M. "Combinatorial approaches for developing upconverting nanomaterials: High-throughput screening, modeling, and applications." *Chemical Society Reviews* 44, no. 6 (2015): 1653–1679.

15. Amato, Alberto, and Vincenzo Di Lecce. "Data preprocessing impact on machine learning algorithm performance." *Open Computer Science* 13, no. 1 (2023): 20220278.

16. Wang, Qingquan, Yujian Liu, Chenchen Li, Bin Xu, Shidang Xu, and Bin Liu. "Machine learning-enhanced nanoparticle design for precision cancer drug delivery." *Advanced Science* 12, no. 30 (2025): e03138.

17. Li, Da, Yongxin Yang, Yi-Zhe Song, and Timothy Hospedales. "Learning to generalize: Meta-learning for domain generalization." In *Proceedings of the Thirty-Second AAAI Conference on Artificial Intelligence*, vol. 32, edited by Sheila A. McIlraith, pp. 3490–3497. AAAI Press, 2018.

18. Salih, Ahmed M., Zahra Raisi-Estabragh, Ilaria Boscolo Galazzo, Petia Radeva, Steffen E. Petersen, Karim Lekadir, and Gloria Menegaz. "A perspective on explainable artificial intelligence methods: SHAP and LIME." *Advanced Intelligent Systems* 7, no. 1 (2025): 2400304.

19. Khan, Ibrahim, Khalid Saeed, and Idrees Khan. "Nanoparticles: Properties, applications and toxicities." *Arabian Journal of Chemistry* 12, no. 7 (2019): 908–931.

20. Nandipati, Mutha, Olukayode Fatoki, and Salil Desai. "Bridging nanomanufacturing and artificial intelligence — a comprehensive review." *Materials* 17, no. 7 (2024): 1621.

21. Mohammadzadeh Kakhki, Roya and Mohammadpoor, Mojtaba. "Machine learning-driven approaches for synthesizing carbon dots and their applications in photoelectrochemical sensors." *Inorganic Chemistry Communications* 159 (2024): 111859.

22. Duman, Ali Nabi, and Almaz S. Jalilov. "Machine learning for carbon dot synthesis and applications." *Materials Advances* 5, no. 18 (2024): 7097–7112.

23. El-Azazy, Marwa, Ahmed I. Osman, Mahmoud Nasr, Yassmin Ibrahim, Nessreen Al-Hashimi, Khalid Al-Saad, Mohammad A. Al-Ghouti et al. "The interface of machine learning and carbon quantum dots: From coordinated innovative synthesis to practical application in water control and electrochemistry." *Coordination Chemistry Reviews* 517 (2024): 215976.

24. Chen, Guohai, and Dai-Ming Tang. "Machine learning as a "catalyst" for advancements in carbon nanotube research." *Nanomaterials* 14, no. 21 (2024): 1688.

25. Booth, Andrew, Anthea Sutton, Mark Clowes, and Marrissa Martyn-St James. *Systematic Approaches to A Successful Literature Review*. 3rd Edition, Sage 2021.
26. Liu, Lei, Zhaolun Zhang, Lihua Cao, Ziyi Xiong, Ying Tang, and Yao Pan. "Cytotoxicity of phytosynthesized silver nanoparticles: A meta-analysis by machine learning algorithms." *Sustainable Chemistry and Pharmacy* 21 (2021): 100425.
27. Lv, Honglin, and Xueye Chen. "Intelligent control of nanoparticle synthesis through machine learning." *Nanoscale* 14, no. 18 (2022): 6688–6708.
28. Karakus, Ceyda Oksel. "Machine learning-assisted navigation of iron oxide nanoparticle synthesis: A meta analysis." *Journal of Biomedical Nanotechnology* 20, no. 10 (2024): 1619–1628.
29. Li, Hao, Yayi Zhao, and Chenjie Xu. "Machine learning techniques for lipid nanoparticle formulation." *Nano Convergence* 12, no. 1 (2025): 35.

https://doi.org/10.1142/9789819830763_0009

CHAPTER 9

On-Line Characterization Techniques in Nanofabrication

Prithvi Basu
Department of Electrical Engineering, Texas A&M University,
College Station, TX 77840, USA
prithvibasu87@gmail.com

Abstract

This chapter explores the vital role of real-time diagnostic tools in monitoring and improving nanofabrication processes. As the demand for precise and complex nanoscale manufacturing grows, on-the-spot, non-destructive techniques are increasingly essential. The chapter covers key on-line characterization methods such as atomic force microscopy, scanning electron microscopy, ellipsometry, optical profilometry and Raman spectroscopy, and their application in monitoring critical factors like surface morphology, film thickness, material composition, and structural integrity during fabrication. It also addresses the challenges of integrating these techniques into automated manufacturing systems and their role in enhancing process control. The chapter concludes by examining the future of on-line characterization, with a focus on advancing these methods to improve the reliability, yield, and performance of nanofabricated devices. By highlighting both established and emerging techniques, it provides a clear understanding of how real-time characterization supports the ongoing evolution of nanofabrication technology.

Keywords: On-line characterization, real-time metrology, nanofabrication, *in situ* monitoring, atomic force microscopy, scanning electron microscopy, ellipsometry, optical profilometry, Raman spectroscopy, thin film characterization, surface morphology, film thickness control, process monitoring, adaptive process control, non-destructive measurement, semiconductor manufacturing, nanoscale precision, automated fabrication systems, process feedback loops, yield and reliability

9.1 Introduction

In recent years, the field of nanofabrication has experienced rapid expansion, driven by the increasing demands of advanced technologies such as semiconductor devices, photonics, MEMS/NEMS, and quantum systems. As feature sizes continue to shrink into the deep nanoscale and structures become more complex — incorporating 3D architecture, heterogeneous materials, and multilayer stacks — the tolerances on critical parameters such as film thickness, surface morphology, material composition, and structural integrity have become exceptionally stringent.[1] In this environment, the ability to monitor and control fabrication in real time, or near real time, is no longer a luxury but a necessity for achieving high precision and reliability.[2]

Traditionally, nanofabrication processes have relied heavily on *ex situ* characterization methods, where measurements are performed off-line after individual fabrication steps to assess outcomes and identify defects. However, as process windows narrow and allowable defect budgets tighten, these post-process analyses often prove insufficient for maintaining the precision and throughput required in advanced manufacturing. To address this limitation, on-line — or in-line — characterization techniques have emerged as essential tools.[3] These techniques integrate diagnostic instruments directly into the fabrication environment, allowing for non-destructive, rapid measurement of key parameters during the process itself. Such real-time access to process data enables immediate feedback and adaptive control, ensuring tighter process regulation and improved yield.[4] Figure 9.1 shows a simplified schematic of a nanofabrication line (e.g., deposition → lithography → etching → inspection → packaging). Arrows show feedback loops from

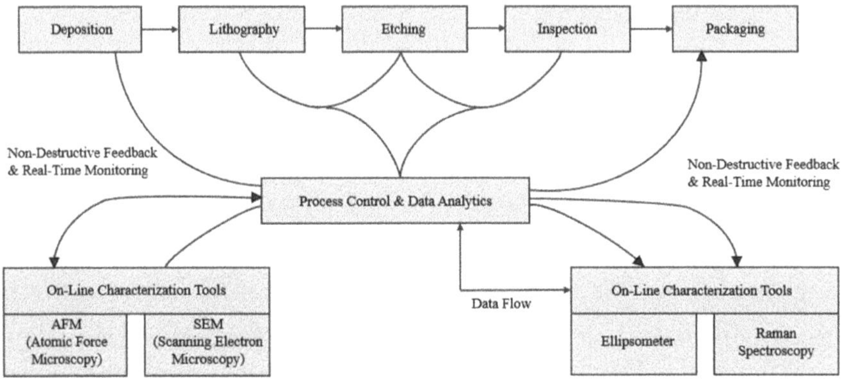

Figure 9.1. Schematic representation of a nanofabrication workflow illustrating integration of on-line characterization tools for real-time process monitoring and control.

on-line characterization tools (atomic force microscopy (AFM), scanning electron microscopy (SEM), ellipsometer, Raman spectroscopy) feeding data back into process control.

The importance of on-line characterization in nanofabrication is underscored by several key factors. At the nanoscale, even slight deviations in surface morphology or material composition can cause substantial performance degradation, manifesting as electrical leakage, optical scattering, or mechanical instability. Real-time monitoring mitigates these risks by providing immediate, actionable data that can be used to fine-tune fabrication parameters before deviations propagate into major defects. Furthermore, the integration of measurement tools into automated or semi-automated process flows enhances manufacturing consistency, scalability, and efficiency which are crucial attributes for transitioning from laboratory-scale demonstrations to industrial production.[5] Optical metrology methods such as scatterometry and ellipsometry, for instance, are increasingly applied in-line within semiconductor fabrication facilities because of their high measurement speed, non-contact operation, and ability to track thin film properties with great accuracy.[6] In contrast, high-resolution surface metrology tools such as AFM and SEM have traditionally been used off-line due to their slower acquisition rates and sample handling requirements. However, recent advancements in instrumentation speed, automation, and hybrid measurement systems are making their integration into on-line environments more feasible.[7]

The main objective of this chapter is to provide a comprehensive overview of on-line characterization techniques applicable to nanoscale fabrication, with particular emphasis on real-time or near-real-time diagnostics that enable process monitoring without disturbing or damaging the sample. Specifically, the chapter aims to present the fundamental principles, capabilities, and limitations of key techniques such as AFM, SEM, ellipsometry, optical profilometry, and Raman spectroscopy in the context of nanofabrication. It further seeks to illustrate how these methods can be used to monitor critical fabrication parameters — including surface morphology, film thickness, material composition, and structural integrity — directly during processing rather than as a post-fabrication evaluation.

Beyond describing the methods themselves, the chapter examines how these diagnostic tools can be integrated into automated manufacturing systems, addressing challenges such as measurement speed, environmental compatibility, data acquisition, and process feedback control. Particular attention is given to practical considerations such as measurement speed, environmental compatibility, system integration, and the ability to generate stable and interpretable data streams for process feedback. In addition, the chapter looks toward future trends in the field, including emerging measurement techniques, the growing role of artificial intelligence and machine learning in real-time data analysis, and the broader integration of on-line metrology into the framework of Industry 4.0 and intelligent nanomanufacturing systems. Overall, the chapter is designed to equip readers with both a strong theoretical understanding of these characterization techniques and a practical appreciation of how they can be implemented in real-world fabrication environments.

9.2 Importance of On-Line Characterization in Nanofabrication

9.2.1 *Precision in Nanofabrication*

In the realm of nanofabrication, achieving the required precision means controlling feature sizes, surface roughness, material composition and interfaces down to the nanometer (and sometimes atomic) scale. On-line characterization plays a crucial role here because it allows one to monitor, in real time or near real time, whether the imposed process conditions are producing the desired nanostructures with the necessary fidelity. For instance, without on-line feedback one might only discover after the fact that lithographic lines are misaligned, surface roughness exceeds specification, or layer thicknesses deviate beyond tolerance. Furthermore, Liddle et al. explicitly stated that future nanomanufacturing may well necessitate nanometer scale metrology which functions real time and on-line.[8]

By integrating metrology (i.e., characterization) during the fabrication steps, deviations can be detected early and corrected — such as by adjusting exposure dose, etch time, deposition rate or mask alignment — rather than only discovering them at end-of-line inspection. This shift from post-process spotting to embedded feedback significantly improves yield, uniformity and device performance. The advantages are multiple: first, process drift (e.g., gradual misregistration, film thickness variation, roughness development) can be identified while there is still opportunity to intervene rather than scrapping entire runs. Second, the ability to gather high-density feedback across the wafer or across runs enables statistical process control at the nanoscale; for example, in-line AFM-based systems have been explored for wafer alignment metrology in nanomanufacturing.[9] Third, on-line metrology supports the adoption of advanced, high-throughput nanomanufacturing methods (such as roll-to-roll nanoimprint, directed self-assembly) by providing the required measurement fidelity and speed to keep pace with production. Indeed, the challenge of achieving high speed and nanometer resolution has been noted as a serious barrier to in-line nanoscale metrology.[10]

Moreover, the choice of characterization technique must itself meet stringent criteria for on-line deployment: non-contact operation (to avoid disturbing delicate nanostructures), high throughput (to avoid becoming the production bottleneck), and sufficient resolution and sensitivity (to detect errors such as line-edge roughness, sidewall angle deviations, interface intermixing).[11] For example, recent reviews of optical metrology for advanced manufacturing emphasize that real-time, embedded optical measurement modules directly in fabrication equipment are now becoming essential.[12] Without such integrated metrology, the fabrication process operates in an open-loop mode, relying on prior calibration or off-line sampling; this increases the risk of undetected shifts or defects accumulating until end-of-line inspection, at which point correction is much more costly (in yield loss or scrap).[13] In short: on-line characterization is a vital enabler of precision in nanofabrication, enabling tighter control, higher yield, and greater reproducibility than would otherwise be achievable.

9.2.2 Non-Destructive Testing

Non-destructive testing techniques are especially important in nanofabrication because many of the structures and devices under fabrication are too delicate or too costly to sacrifice. Non-destructive methods allow inspection and characterization without damaging or altering the material or device, thus preserving its integrity for subsequent processing or use. For example, in the review by She et al., it is shown how non-destructive testing methods are used in nanocomposite additive manufacturing to detect defects such as porosity, delamination or particle agglomeration — without slicing or destroying the part.[14]

The key advantages include preserving the fabricated item for further processing or testing; enabling serial inspections over time (for example during lifetime testing); and reducing waste and cost associated with destructive test samples. In the context of nanofabrication, where layers may be extremely thin, interfaces critical, and the structures complex, the ability to apply non-destructive characterization means one can detect hidden defects, interface roughness, contamination, or layer delamination without compromising the sample for further steps. For instance, advanced metrology roadmaps identify the "non-destructive characterization of buried or embedded interfaces" as a key challenge in nano-metrology.[15]

Moreover, non-destructive testing methods can often be integrated into real-time or *in situ* monitoring of fabrication and service conditions. This means one can track changes in the structure over time (for example under thermal cycling or mechanical load) and detect the onset of degradation or defect growth before failure. Reviews of non-destructive testing for advanced composite or hybrid structures emphasize that as materials become more complex (multi-material, nano-enabled, layered), the demand for high-resolution, non-destructive and often automated inspection is rising.[16]

It is also worth noting the technical challenges: Traditional non-destructive testing methods (ultrasound, X-ray radiography, thermography) were developed for bulk materials and larger defects; in the nano- and micro-fabrication regime one often needs sub-micron-scale sensitivity, minimal disturbance of the sample, and high throughput.[17] For nanofabrication lines, the non-destructive testing tool must often be non-contact (to avoid damaging fragile structures), very fast (to not bottleneck production) and capable of detecting very small defects (for example voids of tens of nanometers, interface mixing, side-wall roughness). For example, a recent review underscores that current non-destructive techniques must evolve to detect tiny defects in insulating materials at ever smaller scales.[18]

In short: non-destructive characterization is essential to maintain both material integrity and throughput in a nanofabrication line. By enabling defect detection without destroying the part, non-destructive testing supports higher yield, lower scrap rates, continuity of manufacturing, and in-service reliability of nano-enabled devices.

9.2.3 *Process Monitoring and Control*

In modern nanofabrication, particularly in high-volume or high-precision environments, maintaining real-time insight into the fabrication process is essential. On-line characterization refers to the continuous or near-continuous monitoring of key parameters — such as layer thickness, surface roughness, feature alignment, composition, stress, and defect density — while fabrication is underway. The purpose is not merely to collect data, but to use that data to control and stabilize the process. By linking characterization tools directly with process equipment, deviations can be detected and corrected before they lead to significant defects or yield losses.

Fang et al. states that additive manufacturing builds parts layer by layer, but without proper feedback control, defects and inconsistencies can easily occur.[19] Their work outlines three key feedback loops — *in situ* defect monitoring, printer fault diagnosis, and closed-loop process control — to improve reliability and part quality. It also highlights the measurement needs, challenges, and future directions for making additive manufacturing processes more consistent and dependable. While these studies focus on additive systems, the same approach applies in nanofabrication processes such as atomic layer deposition (ALD), electron beam lithography, or nanoimprint lithography — where even a nanometer-scale deviation can affect device performance.[20]

Recent reviews highlight that on-line characterization enables "zero-defect" manufacturing by providing feedback to correct errors as they develop rather than after production is completed.

For instance, in laser-based and nanoscale fabrication, *in situ* optical, acoustic, or X-ray monitoring has been used to detect layer non-uniformity, micro-crack formation, and chemical inhomogeneity, feeding this information into adaptive control loops.[21] In nanofabrication, comparable systems monitor film growth rates, etch uniformity, and pattern registration during processing, automatically adjusting variables such as deposition rate, gas composition, or beam focus to maintain desired specifications.

The advantages of this approach are clear. First, it transforms the process from an open-loop to a closed-loop system, reducing reliance on pre-set parameters and manual inspection. Second, it minimizes scrap and rework by catching errors early. Third, it supports the generation of statistical process data that can be used for predictive modeling and machine learning-based process optimization.[22] For example, Yao et al. describe wafer-scale AFM and optical scatterometry systems capable of in-line surface topography measurements with nanometer-level precision, offering the feedback needed for adaptive nanomanufacturing.[9]

However, integrating on-line metrology into nanofabrication remains technically demanding. The measurement systems must be non-invasive, high-resolution, and fast enough not to hinder production throughput. They must also operate under challenging conditions — for instance, within vacuum chambers, high-temperature environments, or chemically reactive atmospheres — while maintaining stability and calibration over long production runs. The review "Lithography, Metrology and Nanomanufacturing" underscores that future nanomanufacturing may well necessitate real-time, on-line metrology to ensure process control at the atomic scale.[8]

In summary, on-line characterization forms the backbone of effective process monitoring and control in nanofabrication. It enables immediate detection of deviations, facilitates closed-loop correction, and maintains consistency across wafers, layers, and production cycles. By embedding metrology directly into fabrication equipment, manufacturers can achieve tighter process control, higher yields, and more reliable nanostructured devices — outcomes essential for both industrial production and advanced research at the nanoscale.

9.3 On-Line Characterization Techniques

Key on-line characterization techniques play a crucial role in understanding and monitoring the physical and chemical properties of materials during and after processing. These advanced methods provide valuable insights into surface morphology, film thickness, optical behavior, and structural composition at the micro- and nanoscale. Techniques such as AFM, SEM, ellipsometry, optical profilometry, and Raman spectroscopy allow researchers to obtain precise and real-time information about material surfaces and interfaces. Together, they form a comprehensive toolkit for ensuring quality control, optimizing fabrication processes, and supporting the development of new materials and technologies.

9.3.1 *Atomic Force Microscopy*

9.3.1.1 *Principles of Atomic Force Microscopy*

An AFM works by bringing a nanometer-sharp tip on a tiny flexible cantilever close to a surface, where atomic-scale forces cause the cantilever to bend or oscillate as shown in Figure 9.2. A laser beam reflected off the back of the cantilever onto a photodiode

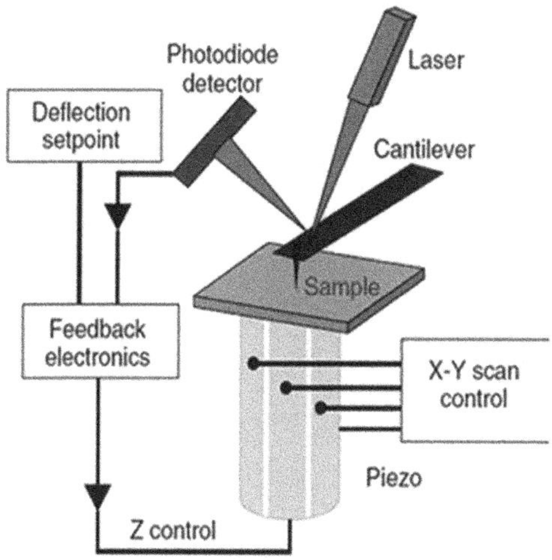

Figure 9.2. Schematic diagram of the AFM. Reprinted with permission from [24] by Frontiers Media.

tracks these small deflections with high sensitivity. As the tip moves across the surface in a controlled scan, a feedback system constantly adjusts the vertical position of the tip or sample to keep the interaction at a set, stable level. The system records the voltages used for these vertical adjustments, and from them reconstructs a highly detailed map of the surface, revealing its topography with nanometer-level resolution.[23]

AFM imaging modes can be broadly grouped:

(a) **Contact mode**, where the tip maintains continuous contact with the surface and the cantilever deflection is used for feedback.[25]
(b) **Tapping (intermittent contact) mode**, where the cantilever is oscillated near its resonant frequency and the amplitude (or phase) is used for feedback. This reduces tip–surface drag and is gentler on soft samples.[26]

(c) **Non-contact mode**, where the tip never quite touches the surface and small changes in oscillation frequency or amplitude are used to sense forces.[27]

In addition to topographic imaging, AFM can carry out force–distance spectroscopy (by placing the tip at a fixed location, indenting or retracting from the surface, and recording force vs. displacement curves) and other advanced modes (e.g., bimodal, peak force tapping, nanomechanical mapping). From these interactions, one can infer mechanical, adhesive, viscoelastic and other local properties of the surface.[29] Figure 9.3(a) displays a typical AFM setup including a sharp tip mounted on a small cantilever, a photodetector that measures how the cantilever bends or vibrates, a piezoelectric scanner that moves the tip or sample with high precision, and electronics that control and stabilize the whole system. Figure 9.3(b) shows the force curve and the three ways an AFM can operate: non-contact, tapping, and contact modes. In non-contact and tapping modes, the cantilever is driven to vibrate near its natural resonance, and the photodetector monitors how the vibration changes as the tip interacts with the surface. In non-contact mode, the tip never touches the sample; instead, it feels only weak attractive forces, which reduce its amplitude and shift its frequency. These changes are used by the feedback electronics to keep the system stable and to create an image of the surface.

Finite element modeling and refined contact mechanics show that even for soft materials or thin layers, the tip–sample geometry, indentation depth, substrate stiffness, and bottom-stiffness effects (for thin films on hard substrates) must

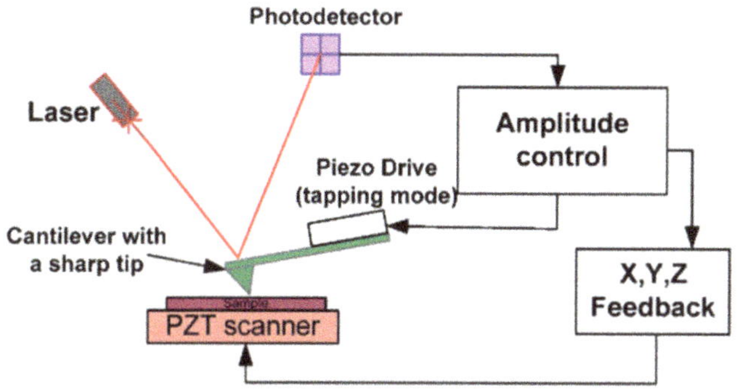

(a)

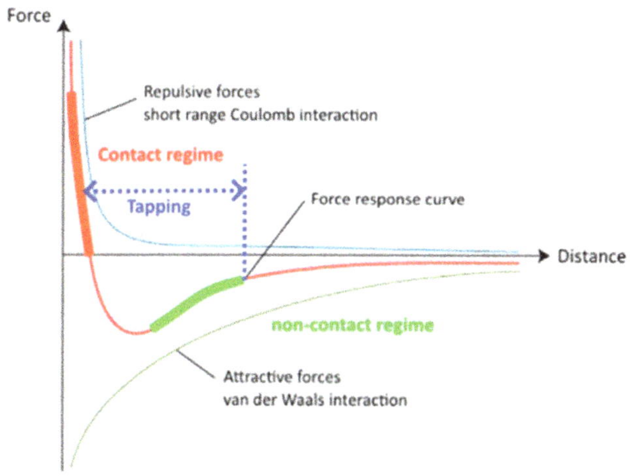

(b)

Figure 9.3. (a) A scheme of tapping mode AFM and (b) force-distance curve between the AFM tip and sample. Reprinted with permission from [28] by MDPI.

be factored for quantitative modulus or stiffness measurement.[30] Thus, AFM is not just a "microscope" but a multimodal measurement platform combining high-resolution imaging of surface morphology with quantitative mechanical and interaction measurements at the nanoscale.

9.3.1.2 *Capabilities in Surface Morphology, Roughness and Nanostructure Features*

Since AFM provides 3D (x,y,z) data of a sample surface with nanometer or even sub-nanometer resolution in height (z) and ~10 nm lateral resolution (x,y) under appropriate conditions, it allows accurate quantitative morphology and roughness analysis. For example, commercial AFM sources note vertical resolution of sub-nanometer and typical lateral resolution of 5–10 nm.[31] In practice, AFM has been used to extract surface roughness parameters (such as Ra, Rq in line profiles, or Sa, Sq, Sz, etc., for areal surface analysis) directly from the height maps. A recent study addressed how particular areal parameters (Sa, Sz, Sq, Sdq, Sdr) behave in AFM measurements, showing that even small instrumental or geometric deviations can influence the computed parameters and stress the need for calibration and uncertainty evaluation.[32]

Beyond mere roughness numbers, AFM height maps allow insight into nanostructural features: islands, grains, steps, pits, nanowires, coatings, textured surfaces, pattern fidelity, etc. For example, in thin films of nanoparticles the variation of root mean square roughness with coverage was modeled and compared to AFM data, showing clear quantitative trends.[33] In many applied studies, AFM has been used to monitor morphological changes induced by processing: e.g., 3D-printed (laser powder bed fusion) surfaces characterized by AFM in a recent 2024 paper found that AFM allowed detailed measurement of the surface structure of hard magnetic Nd–Fe–B parts and comparison with other imaging methods.[34]

Similarly, AFM has been used to measure sidewall roughness of vertical structures (in semiconductor metrology) using a 3D

AFM system with fully automated analysis and high reproducibility (relative standard deviation <2%) for sidewall roughness measurement.[35] Finally, AFM excels for morphology as it allows high-resolution imaging of surface features, quantification of roughness (both profile- and area-based), analysis of nanostructures (size, spacing, coverage, distribution), and linking topography to function (e.g., in coatings, films, biological surfaces).

9.3.1.3 *Applications for Monitoring Roughness, Nanostructure Features and Mechanical Properties*

As AFM can acquire both topographic and mechanical/interaction data, it has seen numerous applications in monitoring roughness evolution, nanostructure feature integrity, and mechanical property mapping.

(a) **Roughness monitoring:** In many manufacturing or material-processing studies, surface roughness is a critical parameter (for adhesion, friction, coating performance, fatigue, wettability). AFM allows the measurement of roughness down to nanoscale. For example, an AFM-based study on nanoparticle film coverings correlated coverage with root mean square roughness and skewness of the height distribution, demonstrating quantitative sensitivity.[32] Another article studied areal topography parameters using AFM (Sa, Sz, etc.) and emphasized metrological rigor in roughness evaluation.[31]

(b) **Nanostructure feature analysis:** AFM is ideal for assessing features such as nanowires, nanotextures, nano-patterned surfaces, grain size, island morphology, and their height/

spacing. For example, an AFM investigation of ZnO nanowires measured mechanical properties (Young's modulus) via tip indentation and correlated nanostructure treatments (annealing) with a tenfold change in modulus.[36] In the context of adhesion and topographic scale, a study showed how nanoscale roughness in diamond coatings (measured by AFM) impacted adhesion forces in hard material contacts; AFM measured the topography across length scales and then adhesion behavior was analyzed.[37]

(c) **Mechanical/interaction property mapping:** Beyond morphology, AFM now allows spatially resolved mapping of mechanical properties such as stiffness (elastic modulus), adhesion, viscoelastic response, and even nano-rheology. A comprehensive review describes AFM-based indentation modes which generate nanomechanical maps by correlating tip forces (or oscillation behavior) with contact mechanics models; the review emphasizes advances since 2019 in high-speed acquisition, improved quantitative accuracy, machine learning for data processing, and new modes such as volumetric imaging of solid–liquid interfaces.[29] For soft materials, an earlier review explains how mapping elastic and viscoelastic properties using AFM has become a key method for soft matter and biomaterials.[38]

This capability is especially useful when one wants to correlate surface morphology (e.g., roughness, texture, nanofeatures) with mechanical response; for example, a nanostructured antibacterial surface with defined peaks/spacings where AFM quantified both the topography and the forces or deformation behavior of the nanostructures under load.[40]

Various 3D nanoscale morphology and a 2D representation of the relative frequency of topographic maps of $GdMnO_3$ thin films are shown in Figure 9.4. $GdMnO_3$ at 650°C and 750°C showed smoother surfaces in Figure 9.4(a,c). The topography of $GdMnO_3$ at 850°C in Figure 9.4(e) is more irregular and has a distribution of sharper peaks. In Figure 9.4(b,d,f), the symmetry parameter Ssk (skewness), and shape of the height distribution Sku (kurtosis) confirm the topographic difference.

AFM provides a highly versatile platform for surface morphology analysis, offering true 3D topography at nanometer resolution plus the capability to map mechanical or interaction properties at each spatial point. For monitoring roughness, AFM allows fine quantification of parameters (e.g., Ra, Rq, Sa, Sq) and can resolve changes induced by processing, coating or nanostructuring. For nanostructure features, AFM allows direct measurement of height, spacing, shape, coverage and defects of nano-objects. For mechanical properties, AFM can map modulus, stiffness, adhesion, viscoelasticity and correlate those with morphology.

However, there are important caveats: quantitative mechanical measurements require careful calibration, proper tip geometry modeling, correction for substrate ("bottom stiffness") effects (especially for thin films), and awareness of how tip-sample interaction may distort the topography itself (especially for soft or deformable samples) as shown by recent finite element modeling.[41] Also, AFM scanning is relatively slow for large areas, and lateral resolution is limited by tip radius and tip convolution, so interpretation of lateral feature size must consider tip shape.

Table 9.1 compares several AFM techniques based on how they operate, what they measure, and where they are commonly

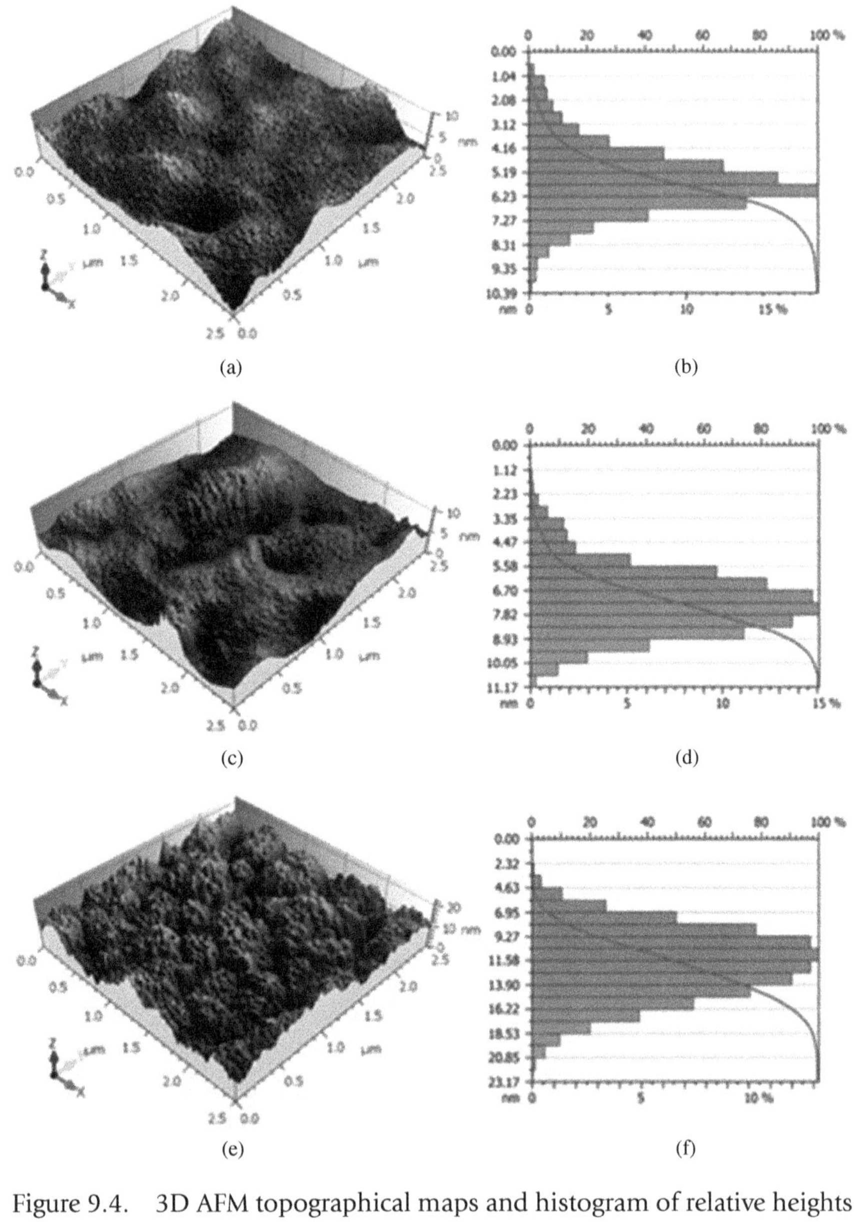

Figure 9.4. 3D AFM topographical maps and histogram of relative heights of $GdMnO_3$ thin films annealed at (a,b) 650°C, (c,d) 750°C, and (e,f) 850°C. Reprinted with permission from [39] by MDPI.

Table 9.1. AFM modes, capabilities, and representative applications.

AFM Mode/ Technique	Principle of Operation	Main Advantages	Limitations	Typical Parameters Measured	Recent Applications
Contact Mode AFM	Tip maintains continuous contact with surface; cantilever deflection used for feedback	High-resolution topography; direct force measurement	Can damage soft samples; lateral forces may distort topography	Height, Ra, Rq, Sa, Sq; friction and adhesion	Used for surface roughness quantification of metal coatings and ceramics[31]; nanoparticle film coverage vs. roughness[32]
Tapping (Intermittent Contact) Mode AFM	Cantilever oscillates near resonance; feedback maintains constant amplitude or phase	Minimizes damage to soft/fragile samples; excellent for polymers, biomaterials	Slightly lower lateral resolution; requires oscillation tuning	3D topography, root mean square roughness, height distributions	Applied to polymer nanocomposites for morphology and roughness[42]; biological membranes in liquid[43]
Non-Contact AFM	Tip oscillates without touching the surface; detects long-range attractive forces	Ideal for clean or ultra-flat surfaces; atomic-scale imaging possible	Sensitive to contamination and environment (humidity, adsorbates)	Atomic lattice imaging, long-range forces, work function variations	Used for atomically resolved imaging of 2D materials and van der Waals heterostructures[44]
Force–Distance Spectroscopy	Tip approaches/ retracts at a single point; force vs. displacement recorded	Quantifies adhesion, stiffness, viscoelasticity	Requires proper contact mechanics modeling	Young's modulus, adhesion energy, yield force	Used to map ZnO nanowire modulus (20–200 GPa range)[35]; soft matter viscoelastic mapping[37]

(*Continued*)

Table 9.1. (*Continued*)

AFM Mode/ Technique	Principle of Operation	Main Advantages	Limitations	Typical Parameters Measured	Recent Applications
PeakForce QNM (Quantitative Nanomechanical Mapping)	Tip oscillates at low frequency; force curve recorded at each pixel	Simultaneous morphology and mechanical mapping; minimal lateral forces	Requires calibration of deflection sensitivity and tip radius	Topography, modulus, adhesion, deformation, energy dissipation	Used for biofilms and soft tissue elasticity mapping[39]; polymer blend phase mapping[45]
Bimodal/ Multifrequency AFM	Simultaneous excitation of two or more resonance modes of the cantilever	Enhance sensitivity to stiffness and adhesion variations	Requires advanced data interpretation	Phase, amplitude, frequency shift (linked to elastic modulus)	Demonstrated for high-speed nanomechanical imaging of heterogeneous surfaces[29]
3D AFM (Tomographic or Volumetric AFM)	Records tip–sample interactions in x, y, z for 3D structure reconstruction	Quantifies sidewall and buried morphology	Complex calibration; slower acquisition	Sidewall roughness, 3D shape reconstruction	Applied in semiconductor metrology for FinFET sidewall roughness (<2% relative standard deviation)[34]
AFM Nanoindentation/ Force Volume Mode	Records full force–displacement curves over a grid of points	Provides spatially resolved mechanical maps	Requires long acquisition time and correction for tip wear	Modulus, hardness, plastic deformation	Used for thin film modulus mapping and polymer nanocomposites[28]

used. Contact mode provides very detailed topography but can damage softer samples, while tapping mode reduces this risk and is often chosen for polymers and biological materials. Non-contact AFM is ideal for very clean, flat surfaces and can even achieve atomic-scale imaging. Techniques like force–distance spectroscopy, PeakForce QNM, and AFM nanoindentation allow researchers to measure mechanical properties such as stiffness, adhesion, and deformation. More advanced methods, including bimodal AFM and 3D AFM, offer enhanced sensitivity or full 3D structural mapping and are increasingly used in semiconductor and materials research.

9.3.2 *Scanning Electron Microscopy*

9.3.2.1 *Principles of Scanning Electron Microscopy*

The SEM directs a focused beam of electrons across a sample surface. As these primary electrons enter the sample, they travel some distance before striking atoms and changing direction — a process called scattering. These repeated interactions create a teardrop-shaped zone beneath the surface where various signals form as shown in Figure 9.5.[46] Secondary electrons are produced when primary electrons inelastically collide with the electrons from the specimen. Since secondary electrons have very low energy, only those generated at or near the surface can escape, while those formed deeper in the interaction volume are absorbed. This shallow origin gives secondary electrons images very fine surface detail. Secondary electron yield increases slightly with the atomic number, making it possible to obtain transmission electron microscope (TEM)-like images from smooth, resin-embedded samples containing heavy metals.[47]

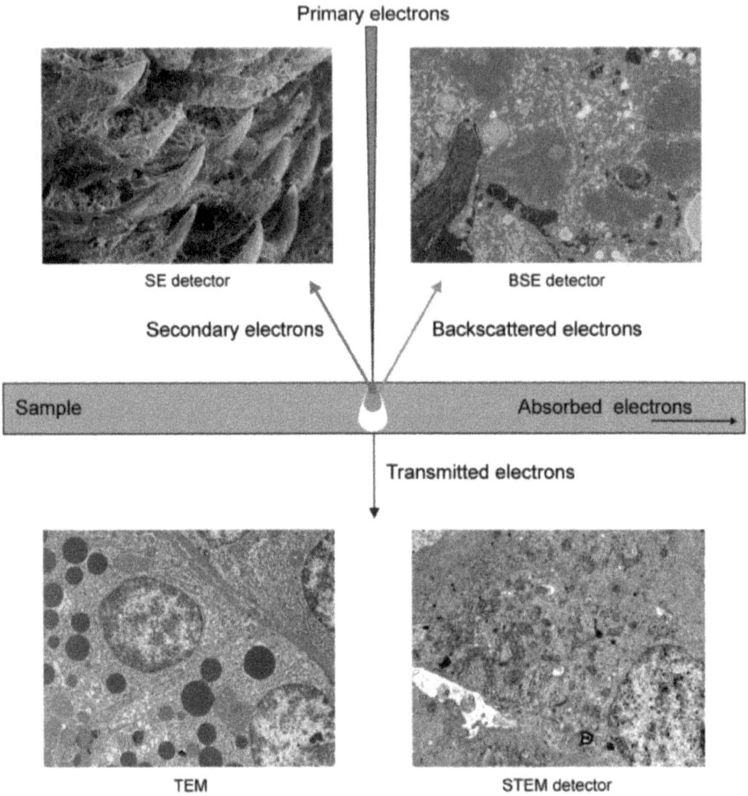

Figure 9.5. Principles of SEM imaging. Reprinted with permission from [52] by MDPI.

A single primary electron can generate multiple secondary electrons, resulting in a strong signal. Secondary electrons are commonly collected using Everhart–Thornley, variable-pressure secondary electron or in-column detectors.[48] Backscattered electrons (BSE) are primary electrons that rebound out of the specimen after elastic scattering. They retain higher energy than secondary electrons and their production strongly depend on atomic number, with heavier elements appearing brighter. As BSE originates from a much larger interaction volume than secondary electrons, the images have lower spatial resolution but provide useful compositional contrast. BSE are detected with extra-column diodes or in-column detectors. They also contribute to secondary electron detector signals and can generate additional secondary electrons as they travel through the sample.[49] The result is an image formation mechanism where the scanning beam position is mapped to detector signals, giving a high-resolution greyscale image of the surface region.[50]

Instrumentally, the SEM consists of an electron gun (thermionic or field emission), condenser and objective lenses (magnetic or electrostatic) to focus and steer the beam, scan coils to raster the beam, the specimen chamber under vacuum, and one or more detectors (secondary electrons, BSE, sometimes energy-dispersive X-ray spectroscopy, cathodoluminescence, etc.).[51] Since the interaction volume of the beam in the solid is typically of nanometer to micrometer dimensions (depending on accelerating voltage and material), SEM can probe details on the nanoscale (often 1–2 nm in good systems) when suitably configured and prepared.[49]

In more advanced systems (e.g., field-emission SEMs, low-voltage SEMs) the probe size is further reduced, and surface-specific imaging (very shallow electron energy, small penetration depth) becomes feasible, which is critical for nanostructure imaging. For example, it has been reported that low landing voltages (1 kV or even ~10 eV) lead to much finer surface detail in nanocrystal imaging.[53] Thus, SEM provides a powerful way to examine the surface morphology, topography, and (to some extent) composition and structural uniformity of materials at high magnification and resolution.

9.3.2.2 *Overview of Scanning Electron Microscopy for High-Resolution Imaging of Nanostructures*

When dealing with nanostructures (e.g., nanowires, nanoparticles, nanopillars, nanosheets), the requirements for imaging become more stringent: small feature sizes, high aspect-ratio morphology, fine surface roughness, alignment of the nanostructures, and detection of defects or deviations from design. SEM is well suited for this because of its large depth of focus (compared to optical microscopy), high resolution and ability to scan large areas relatively rapidly.[54]

In practice, the imaging workflow for nanostructures might involve choosing appropriate accelerating voltage (lower voltages reduce penetration and improve surface sensitivity, but increase probe size and lower signal), selecting proper detectors (e.g., in-column secondary electrons for highest resolution, BSE for contrast of composition or alignment features), optimizing working distance (shorter working distance improves resolution) and ensuring good sample preparation (e.g., conductive coating for insulating nanostructures, minimization of charging, avoidance of beam damage).[55] Figure 9.6 shows an SEM image

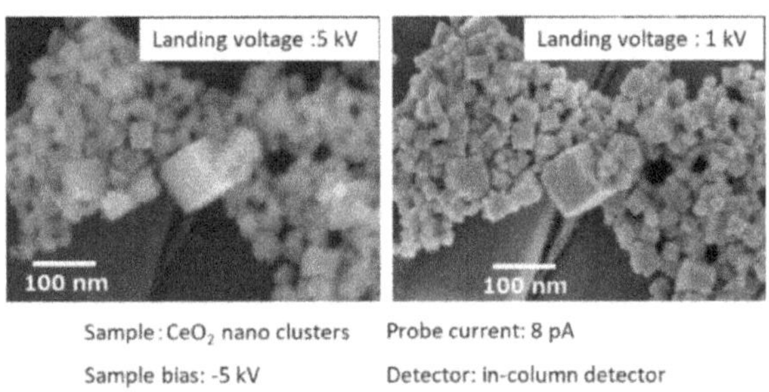

Figure 9.6. SEM images of CeO_2 nanoclusters with different electron landing voltages. Reprinted with permission from [53] by MDPI.

of CeO_2 nanoclusters taken at a 1 kV landing voltage (sample bias −5 kV, 8 pA probe current, in-column detector). The overall shape and size of the nanocrystals can be seen at both tested landing voltages, but the 1 kV image reveals much finer surface details. This improved clarity comes from the reduced electron penetration depth at the lower landing voltage, which makes surface steps easier to distinguish.

Moreover, when imaging nanostructures, the spatial resolution needed is often at or below the few-nanometer scale, and the signal-to-noise ratio of the images becomes critical. A recent review emphasizes that signal-to-noise ratio is a limiting factor in SEM imaging of nanoscale features.[56] Also, the imaging of nanostructures often requires careful consideration of the material–beam interactions, charging effects, drift, vibration and beam damage — all of which can degrade the fidelity of the captured image. The review on mineral characterization by SEM comments on vacuum requirements, coating for non-conductive samples, and the trade-off between resolution and sample preparation complexity.[45]

SEM is increasingly used not just qualitatively but quantitatively for nanostructure characterization: measuring feature sizes, surface roughness, morphology of nanofibers, pore structure in scaffolds, dispersion of nanoparticles in composite matrices, alignment of nanocrystals in films, and so on. For example, in one biopolymer nanocomposite study, SEM images were used to verify alignment of cellulose nanocrystals: high-magnification SEM of a fractured cross-section revealed that the helicoid axis changed through the film.[49]

SEM offers an indispensable tool for high-resolution imaging of nanostructures, bridging the gap between optical microscopy (insufficient resolution) and TEM (which often requires ultra-thin samples and more complicated preparation). The ability to image surfaces and near-surface features *in situ*, with ease of sample preparation relative to TEM, makes SEM a workhorse for nanostructure characterization.

9.3.2.3 *Applications: Imaging Surface Features, Alignment and Defect Detection*

(a) **Surface features/topography:** A primary use of SEM in nanoscience is the examination of the surface morphology of nanostructured materials — this includes the size, shape, distribution, orientation, surface roughness, porosity, grain boundaries, coating coverage, etc. For instance, in the review of optimal mechanical properties and high cellular proliferation on hybrid PCL/HaP/Col/SA scaffolds for tissue engineering, SEM images demonstrated that each type of hydroxyapatite produced a noticeably different surface appearance. All the scaffolds showed a generally porous, cross-linked structure, but the size and shape of these features varied depending on the type of HaP added.[58]

(b) **Alignment of nanostructures:** In many nanotechnology applications (e.g., nanowires for electronics or sensors, anisotropic nanoparticles, aligned nanotubes), the orientation and ordering of the nanostructures on a surface or in a film can significantly impact performance. SEM is used to visualize and assess such alignment. Figure 9.7(a) shows the SEM image of a Si nanowire from bottom-up fabrication method. Figure 9.7(b,c,d) shows 2D schematic diagrams of diameter, length, pitch of nanowire and cylindrical- and tapered-shaped nanowire respectively. For example, a study

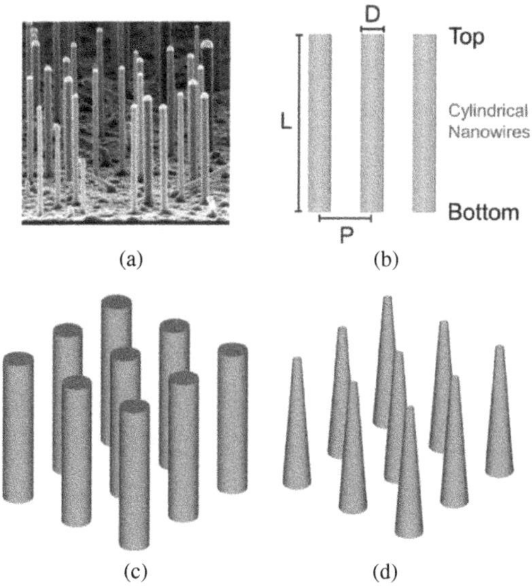

Figure 9.7. SEM image of a nanowire array, illustration of geometry parameters, and nanowire morphology schematics. (a) SEM image of a typical Si nanowire array grown using gold-catalyzed bottom-up fabrication technique. (b) 2D diagram defining key geometric parameters of the nanowires: diameter (D), length (L), and pitch (P). (c) Schematic illustration of cylindrical and (d) tapered nanowires. Reprinted with permission from [57] by Springer.

showed that 1D nanostructure (nanowire) alignment on surfaces can be quantitatively assessed via SEM imaging: an automated method extracted alignment metrics from the SEM image data to compare orientational order.[59]

In another review article on metal nanowires, the authors emphasize that good alignment and patterning of nanowire arrays is critical for device performance and SEM imaging is a key tool to verify alignment and ordering of such arrays.[60] A specific example: silicon nanowires aligned by an off-center spin-coating method were characterized by SEM and then used in a field effect transistor device.[61] Thus, SEM enables not only qualitative visual confirmation of alignment (e.g., wires oriented parallel) but also quantitative metrics (angle distributions, spacing, spacing uniformity) when combined with image processing.

(c) **Defect detection:** In the context of nanostructures (and especially in semiconductor manufacturing, nanoscale devices and nanofabrication), defects (e.g., missing wires, bridging wires, breaks, voids, particulate contamination, misalignments) must be detected and characterized. SEM is widely used for defect review and failure analysis: it provides high resolution to locate tiny defects, and the contrast from surface/backscattered electrons helps identify anomalies. Amato et al. describes how SEM-based defect detection in semiconductor manufacturing helps link defects found by SEM to yield outcomes of wafers.[62] A systematic review of SEM images for automated semiconductor defect inspection summarizes many recent efforts to apply machine learning to SEM images for defect classification.[63]

In nanostructure research (outside pure semiconductor fab) SEM is similarly used: e.g., to spot misoriented nanowires, clustering or agglomeration of nanoparticles, incomplete coating, voids in nanocomposites, or cracks in nanofiber scaffolds. Through such imaging, one may detect defects such as broken or bent nanowires, loose composites, surface pits, or delamination. Since SEM can resolve such surface defects down to a few nanometers (depending on system), it's highly valuable. Furthermore, in the context of alignment and defect detection together: achieving proper alignment often requires verification via SEM, and any misalignment or defect (e.g., misplaced nanowire, gap, extra bridging) is visualized and quantified via SEM. For example, the automated quantification study of 1D nanostructure alignment via SEM images noted that overlap or misorientation influences the alignment metrics significantly.[58]

9.3.3 *Ellipsometry*

9.3.3.1 *Principle of Ellipsometry in Measuring Film Thickness and Optical Properties*

Ellipsometry is an optical measurement technique that exploits the change in polarization of light upon reflection from (or transmission through) a sample to infer film thickness, refractive index (n) and extinction coefficient (k) of thin films and layered materials. In a typical reflection geometry, a beam of light with known polarization strikes the surface at an oblique angle (Θ), and the reflected light becomes elliptically polarized.[64] The change can be described by two parameters, often denoted Ψ and Δ, as shown in Figure 9.8, which relate to the ratio of the

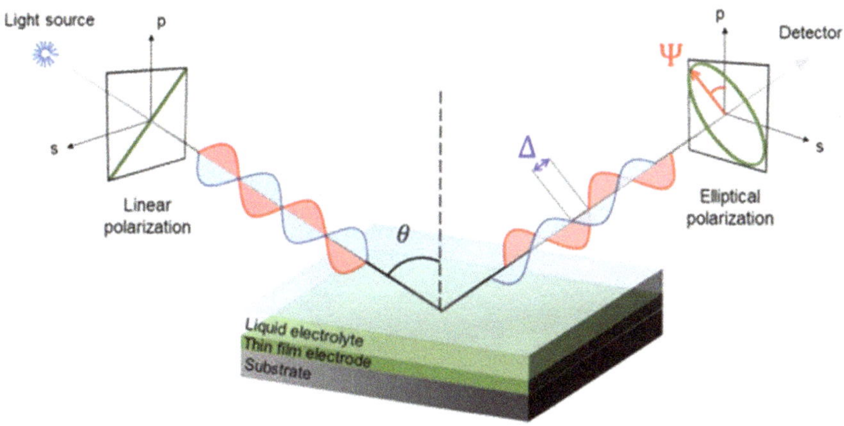

Figure 9.8. Spectroscopic ellipsometry configuration. Reprinted with permission from [65] by IOP Science.

complex reflection coefficients r_p and r_s for p- and s-polarized light respectively, via:

$$\rho = r_p / r_s = \tan \Psi e^{i\Delta}. \quad (1)$$

This fundamental relation means that by measuring Ψ and Δ (often at multiple wavelengths, i.e., spectroscopic ellipsometry), one can construct an optical model of the film/substrate system to extract the thickness of the film and the wavelength-dependent optical constants (n, k) or equivalently the complex dielectric function $\varepsilon(\omega)$. For example, a recent review states that spectroscopic ellipsometry is capable of deriving parameters such as thin film thickness, n, and k in a non-invasive way for a variety of materials including dielectrics, semiconductors and metals.[66]

In practice, one builds a layered optical model: substrate (known), film(s) (unknown thickness, optical constants), maybe interface or rough surface layer (effective medium). The measured $\Psi(\lambda)$, $\Delta(\lambda)$ (or at one wavelength) are fit to the model by adjusting thickness and optical constants.[67] As the path difference and phase shift depends on film thickness relative to wavelength, ellipsometry can often measure thicknesses ranging from sub-nanometer to micrometer scales with high precision (sub-nm thickness resolution, and high sensitivity to changes in optical constants). Mukherjee et al. notes that real-time ellipsometry can determine thickness with precisions down to sub-nanometer and dielectric function with precision to $\sim 10^{-4}$.[68]

The strong sensitivity arises because the phase information (Δ) is very sensitive to small changes in optical path length, and the amplitude ratio (Ψ) responds to changes in reflectivities tied to film optical contrast. In addition, by employing spectroscopic data (varying λ) one gains more information which helps decouple thickness and optical constants — particularly when the film is absorbing or when several layers are present. Ma et al., in the review of measurement for perovskite solar cells, emphasizes that *in situ* ellipsometry measurement can track the dynamic variation of complex optical constants with thickness change during the preparation of perovskite films.[69] Thus, ellipsometry is extremely useful for characterizing thin films: one obtains not only how thick a film is, but also how its optical behavior (n, k or ε) changes — which often correlates with film density, void content, crystallinity, morphology or composition changes.

9.3.3.2 *Real-Time Monitoring of Thin Film Deposition Processes*

Beyond just *ex situ* measurement, ellipsometry has been deployed in real-time, *in situ* monitoring of thin film growth and deposition processes — providing dynamic measurement of film thickness and optical constants as the film is being formed. Since many deposition systems (e.g., sputtering, evaporation, atomic layer deposition, molecular beam epitaxy) can accommodate a monitoring window or port, real-time spectroscopic ellipsometry (RTSE) has become a valuable tool for process control, diagnostics, and optimization. For example, a recent article describes how RTSE was used in the multi-source co-evaporation of $CuInSe_2$ thin films. In that work, the instantaneous effective thickness, deposition rate, and even bulk void volume fraction of the metal layer were tracked in real time via ellipsometry measurements.[71]

In practice, using ellipsometry *in situ* allows the monitoring of several key aspects of film deposition:

- **Thickness evolution in real time:** The film thickness, as it accumulates on the substrate, can be monitored continuously (or with a high temporal resolution). Changes in growth rate, onset of island/coalescence modes, or drift in deposition parameters are visible as changes in the thickness vs. time curve. For instance, one early study on the deposition of hydrogenated amorphous carbon used RTSE with time resolution of ~3 s during growth to deduce both film thickness and evolution of an interface layer.[72] Leigh et al. used spectroscopic ellipsometry to track the earliest stages of polycrystalline diamond growth with high precision.[70] An optical model was first built using *ex situ* spectroscopic ellipsometry spectrum, then applied to *in situ* measurements to follow changes in film thickness, composition, and surface structure during growth as shown in Figure 9.9. Spectroscopic ellipsometry revealed key transitions such as the coalescence of diamond crystallites, marked by decreasing void content and increases in sp^2 carbon and roughness. Overall, spectroscopic ellipsometry provided real-time insight and quantitative analysis for understanding and optimizing early diamond film formation.

- **Optical constant evolution:** RTSE measures how the optical constants of a material, specifically the complex dielectric function, change over time. In the study, these measurements track how perovskite thin films respond to different humidity levels and temperatures. Changes in the dielectric function indicate variations in the film's structure and stability. Tumusange et al. showed that some compositions like $MAPbI_3$ and $MA_{0.7}FA_{0.3}PbI_3$ are more stable on fluorine-doped tin oxide substrates, showing smaller changes in optical constants. But films like $(FAPbI_3)_{0.95}(MAPbBr_3)_{0.05}$ maintain stable optical properties even on less ideal substrates due to their composition and preparation method. In short, monitoring the evolution of optical constants in RTSE reveals how the

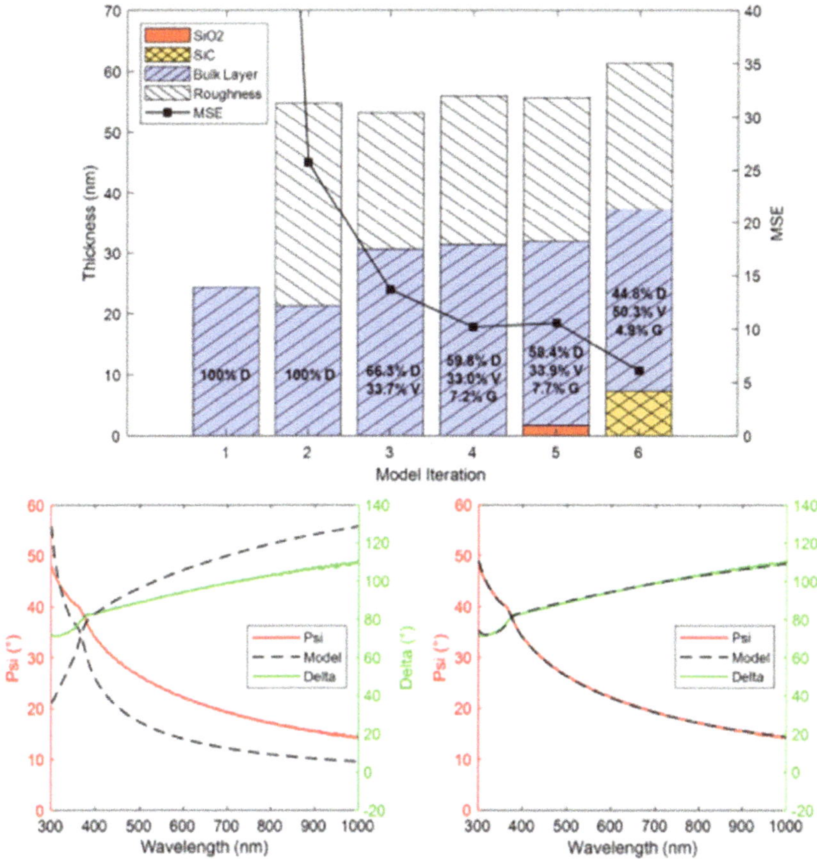

Figure 9.9. Above: The figure shows how the model gradually improved for a 20-minute diamond growth on a Si sample. With each step, the error (MSE) decreased, the layer thickness values were updated, and the predicted amounts of diamond, voids, and glassy carbon in the main layer changed as the model became more accurate. Below: Comparison of the modeled parameters with measured parameters at an incidence angle of 70° for the first iteration (left) and final iteration (right). Reprinted with permission from [70] by Elsevier.

material's structure and stability change under environmental conditions.[73]

- **Deposition rate calibration and control:** Ellipsometry plays a key role in calibrating and controlling deposition rates because it measures film growth in real time with high sensitivity. Sapkota et al. studied how RTSE is used to track the buildup of Cu and In_2Se_3 layers as the evaporation source temperatures change. By directly measuring the effective thickness of each layer, including changes caused by roughness, ellipsometry provides accurate deposition rates for both materials. These rates make it possible to estimate the true atomic flux reaching the substrate and to correct for differences between ideal crystal composition and the actual films. With these corrected values, a reliable calibration map is created that links source temperatures to film composition and growth rate. This allows precise control of the Cu/In ratio and the overall deposition rate during co-evaporation, which is essential for making high-quality copper indium selenide absorber layers for solar cells.[71]

- **Interface and layer structure monitoring:** Ellipsometry is vital for monitoring interfaces and layer structures because it provides a sensitive, non-destructive way to detect early changes in thin films under operating conditions. Minenkov et al. in their study demonstrates that operando ellipsometry tracks how the optical properties of indium tin oxide evolve during electrochemical cycling, revealing the onset of reduction and surface degradation well before electrical failure occurs. As hydrogen evolution triggers the formation of metallic indium and tin at the solid–liquid interface, ellipsometry captures the resulting darkening and loss of transparency, showing how degradation progresses from the surface into the film. By distinguishing between a deteriorating surface layer and an underlying region that still conducts, ellipsometry explains why conductivity remains largely intact despite optical damage. Overall, it provides real-time insight into structural and compositional changes, making it a powerful tool for evaluating thin film stability and identifying electrochemical failure points.[74]

- **Adaptive process optimization:** As real-time data is available, the deposition process can be modified on the fly — e.g., change source power, gas flow, substrate temperature, shutter times — to achieve desired film optical properties or thickness. A study on anti-reflective coatings for CIGS solar cells used *in situ* spectroscopic ellipsometry combined with optical modeling to optimize anti-reflective coating thickness and improved device performance by ~6%.[75]

Beyond these functional benefits, the instrumentation and modeling of *in situ* ellipsometry have been advancing where faster detectors, multi-wavelength or spectroscopic capability, integration with deposition chamber environments (vacuum, high temperature), and advanced modeling (including machine learning-assisted modeling for ultra-thin films) are being explored. For example, a recent study used a machine learning method that solves ellipsometry problems automatically and without ambiguity. The approach is tested on many types of films, including metals, semiconductors, and dielectrics. It works with existing ellipsometers and enables fast, high-throughput optical characterization.[76]

Nevertheless, implementing real-time ellipsometry does involve challenges: accurate optical modeling of a growing film (especially early nucleation stages), handling of surface roughness or voids, ensuring stable measurement geometry inside a deposition chamber, managing beam heating or substrate movement, and decoupling thickness vs. optical constant changes when both evolve simultaneously. The study by Maudet et al. highlights that for ultra-thin films where thickness and optical constant are strongly correlated, a careful multi-step modeling approach is needed to unambiguously extract both.[77]

Table 9.2 shows three case studies of using *in situ* ellipsometry (and allied optical monitoring) during thin film deposition. Across these studies, *in situ* spectroscopic ellipsometry proved valuable because it can track film growth in real time, revealing small changes in thickness, roughness, and optical properties that occur during each stage of deposition. It allowed researchers to follow individual ALD cycles, control ultra-thin metal films only a few nanometers thick, and identify key transitions such as island coalescence and the shift from discontinuous to continuous films. At the same time, the work highlights challenges such as accurately modeling very thin or rough films, separating different contributions in the optical signal, and maintaining reliable measurements inside a deposition chamber. Overall, the studies show that ellipsometry provides detailed, time-resolved insight into film growth that would be missed with only after-growth measurements.

9.3.4 *Optical Profilometry*

9.3.4.1 *Principles of Optical Profilometry for 3D Surface Topography*

Optical profilometry is a non-contact surface metrology technique of increasing importance in micro- and nanofabrication, where 3D surface topographies must be measured quickly, accurately and without physically touching the sample. The basic idea is that optical reflection or interferometric phenomena are utilized to map the height (z-axis) variation of a surface over

Table 9.2. Case studies of using *in situ* ellipsometry (and allied optical monitoring) during thin film deposition.

Study	**Deposition Method and Film Material**	***In situ* Ellipsometry/Monitoring Details**	**Key Insights and Outcomes**
Study A – Precursor-surface interactions revealed during plasma-enhanced ALD of metal oxide thin films by *in situ* spectroscopic ellipsometry[78]	Plasma-enhanced ALD of ultra-thin TiO_2 and WO_3 films	*In situ* spectroscopic ellipsometry during deposition; a "dynamic dual-box model" was applied to resolve thickness and roughness evolution cycle by cycle	The ellipsometry data revealed cycle-by-cycle roughening and smoothing of the surface, producing thickness oscillations that were accurately captured by the dynamic dual-box model, enabling precise real-time tracking of metal oxide ALD growth
Study B – Growth of magnetron-sputtered ultra-thin chromium films: *in situ* monitoring and *ex situ* film properties[79]	Magnetron sputtering of ultra-thin chromium films (2.6 nm up to ~57 nm) on fused silica	*In situ* optical broadband monitoring (transmission) plus optical emission spectroscopy; compared with *ex situ* ellipsometry, reflectance/transmission; thickness/time evolution tracked	Showed that *in situ* optical monitoring can reliably control ultra-thin metal films, and allowed correlation of thickness, optical constants (n, k), morphology, electrical sheet resistance; useful for ultra-thin metal film tuning
Study C – *In situ* and real-time nanoscale monitoring of ultra-thin metal film growth using optical and electrical diagnostic tools[80]	Various ultra-thin metal films (noble metals, transition metals) deposited via sputtering	Combined *in situ* spectroscopic ellipsometry + wafer curvature + electrical probes; real-time monitoring of morphological transitions (percolation thresholds, coalescence) during early growth	Key result: the early nucleation/growth-mode transitions (e.g., 2D → 3D islands to continuous film) can be captured, and ellipsometry gives direct real-time insights into critical thicknesses where film becomes continuous or changes its growth mode

a lateral region (x–y) to produce a full 3D surface profile. For example, many systems adopt white light interferometry, focus variation, confocal scanning, or fringe projection techniques, depending on the scale and surface nature.[81]

In a typical white light interferometric optical profiler, broadband light is directed onto the surface; reflections from the surface and a reference form interference fringes whose constructive/destructive pattern changes as the focus (or path length) is varied. The vertical displacements across the surface can then be extracted from the local phase or fringe contrast. Since the technique sees the surface as a continuous reflectance map, it can resolve vertical height variation on the order of nanometers (for smooth surfaces) and over lateral fields of tens to hundreds of micrometers (or larger).[82]

Figure 9.10 shows the arrangement of a laser profilometer which is based on the presence of a laser beam emitted by an optical source of electromagnetic radiation and an image-reading

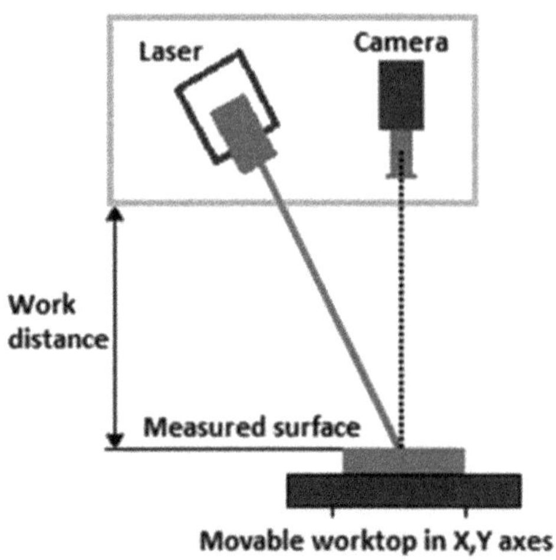

Figure 9.10. Principle of an optical profilometer. Reprinted with permission from [84] by MDPI.

device in the form of a camera, usually based on a charge-coupled device or complementary metal oxide semiconductor sensing chip. Another variant is focus variation or confocal optical profilometry, where instead of interferometric fringes one uses the point of best focus (sharpest image) to determine height, scanning across the surface laterally. These methods are robust for surfaces with some slope or roughness and are used when the reflectivity is variable.[83]

Since the surface topography map is area-based rather than line-based (as with many stylus profilometers), optical profilometry offers several distinct advantages:

(a) It is non-contact, which means delicate or soft surfaces can be mapped without risk of damage or altering the morphology.
(b) It can sample full area maps rapidly rather than scan a single line, which gives access to true areal roughness statistics (e.g., Sa, Sq, Sz) rather than just linear profiles. For example, surface roughness parameters such as Sa (arithmetical mean height) and Sq (root mean square height) can be derived from the 3D height map.[85]
(c) It allows measurement of non-planar features, steps, trenches, and pits, giving a clear context of surface variation (e.g., step height, surface waviness plus roughness).

However, there are also practical limitations, such as reflectivity variations (specular vs. diffuse surfaces), steep slopes, shadowing, and calibration issues. For example, one study on steep-slope micro-optical surfaces noted that standard reflection-based optical profilometry may fail for high slope or curved surfaces because of missed light or diffraction effects and proposed a fluid-based absorption technique to overcome that.[86] In addition, surfaces with large reflectivity contrast (e.g., a semiconductor wafer with metal bumps and diffuse regions) can present challenges for structured light or fringe projection profilometers, and dual-sensor approaches have been proposed to capture both specular and diffuse reflections simultaneously.[87] Optical profilometry provides a versatile and high-resolution method for mapping 3D surface morphology, enabling quantitative topography and roughness metrics over a spatial area, which is far superior to single-line stylus methods for many modern fabrication needs.

9.3.4.2 *Application in Tracking Surface Roughness and Uniformity During Fabrication*

In fabrication processes — for example thin film deposition, etching, lithography, nano-patterning, coating, MEMS[88] manufacture — surface roughness, step heights, layer uniformity and defect topography are key metrics that affect functionality, yield, and reliability. Optical profilometry has increasingly been used in these contexts to monitor and assure surface quality, uniformity, and process control.

(a) **Surface roughness monitoring:** Optical profilometry produces a full 3D map, hence surface roughness can be quantified by areal parameters (e.g., Sa, Sq, Sz) rather than just linear Ra/Rq from stylus traces. For example, one work on circulated coins applied high-accuracy 3D optical profilometry to quantify wear and surface condition by measuring roughness parameters derived from surface height maps.[89] Though that is a wear example, the same principle applies to fabrication: roughness evolution during film growth, polishing, and etching can be tracked. Another application in machined metallic surfaces used fringe projection profilometry (which is a form of optical profilometry) to quantify surface roughness of ground, milled and electrical discharge machining surfaces, demonstrating that the technique can deliver high-speed, accurate roughness measurement over micro-machined textures.[90] Figure 9.11 shows the surface roughness image and the measurements by an optical profilometer on a cast model which was 3D-printed/fabricated using a stereolithography apparatus. The profilometry data shows clear layer-by-layer features, with an average surface roughness of about 0.55 μm and a maximum of 3.54 μm, mainly along the Y direction. Overall, the stereolithography printed cast produced a much smoother surface than fused filament fabrication, which typically has an average roughness of around 2 μm.

(b) **Uniformity and step/feature height measurement:** When fabricating surfaces, one often desires uniform thickness or

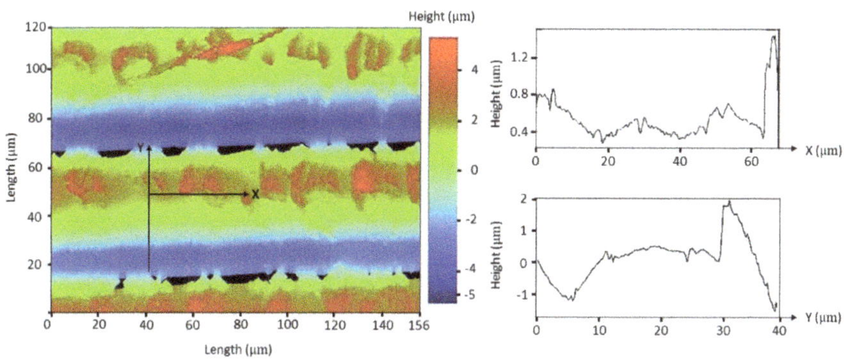

Figure 9.11. Surface roughness image (left) and measurements by optical profilometer (right). Reprinted with permission from [91] by Taylor & Francis.

uniform roughness across an area, or specific step heights (e.g., patterned etch depths, film step features). Optical profilometry can map entire fields and identify spatial variations: e.g., non-uniform etch depths, coating thickness variations, step height non-uniformity or surface warpage.

(c) **In-process/in-line monitoring:** While many profilometry uses are ex-post (after fabrication), optical profilometry is increasingly being considered for in-line or rapid process monitoring because of its non-contact nature and ability to acquire full 3D maps relatively quickly. The recent announcement in the optics field of a method for 3D surface measurement emphasized the need for speed and accuracy in industrial inspection applications (e.g., printed circuit boards, batteries) where surface topography and defect detection are critical.[92] This suggests that optical profilometry (or allied optical 3D surface metrology) is viable for process control, defect detection and uniformity assurance.

(d) **Detecting defects and surface anomalies:** Optical profilometry helps detect defects and surface irregularities on semiconductor wafers by creating a detailed map of the wafer's top surface. In a recent study, this surface data is used to spot features that may lead to poor device performance, such as low breakdown voltage in GaN devices. Small bumps, pits, scratches, or uneven growth patterns can signal deeper structural problems in the wafer. By feeding these surface measurements into machine learning models, it becomes possible to link specific surface anomalies to the likelihood of a wafer failing to meet electrical performance requirements. In this way, optical profilometry serves as an early, non-contact method to identify wafers that may have hidden defects and are at risk of failing during device operation.[93]

(e) **Quantitative process feedback:** By periodically measuring surfaces during or after fabrication, optical profilometry can feed quantitative metrics back into the process: e.g., roughness trending, uniformity maps, step height deviations, or morphological transitions. This allows process engineers to make adjustments (e.g., modify deposition parameters, polishing time, etch duration) to maintain tight surface quality. Although literature explicitly describing real-time in-line optical profilometry in thin film fabrication is less abundant than for ellipsometry or SEM, the technological capabilities (non-contact, full area mapping, nm-scale sensitivity) make it well suited for tracking fabrication quality.

(f) **Practical considerations:** In applying optical profilometry in a fabrication environment there are some key practicalities: the lateral field of view versus resolution trade-off (smaller fields provide higher resolution; larger fields give better area coverage), and dealing with surface slopes or steep features (may cause shadowing or loss of fringe contrast) as noted earlier. Surfaces with varying reflectivity (specular and diffuse spots) require careful setup or dual sensor approaches for reliable profiling. Calibration and stability (vibration control, environmental control) matter for high precision. Du et al. explains that multi-exposure techniques help solve this problem of different surface reflectivity by taking several images at different exposure levels and combining them with a complete 3D shape. However, choosing the right settings for this process usually depends on trial and error. The work described improves this method by giving a clear, guided way to pick the best parameters, such as how wide the Gaussian window should be or which grayscale range produces the best results. With these guided adjustments, the optical profilometer can capture surface features much more completely, raising the coverage of both bright and dark areas to over 99%. This makes it far easier to measure challenging surfaces that reflect light unevenly.[94]

Table 9.3 shows some recent studies relevant to optical profilometry in thin film deposition metrology and characterization

Table 9.3. Recent studies of optical profilometry in thin film deposition characterization and other techniques.

Study	Technique	Key Findings
Method for film thickness mapping with an astigmatic optical profilometer[95]	Astigmatic optical profilometer used for transparent thin films	Demonstrated that an astigmatic optical profilometer — typically used for surface morphology — can be adapted for thickness mapping of transparent films (tens of μm thick); shows ability to map 3D morphology + subsurface interface via optical profilometry
Comparison of optical and stylus methods for surface texture characterization in industrial quality assurance of post-processed laser metal additive Ti-6Al-4V[96]	Optical vs. stylus roughness/ texture measurement in additive manufacturing	Shows the optical (non-contact) methods (including optical profilometry) provide reliable areal roughness and surface texture metrics over large and complex surfaces — important for surface uniformity and roughness tracking in fabrication

Table 9.3. (*Continued*)

Study	Technique	Key Findings
Determination of optical and structural parameters of thin films with differently rough boundaries[97]	Optical characterization of thin films with rough boundaries	Though more focused on optical characterization (ellipsometry/reflectometry) of films with rough boundaries, it explicitly accounts for boundary roughness (i.e., topography) in modeling; relevant because it ties roughness/height variation into optical metrology — indicative of how surface profile matters
Rapid full-field surface topography measurement of large-scale wafers using interferometric imaging[98]	Interferometric full-field topography for large wafers	Demonstrates a method for large-area wafer topography measurement (full-field) using interferometric imaging; highly relevant for surface uniformity tracking and defect/roughness mapping over large areas in fabrication

of other techniques. These examples highlight different uses of optical profilometry and 3D surface metrology. Some focus on mapping film thickness, while others compare roughness or model complex surface boundaries. High-speed, full-field methods are especially useful for tracking how roughness changes during fabrication. Large-area techniques help monitor surface uniformity across an entire substrate. Even when the tools are not classic profilometers, they still represent modern, non-contact ways to measure 3D surfaces in manufacturing environments.

9.3.5 *Raman Spectroscopy*

9.3.5.1 *Principle of Raman Spectroscopy for Material Composition and Structural Analysis*

Raman spectroscopy rests on the phenomenon of inelastic scattering of light by matter. When a monochromatic laser beam is focused onto a material, most photons scatter elastically (Rayleigh scattering) and retain their original energy.[99] A small fraction of photons, however, interact with the vibrational (or phonon) modes of the material, losing or gaining energy in the process. Rayleigh scattering does not change the molecule's vibrational energy, so the scattered light has the same frequency as the incident light and cannot identify molecules. In Stokes Raman scattering, the molecule gains vibrational energy, and the scattered light has a lower frequency. In anti-Stokes Raman scattering, the molecule loses energy, and the scattered light has a higher frequency. Unlike infrared absorption, Raman scattering involves energy changes due to photon interaction, not direct photon absorption.[100] Figure 9.12 shows that when a molecule is exposed to a monochromatic light beam, it can scatter light in three ways: Rayleigh scattering, Stokes Raman scattering, and anti-Stokes Raman scattering.

In a Raman spectrum, one plots intensity versus Raman shift (often in cm^{-1}) and the positions, intensities and widths of the peaks provide a "fingerprint" of the material's composition, molecular bonding environment, crystal structure, presence of defects, strain, and so on.[101] The vibrational mode frequencies depend on mass of atoms, bonding strength, crystallographic symmetry, lattice constant, and local environment. So changes in composition, crystallinity (order vs. disorder), or strain in

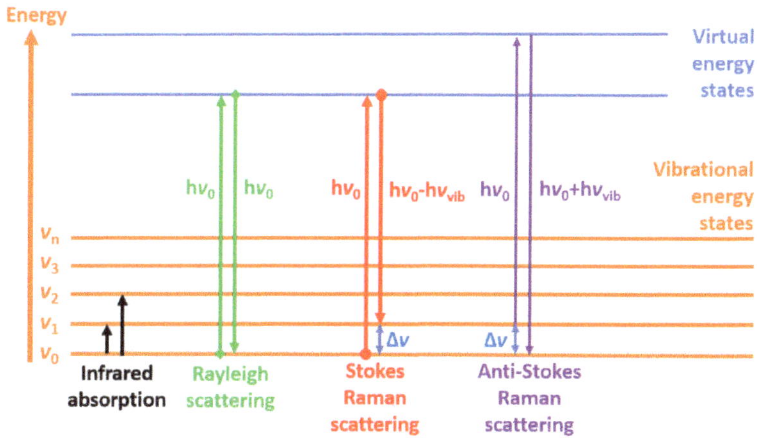

Figure 9.12. Types of scattering and their energy transition due to monochromatic light excitation. Reprinted with permission from [104] by MDPI.

the material will often shift the Raman peak positions (and sometimes broaden them), alter their intensities, or even activate otherwise forbidden modes.[102] From a practical viewpoint, Raman spectroscopy is non-destructive, requires little sample preparation (no vacuum required in many cases), and can be applied *in situ* or on functional devices or films.[103] Thus, it is widely used in materials research to identify phases, assess crystallinity, detect foreign impurities, evaluate strain or stress, and map spatial heterogeneity of composition or structure.

9.3.5.2 *Application in Monitoring Material Composition, Crystallinity and Strain*

(a) **Material composition:** Raman spectroscopy is excellent for identifying what chemical or phase components are present in a material. For example, distinct Raman bands correspond to specific molecular groups or crystal lattice vibrations, so by comparing observed peaks to reference spectra one can determine composition. A recent review highlights that Raman can rapidly classify materials and detect impurities or mixture components, especially when augmented with machine learning methods.

In thin film processing, Raman has been used to detect residual precursor organic materials embedded in a film. For instance, in a study of thin CeO_2 (ceria) spray pyrolysis films, Raman spectra around the F_{2g} mode were analyzed for films deposited and annealed at different temperatures. The F_{2g} peak appeared in all films except the as-deposited film at 400 °C, likely due to its low thickness and weak signal as shown in Figure 9.13. Variations in film thickness and measurement focus caused fluctuations in peak intensity, preventing meaningful conclusions from intensity values.[105]

(b) **Crystallinity/structural order:** Raman is very sensitive to crystallinity because well-ordered lattices have sharp, well-defined phonon modes, whereas disordered, defective or amorphous regions lead to peak broadening, intensity changes, or even the appearance/disappearance of modes. A recent study on Raman spectral mapping of flax cellulose reveals how UV and extreme UV radiation break chemical bonds, causing oxidation, dehydration, and accelerated aging. The spatially resolved spectra also show the role of oxygen in these molecular changes, helping to quantify reactions that lead to cellulose degradation and yellowing.[106]

In polycrystalline inorganic materials, Raman micro-spectroscopy has been used to map grain-by-grain variations in crystallinity, or to detect minor second phases. Another article states that Raman mapping of polycrystalline VO_2 thin films, combined with phase transition measurements using

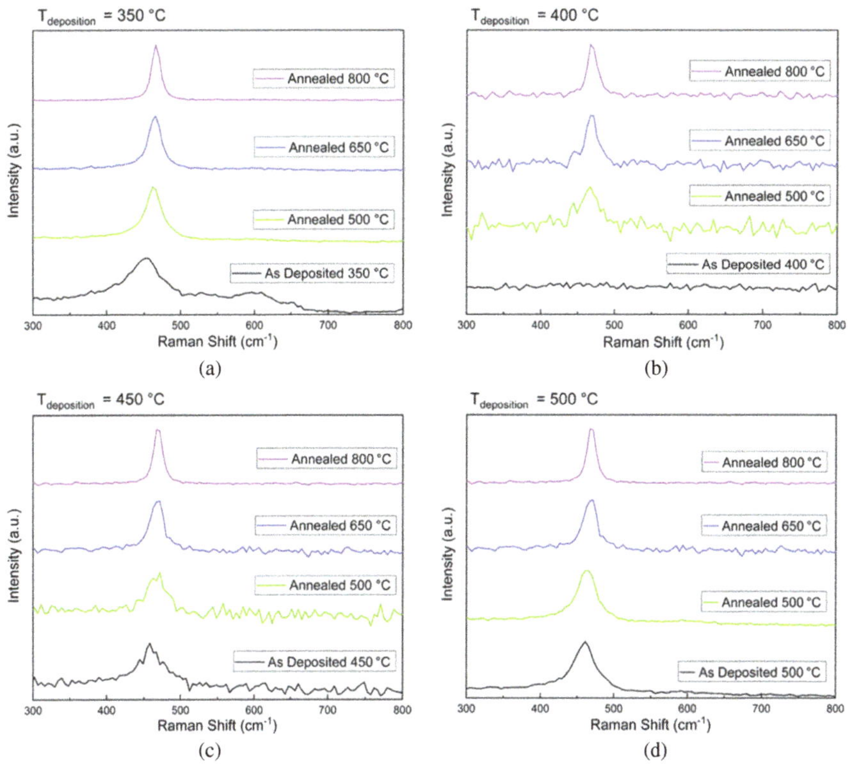

Figure 9.13. Raman spectra of the F_{2g} peak of as-deposited and annealed films deposited at (a) 350 °C, (b) 400 °C, (c) 450 °C, and (d) 500 °C. Reprinted with permission from [105] by Royal Society of Chemistry.

continuous-wave optical excitation, provides a fast, purely optical way to assess structural, morphological, and thermal properties. This approach also reveals details of hysteresis behavior, such as stepped transitions likely caused by the gradual formation of metallic rutile domains within the semiconducting matrix.[107]

(c) **Strain/structural deformation:** Vibrational frequencies depend on interatomic distances and bonding strength so strain (compressive or tensile) in a crystal lattice shifts Raman peak positions. For example, the Raman peak might move to higher or lower wavenumber depending on how the lattice is distorted. This effect allows Raman spectroscopy to act as a non-contact stress/strain probe. For example, an article on Raman micro-spectroscopy noted that by evaluation of relative band intensities and exact band positions, stress/strain can be derived from Raman spectra.[108]

Table 9.4 shows various applications of Raman spectroscopy in thin films or nanostructured film materials, focusing on composition, crystallinity or strain monitoring. These studies show that Raman spectroscopy can reveal crystallinity, preferred orientation, and strain, as seen in CeO_2, Gd-doped CeO_2, and GaN films. It also probes defects and composition, such as oxygen vacancies in perovskites, and confirms material quality in composites like CZTS/PEDOT:PSS. Additionally, Raman peak intensity and sharpness correlate with film thickness and precursor concentration, reflecting structural quality, as shown in WO_3 films. Overall, it provides key insights into structure, stress, and composition in thin films.

9.4 Integration of On-Line Characterization in Automated Systems

9.4.1 *Challenges in Integration*

Incorporating advanced material and process characterization methods (such as SEM, ellipsometry, optical profilometry, Raman spectroscopy) into automated fabrication lines is far from straightforward. Several technical and logistical hurdles arise. First, many characterization techniques were originally designed for standalone laboratory use, not for continuous,

Table 9.4. Applications of Raman spectroscopy in nanostructured films.

Study	Material Deposition Process	Measurements	Key Insight
Deposition parameters and Raman crystal orientation measurements of ceria thin films deposited by spray pyrolysis[104]	CeO_2 (ceria) thin films by spray pyrolysis	Raman F_{2g} mode position vs. film texture/orientation	Found that the Raman peak shift above the bulk value correlates with preferred crystallographic orientation (texture) in the films; thus Raman can sense crystallinity/orientation
The elastic component of anisotropic strain dominates the observed shift in the F_{2g} Raman mode of anelastic ceria thin films[109]	Gd-doped CeO_2 thin films (membranes)	Raman peak shifts (F_{2g}) due to strain (substrate vs. freestanding)	Demonstrated that the observed Raman shift is dominated by the elastic component of anisotropic strain — hence Raman is sensitive to strain in thin films
Raman spectra and defect chemical characteristics of Sr (Ti, Fe) O_{3-y} solid solution of bulk pellets vs. thin films[110]	Sr (Ti, Fe) $O_{3-\gamma}$ perovskite solid-solution thin films and bulk	Raman vibrational modes, defect chemistry (oxygen vacancies, Fe valence)	Showed how Raman can non-destructively probe defect chemistry (oxygen non-stoichiometry, cation valence) in thin films vs. bulk, thus composition/defects via Raman

(*Continued*)

Table 9.4. (*Continued*)

Study	Material Deposition Process	Measurements	Key Insight
Raman spectroscopy and thermoelectric characterization of composite thin films of Cu_2ZnSnS_4 nanocrystals embedded in a conductive polymer PEDOT:PSS[111]	CZTS nanocrystals embedded in polymer thin films	Raman spectral peaks of CZTS + polymer matrix, crystallinity	Raman confirmed material composition and structural features of the composite thin film, showing utility for composite film quality control
Raman characterization of the in-plane stress tensor of gallium nitride[112]	GaN thin films	Raman shifts used to determine in-plane stress tensor	Showed that Raman spectroscopy can quantify in-plane stress (strain) in GaN thin films — direct application to strain monitoring
Raman study of novel nanostructured WO_3 thin films grown by spray deposition[113]	WO_3 thin films by spray deposition	Raman peak intensity and sharpness vs. film thickness and precursor concentration	Demonstrated that Raman peak intensity/sharpness improved with increased film thickness and precursor concentration, indicating higher crystallinity; thus, Raman probes crystallinity and structural quality

high-throughput integration on a production line. For example, *in situ* sensors in additive manufacturing must tolerate harsh environments (high temperature, powder, laser spatter) and deliver reliable signals.[114] Secondly, the logistics of embedding sensors, probes or optics into a moving production line bring mechanical alignment, vibration isolation, optical access (windows, view paths), timing and throughput issues. Third, the data-handling infrastructure must be upgraded: integrated sensors across a line generate large volumes of heterogeneous data (images, spectra, height maps, time signals), with high velocity and variety. A review of data-driven manufacturing points out that "the explosion of data amount" raises major challenges in veracity (data quality), volume and the ability to turn those data into actionable decisions.[115] Fourth, there is a challenge of process integration with control systems: the characterization must link meaningfully to control logic (programmable logic controllers, distributed control systems, manufacturing execution systems) in order to trigger corrective actions, and this demands protocols, machine-to-machine communications, real-time latency guarantees, and reliable interfaces. For example, implementing "single source of truth" data systems for process control were shown to require careful integration of data acquisition and control blocks.[116] Finally, there is the issue of standardization and traceability: calibration of sensors in harsh environments, real-time modeling of what the measurement means for the material state, linking measurement to defect or yield metrics, and validating the measurement benefit all require significant investment of metrology science.[117] In short, while on-line characterization promises improved yield and process control, realizing it in an automated fabrication context involves aligning sensors, data systems, mechanical/optical integration, communication interfaces, modeling and control loops — each of which represents a non-trivial challenge.

9.4.2 Real-Time Data Acquisition and Analysis

To make on-line characterization effective in automated systems, real-time data acquisition and rapid analysis are essential. The requirements here span hardware, software, communications and control feedback. On the acquisition side, sensors or instruments must sample at rates compatible with the fabrication process (for example layer by layer in additive manufacturing, or wafer by wafer in semiconductor fabs) and deliver data with minimal latency. The "velocity" of data must match the dynamics of the process. After acquisition, the raw data must be pre-processed (noise filtering, drift correction, alignment), then analyzed (feature extraction, anomaly detection, model inference), and finally fed into decision logic with minimal delay. A review in the biopharmaceutical domain (which faces similar real-time challenges) cites that real-time release strategies require automated data piping, analysis, aggregation, visualization, and smart utility of data for advanced analytics in real time.[118] For example, researchers introduced a knowledge-guided deep learning framework for monitoring melt pools in additive manufacturing. By integrating engineering principles with learned melt-pool features and modeling their spatiotemporal neighborhoods in 3D, the approach captures process–quality relationships more effectively than pointwise or black box methods. This leads to significantly improved prediction of internal density variations and more interpretable quality-related features. The next stage is integration with the control system: the analyzed data must trigger feedback or feed-forward actions (e.g., adjust power, feed rate, shutter time, or reject part).[119]

The linkage between the measurement system and the control actuation system must be robust, low-latency, and fault-tolerant. For instance, the U.S. National Institute of Standards and Technology roadmap for real-time monitoring notes that while open- and closed-loop control based on *in situ* sensing is the goal, machine manufacturers often hinder access to modify control systems, making implementation difficult. Moreover, data infrastructure must handle storage, versioning, traceability, and linking of measurement data with process parameters and final quality data. The "three Vs" of big data (volume, velocity, variety) dominate this domain and require careful architecture for retention, retrieval and decision logic.[120] In sum: real-time acquisition must be fast and synchronized with the fabrication process; analysis must be predictive or diagnostic in near-real time; and the interface with control systems must allow corrective action or decision making within process windows.

9.4.3 *Synergy Between Characterization Techniques*

While each characterization technique (e.g., ellipsometry, Raman, profilometry, SEM) has specific strengths and limitations, combining multiple complementary techniques offers a far more comprehensive view of a fabrication process and yields stronger process control. For example, imaging methods (optical profilometry) might deliver rapid topography and roughness maps across a surface, while local spectroscopy (Raman) may provide chemical or strain information, and ellipsometry reveals optical/film thickness or composition changes. Linking them in an on-line system enables earlier defect detection, cross-validation of measurement results, and richer feedback to the control system. One application domain is additive manufacturing, where multi-sensor/multi-modal monitoring of the melt pool (vision, thermal, acoustic) combined with data analytics has enabled defect detection with high accuracy (e.g., over 96% accuracy in one study).[121]

Similarly, in a pharmaceutical manufacturing setting, integrating multiple process analytical technology tools (spectroscopy, chromatography, sensors) into a unified data system allowed synchronized measurements and multivariate predictive control of quality attributes.[122] When applied to thin film fabrication, such synergy might look like: real-time ellipsometry tracking film thickness and optical constants during deposition, coupled with in-line profilometry scanning for surface uniformity/roughness on the same substrate, and Raman spectroscopy mapping for crystallinity or composition changes before the film goes into further processing.

The feedback loop could then adjust deposition rate, substrate temperature or post-anneal conditions based on combined measurements. This integrated approach yields a more robust control of yield and uniformity than any single measurement alone. The literature on "immersive digital twins" for manufacturing execution systems further illustrates how multi-sensor, multi-modal data fused into a cyber-physical system enhance process monitoring and control.[123] In essence, synergy means measurement redundancy (for confidence), complementary data (for multiple attributes: morphological, chemical, structural), and cross-domain feedback (for control loops). The outcome is higher process yield, fewer defects, tighter uniformity, and potentially shorter qualification cycles.

9.5 Future Trends in On-Line Characterization

9.5.1 *Emerging Techniques and Advancements*

As fabrication processes push toward ever finer feature sizes, complex materials, and higher throughput, on-line characterization is evolving beyond the traditional optical/SEM/metrology toolbox into new frontiers. For example, hyperspectral imaging (covering hundreds of wavelengths) is becoming viable on the production line, enabling rapid chemical/phase mapping of surfaces or films in real time.[124] Meanwhile, *in situ* TEM and other advanced microscopy is moving from lab-only settings toward integration with fabrication or deposition environments, enabling atomic-scale monitoring of growth, defects or interface formation.[125] Additionally, metrology tools with greater resolution, speed and sensitivity are being developed — for instance through physics-informed imaging, adaptive optics, and novel sensor materials.

A review of nanofabrication tools notes that emerging micro- and nanochannel fabrication methods are paired with novel measurement instrumentation to monitor minute structural features.[126] These developments mean that on-line characterization will increasingly encompass 3D, multi-modal, high-speed metrology rather than only simple spot checks. The key advancement is that measurement tools are becoming faster, more sensitive, more integrated into the process environment, and capable of delivering richer data (chemical, structural, spatial) in real time or near real time.

9.5.2 Integration with Artificial Intelligence and Machine Learning

A major trend is the marriage of on-line characterization with artificial intelligence and machine learning to manage, interpret and act on the vast data streams that modern sensors produce. For instance, in the semiconductor and optical metrology domain, a recent review describes how artificially intelligent optical spectroscopy combinations are used for forward surrogate modeling, inverse prediction and physics-informed neural networks to address challenges such as tool-to-tool matching and characterization of high-aspect-ratio structures.[127] Similarly, in the broader nanomanufacturing realm, a comprehensive recent review outlines how artificial intelligence and deep learning are being used to analyze nanoscale images, optimize nanomaterial synthesis, and perform nano-quality assurance.[128] The implication for on-line characterization is clear: sensors generate huge volumes of data; artificial intelligence/machine learning algorithms will shift, interpret, correlate such data with process parameters and product quality, and then drive feedback decisions. Real-time image recognition, anomaly detection, predictive modeling of defects or drift, and automated calibration of measurement tools all become possible when characterization is coupled with artificial intelligence. This evolution will enable not just measurement but smart measurement: the system can learn from past data, anticipate issues, and proactively adjust process or measurement conditions.

9.5.3 *The Role of On-Line Characterization in Future Nanofabrication Technologies*

As nanofabrication technologies advance — for example quantum devices, photonics, MEMS/NEMS, advanced sensors, 2D materials, and heterogeneous integration — the demands on characterization escalate smaller features, novel materials, more complex architectures, multi-material stacks, interfaces, strain/defect control and high yield. In this environment, on-line characterization becomes indispensable. For instance, when fabricating quantum devices with nanometer-scale precision, monitoring of film thickness, uniformity, surface roughness, strain, and defect density must happen in near-real time to maintain yield.[129] A recent review highlights how emerging computational tools — particularly machine learning, genetic algorithms, fuzzy logic, and hybrid soft-computing approaches — can greatly enhance nanofabrication by enabling rapid, data-driven characterization and prediction of material behavior. These methods reduce reliance on slow, experimental trial and error and can support real-time or online characterization by efficiently exploring design spaces and forecasting nanoscale properties. Together, they provide a roadmap for integrating intelligent computational frameworks into future nanofabrication workflows to accelerate the development of high-performance nanomaterials.[130] In other words, as fabrication moves toward smaller scales and higher complexity, on-line characterization will shift from optional quality check to critical process enabler. It will support real-time decision making, adaptive process correction, and ensure that sophisticated nanofabricated devices meet specifications without excessive post-process inspection or rework.

9.5.4 *Towards Industry 4.0 in Nanofabrication*

Finally, on-line characterization is central to the Industry 4.0 vision for manufacturing: a fully connected, data-driven, automated production environment where sensors, machines, analytics and control systems collaborate to optimize output, quality and flexibility.[131] In the nanofabrication context, this means that metrology tools become part of the cyber-physical system: data flows are measured in real time to control systems, analytics modules detect deviations or trends, corrective actions are taken autonomously (or semi-autonomously), and the system continuously improves via machine learning. A recent review outlines how artificial intelligence and machine learning are becoming central to Industry 4.0-driven nanomanufacturing by enhancing computational design, process optimization, predictive maintenance, and real-time quality assurance. As nanoscale production grows more complex, artificial intelligence-enabled tools offer scalable, data-driven solutions that improve efficiency, reduce costs, and accelerate innovation across the full manufacturing lifecycle. The work also highlights challenges, including implementation, data needs, and regulatory considerations that must be addressed to fully realize intelligent, next-generation nanomanufacturing.[132] The shift is from isolated metrology stations toward integrated digital twins, sensor networks, closed-loop feedback and autonomous control. In the future, on-line characterization data will not only report what is, but actively drive what should be, enabling higher yields, reduced defects, shorter cycle time, and more responsive manufacturing. The challenge, of course, is managing data integrity, standardizing interfaces, ensuring interoperability, safeguarding cybersecurity, and training the workforce to operate in a more autonomous environment.

9.6 Conclusion

On-line (or in-process) characterization in nanofabrication has emerged as a central enabler of high-precision, high-yield device manufacturing. By embedding metrology tools that monitor film thickness, surface topography, composition, crystallinity, strain and defect status *during* the fabrication flow, manufacturers gain visibility into what is happening in real time, not just after the fact. This allows corrective actions, process optimization, and improved control of critical nanoscale features. As described in earlier sections, advanced measurement tools (e.g., optical profilometry for topography, ellipsometry for optical/thickness, Raman for structural/compositional information) provide rich data. When tied directly into control systems, these tools help ensure that each layer/step meets specification, that defects are flagged early, and that throughput and yield are improved. As one recent quality control review shows, the push toward in-line, real-time metrology is becoming integral to micro- and nano-technology manufacturing. In short, the role of on-line characterization is no longer optional but foundational for nano-precision fabrication.

Implementing on-line characterization is far from trivial, and several challenges remain. Technically, achieving the required resolution, speed and repeatability in a production environment is demanding and tools must be fast enough to keep up with processing. For example, it has been pointed out that while nanometrology tools exist in labs, realizing atomic-scale imaging at speeds high enough to provide the required process control information remains a key barrier. Logistically, integration into tool flows, synchronization with production devices, calibration, data management, and linking measurements to control loops all add complexity. The merits are tighter process control, fewer defects, reduced scrap, faster cycle times, and enablement of advanced device architectures (2D materials, MEMS/NEMS, quantum devices) that demand nanometer precision. Moreover, quality control and reliability frameworks are increasingly recognizing the need for in-line metrology, advanced analytics and multi-modal data fusion. As manufacturing moves to smaller nodes, more heterogeneous materials and more complex stacks, the opportunity for on-line characterization to deliver value is enormous. The key is to overcome the cost-throughput-precision tradeoff, standardize interfaces, and build the data ecosystems to turn measurement into actionable process control.

Looking ahead, the evolution and integration of advanced characterization techniques will be a defining factor in the performance and reliability of next-generation nanofabricated devices. As fabrication moves into domains such as quantum components, nanophotonics, heterogeneous integration, and flexible electronics, the demand for real-time, high-precision, richly informative metrology only grows. On-line characterization will shift from being an accessory to being embedded deeply into the process fabric: measurement, decision, and correction will become tightly interwoven. The manufacturing systems of the future will not wait for roll-out inspections; they will monitor continuously, learn from data, adapt on the fly and steer the process to meet specs with minimal intervention. Some recent reviews of advanced manufacturing describe exactly this transition toward self-adaptive, sensor-rich, data-driven fabrication systems. In essence, the promise is that measurement will no longer simply tell us what went wrong — it will become part of what makes things go right.

Acknowledgments

The author would like to express his gratitude to Dr. Ampere A. Tseng, Distinguished Professor and Director Emeritus at Arizona State University, for his valuable information and suggestions in preparing this manuscript.

References

1. Chen, Yong and Anne Pépin. "Nanofabrication: Conventional and Non-conventional Methods." *Electrophoresis* 22, no. 2 (2001): 187–207.
2. Gates, Byron D., Qiaobing Xu, J. Christopher Love and Daniel B. Wolfe. "Unconventional Nanofabrication." *Annual Review of Materials Research* 34, no. Volume 34, 2004 (2004): 339–372.
3. Mourdikoudis, Stefanos, Roger M. Pallares and Nguyen T. K. Thanh. "Characterization Techniques for Nanoparticles: Comparison and Complementarity upon Studying Nanoparticle Properties." *Nanoscale* 10, no. 27 (2018): 12871–12934.
4. Hergeth, Wolf-Dieter. "On-Line Characterization Methods." in *Polymeric Dispersions: Principles and Applications*, edited by José M. Asua, pp. 267–288. Springer Nature, 1997.
5. Heinrich, Harald, Arthur Deutschländer, Feryel Zoghlami and Okan Kamil Sen. "The Long Journey From Standardization to Fully Fab Automation and More." *IEEE Transactions on Semiconductor Manufacturing* 32, no. 4 (2019): 408–414.
6. Prasad, Rai Dhirendra, Shivanand B. Teli, Rai Surendra Prasad, Rai Bishvendra Prasad, Saurabh R. Prasad, Pramode Sinha, Amit Sinha, Preety Sinha, Mamata Saxena, Rai Rajnarayan. Prasad, R. S. Pande, Naresh Charmode, K. G. Deshmukh, Prashant D. Sarvalkar, U. D. Kadam, Chandrasekhar Chiplunkar, Nirmala Prasad, M.V. Padvi and Zhanhu Guo. "A Review on Thin Film Technology and Nanomaterial Characterization Techniques." *ES Materials and Manufacturing* 25, no. 0 (2024): 1198.
7. Radnik, Jörg, Vasile-Dan Hodoroaba, Harald Jungnickel, Jutta Tentschert, Andreas Luch, Vanessa Sogne, Florian Meier, Loïc Burr, David Schmid, Christoph Schlager, Tae Hyun Yoon, Ruud Peters, Sophie M Briffa and Eugenia Valsami-Jones. "Automation and Standardization — A Coupled Approach towards Reproducible Sample Preparation Protocols for Nanomaterial Analysis." *Molecules* 27, no. 3 (2022): 985.

8. Liddle, James Alexander and Gregg M. Gallatin. "Lithography, Metrology and Nanomanufacturing." *Nanoscale* 3, no. 7 (2011): 2679–2688.
9. Yao, Tsung-Fu, Andrew Duenner, and Michael Cullinan. "In-Line Dimensional Metrology in Nanomanufacturing Systems Enabled by a Passive Semiconductor Wafer Alignment Mechanism." *Journal of Micro and Nano-Manufacturing* 5 (2016): 011001.
10. Yao, Tsung-Fu, Andrew Duenner and Michael Cullinan. "In-Line Metrology of Nanoscale Features in Semiconductor Manufacturing Systems." *Precision Engineering* 47 (2017): 147–157.
11. Ivanov, Aleksandar P., Kevin J. Freedman, Min Jun Kim, Tim Albrecht and Joshua B. Edel. "High Precision Fabrication and Positioning of Nanoelectrodes in a Nanopore." *ACS Nano* 8, no. 2 (2014): 1940–1948.
12. Zhao, Fangyuan, Hanyao Tang, Xuerong Zou, and Xinghui Li. "A Review of Optical Metrology Techniques for Advanced Manufacturing Applications." *Micromachines* 16, no. 11 (2025): 1224.
13. Cui, Mengjia, Wei Yang, Yingchun Guan and Zhen Zhang. "Fabrication of High Precision Grating Patterns with a Compliant Nanomanipulator-Based Femtosecond Laser Direct Writing System." *Precision Engineering* 78, no. 19 (2022): 60–69.
14. She, Yulong, Jie Tang, Chaoyang Wang, Chaoyang Wang, Zhicheng Wang, Zhengren Huang, and Yong Yang. "Nano-Additive Manufacturing and Non-Destructive Testing of Nanocomposites." *Nanomaterials* 13, no. 20 (2023): 2741.
15. Postek, Michael T. and Kevin Lyons. "Instrumentation, Metrology, and Standards: Key Elements for the Future of Nanomanufacturing." *Instrumentation, Metrology, and Standards for Nanomanufacturing* 6648 (2007): 11–17.
16. Abdollahi-Mamoudan, Farima, Clemente Ibarra-Castanedo and Xavier P V Maldague. "Non-Destructive Testing and Evaluation of Hybrid and Advanced Structures: A Comprehensive Review of Methods, Applications, and Emerging Trends." *Sensors* 25, no. 12 (2025): 3635.
17. Adeleke, Adeniyi Kehinde, Thompson Odion Igunma and Zamathula Sikhakhane Nwokediegwu. "Developing Nanoindentation and Non-Contact Optical Metrology Techniques for Precise Material Characterization in Manufacturing." *International Journal of Multidisciplinary Research and Growth Evaluation* 3, no. 1 (2022): 720–734.

18. Zhang, Xinlong, Li Cheng, Yunfan Liu, Bo Tao, Jiuyi Wang and Ruijin Liao. "A Review of Non-Destructive Methods for the Detection Tiny Defects within Organic Insulating Materials." *Frontiers in Materials* 9 (2022): 995516.

19. Fang, Qihang, Gang Xiong, MengChu Zhou, Tariku Sinshaw Tamir, Chao-Bo Yan and Huaiyu Wu. "Process Monitoring, Diagnosis and Control of Additive Manufacturing." *IEEE Transactions on Automation Science and Engineering* 21, no. 1 (2024): 1041–1067.

20. Basu, Prithvi, Jyoti Verma, Vishnuram Abhinav, Ratneshwar Kumar Ratnesh, Yogesh Kumar Singla and Vibhor Kumar. "Advancements in Lithography Techniques and Emerging Molecular Strategies for Nanostructure Fabrication." *International Journal of Molecular Sciences* 26, no. 7 (2025): 3027.

21. Chen, Lequn, Guijun Bi, Xiling Yao, Jinlong Su, Chaolin Tan, Wenhe Feng, Michalis Benakis, Youxiang Chew and Seung Ki Moon. "In-Situ Process Monitoring and Adaptive Quality Enhancement in Laser Additive Manufacturing: A Critical Review." *Journal of Manufacturing Systems* 74, no. 1 (2024): 527–574.

22. Venkatesan, Shashank, Michael A. Cullinan and Michael Baldea. "Recent Advances in Continuous Nanomanufacturing: Focus on Machine Learning-Driven Process Control." *Reviews in Chemical Engineering* 41, no. 4 (2025): 311–331.

23. Variola, Fabio. "Atomic Force Microscopy in Biomaterials Surface Science." *Physical Chemistry Chemical Physics* 17, no. 5 (2015): 2950–2959.

24. Pang, Dalong, Alain R Thierry and Anatoly Dritschilo. "DNA Studies Using Atomic Force Microscopy: Capabilities for Measurement of Short DNA Fragments." *Frontiers in Molecular Biosciences* 2, no. 1 (2015).

25. Liu, Chak-Ming, Wen-Hsiang Chang, En-De Chu, Chun-Han Yin, Yann-Wen Lan, Hsiang-Chih Chiu and Wen-Chin Lin. "Photoluminescence Patterns via Single-Layer Exfoliation in Gr/MoS_2 Heterostructure Using Contact-Mode Atomic Force Microscopy." *Applied Surface Science* 644, no. 11 (2024): 158778.

26. Song, Peijie, Xiaojuan Li, Jianjun Cui, Kai Chen and Yandong Chu. "Mixed-Mode Oscillations of an Atomic Force Microscope in Tapping Mode." *Chaos: An Interdisciplinary Journal of Nonlinear Science* 34, no. 6 (2024): 063138.

27. Hao, Xiaoyu, Yan Li, Yan Li, Hongyan Ji, Tingting Wang, Haolong Fan, Quanzhen Zhang, Huixia Yang, Liwei Liu, Teng Zhang and Yeliang Wang.

"Direct Visualization of Organometallic Intermediates on Cu (111) with Bond-Resolving Non-Contact Atomic Force Microscopy." *Surfaces* 7, no. 3 (2024): 529–536.

28. Yasakau, Kiryl. "Application of AFM-Based Techniques in Studies of Corrosion and Corrosion Inhibition of Metallic Alloys." *Corrosion and Materials Degradation* 1, no. 3 (2020): 345–372.
29. Nguyen, Cam-Phu Thi, Peggy Schoenherr, Lyman Juli and Jan Seidel. "In-Depth Atomic Force Microscopy Investigation of Nanoscale Mechanical Properties of Pāua Nacre." *Nanoscale* 13, no. 40 (2021): 16959–16967.
30. Garcia, Ricardo and Jaime R. Tejedor. "Advances in Nanomechanical Property Mapping by Atomic Force Microscopy." *Nanoscale Advances* 7, no. 20 (2025): 6286–6307.
31. Petrov, Mikhail, Daniel Canena, Nikita Kulachenkov, Nishant Kumar, Pierre Nickmilder, Philippe Leclère and Igor Sokolov. "Mechanical Spectroscopy of Materials Using Atomic Force Microscopy (AFM-MS)." *Materials Today* 80, no. 7 (2024): 218–225.
32. Alar, Vesna, Andrej Razumić, Biserka Runje, Ivan Stojanović, Marin Kurtela and Branko Štrbac. "Application of Areal Topography Parameters in Surface Characterization." *Applied Sciences* 15, no. 12 (2025): 6573.
33. Adamczyk, Zbigniew, Marta Sadowska and Małgorzata Nattich-Rak. "Quantifying Nanoparticle Layer Topography: Theoretical Modeling and Atomic Force Microscopy Investigations." *Langmuir* 39, no. 42 (2023): 15067–15077.
34. Pelevin, Ivan A., Tatiana P. Kaminskaya, Stanislav V. Chernyshikhin, Kirill B. Larionov and Ella L. Dzidziguri. "Atomic Force Microscopy's Application for Surface Structure Investigation of Materials Synthesized by Laser Powder Bed Fusion." *Compounds* 4, no. 3 (2024): 562–570.
35. Yoo, Su-Been, Seong-Hun Yun, Ah-Jin Jo, Sang-Joon Cho, Haneol Cho, Jun-Ho Lee and Byoung-Woon Ahn. "Automated Measurement and Analysis of Sidewall Roughness Using Three-Dimensional Atomic Force Microscopy." *Applied Microscopy* 52, no. 1 (2022): 1.
36. Pruchnik, Bartosz C., Janusz D. Fidelus, Ewelina Gacka, Krystyna Mika, Leszek Zaraska, Grzegorz D. Sulka and Teodor P. Gotszalk. "Atomic Force Microscopy in Mechanical Measurements of Single Nanowires." *Ultramicroscopy* 263 (2024): 113985.
37. Thimons, Luke A., Abhijeet Gujrati, Antoine Sanner, Lars Pastewka and Tevis D. B. Jacobs. "Hard-Material Adhesion: Which Scales of Roughness Matter?." *Experimental Mechanics* 61, no. 7 (2021): 1109–1120.

38. Garcia, Ricardo. "Nanomechanical Mapping of Soft Materials with the Atomic Force Microscope: Methods, Theory and Applications." *Chemical Society Reviews* 49, no. 16 (2020): 5850–5884.

39. Romaguera-Barcelay, Yonny, Ştefan Ţălu, Robert Saraiva Matos, Rosane Maria Pessoa Betânio Oliveira, Joaquim Agostinho Moreira, Javier Perez de Cruz and Henrique Duarte da Fonseca Filho. "Fractal-Stereometric Correlation of Nanoscale Spatial Patterns of $GdMnO_3$ Thin Films Deposited by Spin Coating." *Applied Sciences* 11, no. 9 (2021): 3886.

40. Wood, Jonathan, Richard Bright, Dennis Palms, Dan Barker and Krasimir Vasilev. "Damage Behavior with Atomic Force Microscopy on Anti-Bacterial Nanostructure Arrays." *Nanomaterials* 14, no. 3 (2024): 253.

41. Giblin-Burnham, Joshua, Yousef Javanmardi, Emad Moeendarbary and Bart W Hoogenboom. "Finite Element Modelling of Atomic Force Microscopy Imaging on Deformable Surfaces." *Soft Matter* 20, no. 47 (2024): 9483–9492.

42. Asih, Geolita Ihsantia Ning, Ande Fudja Rafryanto, Sri Hartati, Xiaoyi Jiang, Alinda Anggraini, Azis Yudhowijoyo, Jizhou Jiang and Arramel. "Recent Advances of Polymer Nanocomposites in Emerging Applications." *Composite Functional Materials* 1, no. 1 (2025): 20250105.

43. Radić, Tea Mišić, Petra Vukosav, Andrea Čačković and Alexander Dulebo. "Insights into the Morphology and Surface Properties of Microalgae at the Nanoscale by Atomic Force Microscopy (AFM): A Review." *Water* 15, no. 11 (2023): 1983.

44. Celano, Umberto, Daniel Schmidt, Carlos Beiti, George Orji, Albert V Davydov and Yaw Obeng. "Metrology for 2D Materials: A Perspective Review from the International Roadmap for Devices and Systems." *Nanoscale Advances* 6, no. 9 (2024): 2260–2269.

45. Park, Yejin, Dokyung Jeong, Uidon Jeong, Hyunbum Park, Sungmin Yoon, Minjae Kang and Doory Kim. "Polarity Nano-Mapping of Polymer Film Using Spectrally Resolved Super-Resolution Imaging." *ACS Applied Materials & Interfaces* 14, no. 40 (2022): 46032–46042.

46. Ali, Asif, Ning Zhang and Rafael M. Santos. "Mineral Characterization Using Scanning Electron Microscopy (SEM): A Review of the Fundamentals, Advancements, and Research Directions." *Applied Sciences* 13, no. 23 (2023): 12600.

47. Trépout, Sylvain, Marie-Laure Sgarra, Sergio Marco and Georg Ramm. "An Introduction to Scanning Transmission Electron Microscopy for the Study of Protozoans." *Molecular Microbiology* 121, no. 4 (2024): 659–670.

48. Sekiguchi, Takashi, Yuanzhao Yao, Ryosuke Sonoda and Yasunari Sohda. "Acceptance Characterization of Electron Detector in SEM Using Stainless Steel Sphere." *Microscopy* 74, no. 2 (2025): 79–85.

49. Prasad, Neeraj R., Sudhir Desai and Mrunalini Jagtap. "A Critical Review on Glimpses of Nanoscience and Technology." *Engineering and Applied Science Letters* 8, no. 1 (2025): 1–37.

50. Venkateshaiah, Abhilash, Vinod V.T. Padil, Malladi Nagalakshmaiah, Stanisław Waclawek, Miroslav Černík and Rajender S. Varma. "Microscopic Techniques for the Analysis of Micro and Nanostructures of Biopolymers and Their Derivatives." *Polymers* 12, no. 3 (2020): 512.

51. Li, Yang, Hongyu Long, Yurong Wang, Bo Wei, Yuemei Tan, Jun Zhou, Xiaoyu Liang, Lei Zhang and Feng Lin. "A Novel and High-Efficiency Dual-Electron Gun Concurrent Preheating Additive Manufacturing Technology: Suppressing Thermal Cracks in Non-Weldable IN738LC Nickel-Based Superalloy by Mitigating Temperature Fluctuations." *Virtual and Physical Prototyping* 20, no. 1 (2025): e2518615.

52. Lewczuk, Bogdan and Natalia Szyryńska. "Field-Emission Scanning Electron Microscope as a Tool for Large-Area and Large-Volume Ultrastructural Studies." *Animals* 11, no. 12 (2021): 3390.

53. Asano, Natsuko, Jinfeng Lu, Shunsuke Asahina and Seiichi Takami. "Direct Observation Techniques Using Scanning Electron Microscope for Hydrothermally Synthesized Nanocrystals and Nanoclusters." *Nanomaterials* 11, no. 4 (2021): 908.

54. Imran, Muhammad, Zhenhua Hu, Fang Ding, Harse Sattar and Guang-Nan Luo. "Controlled Ablation of Impurity Deposition in Fusion Devices by Using Variable Laser Fluence within Depth-of-Focus of Focusing Lens." *Applied Physics B* 130, no. 2 (2024): 27.

55. Tomc, Blaž, Marjan Bele, Ana Rebeka Kamšek, Milena Martins, Aleš Marsel, Miha Hotko, Stefan Popović, Gregor Kapun, Črtomir Donik, Mitja Kostelec, Matjaž Godec, Nejc Hodnik and Luka Suhadolnik "Workflow and Practical Guidance for Identical Location Scanning Electron Microscopy: Reliable Tracking of Localized Transformations." *Small Methods* 9, no. 9 (2025): e01290.

56. Nyaburi, Kevin Mogere, Surya Kamal and Richard K Hailstone. "Scanning Electron Microscopy Noise Classification Using Machine Learning." *Microscopy and Microanalysis* 31, no. Supplement_1 (2025): ozaf048.236.

57. Jäger, Stefan T. and Steffen Strehle. "Design Parameters for Enhanced Photon Absorption in Vertically Aligned Silicon Nanowire Arrays." *Nanoscale Research Letters* 9, no. 1 (2014): 511.

58. González-Alva, Patricia, Karla Karina Gómez-Lizárraga, Jesus Rafael González-Parra, Arturo Barba, J. G. Bañuelos and Miguel Hernández. "Optimal Mechanical Properties and High Cellular Proliferation on Hybrid PCL/HaP/Col/SA Scaffolds for Tissue Engineering." *Surface and Interface Analysis* 57, no. 8 (2025): 659–665.

59. Dong, Jianjin, Irene A. Goldthorpe and Nasser Mohieddin Abukhdeir. "Automated Quantification of One-Dimensional Nanostructure Alignment on Surfaces." *Nanotechnology* 27, no. 23 (2016): 235701.

60. Chen, Ying, Tianwei Liang, Lei Chen, Yaofei Chen, Bo-Ru Yang, Yunhan Luo and Gui-Shi Liu. "Self-Assembly, Alignment, and Patterning of Metal Nanowires." *Nanoscale Horizons* 7, no. 11 (2022): 1299–1339.

61. Lee, Giwon, Haena Kim, Seon Baek Lee, Daegun Kim, Eunho Lee, Seong Kyu Lee and Seung Goo Lee. "Tailored Uniaxial Alignment of Nanowires Based on Off-Center Spin-Coating for Flexible and Transparent Field-Effect Transistors." *Nanomaterials* 12, no. 7 (2022): 1116.

62. Amato, Umberto, Anestis Antoniadis, Italia De Feis, Anastasiia Doinychko, Irène Gijbels, Antonino La Magna, Daniele Pagano, Francesco Piccinini, Easter Selvan Suviseshamuthu, Carlo Severgnini, Andres Torres and Patrizia Vasquez. "Detecting Important Features and Predicting Yield from Defects Detected by SEM in Semiconductor Production." *Sensors* 25, no. 13 (2025): 4218.

63. Dehaerne, Enrique, Bappaditya Dey, Victor Blanco, Jesse Davis. "Scanning Electron Microscopy-Based Automatic Defect Inspection for Semiconductor Manufacturing: A Systematic Review." *Journal of Micro/Nanopatterning, Materials, and Metrology* 24, no. 2 (2025): 020901.

64. Losurdo, Maria, Michael Bergmair, Giovanni Bruno, Denis Cattelan, Christoph Cobet, Antonello de Martino, Karsten Fleischer, Zorana Dohcevic-Mitrovic, Norbert Esser, Melanie Galliet, Rados Gajic, Dušan Hemzal, Kurt Hingerl, Josef Humlicek, Razvigor Ossikovski, Zoran V. Popovic and Ottilia Saxl. "Spectroscopic Ellipsometry and Polarimetry for Materials and Systems Analysis at the Nanometer Scale: State-of-the-Art, Potential, and Perspectives." *Journal of Nanoparticle Research* 11, no. 7 (2009): 1521–1554.

65. Soult, Marta Cazorla, Valerie Siller, Xinhua Zhu, Robert Gehlhaar, Pawel J. Wojcik, Alex Morata, Albert Tarancón, Philippe M. Vereecken and Annick Hubin. "Spectroscopic Ellipsometry for Operando Monitoring of (De)Lithiation-Induced Phenomena on $LiMn_2O_4$ and $LiNi_{0.5}Mn_{1.5}O_4$ Electrodes." *Journal of The Electrochemical Society* 169, no. 4 (2022): 040501.

66. Politano, Grazia Giuseppina and Carlo Versace. "Spectroscopic Ellipsometry: Advancements, Applications and Future Prospects in Optical Characterization." *Spectroscopy Journal* 1, no. 3 (2023): 163–181.

67. Rauch, Nikolaus, E. Andersen, Ignacio Gabriel Vicente-Gabás, Jiri Duchoslav, Alexey Minenkov, Jacek Gasiorowski, Christoph Flötgen, Kurt Hingerla and Heiko Groiss. "A Model for Spectroscopic Ellipsometry Analysis of Plasma-Activated Si Surfaces for Direct Wafer Bonding." *Applied Physics Letters* 121, no. 8 (2022): 081603.

68. Mukherjee, Deshabrato and Peter Petrik. "Real-Time Ellipsometry at High and Low Temperatures." *ACS Omega* 8, no. 4 (2023): 3684–3697.

69. Ma, Liyuan, Xipeng Xu, Changcai Cui, Tukun Li, Shan Lou, Paul J. Scott, Xiangqian Jiang and Wenhan Zeng. "A Review of Measurement and Characterization of Film Layers of Perovskite Solar Cells by Spectroscopic Ellipsometry." *Nanomaterials* 15, no. 4 (2025): 282.

70. Leigh, William G. S., Evan L.H. Thomas, Jerome, A. Cuenca, Soumen Mandal and Oliver A. Williams. "*In-Situ* Monitoring of Microwave Plasma-Enhanced Chemical Vapour Deposition Diamond Growth on Silicon Using Spectroscopic Ellipsometry." *Carbon* 202 (2023): 204–212.

71. Sapkota, Dhurba R., Balaji Ramanujam, Puja Pradhan, Mohammed A. Razooqi Alaani, Ambalanath Shan, Michael J. Heben, Sylvain Marsillac, Nikolas J. Podraza and Robert W. Collins. "Real-Time Spectroscopic Ellipsometry for Flux Calibrations in Multi-Source Co-Evaporation of Thin Films: Application to Rate Variations in $CuInSe_2$ Deposition." *Materials* 17, no. 16 (2024): 4048.

72. Cong, Yue, IIsin An, K. Vedam, and Robert W. Collins. "Optical Characterization of a Four-Medium Thin Film Structure by Real Time Spectroscopic Ellipsometry: Amorphous Carbon on Tantalum." *Applied Optics* 30, no. 19 (1991): 2692–2703.

73. Tumusange, Marie Solange, Biwas Subedi, Cong Chen, Maxwell M. Junda, Zhaoning Song, Yanfa Yan and Nikolas J. Podraza. "Impact of Humidity and Temperature on the Stability of the Optical Properties and Structure of $MAPbI_3$, $MA_{0.7}FA_{0.3}PbI_3$ and $(FAPbI_3)_{0.95}(MAPbBr_3)_{0.05}$ Perovskite Thin Films." *Materials* 14, no. 14 (2021): 4054.

74. Minenkov, Alexey, Sophia Hollweger, Jiri Duchoslav, Otgonbayar Erdene-Ochir, Matthias Weise, Elena Ermilova, Andreas Hertwig, Manuela Schiek. "Monitoring the Electrochemical Failure of Indium Tin Oxide Electrodes via Operando Ellipsometry Complemented by Electron Microscopy and Spectroscopy." *ACS Applied Materials & Interfaces* 16, no. 7 (2024): 9517–9531.

75. Rajan, Grace, Shankar Karki, Robert W. Collin, Nikolas J. Podraza and Sylvain Marsillac. "Real-Time Optimization of Anti-Reflective Coatings for CIGS Solar Cells." *Materials* 13, no. 19 (2020): 4259.

76. Liu, Jinchao, Di Zhang, Dianqiang Yu, Mengxin Ren and Jingjun Xu. "Machine Learning Powered Ellipsometry." *Light: Science & Applications* 10, no. 1 (2021): 55.

77. Maudet, Florian, Charlotte Van Dijck, Muhammad Hamid Raza, Catherine Dubourdieu. "A Fully Automatized Method for the Unambiguous Wavelength-by-Wavelength Determination of the Thickness and Optical Property of a Very Thin Film with a Transparent Range." *Journal of Applied Physics* 134, no. 4 (2023): 045301.

78. Kilic, Ufuk, Alyssa Mock, Derek Sekora, Simeon Gilbert, Shah Valloppilly, Giselle Melendez, Natale Ianno, Marjorie Langell, Eva Schubert and Mathias Schubert. "Precursor-Surface Interactions Revealed during Plasma-Enhanced Atomic Layer Deposition of Metal Oxide Thin Films by in-Situ Spectroscopic Ellipsometry." *Scientific Reports* 10, no. 1 (2020): 10392.

79. Belosludtsev, Alexandr, Anna Sytchkova, Kazimieras Baltrusaitis, Viktoras Vaicikauskas, Vitalija Jasulaitiene and Tatjana Gric. "Growth of Magnetron-Sputtered Ultrathin Chromium Films: In Situ Monitoring and Ex Situ Film Properties." *Coatings* 13, no. 2 (2023): 347.

80. Colin, Jonathan, Andreas Jamnig, Clarisse Furgeaud, Anny Michel, Nikolaos Pliatsikas, Kostas Sarakinos and Gregory Abadias. "In Situ and Real-Time Nanoscale Monitoring of Ultra-Thin Metal Film Growth Using Optical and Electrical Diagnostic Tools." *Nanomaterials* 10, no. 11 (2020): 2225.

81. Nouira, Hichem, Jose A. Salgado, Nadim El-Hayek, Sebastien Ducourtieux, Alexandra Delvallée and Nabil Anwer. "Setup of a High-Precision Profilometer and Comparison of Tactile and Optical Measurements of Standards." *Measurement Science and Technology* 25, no. 4 (2014): 044016.

82. Debnath, Sanjit K. and Mahendra P. Kothiyal. "Optical Profiler Based on Spectrally Resolved White Light Interferometry." *Optical Engineering* 44, no. 1 (2005): 013606.

83. Lomanovskaya, Tatiana A., Gennadii A. Piavchenko, Vladislav O. Soldatov, Artem A. Venediktov, Egor A. Kuzmin, Natalia L. Kartashkina, Svetlana G. Mukhamedova, Tatiana V. Boronikhina, Aleksandr G. Markov, Dmitry V. Telyshev, Igor Meglinski and Alexander N. Yatskovskiy. "Structural Changes of Erythrocyte Membrane Revealed by 3D Confocal Optical Profilometer." *Journal of Biophotonics* 16, no. 1 (2023): e202200222.

84. Ružbarský, Juraj. "Roughness Control of Surfaces Using a Laser Profilometer with the Selected Material Cutting Technology." *Materials* 16, no. 11 (2023): 4109.

85. Ekici, Ömer, Kubilay Aslantaş, Özgür Kanık and Ali Keleş. "Evaluation of Surface Roughness after Root Resection: An Optical Profilometer Study." *Microscopy Research and Technique* 84, no. 4 (2021): 828–836.

86. Antón, Juan Carlos Martínez, Jose Alonso, and Jose Antonio Gómez Pedrero. "Topographic Optical Profilometry of Steep Slope Micro-Optical Transparent Surfaces." *Optics Express* 23, no. 7 (2015): 9494–9507.

87. Chen, Liang-Chia, Duc-Hieu Duong and Chin-Sheng Chen. "Optical 3-D Profilometry for Measuring Semiconductor Wafer Surfaces with Extremely Variant Reflectivities." *Applied Sciences* 9, no. 10 (2019): 2060.

88. Abhinav, Vishnuram, Prithvi Basu, Shikha Supriya Verma, Jyoti Verma, Atanu Das, Savita Kumari, Prateek Ranjan Yadav and Vibhor Kumar. "Advancements in Wearable and Implantable BioMEMS Devices: Transforming Healthcare Through Technology." *Micromachines* 16, no. 5 (2025): 522.

89. Kapłonek, Wojciech, Tadeusz Mikolajczyk, Danil Yurievich Pimenov, Munish Kumar Gupta, Mozammel Mia, Shubham Sharma, Karali Patra and Marzena Sutowska. "High-Accuracy 3D Optical Profilometry for Analysis of Surface Condition of Modern Circulated Coins." *Materials* 13, no. 23 (2020): 5371.

90. Khavieya, Anandhan, G. V. BhanuTeja, Sohini Chowdhury, Nara Arunachalam and A. R. Ganesan. "Quantification of Surface Roughness Using Fringe Projection Profilometry for Metallic Machined Surfaces." *Journal of Tribology* 147, no. 114202 (2025): 1–41.

91. Asgari, Mahdi, Francesco Lucci, Jakub Bialek, Barthélemy Dunan, Gaëlle Andreatta, Rita Smajda, Sébastien Lani, Nicolas Blondiaux, Shoaib Majeed, Sandro Steiner, Jean-Pierre Schaller, Stefan Frentzel, Julia Hoeng and Arkadiusz K. Kuczaj. "Development of a Realistic Human Respiratory Tract Cast Representing Physiological Thermal Conditions." *Aerosol Science and Technology* 53, no. 8 (2019): 860–870.

92. Engel, Thomas. "3D Optical Measurement Techniques." *Measurement Science and Technology* 34, no. 3 (2022): 032002.

93. Gallagher, James C., Michael A. Mastro, Alan G. Jacobs, Robert. J. Kaplar, Karl D. Hobart and Travis J. Anderson. "Detecting Defects That Reduce Breakdown Voltage Using Machine Learning and Optical Profilometry." *Scientific Reports* 14, no. 1 (2024): 7440.

94. Du, Junlin, Fan Yang, Hong Guo, Jiangping Zhu and Pei Zhou. "Parameter Selection on a Multi-Exposure Fusion Method for Measuring Surfaces with Varying Reflectivity in Microscope Fringe Projection Profilometry." *Applied Optics* 63, no. 13 (2024): 3506–3517.

95. Liao, Hsien-Shun, Shih Han Cheng and En-Te Hwu. "Method for Film Thickness Mapping with an Astigmatic Optical Profilometer." *Sensors* 22, no. 8 (2022): 2865.

96. Buchenau, Theresa, Tobias Mertens, Hubertus Lohner, Hauke Bruening and Marc Amkreutz. "Comparison of Optical and Stylus Methods for Surface Texture Characterisation in Industrial Quality Assurance of Post-Processed Laser Metal Additive Ti-6Al-4V," *Materials* 16, no. 13 (2023): 4815.

97. Ohlídal, Ivan, Jiří Vohánka, Jan Dvořák, Vilma Buršíková and Petr Klapetek. "Determination of Optical and Structural Parameters of Thin Films with Differently Rough Boundaries." *Coatings* 14, no. 11 (2024): 1439.

98. Ye, Ruifang, Jiarui Zeng, Heyan Zhang, Yujie Su and Huihui Li. "Rapid Full-Field Surface Topography Measurement of Large-Scale Wafers Using Interferometric Imaging." *Photonics* 12, no. 9 (2025): 835.

99. Placzek, George. *The Rayleigh and Raman Scattering*. Lawrence Radiation Laboratory, 1959.

100. Cialla-May, Dana, Michael Schmitt and Jürgen Popp. "Theoretical Principles of Raman Spectroscopy." *Physical Sciences Reviews* 4 (2019): 6.

101. Yogurtcu, Burak, Nur Cebi, Anıl Tevfik Koçer and Azime Erarslan. "A Review of Non-Destructive Raman Spectroscopy and Chemometric Techniques in the Analysis of Cultural Heritage." *Molecules* 29, no. 22 (2024): 5324.

102. Nogueira, Bernardo A., and Chiara Castiglioni. "Raman Spectroscopy of Crystalline Materials and Nanostructures." *Crystals* 14, no. 3 (2024): 251.

103. Zhao, Na, Xiao Chang, Xianghong Liu, Ji Li, Wei Zheng and Jun Zhang. "Probing the Structural Dynamics of In_2O_3 Using in Situ Raman Spectroscopy: Bridging Material Dynamics and Sensor Functionality." *Angewandte Chemie International Edition* 64, no. 45 (2025): e202512808.

104. Zhang, Shuyan, Yi Qi, Sonia Peng Hwee Tan, Renzhe Bi, Malini Olivo. "Molecular Fingerprint Detection Using Raman and Infrared Spectroscopy Technologies for Cancer Detection: A Progress Review." *Biosensors* 13, no. 5 (2023): 557.

105. Simons, Philipp, Kierstin P. Torres and Jennifer L. M. Rupp. "Deposition Parameters and Raman Crystal Orientation Measurements of Ceria Thin

Films Deposited by Spray Pyrolysis." *Journal of Materials Chemistry A* 10, no. 16 (2022): 8898–8910.

106. Botti, S., Paolo Di Lazzaro, Francesco Flora, L. Mezi and Daniele Murra. "Raman Spectral Mapping Reveal Molecular Changes in Cellulose Aging Induced by Ultraviolet and Extreme Ultraviolet Radiation." *Cellulose* 31, no. 2 (2024): 749–758.

107. Mussi, Valentina, Fabio Antonio Bovino, R. Falsini, Davide Daloiso, Federico Vittorio Lupo, Rincy Kunjumon, Roberto Li Voti, T. Cesca, Roberto Macaluso, C. Sibilia and Giovanni Mattei. "Polarized Raman Mapping and Phase-Transition by CW Excitation for Fast Purely Optical Characterization of VO_2 Thin Films." *Scientific Reports* 14, no. 1 (2024): 19338.

108. Schmid, Thomas, Norbert Schäfer and Daniel Abou-Ras. "Raman Microspectroscopy Provides Access to Compositional and Microstructural Details of Polycrystalline Materials." *Spectroscopy Europe* 28 (2016): 5.

109. Freidzon, Daniel, Olga Kraynis, Ellen Wachtel, Igor Lubomirsky and Tsachi Livneh. "The Elastic Component of Anisotropic Strain Dominates the Observed Shift in the F_{2g} Raman Mode of Anelastic Ceria Thin Films." *Physical Chemistry Chemical Physics* 25, no. 44 (2023): 30563–30571.

110. Sediva, Eva and Jennifer L. M. Rupp. "Raman Spectra and Defect Chemical Characteristics of Sr (Ti, Fe) O_{3-y} Solid Solution of Bulk Pellets vs. Thin Films." *Journal of Materials Chemistry A* 11, no. 48 (2023): 26752–26763.

111. Havryliuk, Yevhenii, Volodymyr Dzhagan, Anatolii Karnaukhov, Oleksandr Selyshchev, Julia Hann and Dietrich R. T. Zahn. "Raman Spectroscopy and Thermoelectric Characterization of Composite Thin Films of Cu_2ZnSnS_4 Nanocrystals Embedded in a Conductive Polymer PEDOT: PSS." *Nanomaterials* 13, no. 1 (2023): 41.

112. Han, Bowen, Mingyuan Sun, Ying Chang, Saisai He, Yuqi Zhao, Chuanyong Qu and Wei Qiu. "Raman Characterization of the In-Plane Stress Tensor of Gallium Nitride." *Materials* 16, no. 6 (2023): 2255.

113. Popescu, Andreea Gabriela Marina, Ioan Valentin Tudose, Cosmin Romanitan, Marian Popescu, Marina Manica, Paul Schiopu, Marian Vladescu, Mirela Petruta Suchea, Cristina Pachiu. "Raman Study of Novel Nanostructured WO_3 Thin Films Grown by Spray Deposition." *Nanomaterials* 14, no. 14 (2024): 1227.

114. Aydogan, Beytullah and Kevin Chou. "Review of In Situ Detection and Ex Situ Characterization of Porosity in Laser Powder Bed Fusion Metal Additive Manufacturing." *Metals* 14, no. 6 (2024): 669.

115. Xu, Ke, Yingguang Li, Changqing Liu, Xu Liu, Xiaozhong Hao, James Gao and Paul G. Maropoulos. "Advanced Data Collection and Analysis in Data-Driven Manufacturing Process." *Chinese Journal of Mechanical Engineering* 33, no. 1 (2020): 43.

116. Vistein, Michael, Monika Mayer, Manuel Endraß and Frederic Fische. "Single Source of Truth: Integrated Process Control and Data Acquisition System for the Development of Resistance Welding of CFRP Parts." *20th International Conference on Informatics in Control, Automation and Robotics, ICINCO 2023* 1 (2023): 592–599.

117. Mustapää, Tuukka, Pekka Nikander, Daniel Hutzschenreuter and Raine Viitala. "Metrological Challenges in Collaborative Sensing: Applicability of Digital Calibration Certificates." *Sensors* 20, no. 17 (2020): 4730.

118. Wasalathanthri, Dhanuka P., Matthew S Rehmann, Yuanli Song, Yan Gu, Luo Mi, Chun Shao, Letha Chemmalil, Jongchan Lee, Sanchayita Ghose, Michael C Borys, Julia Ding and Zheng Jian Li. "Technology Outlook for Real-Time Quality Attribute and Process Parameter Monitoring in Biopharmaceutical Development-A Review." *Biotechnology and Bioengineering* 117, no. 10 (2020): 3182–3198.

119. Zhang, Siqi, Hui Yang and Yan Lu. "Engineering-Guided Deep Learning of Melt-Pool Dynamics for Additive Manufacturing Quality Monitoring." *Journal of Computing and Information Science in Engineering* 24, no. 10 (2024): 101002.

120. Seenivasan, Raagul, Jey Kumar Pachiyappan, Murthannagari Vivek Reddy and GNK Ganesh. "A Systematic Review: Exploration of Process Analytical Technology Techniques (PAT) and their Multifacated Advantages in Industrial Processes." *International Journal of Applied Pharmaceutics* 16, no. 2 (2024): 44–51.

121. Ahmmed, Md Sazol, Laraib Khan, Muhammad Arif Mahmood and Frank Liou. "Digital Twins, AI, and Cybersecurity in Additive Manufacturing: A Comprehensive Review of Current Trends and Challenges." *Machines* 13, no. 8 (2025): 691.

122. Maruthamuthu, Murali K., Scott R Rudge, Arezoo M Ardekani, Michael R Ladisch and Mohit S Verma. "Process Analytical Technologies and Data Analytics for the Manufacture of Monoclonal Antibodies." *Trends in Biotechnology* 38, no. 10 (2020): 1169–1186.

123. Caiza, Gustavo and Ricardo Sanz. "An Immersive Digital Twin Applied to a Manufacturing Execution System for the Monitoring and Control of Industry 4.0 Processes." *Applied Sciences* 14, no. 10 (2024): 4125.

124. Chen, Xiaoyun, Jin Wang, Christopher Thurber, Matthew Benedict, Kurt Olson, Eric Marchbanks, Hyunwoo Kim, and Michael Bishop. "Real-Time Mapping of Polymer Film Thickness Using Near-Infrared Hyperspectral Imaging." *Applied Spectroscopy* 79, no. 10 (2025): 1518–1526.

125. Abdellah, Ahmed M., Kholoud E. Salem, Liza-Anastasia DiCecco, Fatma Ismail, Amirhossein Rakhsha, Kathryn Grandfield and Drew Higgins. "In Situ Transmission Electron Microscopy of Electrocatalyst Materials: Proposed Workflows, Technical Advances, Challenges, and Lessons Learned." *Small Methods* 9, no. 1 (2025): 2400851.

126. Karimi, Koosha, Ali Fardoost, Nikhil Mhatre, Jay Rajan, David Boisvert and Mehdi Javanmard. "A Thorough Review of Emerging Technologies in Micro- and Nanochannel Fabrication: Limitations, Applications, and Comparison." *Micromachines* 15, no. 10 (2024): 1274.

127. Xu, Weiwang, Houdao Zhang, Lingjing Ji and Zhongyu Li. "AI-Powered Next-Generation Technology for Semiconductor Optical Metrology: A Review." *Micromachines* 16, no. 8 (2025): 838.

128. Nandipati, Mutha, Olukayode Fatoki and Salil Desai. "Bridging Nanomanufacturing and Artificial Intelligence — A Comprehensive Review." *Materials* 17, no. 7 (2024): 1621.

129. Kwoen, Jinkwan and Yasuhiko Arakawa. "In Situ Monitoring of Quantum Dot Growth Using a Magnification Inferred Curvature Method." *Applied Physics Letters* 125, no. 17 (2024): 172108.

130. Das, Hirendra and Mridusmita Sharma. "Transforming Computational Nanotechnology: Accelerating Material Discovery, Design, and Property Prediction through Soft Computing Techniques." *ACS Applied Electronic Materials* 7, no. 13 (2025): 5757–5787.

131. Ali, Muhammad A., Muhammad S. Irfan, Tayyab Khan, Muhammad Y. Khalid and Rehan Umer. "Graphene Nanoparticles as Data Generating Digital Materials in Industry 4.0." *Scientific Reports* 13, no. 1 (2023): 4945.

132. Yadavalli, Vamsi K. "The Convergence of Nanomanufacturing and Artificial Intelligence: Trends and Future Directions." *Nanotechnology* 36, no. 22 (2025): 222001.

https://doi.org/10.1142/9789819830763_0010

CHAPTER 10

Microwave-Assisted Scanning Probe Microscopy from Near-Field Sensing to Nanofabrication Applications

Filippo Ozino Caligaris, Lorenzo Andrea Magnarin, Sakib Ahmed, Sanchari Sen, Umberto Celano*

School of Electrical, Computer, and Energy Engineering, Arizona State University, Tempe, Arizona 85281, USA

*Corresponding author: umberto.celano@asu.edu

Abstract

Since its inception, atomic force microscopy has evolved beyond high-resolution topographic imaging to include a diverse range of "secondary modes" capable of characterizing electrical, magnetic, and mechanical material properties. This chapter focuses on the advanced integration of electromagnetic radiation, specifically within the microwave regime, into the atomic force microscopy architecture for nanofabrication and characterization. We examine two distinct imaging methodologies leveraging this integration: scanning near-field microwave microscopy, which utilizes metallic probes to detect variations in permittivity and subsurface conductivity; and quantum sensing via optically detected magnetic resonance, where microwaves drive spin transitions in engineered probes to map magnetic and spintronic features with non-perturbative high sensitivity. Following a review of the historical and technical developments of these techniques, we detail their respective operational physics, experimental setups, and contrast mechanisms. Finally, the chapter highlights the critical role of these microwave-assisted modes in contemporary nanoscale characterization and nanofabrication, particularly within semiconductor and biomaterial applications.

Keywords: atomic force microscopy, optically detected magnetic resonance, scanning microwave impedance microscopy, near-field scanning microwave microscopy

10.1 Introduction

Since their introduction, atomic force microscopy (AFM) methods have offered a platform to explore materials with an unprecedented level of detail and flexibility. While most of the merits of high-resolution imaging can be found in the use of nanosized probes, often manufactured with atomic precision, the inherent flexibility to extend AFM modes to measure properties beyond surface topography, such as electrical, magnetic, and mechanical characteristics, arises from the ability to sense short- and long-range forces acting on the tip–sample junction. The inherent flexibility to extend the AFM modes to measure properties beyond surface topography such as electrical, magnetic, and mechanical characteristics arise from the measurement ability to sense short- and long-range forces acting on the tip-sample junction. These specialized techniques, often referred to as secondary AFM modes, were introduced immediately after the basic contact and tapping modes and have continued to grow in the last four decades. The first generation of secondary modes generally explored some kind of biased conducting AFM tips, in conjunction with DC and AC signals, applied between the tip and the sample. Contact-mode methods yielded multiple variations of electrical two-probe experiments, including techniques for the characterization of local conductivity in thin films, electronic leakage in thin dielectrics, active carrier in semiconductors, and mapping of ferroelectric domains, to name a few. Non-contact methods yielded the ability to probe magnetic, electrostatic, and mechanical properties associated with long-range forces between the tip and the sample. In parallel with the use of electrical stimulation, early AFM probes were customized to create tiny apertures at the end of the cantilever that allowed for optical imaging with sub-wavelength resolution. Here, laser light is illuminated through the aperture, which is brought into the near-field operation regime of the sample surface (0–10 nm). The use of different wavelengths and modes of operation such as transmission, reflection, and collection gave rise to the field of near-field imaging, overcoming the diffraction limit of light to achieve resolution down to tens of nanometers. Beyond sensing additional information about the sample surface, these modes have distinguished themselves for their ability to manipulate matter at the nanoscale. Among notable examples, we report on the use of tip-assisted localized electrochemical oxidation for patterning, stress-assisted chemical reaction, and pressure-induced surface modification for tomographic analysis, to name but a few.

With the advent of more sophisticated instruments, a new generation of secondary modes appeared, leveraging the improved scanning stability of scanners and tip placement in order to couple electromagnetic radiation to the AFM probe in a wide spectral range that includes microwaves, visible, terahertz, infrared, and ultraviolet radiation. These modes allow the detection of polarization-resolved field mapping, nanoscale spectroscopy, as well as estimation of the complex optical constants (n, k) of the material. However, the case of microwave radiation is quite unique since it is exploited in various ways by different probing methods. For example, by the use of metallic probes, microwaves can be confined and coupled to the tip-sample junction, thus creating a hybrid method that is microwave-based and a near-field-assisted scanning probe technique. This implementation has demonstrated sensitivity to variations in permittivity and conductivity of the sample surface, with the addition of a deep microwave interaction volume, which makes it particularly effective for probing subsurface electrical features, dopant distributions, and buried interfaces in semiconductor devices.

Alternatively, a microwave field can be used to drive transitions between spin sublevels of spin-active defects contained in a specially engineered AFM probe. Here, optically detected magnetic resonance (ODMR) bridges optical excitation with microwave-driven spin resonance to probe general properties that affect the spin of electrons, i.e., magnetic properties at the nanoscale, electric fields, or even temperature. In essence, when the microwave frequency matches the energy difference between spin states, resonant spin transitions occur, altering the optical emission intensity. This implementation has demonstrated ultra-high spatial resolution and magnetic field sensitivity while remaining non-perturbative. For these properties, this class of techniques is particularly effective for qualifying the fabrication of magnetic and spintronic devices that leverage antiferromagnets, multiferroics, and skyrmions.

Both approaches, namely the use of microwave field as a direct probe or as a driver for ODMR, are key for materials characterization and nanoscale fabrication. These methods offer an unparalleled level of sensitivity in their respective fields such that without them, the accurate level of process control would be impossible in multiple technology-relevant applications, e.g., for the detection of impurities, buried doping layers, and ultra-confined magnetic features. For their instrumental role in the nanofabrication of semiconductor devices, biomaterials, polymers, and more, these two techniques are described in this chapter, along with a brief schematic presented in Figure 10.1 combining both techniques. First, we review the history and the main technical developments that transformed the original concepts into imaging methodologies. Here, we revise the differences between the physics of operation and the existing setups. Later, we introduce the principles of

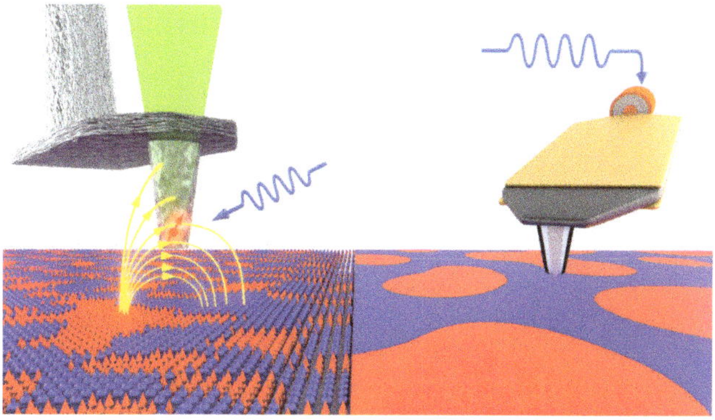

Figure 10.1. Application of microwave fields to manipulate spin sublevels of color centers placed on an AFM probe to measure nanoscale magnetic, electrical, and thermal properties of certain samples, along with the use of metallic probes to confine and couple microwaves to the tip-sample junction in order to implement a microwave-based, near-field-assisted scanning probe technique.

contrast mechanisms for each method, discussing the strengths and limitations. Finally, we report on the main area of application in which these modes have been instrumental for nanoscale characterization and nanofabrication.

10.2 Microwave-Based Near-Field Assisted Imaging Using Scanning Probe Methods

Microwave-based near-field assisted imaging using scanning probes represents a significant advance in AFM, providing a means of probing local electronic and dielectric properties through the detection of reflected microwave signals. Commonly referred to as scanning microwave impedance microscopy (sMIM) — the term adopted hereafter — and alternatively known as scanning microwave microscopy or near-field scanning microwave microscopy, this technique was first introduced by K. Lai in

2007[1] and builds upon the principle of transmitting a microwave signal through an AFM tip in contact with a sample, followed by analysis of the reflected response. In Lai's initial implementation, ~1 GHz microwaves were guided through the cantilever and tip. At the same time, the reflected signal was decomposed into in-phase and out-of-phase components using a 50 Ω matching network, enabling direct determination of the complex impedance of the tip–sample system.[1] One of the main advantages of sMIM lies in its high sensitivity and spatial resolution. Charge resolution can reach the sub-attofarad regime, with lateral resolution between 10 and 100 nm depending on the probe geometry.[2] Recent advancements in the technique have enhanced its sensitivity to the point of Johnson noise-limited performance, extending its detection capability to the sub-zeptofarad range.[3] Because the microwave spectrum spans a wide frequency range (300 MHz–300 GHz),[4] selecting an appropriate operating frequency is essential. This choice can be guided by considering the equivalent circuit that describes the tip–sample junction in contact-mode AFM. Given that the tip apex is several orders of magnitude smaller than the wavelength of the applied radiation, the interaction can be modeled using lumped elements: a resistance and capacitance connected in parallel.[4] In practice, a thin oxide layer at the sample surface is often present, which extends the model to include a series capacitance representing the oxide in series with the parallel resistor-capacitor circuit.[5] In a typical measurement, sMIM provides multiple imaging channels, usually five and up to seven in extended configurations. Besides the topography channel, both the real and imaginary components of the reflected microwave signal are recorded, commonly interpreted as resistance- and capacitance-related contributions, respectively. When both a lock-in amplifier and an AC voltage source are incorporated into the setup, the capacitance channel can be extended to yield differential capacitance (*dC*/*dV* amplitude and phase). With an additional lock-in amplifier, analogous differential resistance (*dR*/*dV*) signals can also be obtained. The versatility of sMIM has already been demonstrated across a broad spectrum of applications. These include the characterization of biological materials,[6] the analysis of devices based on novel materials and oxides,[7,8] the investigation of quantum transport phenomena in carbon nanotubes,[9] the study of grain boundary effects in next-generation photovoltaic materials,[10,11] and diverse applications in condensed matter physics.[12,13]

10.2.1 *Setup and Technical Components*

The instrumentation required for sMIM is based on a conventional AFM platform, augmented with an electronic circuit that generates, couples, and separates incident and reflected microwave signals. The vertical (z-axis) movement of the piezoelectric scanner is still regulated via the AFM's optical beam deflection system, as in standard AFM operation. However, the microwave circuitry introduces additional components essential to the measurement setup. These include a microwave source, a directional coupler for guiding the signal into the sMIM probe, a cancellation circuit for suppressing background contributions, an impedance matching network to optimize signal transfer between the tip–sample junction and the transmission line, and finally a mixer for demodulating the reflected signal into its real and imaginary components.[5,14] A schematic representation of the sMIM setup is depicted in Figure 10.2(a). The probe itself can be modeled electrically as comprising two admittance pathways: the tip–sample admittance, Y_{ts},

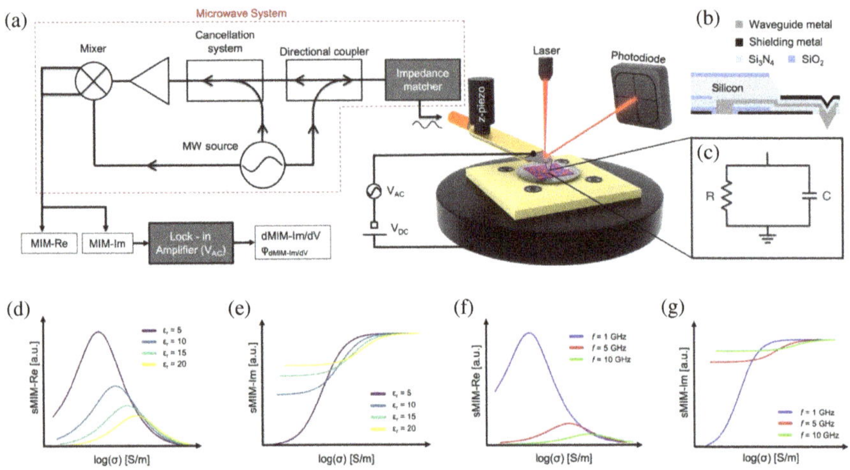

Figure 10.2. Schematic and modeling of the sMIM experimental setup and signal response. (a) Schematic of the sMIM system integrated with an AFM, including a 3D rendering of the AFM setup. (b) Cross-sectional view of the shielded tip structure. (c) Lumped-element model representing the tip–sample junction. (d) Simulated real part of the S_{11} response as a function of sample conductivity for various permittivity values. (e) Simulated imaginary part of the S_{11} response as a function of sample conductivity for various permittivity values. (f) Simulated real part of the S_{11} response as a function of sample conductivity for different microwave frequencies. (g) Simulated imaginary part of the S_{11} response as a function of sample conductivity for different microwave frequencies.

which varies dynamically during a measurement, and the probe's intrinsic admittance, Y_{probe}, arising from the cantilever and waveguide structures excluding the tip apex. Since Y_{probe} is essentially constant, the total admittance can be written as:

$$Y_{total} = Y_{ts} + Y_{probe} \tag{10.1}$$

To ensure efficient transmission, the tip–sample admittance must be matched to that of the transmission line, typically fixed at $Y_0 = 1/50\ \Omega^{-1}$. Without such impedance matching, reflections would occur, degrading the signal-to-noise ratio.[15,16] Matching is commonly achieved using resonant elements such as a quarter-wave ($\lambda/4$) transmission-line resonator.[17] In commercial sMIM systems, more complex and precise resonator systems are used, often using an array of lumped elements to improve the signal-to-noise ratio. Additionally, a resonator enables the selection of an optimal operation frequency with the lowest reflected-to-transmitted signal ratio. The cancellation circuit provides a further increase in signal-to-noise ratio. Since both incident and reflected signals contain a significant background component that can obscure the small differential signal arising from local material properties, cancellation is achieved by introducing a phase-shifted (180°) copy of the baseline signal into the reflected path. This destructive interference effectively removes the baseline, isolating the material-dependent contribution for subsequent amplification.[3] Such cancellation is critical for achieving the sub-attofarad resolution that makes sMIM attractive for nanoscale device characterization and failure analysis.[17] The mixer represents the final stage of the electronic system, demodulating the amplified reflected signal into orthogonal components. By selecting an appropriate demodulation phase, the real and imaginary contributions can be separated into the MIM-Re and MIM-Im channels. This splitting of the signal is particular to sMIM and separates the technique from a similar technique, named scanning microwave microscopy.[18] In sMIM, the variation in the tip-sample admittance is measured, allowing for this separation due to the complex nature of the admittance, which will be described at a later point in this chapter. In contrast, scanning microwave microscopy only returns the amplitude and signal of the reflected wave, yielding similar results but at the cost of additional calibration to quantify the signal.[18] Because phase shifts vary with probe geometry and coupling conditions, calibration is required whenever a probe is changed. This is typically performed using a purely capacitive reference sample. By adjusting the demodulation phase until the MIM-Re signal is minimized, the resistive (MIM-Re) and capacitive (MIM-Im) channels can be reliably distinguished. Operation in the near-field regime introduces the problem of parasitic reflections, making probe design critical. An sMIM probe must both guide microwaves efficiently to the apex and suppress spurious reflections.[19] A successful approach was demonstrated by Yang et al. in 2012,[20] who fabricated cantilever-based probes with sharp pyramidal tips using a ten-step microfabrication process. Their design incorporates a waveguide structure extending the length of the cantilever, shielded on both sides by metal layers, with a dielectric spacer (e.g., Si_3N_4) separating the conductive layers, as shown in Figure 10.2(b).[20] Since the probes are used as the primary waveguide for the microwaves, material and geometrical considerations have to be done when designing sMIM-compatible probes. The tip radius heavily influences the minimal feature size that can be resolved using the technique.[21] This phenomenon led to the fabrication and alteration of sMIM tips, generally achieved either via chemical manipulation of the tip's apex or via focused ion beam milling, to improve sMIM's lateral resolution.[22] Tuning fork sMIM-compatible probes, which do not necessitate a laser-based optical feedback mechanism typical for AFM modes, have also been presented in the past, allowing sMIM measurements in complex environments where an optical setup is not possible or unwanted.[23]

10.2.2 *Imaging Contrast Mechanism*

While scanning capacitance microscopy measures variations in the capacitance of a metal–oxide–semiconductor (MOS) structure under different bias conditions, sMIM obtains contrast from the evanescent electric field generated beneath the probe tip.[5] In the case of sMIM, the response of the material to the applied microwave field is governed by its dielectric and conductive properties, which are described by the complex permittivity:

$$\hat{\varepsilon} = \varepsilon' + j\left(\varepsilon'' + \frac{\sigma}{\omega}\right) \tag{10.2}$$

where ε' denotes the real part of the dielectric permittivity, ε'' corresponds to dielectric losses, σ is the electrical conductivity, and ω represents the angular frequency of the probing microwave field.[5] By separating the reflected signal into its real and imaginary components, sMIM provides simultaneous access to the conductive and dielectric contributions of the material. These signals are conventionally referred to as MIM-Re (or MIM-R) and MIM-Im (or MIM-C), respectively. When an alternating bias is applied to the sample, the dielectric channel (MIM-Im) can be further analyzed to extract the amplitude and phase of the differential response (dMIM-Im/dV), analogous to the dC/dV signals employed in scanning capacitance microscopy.[5] Reliable interpretation of sMIM measurements generally requires numerical simulations, as the extracted signals depend on a number of experimental parameters that shape the tip–sample interaction. The most relevant include: the probing frequency (determined by the microwave source), the microwave power (which influences the probing volume), the scan speed (affecting lateral resolution), the probe setpoint (controlling tip–sample force, and thereby resolution and tip wear), and the lock-in amplifier settings (AC bias amplitude and reference frequency). The measured contrast in sMIM ultimately reflects the sample's response to a microwave field. As the relevant signal is the demodulated impedance of the reflected wave relative to the incident wave,[5] it is often convenient to describe the tip–sample interaction in terms of the complex admittance of a parallel resistor-capacitor circuit, as depicted in Figure 10.2(c):

$$Y_{ts} = G + j\omega C \tag{10.3}$$

where G denotes the conductance (loss-related component), C is the effective tip–sample capacitance, and ω is the angular frequency. The variation of this admittance under scanning gives rise to the measured MIM-Re and MIM-Im channels. The characteristic response of these channels can be illustrated by considering different classes of materials. For insulators, where mobile charge carriers are scarce, the electric field penetrates deeply into the material, producing only minor perturbations in both MIM-Re and MIM-Im signals. In semiconductors, the increased carrier density introduces dielectric losses, leading to energy dissipation and an enhanced MIM-Re signal (through the conductance term G).[5] Simultaneously, partial field screening by mobile charges increases the effective capacitance, and therefore the MIM-Im signal. For metals, the abundance of free carriers causes nearly complete field screening, analogous to a plate-capacitor configuration. In this regime, the effective capacitance becomes very large, driving the MIM-Im response to saturation, while the MIM-Re signal diminishes back toward zero as dissipative losses vanish.[14] The MIM-Re response is small for both insulating and metallic samples, peaking only at intermediate conductivities typical of semiconductors, where dielectric losses are most significant. By contrast, the MIM-Im response increases monotonically with conductivity, ranging from near zero in insulators (deep field penetration) to saturation in metals (complete screening).[14] Although the general shape of these curves is preserved, their magnitude and position can be modulated by parameters such as the probing frequency and the dielectric permittivity of the sample.[14] Yong-Tao Cui and Eric Yue Ma have shown that the optimal working frequency is achieved when the real and imaginary components of the admittance are equal, i.e., when $\omega C \approx 1/R$. This condition can be further simplified to $\omega\varepsilon \approx \sigma$, where $\omega = 2\pi f$ is the angular frequency, ε the dielectric constant, and σ the conductivity.[5] For a typical semiconductor resistivity of $\sim 10^2\ \Omega$ cm, the optimal operating frequency is approximately 2 GHz.[5] To probe a wide range of materials with different electrical properties, tuning the frequency range can be of great use. In addition to characterization, microwave-induced fabrication can also benefit from such a setup, allowing for variations in the microwave-sample interaction process. The main downside of such advanced setups is the increased requirements for the waveguides, resonator, and mixer. The real and imaginary parts of the sMIM signal were modeled using the COMSOL AC module and are shown in Figure 10.2(d–g) as a function of sample conductivity, for various dielectric constants and operating frequencies.

10.2.3 *Modes of Operation*

Several operational modes of sMIM are available, each tailored to specific measurement needs.[5] In direct MIM, the standard configuration, the tip is raster scanned, and the MIM-Re and MIM-Im channels are recorded. Differential MIM, analogous to AFM tapping mode, involves modulating the tip–sample distance z at a set frequency; demodulation at this frequency yields $d\mathrm{MIM}/dz$ signals, providing absolute measurements of dY_{ts}/dz with improved drift stability. Multi-frequency MIM further extends the technique's versatility by probing the sample at several microwave frequencies. This approach expands the accessible conductivity range and is particularly beneficial for heterogeneous samples, where different regions may require distinct probing conditions. However, implementing this mode necessitates more sophisticated microwave sources, guiding structures, and matching circuits capable of long, drift-free measurements, thereby increasing system complexity, cost, and susceptibility to failure.[5] Such multi-frequency operation enables spatially and spectrally resolved measurements, offering a powerful means of investigating phase transitions and related inhomogeneities.[24,25] Finally, ultrahigh-resolution sMIM has been demonstrated by Lee et al., who exploited intentional tip modification through debris accumulation to reshape the apex geometry, thereby enhancing spatial resolution to ~5 nm.[26] While reminiscent of probe modification strategies in AFM[27] and conductive AFM,[28,29] this approach remains non-standard and yields variable results, though it highlights the continued potential for resolution improvements through probe engineering. Together, these instrumentation advances, probe designs, and operational modes define the versatility of sMIM, enabling its application across a wide range of materials and device characterization challenges at the nanoscale.

10.2.4 *Characterization of Nanofabricated Features with Scanning Microwave Impedance Microscopy*

The spectrum of materials accessible to investigation by sMIM is vast. The two principal measurement channels, MIM-C and MIM-R, exhibit distinct responses when probing dielectrics, semiconductors, and metals, thereby enabling material-specific contrast. When combined with lock-in-based AC modulation, the technique further facilitates the extraction of dopant distributions and other semiconductor properties. Owing to its inherently non-destructive nature, sMIM is particularly well-suited for the characterization of fragile or sensitive systems, including carbon allotropes and 2D materials. The subsequent survey of sMIM applications is intended to highlight the diversity of research directions and characterization opportunities afforded by this technique.

10.2.4.1 *Semiconductors*

With the continuous scaling and increasing architectural complexity of semiconductor devices, the demands on failure analysis methodologies have grown correspondingly. In this context, sMIM holds considerable promise as a next-generation characterization technique, with the potential to surpass the prevailing AFM method for such purposes, namely scanning capacitance microscopy. Early demonstrations of sMIM on dopant calibration structures, such as those provided by Imec, revealed straightforward calibration procedures and the reliable extraction of doping profiles.[30,31] These advances enabled the subsequent transfer of dopant quantification protocols to technologically relevant devices, including CMOS and power electronics, underscoring the technique's strong relevance to the semiconductor industry.[31–33] Beyond dopant profiling, sMIM has also been employed to investigate the electrical properties of materials such as gallium nitride (GaN) and silicon, further illustrating its versatility and applicability across a broader range of device platforms.[34] As shown in Figure 10.3, sMIM has been effectively applied to semiconductor-relevant 2D systems such as monolayer tungsten ditelluride (WTe_2), enabling direct visualization of spatial variations in local conductivity. These measurements reveal insulating regions near metal contacts and distinct features associated with structural defects, strain, and magnetic field effects — factors that are also critical in determining device performance in advanced semiconductor architectures. In addition, Figure 10.3 highlights the capability of near-field microwave imaging to resolve Moiré superlattice patterns in twisted bilayer graphene (TBG), capturing nanoscale conductance variations arising from interlayer coupling and strain-induced reconstruction. Such high-resolution electrical mapping techniques underscore the potential of MIM-based approaches for probing interface quality, contact integrity, and electronic uniformity in next-generation semiconductor materials and devices.

10.2.4.2 *Dielectrics*

In addition to dopant quantification, sMIM can be employed to directly probe dielectric properties, most notably the dielectric constant of oxide layers.[35] When combined with precise dopant profiling, this capability enables simultaneous characterization of advanced transistor architectures, particularly at lower technology nodes. Such measurements provide insight into both

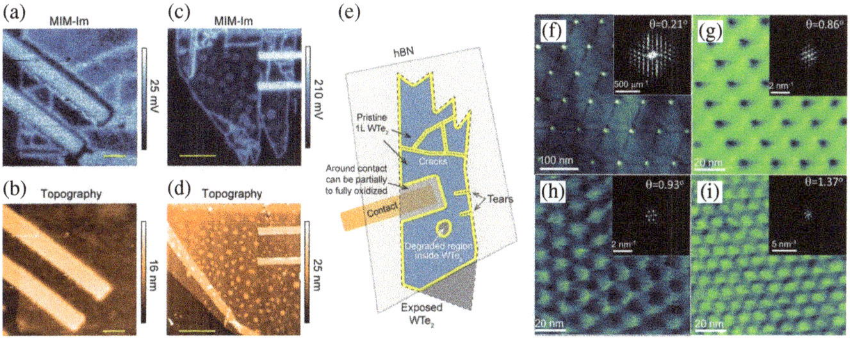

Figure 10.3. Conductivity features near contacts and around defects in monolayer WTe_2, and Moiré superlattice patterns in TBG. (a,b) MIM-Im and corresponding topography images of a monolayer WTe_2 device, measured at T = 10 K, $V_{(g)}$ = 3.3 V, and B = 0 T. In the topography, the flake appears continuous, but MIM imaging reveals that regions surrounding the metal contacts are highly insulating (dark contrast). Scale bars, 1 μm. (c,d) MIM-Im and topography images of another monolayer WTe_2 device, measured at T = 10 K, $V_{(g)}$ = 0 V, and B = 9 T. The small ring-shaped features in the MIM-Im image correspond to topographic blisters. Scale bars, 3 μm. (e) Schematic illustration showing different conductivity features identified in the experiments, including pristine and degraded regions, cracks, tears, and areas around the contacts where oxidation may occur. Adapted under CC BY 4.0 license.[38] (f) 400 × 400 nm conductance image of a TBG:hBN:glass device with a twist angle $\theta = 0.21°$, showing a strain-soliton network resulting from surface reconstruction. The false color corresponds to reflected signal intensity (white = higher conductance). Inset: Fourier transform of the Moiré pattern. (g-i) 100 × 100 nm conductance images showing Moiré patterns at various twist angles and substrates: (g) TBG:hBN:glass, (h) TBG:SiO_2:Si , and (i) TBG:glass. Insets: corresponding Fourier transforms and measured twist angles ($\theta = 0.86°$, $0.93°$, $1.37°$). Adapted under CC BY 4.0 license.[39]

the dielectric components, such as gate oxides and spacers, and the active regions of the device, including the channel, source, and drain. When integrated with finite element modeling, sMIM further allows for the reasonably reliable quantitative extraction of dielectric properties, whether in terms of layer thickness or dielectric constant.[36,37] However, achieving high accuracy in such measurements requires careful consideration of additional factors, including the influence of the interfacial water layer between tip and sample, as well as the geometrical characteristics of the probe tip. Both of these parameters must be precisely controlled experimentally and rigorously implemented in the modeling framework to ensure reliable interpretation.

10.2.4.3 *Quantum Materials*

The rapid growth of research on 2D materials, ranging from graphene to transition metal dichalcogenides and other quantum materials such as carbon nanotubes, has been driven by their promising applications in semiconductor device technologies including transistors[40,41] and photodetectors.[42,43] Figure 10.4 illustrates the versatility of sMIM techniques, from subsurface imaging of conventional semiconductor architectures to the visualization of edge conductivity in quantum materials like monolayer WTe_2. Within this context, sMIM has demonstrated clear advantages over other electrical AFM-based techniques, particularly for the characterization of 2D systems.[44] One of its most significant strengths lies in its minimal sample preparation requirements; notably, sMIM does not require a back contact for electrical characterization, a feature that is especially advantageous given the practical difficulties of implementing such contacts in 2D materials. Recent studies have reported quantitative analysis of molybdenum disulfide (MoS_2) layers using sMIM, achieving a substantially higher signal-to-noise ratio compared to electrostatic force microscopy.[44] Similarly, quantitative measurements on the ferroelectric material α-phase indium selenide

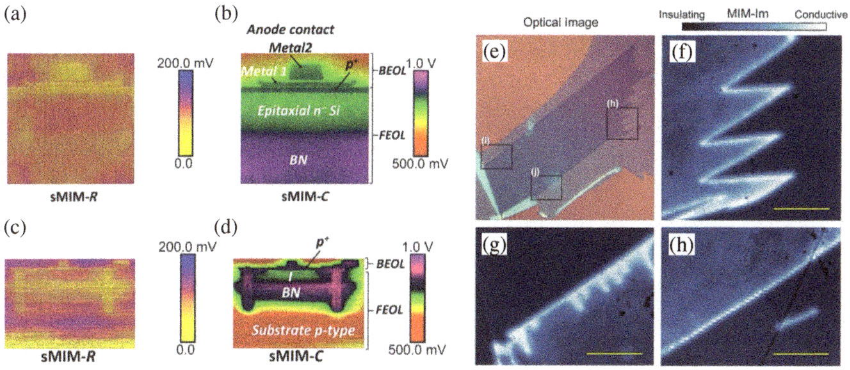

Figure 10.4. Subsurface sMIM mapping of a PIN diode structure and edge conductivity imaging in monolayer WTe_2. (a,b) sMIM-R and sMIM-C maps acquired over a 25 μm × 25 μm area at the PIN anode. All front-end-of-line and back-end-of-line layers are resolved, including the epitaxial n^--Si, BN, and metal contacts. (c,d) Large-area sMIM-R and sMIM-C maps (83 μm × 54 μm) acquired at V_{DC} = 0 V on the same PIN diode device. The contrast clearly delineates the spatial positions of each device layer, including the BN barrier and p^+ regions. Adapted under CC BY 4.0 license.[48] (e) Optical image of a WTe_2 monolayer exfoliated onto SiO_2 and encapsulated by a 10-nm hBN layer. (f-h) MIM-Im images of the selected regions, measured at T = 8 K and B = 0 T. The bright edges correspond to highly conductive boundaries of the WTe_2 flake. Scale bars, 5 μm. Adapted under CC BY 4.0 license.[38]

(α-In_2Se_3) revealed MOS capacitor-like behavior, while sMIM characterization of α-In_2Se_3 field effect transistor (FET) channels demonstrated its capability for nanoscale imaging of ferroelectric domains.[45] Beyond these examples, sMIM has also been applied to graphene-based systems, including the study of graphene encapsulated in hexagonal boron nitride (hBN)[46] as well as the imaging of Moiré patterns in bilayer graphene.[47] As research on 2D materials continues to expand, sMIM is poised to become an essential technique for their quantitative electrical characterization, providing insights into both fundamental properties and device-relevant functionalities. Microwave-assisted imaging of carbon nanotubes has also been portrayed as a promising approach, as it allows for gaining insights into the electronic and quantum behavior of the samples under investigation.[9]

10.2.5 *Benefits and Limitations*

sMIM offers several advantages over conventional AFM modes, particularly for the characterization of nanofabricated structures. The technique has already been successfully applied to image nanoscale features such as nanocapacitors,[22] buried structures including SiO_2 plateaus embedded within Si_3N_4 layers,[49] and doped subsurface regions such as highly doped Si beneath silicon overlayers.[21] In addition to buried features, sMIM is well-suited for fragile samples that must be examined in non-contact mode. However, non-contact operation introduces a reduction in dynamic range, primarily due to the absence of the water meniscus at the tip–sample junction, as confirmed by finite element modeling simulations.[22] Another distinct advantage of sMIM is that it does not require a metallic back contact, enabling the characterization of samples where such a contact is either impractical or impossible to fabricate — an ability it shares with other near-field imaging approaches. Given its intended application in semiconductor metrology, high sensitivity to dopant concentration is of particular importance. sMIM has been employed for the characterization of devices such as PIN diodes (i.e., undoped intrinsic semiconductor region between a p-type semiconductor and an n-type semiconductor region) and silicon carbide (SiC) MOSFETs, demonstrating its utility for both device characterization and failure analysis.[50,51] Comparative studies with established dopant profiling techniques, including scanning spreading resistance microscopy and scanning capacitance microscopy, consistently highlight the advantages of sMIM. Specifically, sMIM enables the simultaneous acquisition of topographical, resistive, capacitive, and dopant-related information, although its lateral resolution remains somewhat limited by the lack of ultrasharp sMIM-compatible probes.

Although sMIM introduces numerous advantages, certain limitations remain, primarily due to the relatively recent development of the technique. Among the primary challenges are the reliance on finite element modeling simulations for the direct extraction of quantitative physical parameters, the insufficient understanding of the influence of the water meniscus — rendering environmental conditions a critical factor for image quality and dynamic range — and the reduced lateral resolution compared to established scanning probe microscopy modalities. The raw output of sMIM measurements is typically expressed in volts (V), whereas scanning microwave microscopy, a closely related technique, produces data in terms of signal amplitude and phase. In both cases, quantitative interpretation requires either extensive simulations or rigorous calibration procedures to determine physical quantities such as sheet resistance, capacitance values, and dielectric constants. At present, however, no standardized reference or calibration samples exist for the direct quantification of sMIM data.[51] The role of the water meniscus at the tip–sample junction has been the subject of ongoing discussion, with experimental and simulation-based studies producing conflicting conclusions.[37] Finite element modeling analyses suggest that the presence of a water layer may enhance microwave coupling, thereby improving both lateral resolution and signal-to-noise ratio.[37] Experimental results from our laboratory, obtained using an sMIM system capable of operating in both air and high vacuum, demonstrate an increased dynamic range when measurements are conducted under vacuum conditions. While a residual water layer is likely still present, its reduced thickness may favorably alter the coupling geometry. Interestingly, partial water re-adsorption following re-exposure to ambient conditions has also been observed to enhance further dynamic range, an effect that persists for several hours after venting. These observations highlight the complex and still poorly understood role of water in governing microwave channeling and its impact on lateral resolution and dynamic range. Finally, the lateral resolution of sMIM remains inferior to that of other dopant profiling techniques, such as scanning spreading resistance microscopy and scanning capacitance microscopy.[50] While improvements have been demonstrated using custom-fabricated ultrasharp or hydrophobic probes, a standardized commercial solution is still lacking, limiting the routine resolution performance of sMIM imaging.[22]

10.2.6 *Future Developments*

Although precise nanofabrication using the sMIM platform has not yet been experimentally demonstrated, several promising directions exist that warrant further development and integration into standard AFM systems. As a relatively recent addition to the scanning probe microscopy family, sMIM remains an emerging technique whose operational mechanisms and full capabilities are still being refined. Consequently, multiple advancements are required before it can evolve into a practical nanofabrication tool comparable to AFM-based nanolithography. At present, the detailed interaction between the microwave field and the sample, as well as the exact spatial extent of the effective interaction volume, remains insufficiently understood. Addressing these limitations will require not only deeper investigations into the contrast-formation mechanisms and fundamental working principles of sMIM but also significant hardware improvements — particularly the development of ultrasharp, sMIM-compatible probes capable of maintaining high spatial fidelity. Among the existing challenges, the limited lateral resolution represents perhaps the most significant bottleneck. Although the theoretical imaging potential of sMIM surpasses that of well-established electrical scanning probe microscopy modes such as scanning capacitance microscopy and scanning spreading resistance microscopy, its current inability to resolve buried nanostructures below a few nm from the surface limits the applicability in advanced semiconductor device characterization. In parallel, further exploration of microwave-assisted fabrication within sMIM frameworks should be pursued. Controlled, localized microwave delivery could enable a variety of high-precision nanofabrication processes, bridging the gap between nanoscale imaging and direct material modification. Such advancements would position sMIM as not merely a diagnostic instrument, but a multifunctional platform for both characterization and active nanoscale fabrication.

10.3 Microwave-Assisted Magnetometry Using Color Centers

10.3.1 *Scanning Nitrogen Vacancy Magnetometry*

Among microwave-excited nanoscale measurement techniques, scanning nitrogen vacancy magnetometry (SNVM) is taking center stage in being critical to nanofabrication flows and identifying defects for several device workflows, especially devices that involve magnetic fields. It was realized that the sensitivity of magnetic sensing can be enhanced by the transfer of the microwave signal to the optical domain, which is realized in the ODMR method.[52] ODMR is a powerful spectroscopic technique that bridges optical excitation with microwave-driven spin resonance to probe general properties that affect the spin of electrons, i.e., magnetic properties at the nanoscale, electric fields, or even temperature. In ODMR, a material containing spin-active defects or paramagnetic centers is illuminated with laser light to generate photoluminescence, while simultaneously being subjected to a microwave field that drives transitions between spin sublevels. When the microwave frequency matches the energy difference between these spin states, resonant spin transitions occur, altering the optical emission intensity. This change in photoluminescence serves as a highly sensitive, non-contact probe of the local magnetic environment.

Unlike conventional electron spin resonance, which detects microwave absorption directly, ODMR translates spin dynamics into optical signals, enabling operation with extremely small ensembles or even single spin centers. Because the optical readout is spatially localized, ODMR allows for nanoscale mapping of magnetic fields, spin populations, and energy-level structures with sub-micron spatial precision. It can be performed in both continuous-wave and pulsed modes, where pulsed ODMR further enables time-resolved and coherent spin manipulation studies. Owing to its non-invasive nature, high sensitivity, and compatibility with room temperature operation, ODMR has become a cornerstone of modern quantum sensing and nanoscale magnetometry, underpinning the development of next-generation spin-based imaging and metrology platforms.

While diamonds contain many defects like Ni, Si, Cr, etc., color centers like the nitrogen vacancy (NV) center are of special interest for their quantum sensing properties. NV centers in diamond show excellent spin coherence times that exceed 50 μs at room temperature, along with photostable photon emission and ambient temperature operation.[53] An NV center consists of a substitutional nitrogen and an adjacent lattice vacancy defect in a diamond crystal. A negatively charged NV (NV^-) has six electrons, with two electrons from nitrogen, one electron from each of the three carbon atoms, and an additional electron from the lattice. These electrons occupy four sp^3 atomic orbitals with electronic spin quantum number $S = 1$. These sp^3 orbitals linearly combine to form four molecular orbitals,[54] comprising the ground electronic configuration. The ground electronic configuration of the NV^- center has a lowest-energy state described as an orbital singlet, spin triplet state $^3A_2$, which exhibits characteristic fine structure (Zeeman splitting). The four molecular orbitals also give rise to electronic excited states: the orbital-singlet spin-triplet ground state $^3A_2$, orbital-doublet spin-triplet excited state ^3E, orbital-doublet spin-singlet ^1E, and orbital-singlet spin-singlet $^1A_1$. The spin-triplet ground state ($^3A_2$) has a zero-field splitting (ZFS) of $D \approx 2.87$ GHz between the $|0\rangle$ and $|\pm1\rangle$ spin sublevels.[55] Excited state ^3E is coupled to the $^3A_2$ ground state by an optical 637-nm zero phonon line (ZPL). The $^3E \leftrightarrow {}^3A_2$ is a radiative transition that generally conserves the electron spin state m_s as a result of weak spin-orbit interaction.[56] The optical transitions $^3E \rightarrow {}^3A_2$ ($^3A_2 \rightarrow {}^3E$) are responsible for both absorption and fluorescence (photoluminescence)[57] shown in Figure 10.5.

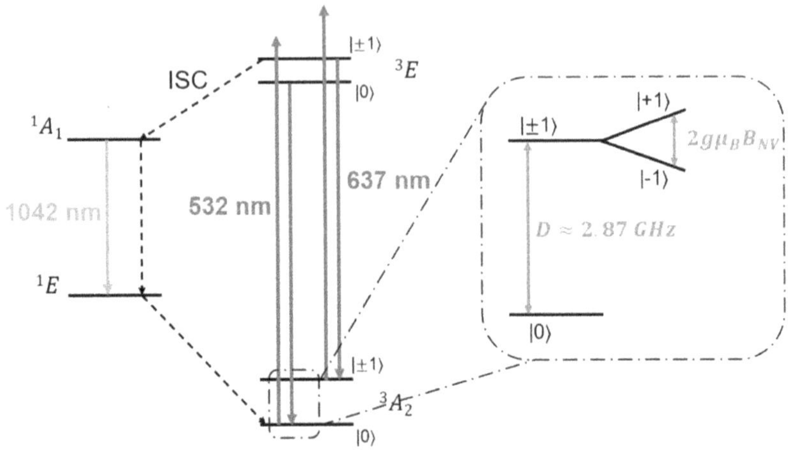

Figure 10.5. Energy level scheme of a typical SNVM operation. The inset shows a closer look at the ground-level spin triplet states.

SNVM is a quantum sensing technique that combines scanning probe microscopy with the quantum spin properties of NV centers in diamond to achieve non-invasive, nanoscale magnetic field imaging under ambient or cryogenic conditions. The NV center functions as a solid-state quantum system whose ground-state spin triplet can be coherently driven by resonant microwave fields in the gigahertz range (~2.87 GHz).[58] Under continuous green laser illumination (typically 532 nm), the NV center is optically excited and polarized into the spin state $|0\rangle$. When microwave radiation is applied at frequencies resonant with the $|0\rangle \leftrightarrow |\pm1\rangle$ spin transitions, it drives coherent population transfer between these sublevels. The resulting changes in spin populations are observed as variations in the photoluminescence intensity, a phenomenon known as ODMR, as shown in Figure 10.6.

The presence of an external magnetic field lifts the degeneracy of the $|\pm1\rangle$ states via the Zeeman effect, producing a frequency

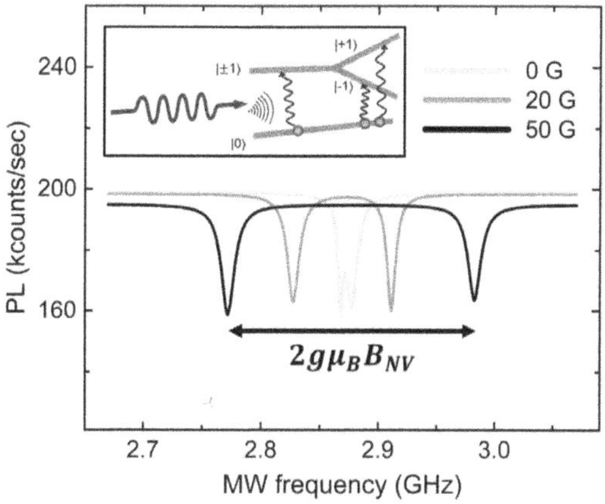

Figure 10.6. ODMR spectrum for variable external magnetic fields. The inset shows the effect of the microwave signal for the transition of electrons from $|0\rangle$ to $|\pm1\rangle$ states as well as the separation of $|\pm1\rangle$ states due to external magnetic fields. The gray spheres denote electrons.

splitting proportional to the magnetic field projection along the NV symmetry axis (splitting = $2g\mu_B B_{NV}$, where constants g and μ_B are the Landé g-factor and the Bohr magneton, respectively, and B_{NV} is the projection of the externally applied magnetic field to the NV defect symmetry axis). This relationship forms the basis for quantitative magnetic field mapping at the nanoscale.[58] Precise control of the microwave frequency, amplitude, and pulse duration is central to SNVM operation, as these parameters define the fidelity of spin-state manipulation and the accuracy of magnetic field readout.

10.3.2 *Modes of Operation*

There are 3 different measurement modes for SNVM, namely, full-B mode, iso-B mode, and dual iso-B mode, as shown in Figure 10.7.

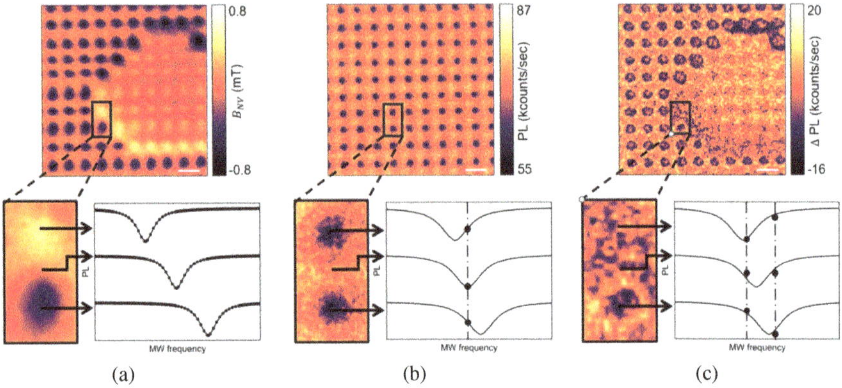

Figure 10.7. Different measurement modes of SNVM are explained on magnetic random access memory bits.[61] (a) Full-B mode. (b) Iso-B mode. (c) Dual iso-B mode. Each mode contains a large area scan of multiple magnetic random access memory bits (top panels). On the bottom left panels, two contrasting bits are shown along with the ODMR responses of the $|+1\rangle$ ground spin state displayed on the bottom right panels. Top panels scale bars, 250 nm.

In full-B mode, at each pixel a full ODMR spectrum is recorded but spanning across a range of microwave frequencies, thereby collecting the resonant frequency. The data is displayed as a map of the value of the magnetic field for each pixel (Figure 10.7(a)). While this mode is critical for the positive to negative magnetic field acquisition of the sample, the data acquisition interval is long.

The iso-B mode reduces the measurement window to a single microwave frequency. The image generated is a map of fluorescence corresponding to a single microwave frequency. The advantage of this mode is faster (~100×) data collection. The iso-B mode is conceptually like the quenching mode, as both produce qualitative magnetic field maps based on fluorescence contrast (Figure 10.7(b)); however, unlike the quenching mode, iso-B imaging requires the application of a finite microwave drive.

Finally, the dual iso-B mode is used if a sample has pronounced topography; dual iso-B mode resembles the full-B mode data but only takes twice as long as the iso-B mode. This mode records the fluorescence differences at two distinct frequency points (Figure 10.7(c)). This approach makes it insensitive to changes in the fluorescence baseline.[59,60]

10.3.3 *Microwave Pulsing Sequences*

Continuous wave microwave excitation allows rapid mapping of magnetic textures. For pulsed microwave sequences, a number of established sensing protocols exist for measuring static (DC), slowly varying, and AC magnetic fields,[62] as shown in Table 10.1.

10.3.3.1 *Continuous Wave ODMR*

Continuous wave ODMR is a robust and simple method that can image the vector components of a magnetic field. Because of its easy implementation, continuous wave ODMR is the

Table 10.1. Optical excitation, readout, and microwave pulsing sequences for the measurement of DC, slowly varying, and AC magnetic fields.

	CW ODMR	Pulsed ODMR	Ramsey	Hahn echo	Dynamical decoupling	Rabi	T_1 relaxometry
Laser							
MW		π	$\pi/2$ $\pi/2$	$\pi/2$ π $\pi/2$	$\pi/2$ $n\pi$ $\pi/2$		π
Readout							
Bias field							
Sample B-field							
Swept parameter	MW frequency	MW frequency	Free precession interval, τ	Spin evolution time, τ	Spin evolution time, τ	MW pulse duration, bias field	Laser pulse duration, bias field

most common technique used for quantum diamond magnetometry applications. Here, continuous laser pumping, microwave driving, and fluorescence readout are typically performed simultaneously.

10.3.3.2 *Ramsey Sequence*

The purpose of this sequence is to collect time-domain phase evolution for DC magnetic field sensing. To conceptualize the microwave pulsing sequence of this technique, we need to visualize the NV center's spin state on the Bloch sphere, which is a geometrical representation of a two-level quantum system, demonstrated in Figure 10.8. A green laser pulse initializes the spin state into the $m_s = 0$ state to begin the sequence. Then there are two $\pi/2$ microwave pulses that are separated by a free precession interval τ during which the NV spin precesses at a frequency determined by the local magnetic field,[62,63] as demonstrated in Table 10.1.

To obtain the magnetic field, a Ramsey pulse sequence is repeated several times, sweeping τ such that each measurement is taken for different τ values. The resultant fluorescence contrast

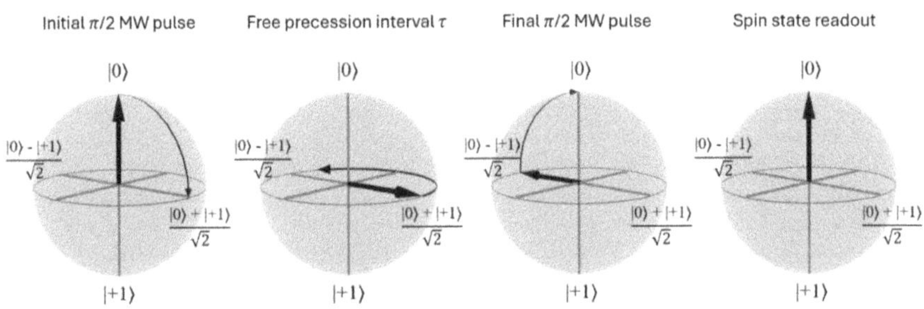

Figure 10.8. The electron spin state is represented as a vector on a Bloch sphere for the Ramsey sequence, with eigenstates pointing in the direction of the poles. Reproduced with permission.[64]

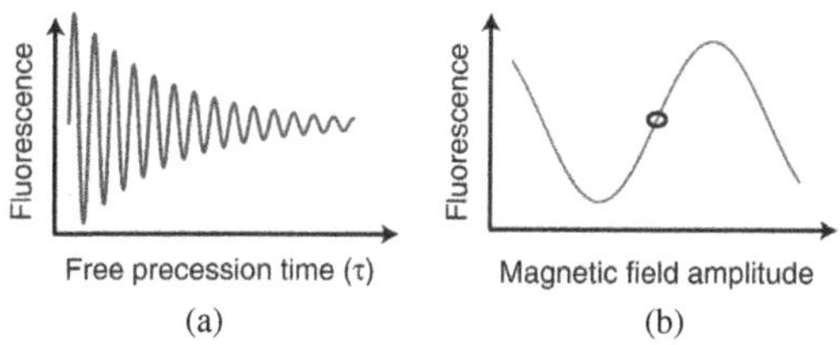

Figure 10.9. (a) Ramsey free induction decay analysis to investigate optimal dephasing time (T_2^*) and sensing time (τ_{sense}). (b) Ramsey magnetometry curve to optimize the sensitivity of the sensed signal. The black circle indicates the free precession time to be fixed to the point of maximum slope of the free induction decay curve closest to T_2^*. The steeper the slope, the more sensitive the scheme. Adapted under open access.[57]

signal as a function of τ is known as the Ramsey fringes, illustrated in Figure 10.9.[57]

10.3.3.3 *Pulsed Optically Detected Magnetic Resonance*

Continuous laser illumination during microwave excitation reduces measurement sensitivity because the simultaneous processes of optical spin initialization and microwave-driven transitions interfere with one another, leading to power broadening of the resonance. To overcome this limitation, pulsed ODMR separates the steps in time: a laser pulse first initializes the NV spin state, followed by a controlled microwave π pulse for spin manipulation, and a subsequent laser pulse for optical readout, as described in Table 10.1. Unlike Ramsey magnetometry, pulsed ODMR is linearly sensitive to spatial and temporal variations in the Rabi frequency, that is, the rate at which the spin coherently oscillates between its two energy states under resonant microwave excitation. When these Rabi frequency variations are small, pulsed ODMR can achieve sensitivities comparable to Ramsey magnetometry, without requiring very high microwave power.[64] When Rabi frequency variations are large, pulsed ODMR becomes less sensitive and less quantitative because the microwave pulses no longer drive consistent spin rotations across the sensing region. This typically occurs near non-uniform microwave fields, metallic or topographic features, or with multi-NV probes, where the coupling to the microwave field varies spatially. Under such conditions, Ramsey magnetometry (which depends on phase, not pulse amplitude) provides more reliable and uniform sensitivity.[64,65]

10.3.3.4 *Hahn-echo Sequence*

For AC magnetic field sensitivity improvements, the Hahn-echo sequence is implemented to extend the decoherence time (T_2) and enhance sensitivity.[63,66–67] The addition of a π pulse into the middle of a Ramsey sequence mitigates environmental perturbations that are slow compared to the free precession interval between pulses, as shown in Table 10.1. The spacing between microwave pulses acts as a narrowband filter in frequency space. The width of this filter is given by the filter response function. A Bloch sphere depiction of the Hahn-echo sequence during the free precession intervals and the effect of the π microwave pulse is shown between the intervals in Figure 10.10.

10.3.3.5 *Dynamical Decoupling Sequences*

Dynamic decoupling sequences[68] have been commonly implemented for extended coherence times and enhancement of magnetic sensitivity. These techniques commonly apply multiple refocusing pulses with spacing determined by the period of the sample field.[69] A general dynamic decoupling pulsing scheme

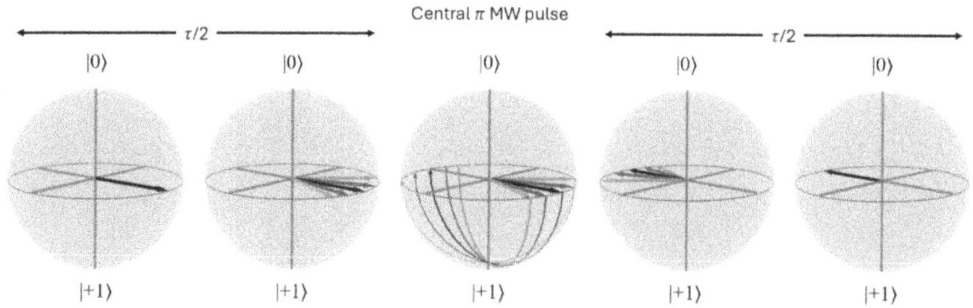

Figure 10.10. Bloch sphere depiction of the Hahn-echo sequence. Reproduced with permission.[64]

is demonstrated in Table 10.1. The magnetic field sensitivity (η) is primarily limited by photon shot noise, spin dephasing (T_2^*), and electron spin resonance contrast, typically reaching values down to a few $\mu T \cdot Hz^{-1/2}$ for single NV centers.[70]

10.3.3.6 *Carr-Purcell-Meiboom-Gill Pulse Sequence*

The Carr-Purcell-Meiboom-Gill (CPMG) pulse sequence is a well-established and commonly implemented dynamical decoupling sequence employing π pulses, which rotate the NV about the same axis as is polarized by the initial $\pi/2$ microwave pulse.[71–72] A comparative analysis of the Hahn-echo scheme and the CPMG scheme is demonstrated in Figure 10.11, where it is clearly visible that the CPMG pulsing sequence improves magnetic field acquisition sensitivity by optimizing the decoherence time T_2.

10.3.3.7 *Rabi Sequence*

Rabi sequence can determine local magnetic fields oscillating at or near GHz.[73] A microwave drive is left on for a varying amount of time. If the microwave drive is in resonance with the NV spin

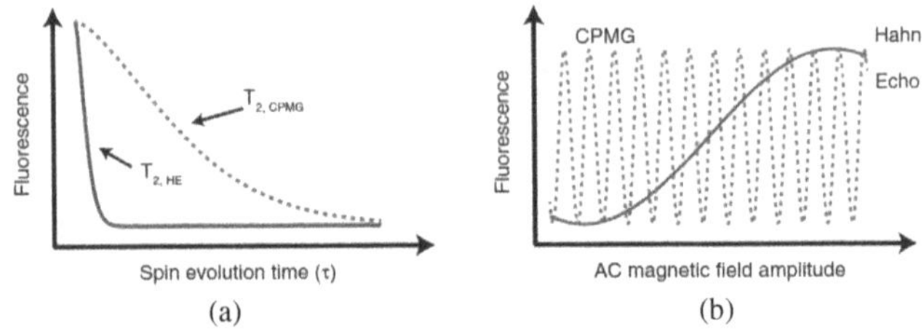

Figure 10.11. (a) T_2 decoherence curves for the Hahn-echo sequence and the CPMG-32 sequence demonstrate extended T_2 for the CPMG sequence compared to the Hahn-echo. (b) Magnetometry curves indicate that extended T_2 for CPMG sequence results in increased magnetic field sensitivity, as evident from the slope of the CPMG magnetometry curve. Adapted under open access.[57]

state transitions, for example, between $m_s = 0$ and $m_s = \pm 1$ states, the population will be driven back and forth between the spin states. The strength of this GHz drive determines the rate at which the transition is driven. This rate is called Rabi frequency, and scales with the square root of the input microwave power.

10.3.3.8 T_1 *Relaxometry*

T_1 is the spin lattice relaxation time, which quantifies how quickly the NV center's spin returns to thermal equilibrium after being polarized (initialized) into its ground spin state. Therefore, T_1 relaxometry provides a direct, non-invasive probe of local magnetic noise density. A comparative analysis of continuous wave ODMR, Rabi sequence, and T_1 relaxometry is presented in Figure 10.12.

Thus, microwave pulsing enables coherent control of the NV spin, allowing selective probing of static and dynamic magnetic fields. By tailoring pulse duration and phase, the NV center's spin states can be manipulated to encode magnetic, electric,

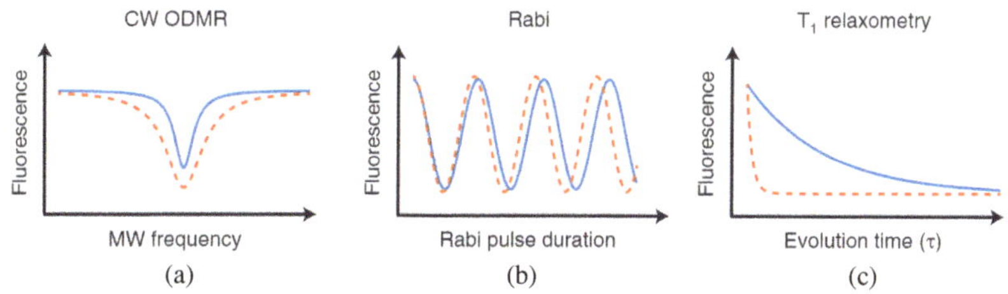

Figure 10.12. (a) Continuous wave ODMR before (blue) and after (red) microwave power increase: increased microwave power will increase both ODMR fluorescence contrast and linewidth. (b) Rabi oscillation before (blue) and after (red) microwave field amplitude increase: increased amplitude of the microwave field will increase the Rabi frequency. (c) T_1 relaxometry before (blue) and after (red) the enhancement of magnetic noise: phonon-limited T_1 decay rate is increased by high-frequency magnetic noise near the NV resonance frequency. Adapted under open access.[57]

or thermal information into measurable photoluminescence contrast. Optimized pulse design enhances sensitivity, extends coherence times, and filters environmental noise, linking quantum spin dynamics to practical nanoscale measurements. Thus, microwave pulsing provides the essential control interface that transforms NV centers into versatile quantum sensors for high-resolution magnetometry.

10.3.4 *Microwave Delivery of Scanning Nitrogen Vacancy Magnetometry*

For SNVM, the microwave signal is typically delivered using a miniature loop antenna or coil (typically 10–100 μm diameter) or micro-coil placed near the NV probe, as depicted in Figure 10.13(a). This is among the most common configurations in commercial and home-built SNVM systems. The popularity of this configuration arises from the fact that, when sample scanning

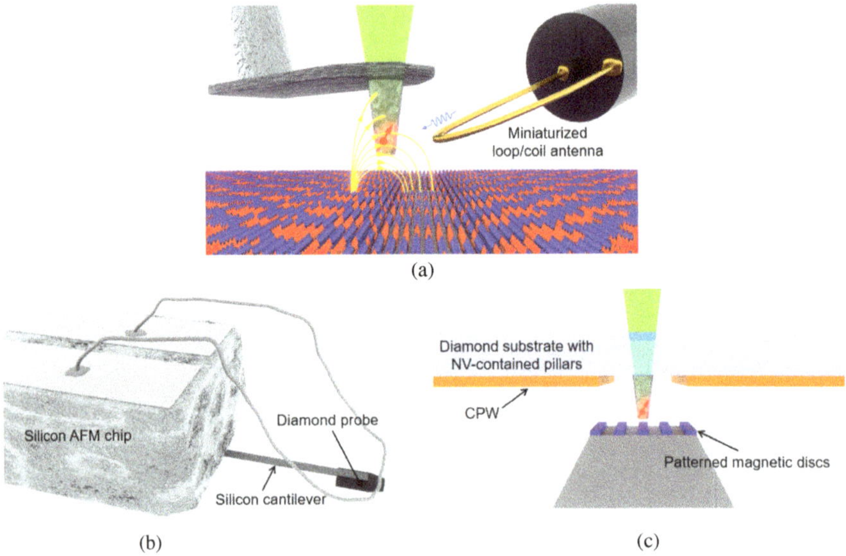

Figure 10.13. Microwave delivery configurations for SNVM. (a) Microwave delivery using a loop/coil antenna. Here, the antenna and the NV center are ensured to be equidistant from each other throughout the scan, while both are different entities. (b) Microwave delivery through a loop antenna mounted directly on the silicon AFM chip that is holding the diamond probe attached to a silicon cantilever. Reproduced with permission.[77] (c) Microwave delivery through CPW, which is evaporated on the bottom surface of a diamond substrate. Reproduced with permission.[78]

is implemented via stage translation, the microwave field at the NV center remains spatially invariant during imaging, ensuring stable excitation conditions. This configuration is generally simple to fabricate, flexible, and efficient at field coupling at close distances;[74] some variation of this includes a metal stripline located near the probe.[75,76] Moreover, an integrated radiofrequency micro-antenna for NV center spin manipulation can be fabricated onto a Si cantilever, making the design versatile and compatible with virtually all AFM instruments,[77] demonstrated in Figure 10.13(b). Furthermore, a coplanar waveguide (CPW) may be fabricated on the surface of the diamond probe, with the apex region left uncoated, to achieve improved proximity and uniform and consistent coupling between the applied microwave field and the NV center,[78] shown in Figure 10.13(c).

10.3.5 *Characterization of Nanofabricated Features with Scanning Nitrogen Vacancy Magnetometry*

SNVM operates in ambient conditions, enabling feature resolution up to tens of nanometers, which makes it a crucial technique to be integrated into nanofabrication stages for closed-loop process control and device validation. While SNVM is not a direct nanofabrication technique, it has emerged as an essential quantum metrology platform integrated into the nanofabrication cycle for spintronic, magnonic, memory, and quantum device development. By combining atomic-scale magnetic sensitivity with nanoscale spatial resolution, SNVM enables post-growth and post-patterning characterization of thin films and nanostructures fabricated through methods such as pulsed laser deposition, sputtering, focused ion beam milling, and electron beam lithography. For example, pulsed laser deposition -grown Cr_2O_3 antiferromagnetic films, relevant to voltage-controlled spintronic memory and magnetoelectric logic devices, have been imaged using SNVM to reveal grain-dependent domain formation and local variations in exchange coupling, directly linking growth conditions to device reliability.[79] Magnetic skyrmions are nanoscale, vortex-like spin configurations that form in certain magnetic materials due to competing magnetic interactions. Generally, they are stabilized in thin film heterostructures by exchange interactions, magnetic anisotropy, and Dzyaloshinskii–Moriya interactions. Their nanometric size, stability, and controllable motion make them promising for spintronic memory and logic. Dovzhenko et al.[78] used single NV imaging to reconstruct the full spin structure of room-temperature skyrmions in sputtered Pt/Co/Ta multilayers, revealing internal Néel-to-Bloch helicity transitions linked to growth-induced asymmetry. Subsequently, Rana et al.[80] visualized stable, zero-field skyrmions (~60 nm) in exchange-biased Pt/Co/NiFe/IrMn stacks fabricated by magnetron sputtering, confirming interfacial tuning of skyrmion stability. Together, these works position SNVM as a key feedback tool connecting thin film growth and interfacial design to functional skyrmion behavior, accelerating the scalable fabrication of skyrmion-based spintronic devices.

Sputtered CoFeB and NiFe thin films, common in magnetic random access memory and domain wall logic devices, have been investigated to visualize domain wall pinning and fabrication-induced inhomogeneities at nanometer scales.[81,82] SNVM's quantitative imaging of out-of-plane magnetic fields in spintronic nanostructures such as CoFeB underscores the mutual interdependence between NV device fabrication and nanoscale metrology.[83] Similarly, focused ion beam and lithographically defined yttrium iron garnet nanostructures, which form the foundation for magnonic and spin wave-based computing architectures, have been analyzed by SNVM to spatially resolve frequency-degenerate spin-wave modes and thermally excited magnon populations.[84] Moreover, NV-based scanning probes have mapped stray magnetic fields in superconducting and topological thin films, aiding the optimization of quantum sensing and cryogenic hybrid device platforms.[85] Collectively, these studies establish SNVM as a non-destructive, feedback-driven metrology technique that bridges material synthesis, nanofabrication, and device-level functionality, providing quantum-level insight crucial for the engineering of next-generation micro- and nanoelectronic systems. Beyond magnetic materials, scanning NV sensing has also been employed to image ferroelectric domains and local polarization fields in epitaxially fabricated piezoelectric ($Pb[Zr_{0.2}Ti_{0.8}]O_3$) and improper ferroelectric ($YMnO_3$) materials through their electric fields for its potential as a ferroelectric, piezoelectric, and magnetoelectric devices.

10.3.6 *Spin Defect Hosts Beyond Diamond*

Although diamond as a host material for NV color centers is well studied and widely employed, there are non-negligible impediments to the proper utilization of diamond as an optimal quantum sensor. Difficulties in scalable and high-quality diamond film growth, costly and complex growth techniques, obscurity in chip-scale integration, and incompatibility with other materials are notable.[86] To address some of these challenges, researchers have been exploring alternative platforms, and SiC is one that is of particular interest due to comparable coherence characteristics of its quantum defects with diamond NV centers, enhanced photoluminescence emission capabilities, and simpler fabrication mechanisms. Among various types of defects in SiC, the most investigated ones for quantum sensing applications are the negatively charged silicon vacancy (V_{Si}^-) (Figure 10.14(a)) and the neutral divacancy ($V_{Si}V_C^0$).[86] There are also some discussions on other defects in SiC which demonstrate quantum sensing capabilities, among which the negatively charged nitrogen vacancy ($N_CV_{Si}^-$) defect is noteworthy. The ZPLs

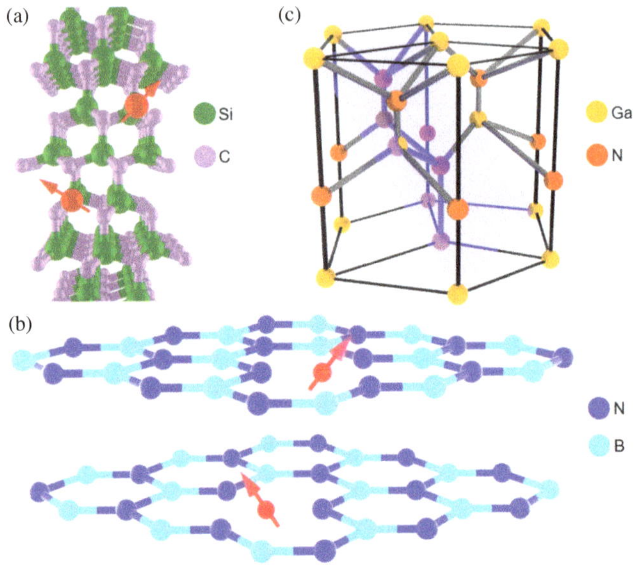

Figure 10.14. Spin defects beyond NV centers in diamond. (a) Negatively charged silicon vacancy defects in 4H-SiC crystals. Reproduced with permission.[87] The red entities each identify as a single defect. (b) Boron vacancy defects in honeycomb lattice structures of hBN single layers. Reproduced with permission.[86] The red entities each indicate a single defect consisting of a missing boron atom surrounded by three nitrogen atoms. (c) Crystal structure of wurtzite GaN, where the blue-shaded plane represents the sample surface containing single photon emitter defects. Reproduced with permission.[88] Images do not indicate scale or proportions.

for optical transition between ground and excited states and the room temperature axial ground-state ZFS parameters (D) for spin state control of these defects using microwave signals are summarized in Table 10.2.

The 3D nature of the host diamond and SiC materials possesses some undesirable impacts on the spatial resolution and sensitivity of the measured entity. It is challenging to place spin defects close to the surface of the hosts due to the dangling bonds

Table 10.2. ZPL and ZFS parameters of spin defect color centers in various host materials.

Host Material	Quantum Defect	ZPL	ZFS (D)
SiC	V_{Si}^-	862 nm – 917 nm[96,97]	2.25 MHz – 64 MHz[87,97,98]
	$V_{Si}V_C^0$	1007 nm – 1140 nm[99–101]	1.22 GHz – 1.42 GHz[99,100,102]
	$N_CV_{Si}^-$	1155 nm – 1289 nm[103–105]	1.23 GHz – 1.37 GHz[104–106]
hBN	V_B^-	773 nm – 820 nm[90,107]	3.48 GHz[108]
	$V_NC_B^-$	500 nm – 800 nm[109–111]	0.4 GHz – 1.1 GHz[109–111]
GaN	Single photon emitters	603 nm – 736 nm[94,95,112]	200 MHz – 389 MHz[112]

on the surface, resulting in a non-trivial sensor-sample offset. Additionally, spin defect sensors that are placed too close to the sample under test often demonstrate inferior spin coherence properties, resulting in reduced sensitivity.[89] A promising candidate to address these problems is the van der Waals materials, as they are intrinsically stable down to the monolayer limit.[86] One of the most studied van der Waals materials containing optically stable quantum sensors is hBN.[90,91] hBN crystals contain negatively charged spin defects such as boron vacancy centers (V_B^-) (Figure 10.14(b)) that demonstrate robust ODMR signals over a wide range of temperature.[92] hBN can be exfoliated down to a few atomic layers and can be placed on the sample under test, ensuring sensor-sample separation of only a few atoms without the consideration of dangling bonds on the surface. Moreover, similar to other van der Waals materials, hBN can be integrated into versatile heterostructures, opening huge potential for unexploited quantum sensing applications. Such a demonstration of V_B^- as active sensors in hBN was first presented by Gottscholl et al.,[92] where the authors imaged temperature and current density of an operational two-terminal graphene device. Apart from V_B^- defects, carbon-related defects in hBN are also gaining interest due to stable thermal and optical characteristics. These defects are also known as negatively charged nitrogen vacancies with a neighboring carbon substitutional atom ($V_N C_B^-$). The ZPL and ZFS parameters of these hBN defects at room temperature are summarized in Table 10.2.

Among the most recent investigations to find new entities for quantum sensing applications, spin defects in GaN are at the forefront of discussions. GaN has a wide direct bandgap and high breakdown field, making it suitable for power electronics applications.[93] Moreover, the capability of GaN to emit light over the entire visible spectrum and excellent photophysical properties make it more appealing for novel quantum sensing schemes.[94] Basically, GaN films contain bright, photostable, and sharp photoluminescence generation capable single photon emitters as defects (Figure 10.14(c)).[95] The most common ZPL and ZFS parameters of these GaN defects at room temperature are summarized in Table 10.2.

10.3.7 *Microwave Antenna Optimization*

Sensitivity of the measured fields can be significantly improved by implementing NV ensembles instead of a single NV center placed on a probe in scanning mode.[113] With the expense of spatial resolution, NV ensembles can also speed up measurements and reduce experimental complexity, adding to enhanced sensitivity.[114] In this scheme, the most common procedure is to place a thin layer containing multiple NV centers on or under a diamond slab and place the desired sample to be measured on or under the slab, depending on the placement of the NV center layer. The photoluminescence signals from the NV centers are then imaged in a wide-field detection scheme, enabling the collection of the entire region of interest at once on a camera instead of scanning point by point. In other words, the electron spin resonance spectrum on each pixel of the image is acquired by sweeping the microwave signal and fitting the ODMR data per pixel.

To optimize the sensitivity, an important step is to improve spin dephasing and coherence times of NV centers, which heavily depend on the configurations of the microwave antenna and associated pulsing signals. For scanning configuration with a single NV center-contained probe, placing a microwave antenna is relatively simpler as shown in Figure 10.13(a). A small single loop of metal wire (preferably Cu or Au) is placed within a few tens of micrometers to the probe (NV center), and by making sure that the antenna is not blocking the photoluminescence emission from the NV center, it is possible to obtain optimal ODMR contrast. However, for wide-field imaging using NV ensembles, there is a trade-off between the homogenous microwave intensity observed by the NV centers and the overall microwave intensity. Consequently, a more specialized and engineered microwave antenna is required to ensure uniform intensity over the region of interest and an optimized signal. Additionally, as mentioned in Section 3.6 and Table 10.2, color centers in different host materials require distinct ZFS parameters to quantify measured entities, resulting in the requirement of the generation capability of precise and wide-band microwave signals. In recent years, there have been a lot of discussions about the improvement of antenna design from the perspective of its geometry, material, placement near the color center ensembles, and various pulsing schemes to improve radiation performance, as these parameters directly impact the precision and efficiency of quantum sensing systems.[115]

10.3.7.1 *Single Straight Wire-Based Antenna*

A single straight wire (mostly Cu) is the simplest form of microwave signal delivery scheme for quantum sensing,[117] as shown in Figure 10.15(a). However, the radiated signal is highly

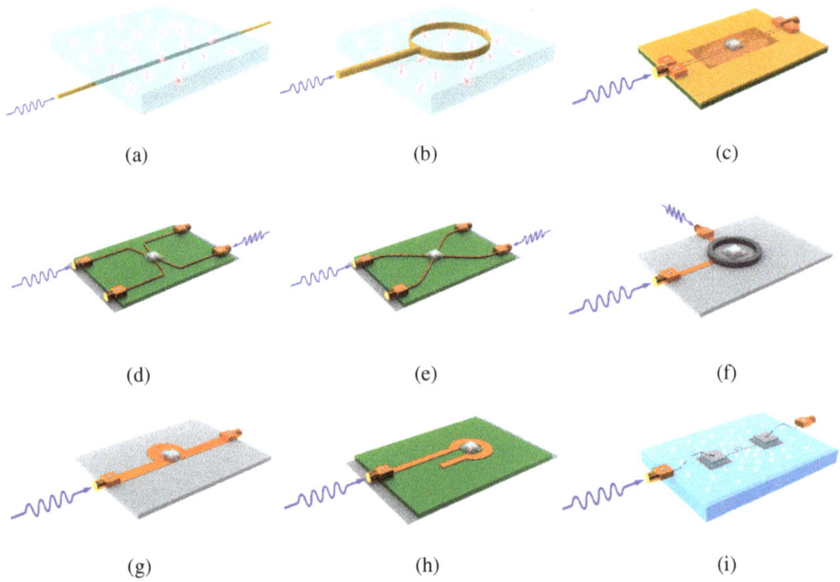

Figure 10.15. Microwave signal delivery schemes for the manipulation of color center ensembles. (a) Single straight wire-based antenna. Reproduced with permission.[116] (b) Loop/coil antenna. Reproduced with permission.[124] (c) Broadband CPW antenna. Reproduced with permission.[124] (d,e) Circularly polarized microstrip antenna. Reproduced with permission.[126,128] (f) Dielectric resonator antenna. Reproduced with permission.[129] (g) Stripline-based, planar antenna. Reproduced with permission.[131] (h) Microstrip patch loop antenna. Reproduced with permission.[132] (i) Diamond-embedded microwave antenna. Reproduced with permission.[133]

inhomogeneous, and the transmission loss of the signal increases noticeably with increasing frequency.[117] Due to these phenomena, a single straight wire-based antenna is not applicable in color center ensemble experiments and is limited to single color center-based sensing.

10.3.7.2 *Loop/Coil Antenna*

The loop/coil antenna is probably the most common microwave signal generation medium for color center ensemble-based applications. Consequently, a number of designs have been demonstrated in the past decade, among which the microwave planar ring antenna,[117] double split-ring microwave resonator-based antenna,[119] helical antenna,[120] and coaxial resonant coil microwave antenna[121] are worth mentioning. More recently, color center ensembles have been employed as quantum sensors for optically detecting nuclear magnetic resonance signals in a wide-field imaging scheme.[122–124] In these experiments, a microwave loop antenna placed on diamond chips containing color centers is implemented to manipulate their spin states. Such an example schematic is demonstrated in Figure 10.15(b). Despite its simplicity and widely acceptable use cases, the loop/coil antenna suffers from low microwave signal intensity and difficulty in tuning the frequency over a wide range.

10.3.7.3 *Broadband Coplanar Waveguide Antenna*

A CPW-based antenna can overcome the limitations of single straight copper wire-based antennas and loop/coil antennas. However, although traditional CPW antennas can achieve higher bandwidth signals, they are susceptible to environmental disturbances and low power-to signal conversion efficiency.

To overcome these limitations, a broadband CPW-based antenna has been demonstrated[125] with a transmitted signal bandwidth up to 15.8 GHz and high power-to- signal conversion efficiency, as shown in Figure 10.15(c). Moreover, in this structure, it is also possible to place the objective lens very close to the device under test, which is not always possible with other antenna designs.

10.3.7.4 *Circularly Polarized Microstrip Antenna*

In most nuclear magnetic resonance and electron spin resonance applications, linearly polarized microwave signals are employed. To increase the effective strength and homogeneity of the signals, circularly polarized signals are good candidates for quantum sensing. Such a microstrip-based antenna has been demonstrated[126] and schematically shown in Figure 10.15(d), where the microwave field frequency is estimated to be in the range of 3 GHz with less power requirements compared to the case of linearly polarized signals.

Circular polarization has also been demonstrated on a two-crossed microstrip resonator-based antenna,[127,128] as shown in Figure 10.15(e). These experiments explain the procedure of the identification of color centers with different orientations in a quicker way, and understand the selective excitation of color center spin states through circular polarization.

10.3.7.5 *Dielectric Resonator Antenna*

A circularly polarized microwave antenna design with the capability of homogeneous magnetic field generation in a high-permittivity dielectric resonator has been demonstrated.[129] To obtain a circular polarization, the design contains two orthogonal $HEM_{11\delta}$ modes, which are excited with signals of the same amplitude and a 90° phase shift in a hollow cylindrical dielectric resonator, as shown in Figure 10.15(f). Although resonator-based antenna designs provide limited bandwidth options, this design is capable of generating uniform microwave signals throughout a large area, making it a decent option for effective color center ensembles spin manipulation. Another S-band tunable loop gap resonator antenna design has been described[130] which is capable of generating strong, homogeneous microwave signals with broadband capabilities up to 80 MHz.

10.3.7.6 *Stripline-Based, Planar Microwave Antenna*

In this stripline-based design (Figure 10.15(g)), a planar Ω-shaped antenna is fabricated on a borosilicate glass substrate with a single crystalline diamond containing color center ensembles to obtain the electron spin resonance spectrum.[131] The numerical and geometrical optimization of the design was performed using a time domain solver. This design seems to outperform the traditional straight-wire antennas, regular stripline antennas, coil antennas, and resonators by producing a high bandwidth and field amplitude microwave signal with expected resonance properties suitable for color center ensemble-based sensing applications. Also, the authors mentioned the competent nature of this antenna design with various pulsed ODMR schemes and complex dynamic decoupling schemes, making it more versatile than the regular antennas.

10.3.7.7 *Microstrip Patch Loop Antenna*

A microstrip technology-based patch loop antenna is a common form of microwave signal generation to meet radiation requirements for the color center ensembles.[132] The authors investigated the antenna design through electromagnetic simulations and analyzed its radiation performance by optimizing antenna geometric parameters to meet the microwave signal uniformity requirements of the color centers' spin resonance (Figure 10.15(h)).

10.3.7.8 *Diamond-Embedded Microwave Antenna*

To ensure localized, efficient, stable, and homogenous microwave fields at color center ensembles, an innovative diamond-embedded Ω-shaped microwave antenna design procedure has been described.[133] Using nanofabrication and microwave plasma chemical vapor deposition technique, the authors incorporated an Ω-shaped metal antenna embedded in a single crystal diamond with color centers, as shown in Figure 10.15(i). This microwave antenna structure offers some distinct advantages as well. It ensures unaffected microwave operation due to the containment of the antenna in the diamond, irrespective of surface operations. The diamond layer also acts as a protective layer for the antenna, establishing a more robust structure for various experiments. Moreover, the diamond-embedded antenna allows for closer proximity to the device under test, ensuring improved sensing efficiency and weak field detection capability.[133]

References

1. Lai, K., M. B. Ji, N. Leindecker, M. A. Kelly, and Z. X. Shen. "Atomic-force-microscope-compatible near-field scanning microwave microscope with separated excitation and sensing probes." *Review of Scientific Instruments* 78, no. 6 (2007): 063702.
2. Wu, Xiaoyu, Zhenqi Hao, Di Wu, Lu Zheng, Zhanzhi Jiang, Vishal Ganesan, Yayu Wang, and Keji Lai. "Quantitative measurements of

nanoscale permittivity and conductivity using tuning-fork-based microwave impedance microscopy." *Review of Scientific Instruments* 89, no. 4 (2018): 043704.

3. Shan, Jun-Yi, Nathaniel Morrison, Su-Di Chen, Feng Wang, and Eric Y. Ma. "Johnson-noise-limited cancellation-free microwave impedance microscopy with monolithic silicon cantilever probes." *Nature Communications* 15 (2024): 5043.
4. Chu, Zhaodong, Lu Zheng, and Keji Lai. "Microwave Microscopy and Its Applications." *Annual Review of Materials Research* 50 (2020): 105–130.
5. Cui, Yong-Tao, and Eric Yue Ma. "Microwave Impedance Microscopy." In *Capacitance Spectroscopy of Semiconductors*: edited by Jian V. Li, and Giorgio Ferrari, pp. 411–435. Jenny Stanford Publishing, 2018.
6. Tselev, Alexander, Jeyavel Velmurugan, Anton V. Ievlev, Sergei V. Kalinin, and Andrei Kolmakov. "Seeing through Walls at the Nanoscale: Microwave Microscopy of Enclosed Objects and Processes in Liquids." *ACS Nano* 10, no. 3 (2016): 3562–3570.
7. Ren, Yuan, Hongtao Yuan, Xiaoyu Wu, Zhuoyu Chen, Yoshihiro Iwasa, Yi Cui, Harold Y. Hwang, and Keji Lai. "Direct Imaging of Nanoscale Conductance Evolution in Ion-Gel-Gated Oxide Transistors." *Nano Letters* 15, no. 7 (2015): 4730–4736.
8. Kim, Joon-Seok, Yingnan Liu, Weinan Zhu, Seohee Kim, Di Wu, Li Tao, Ananth Dodabalapur, Keji Lai, and Deji Akinwande. "Toward air-stable multilayer phosphorene thin-films and transistors." *Scientific Reports* 5, no. 1 (2015): 8989.
9. Seabron, Eric, Scott MacLaren, Xu Xie, Slava V. Rotkin, John A. Rogers, and William L. Wilson. "Scanning probe microwave reflectivity of aligned single-walled carbon nanotubes: Imaging of electronic structure and quantum behavior at the nanoscale." *ACS Nano* 10, no. 1 (2016): 360–368.
10. Chu, Zhaodong, Mengjin Yang, Philip Schulz, Di Wu, Xin Ma, Edward Seifert, Liuyang Sun, Xiaoqin Li, Kai Zhu, and Keji Lai. "Impact of grain boundaries on efficiency and stability of organic-inorganic trihalide perovskites." *Nature Communications* 8, no. 1 (2017): 2230.
11. Tuteja, Mohit, Prakash Koirala, Vasilios Palekis, Scott MacLaren, Chris S. Ferekides, Robert W. Collins, and Angus A. Rockett. "Direct Observation of CdCl2 Treatment Induced Grain Boundary Carrier Depletion in CdTe Solar Cells Using Scanning Probe Microwave Reflectivity Based Capacitance Measurements." *Journal of Physical Chemistry C* 120, no. 13 (2016): 7020–7024.

12. Lai, Keji, Worasom Kundhikanjana, Michael A. Kelly, Zhi-Xun Shen, Javad Shabani, and Mansour Shayegan. "Imaging of Coulomb-Driven Quantum Hall Edge States." *Physical Review Letters* 107, no. 17 (2011): 176809.

13. Ma, Eric Yue, Yong-Tao Cui, Kentaro Ueda, Shujie Tang, Kai Chen, Nobumichi Tamura, Phillip M. Wu, Jun Fujioka, Yoshinori Tokura, and Zhi-Xun Shen. "Mobile metallic domain walls in an all-in-all-out magnetic insulator." *Science* 350, no. 6260 (2015): 538–541.

14. Barber, Mark E., Eric Yue Ma, and Zhi-Xun Shen. "Microwave impedance microscopy and its application to quantum materials." *Nature Reviews Physics* 4, no. 1 (2022): 61–74.

15. Kim, Nam-Tae. "Synthesis of impedance-matching networks for RF-power amplifier applications." *Microwave and Optical Technology Letters* 39, no. 3 (2003): 207–211.

16. Hoffmann, Johannes, Sophie de Préville, Bruno Eckmann, Hung-Ju Lin, Benedikt Herzog, Kamel Haddadi, Didier Théron, Georg Gramse, Damien Richert, José Morán-Meza, and François Piquemal. "Comparison of Impedance Matching Networks for Scanning Microwave Microscopy." *IEEE Transactions on Instrumentation and Measurement* 73 (2024): 1–9.

17. Lai, K., W. Kundhikanjana, M. Kelly, and Z. X. Shen. "Modeling and characterization of a cantilever-based near-field scanning microwave impedance microscope." *Review of Scientific Instruments* 79, no. 6 (2008): 063703.

18. Gramse, G., M. Kasper, L. Fumagalli, G. Gomila, P. Hinterdorfer, and F. Kienberger. "Calibrated complex impedance and permittivity measurements with scanning microwave microscopy." *Nanotechnology* 25, no. 14 (2014): 145703.

19. Lai, Keji, Worasom Kundhikanjana, Michael A. Kelly, and Zhi-Xun Shen. "Nanoscale microwave microscopy using shielded cantilever probes." *Applied Nanoscience* 1, no. 1 (2011): 13–18.

20. Yang, Yongliang, Keji Lai, Qiaochu Tang, Worasom Kundhikanjana, Michael A Kelly, Kun Zhang, Zhi-xun Shen, and Xinxin Li. "Batch-fabricated cantilever probes with electrical shielding for nanoscale dielectric and conductivity imaging." *Journal of Micromechanics and Microengineering* 22, no. 11 (2012): 115040.

21. Scrymgeour, D. A., A. Baca, K. Fishgrab, R. J. Simonson, M. Marshall, E. Bussmann, C. Y. Nakakura, M. Anderson, and S. Misra. "Determining the resolution of scanning microwave impedance microscopy using atomic-precision buried donor structures." *Applied Surface Science* 423 (2017): 1097–1102.

22. Wang, Fei, Nicolas Clément, Damien Ducatteau, David Troadec, Hassan Tanbakuchi, Bernard Legrand, Gilles Dambrine, and Didier Théron. "Quantitative impedance characterization of sub-10 nm scale capacitors and tunnel junctions with an interferometric scanning microwave microscope." *Nanotechnology* 25, no. 40 (2014): 405703.

23. Cui, Yong-Tao, Eric Yue Ma, and Zhi-Xun Shen. "Quartz tuning fork based microwave impedance microscopy." *Review of Scientific Instruments* 87, no. 6 (2016): 063711.

24. Shan, Junyi, Nathaniel Morrison, and Eric Y. Ma. "Advanced Microwave Impedance Microscopy for Emerging Materials and Devices." In *2024 IEEE/MTT-S International Microwave Symposium — IMS 2024*: 994–997. 2024.

25. Kitano, Haruhisa, Takeyoshi Ohashi, Hibiki Ryuzaki, Atsutaka Maeda, and Ichiro Tsukada. "Frequency-dependent microwave conductivity of La2–xSrx-CuO4 thin films." *Physica C: Superconductivity* 412–414 (2004): 130–133.

26. Lee, Kyunghoon, M. Iqbal Bakti Utama, Salman Kahn, Appalakondaiah Samudrala, Nicolas Leconte, Birui Yang, Shuopei Wang, Kenji Watanabe, Takashi Taniguchi, M. Virginia P. Altoé, Guangyu Zhang, Alexander Weber-Bargioni, Michael Crommie, Paul D. Ashby, Jeil Jung, Feng Wang, and Alex Zettl. "Ultrahigh-resolution scanning microwave impedance microscopy of moiré lattices and superstructures." *Science Advances* 6, no. 50 (2020): eabd1919.

27. Gross, Leo, Fabian Mohn, Nikolaj Moll, Peter Liljeroth, and Gerhard Meyer. "The Chemical Structure of a Molecule Resolved by Atomic Force Microscopy." *Science* 325, no. 5944 (2009): 1110–1114.

28. Soh, Eugene J. H., Syed Ghazi Sarwat, Giulio Mazzotta, Benjamin F. Porter, Moritz Riede, Robin Nicholas, Judy S. Kim, and Harish Bhaskaran. "Filamentary High-Resolution Electrical Probes for Nanoengineering." *Nano Letters* 20, no. 2 (2020): 1067–1073.

29. Hong, Seung Sae, Judy J. Cha, and Yi Cui. "One Nanometer Resolution Electrical Probe via Atomic Metal Filament Formation." *Nano Letters* 11, no. 1 (2011): 231–235.

30. Amster, Oskar, Fred Stanke, Stuart Friedman, Yongliang Yang, St. J. Dixon-Warren, and B. Drevniok. "Practical quantitative scanning microwave impedance microscopy." *Microelectronics Reliability* 76–77 (2017): 214–217.

31. Amster, Oskar, Yongliang Yang, B. Drevniok, Stuart Friedman, Fred Stanke, and St. J. Dixon-Warren. "Practical quantitative scanning microwave impedance microscopy of semiconductor devices." In *2017 IEEE 24th*

International Symposium on the Physical and Failure Analysis of Integrated Circuits (IPFA): 1–4. 2017.

32. Amster, Oskar, Kurt Rubin, Yongliang Yang, Dorai Iyer, A. Messinger, and R. Crowder. "Nano C-V imaging of Semiconductor Devices with Scanning Microwave Impedance Microscopy." In *2018 IEEE International Symposium on the Physical and Failure Analysis of Integrated Circuits (IPFA)*: 1–5. 2018.

33. Amster, O., K. A. Rubin, Y. Yang, D. Iyer, and A. Messinger. "Application of Scanning Microwave Microscopy nano-C-V to investigate dopant defect under a poly gate device." *Microelectronics Reliability* 88–90 (2018): 250–254.

34. Friedman, Stuart, Oskar Amster, Yongliang Yang, and Fred Stanke. "Nanoscale Capacitance and Capacitance-Voltage Curves for Advanced Characterization of Electrical Properties of Si and GaN Structures Using Scanning Microwave Impedance Microscopy." In *ISTFA 2016: Proceedings from the 42nd International Symposium for Testing and Failure Analysis* (2016): 449–453.

35. Iyer, D., A. Messinger, R. Crowder, Y. Zhang, O. Amster, S. Friedman, Y. Yang, and F. Stanke. "Measurement of Dielectric Constant and Doping Concentration of a Cross-Sectioned Device by Quantitative Scanning Microwave Impedance Microscopy." In *ISTFA 2017: Proceedings from the 43rd International Symposium for Testing and Failure Analysis* (2017): 613–617.

36. Rubin, K. A., W. Jolley, and Y. Yang. "Finite-Element Modeling and Quantitative Measurement Using Scanning Microwave Microscopy to Characterize Dielectric Films." In *ISTFA 2018: Proceedings from the 44th International Symposium for Testing and Failure Analysis* (2018): 561–565.

37. Tami, Diego, Douglas A. A. Ohlberg, Cássio Gonçalves do Rego, Gilberto Medeiros-Ribeiro, and Jhonattan C. Ramirez. "Scanning microwave impedance microscopy and its applications: A review." *APL Materials* 13, no. 1 (2025): 010602.

38. Shi, Yanmeng, Joshua Kahn, Ben Niu, Zaiyao Fei, Bosong Sun, Xinghan Cai, Brian A. Francisco, Di Wu, Zhi-Xun Shen, Xiaodong Xu, David H. Cobden, and Yong-Tao Cui. "Imaging quantum spin Hall edges in monolayer WTe2." *Science Advances* 5, no. 2 (2019): eaat8799.

39. Ohlberg, Douglas A. A., Diego Tami, Andreij C. Gadelha, Eliel G. S. Neto, Fabiano C. Santana, Daniel Miranda, Wellington Avelino, Kenji Watanabe, Takashi Taniguchi, Leonardo C. Campos, Jhonattan C. Ramirez, Cássio Gonçalves do Rego, Ado Jorio, and Gilberto Medeiros-Ribeiro. "The limits of near field immersion microwave microscopy evaluated by imaging bilayer graphene moiré patterns." *Nature Communications* 12, no. 1 (2021): 2980.

40. Iannaccone, Giuseppe, Francesco Bonaccorso, Luigi Colombo, and Gianluca Fiori. "Quantum engineering of transistors based on 2D materials heterostructures." *Nature Nanotechnology* 13, no. 3 (2018): 183–191.

41. Giannazzo, Filippo, Giuseppe Greco, Fabrizio Roccaforte, and Sushant S. Sonde. "Vertical Transistors Based on 2D Materials: Status and Prospects." *Crystals* 8, no. 2 (2018): 70.

42. Jiang, Jian, Yao Wen, Hao Wang, Lei Yin, Ruiqing Cheng, Chuansheng Liu, Liping Feng, and Jun He. "Recent Advances in 2D Materials for Photodetectors." *Advanced Electronic Materials* 7, no. 7 (2021): 2001125.

43. Long, Mingsheng, Peng Wang, Hehai Fang, and Weida Hu. "Progress, Challenges, and Opportunities for 2D Material Based Photodetectors." *Advanced Functional Materials* 29, no. 19 (2019): 1803807.

44. Lei, Le, Rui Xu, Shili Ye, Xinsheng Wang, Kunqi Xu, Sabir Hussain, Yan Jun Li, Yasuhiro Sugawara, Liming Xie, Wei Ji, and Zhihai Cheng. "Local characterization of mobile charge carriers by two electrical AFM modes: multi-harmonic EFM versus sMIM." *Journal of Physics Communications* 2, no. 2 (2018): 025013.

45. Wang, Lin, Han Chen, Mingfeng Chen, Yinfeng Long, Kai Liu, and Kian Ping Loh. "A Scanning Microwave Impedance Microscopy Study of α-In2Se3 Ferroelectric Semiconductor." *Advanced Functional Materials* 34, no. 28 (2024): 2316583.

46. Bargas, Gabriel, Douglas A. A. Ohlberg, Kenji Watanabe, Takashi Taniguchi, Leonardo C. Campos, and Gilberto Medeiros-Ribeiro. "Scanning Microwave Impedance Microscopy for Characterization of Graphene Systems Encapsulated by Hexagonal Boron Nitride." *physica status solidi (b)* 262, no. 7 (2025): 2400548.

47. Ohlberg, Douglas A. A., Andreij C. Gadelha, Diego Tamia, Eliel G. S. Neto, Daniel A. Miranda, Jessica S. Lemos, Fabiano C. Santana, Leonardo C. Campos, Jhonattan C. Ramírez, Cássio Gonçalves do Rego, Ado Jorio, and Gilberto Medeiros-Ribeiro. "Observation of moiré superlattices on twisted bilayer graphene by scanning microwave impedance microscopy." In *Low-Dimensional Materials and Devices 2020*: pp. 31–37. SPIE, 2020.

48. Germanicus, Rosine Coq, Peter De Wolf, Florent Lallemand, Catherine Bunel, Serge Bardy, Hugues Murray, and Ulrike Lüders. "Mapping of integrated PIN diodes with a 3D architecture by scanning microwave impedance microscopy and dynamic spectroscopy." *Beilstein Journal of Nanotechnology* 11, no. 1 (2020): 1764–1775.

49. Rubin, Kurt A., Yongliang Yang, Oskar Amster, David A. Scrymgeour, and Shashank Misra. "Scanning Microwave Impedance Microscopy (sMIM) in Electronic and Quantum Materials." In *Electrical Atomic Force Microscopy for Nanoelectronics*: edited by Umberto Celano, pp. 385–408. Springer International Publishing, 2019.

50. Zhang, Ting, Yan Liu, Anli Yang, Yupu Wang, Ke Xiao, Bing Yang, Qichao Ding, and Mai Zhihong. "Dopant Analysis of SiC MOSFET with Scanning Probe Technique." In *2024 IEEE International Symposium on the Physical and Failure Analysis of Integrated Circuits (IPFA)*: pp. 01–05. 2024.

51. Liu, Xiaomeng, Xiangsheng Wang, Xinyou Liu, Yanpeng Song, Yiwen Zhang, Hailing Wang, Ying Zhang, Guilei Wang, and Chao Zhao. "Application of atomic force microscopy technology in doping characterization of semiconductor materials and devices." *Microelectronic Engineering* 297 (2025): 112310.

52. Köhler, J. "Magnetic resonance of a single molecular spin." *Physics Reports* 310, no. 5–6 (1999): 261–339.

53. Epstein, R. J., F. M. Mendoza, Y. K. Kato, and D. D. Awschalom. "Anisotropic interactions of a single spin and dark-spin spectroscopy in diamond." *Nature Physics* 1, no. 2 (2005): 94–98.

54. Goldman, M. L., M. W. Doherty, A. Sipahigil, N. Y. Yao, S. D. Bennett, N. B. Manson, A. Kubanek, and M. D. Lukin. "State-selective intersystem crossing in nitrogen-vacancy centers." *Physical Review B* 91, no. 16 (2015): 165201.

55. Chernobrod, Boris M., and Gennady P. Berman. "Spin microscope based on optically detected magnetic resonance." *Journal of Applied Physics* 97, no. 1 (2005): 014903.

56. Batalov, A., V. Jacques, F. Kaiser, P. Siyushev, P. Neumann, L. J. Rogers, R. L. McMurtrie, N. B. Manson, F. Jelezko, and J. Wrachtrup. "Low Temperature Studies of the Excited-State Structure of Negatively Charged Nitrogen-Vacancy Color Centers in Diamond." *Physical Review Letters* 102, no. 19 (2009): 195506.

57. Levine, Edlyn V., Matthew J. Turner, Pauli Kehayias, Connor A. Hart, Nicholas Langellier, Raisa Trubko, David R. Glenn, Roger R. Fu, and Ronald L. Walsworth. "Principles and techniques of the quantum diamond microscope." *Nanophotonics* 8, no. 11 (2019): 1945–1973.

58. Taylor, J. M., P. Cappellaro, L. Childress, L. Jiang, D. Budker, P. R. Hemmer, A. Yacoby, R. Walsworth, and M. D. Lukin. "High-sensitivity diamond magnetometer with nanoscale resolution." *Nature Physics* 4, no. 10 (2008): 810–816.

59. Rondin, L., J.-P. Tetienne, P. Spinicelli, C. Dal Savio, K. Karrai, G. Dantelle, A. Thiaville, S. Rohart, J.-F. Roch, and V. Jacques. "Nanoscale magnetic field mapping with a single spin scanning probe magnetometer." *Applied Physics Letters* 100, no. 15 (2012): 153118.

60. Maletinsky, P., S. Hong, M. S. Grinolds, B. Hausmann, M. D. Lukin, R. L. Walsworth, M. Loncar, and A. Yacoby. "A robust scanning diamond sensor for nanoscale imaging with single nitrogen-vacancy centres." *Nature Nanotechnology* 7, no. 5 (2012): 320–324.

61. Borràs, Vicent J., Robert Carpenter, Liza Žaper, Siddharth Rao, Sebastien Couet, Mathieu Munsch, Patrick Maletinsky, and Peter Rickhaus. "A quantum sensing metrology for magnetic memories." *npj Spintronics* 2, no. 1 (2024): 14.

62. Balasubramanian, Gopalakrishnan, Philipp Neumann, Daniel Twitchen, Matthew Markham, Roman Kolesov, Norikazu Mizuochi, Junichi Isoya, Jocelyn Achard, Johannes Beck, Julia Tissler, Vincent Jacques, Philip R. Hemmer, Fedor Jelezko, and Jörg Wrachtrup. "Ultralong spin coherence time in isotopically engineered diamond." *Nature Materials* 8, no. 5 (2009): 383–387.

63. Maze, J. R., P. L. Stanwix, J. S. Hodges, S. Hong, J. M. Taylor, P. Cappellaro, L. Jiang, M. V. Gurudev Dutt, E. Togan, A. S. Zibrov, A. Yacoby, R. L. Walsworth, and M. D. Lukin. "Nanoscale magnetic sensing with an individual electronic spin in diamond." *Nature* 455, no. 7213 (2008): 644–647.

64. Barry, John F., Jennifer M. Schloss, Erik Bauch, Matthew J. Turner, Connor A. Hart, Linh M. Pham, and Ronald L. Walsworth. "Sensitivity optimization for NV-diamond magnetometry." *Reviews of Modern Physics* 92, no. 1 (2020): 015004.

65. Doherty, Marcus W., Neil B. Manson, Paul Delaney, Fedor Jelezko, Jörg Wrachtrup, and Lloyd C.L. Hollenberg. "The nitrogen-vacancy colour centre in diamond." *Physics Reports* 528, no. 1 (2013): 1–45.

66. Findler, Christoph, Rémi Blinder, Karolina Schüle, Priyadharshini Balasubramanian, Christian Osterkamp, and Fedor Jelezko. "Detecting nitrogen-vacancy-hydrogen centers on the nanoscale using nitrogen-vacancy centers in diamond." *Physical Review Materials* 8, no. 2 (2024): 026203.

67. Hahn, E. L. "Spin Echoes." *Physical Review* 80, no. 4 (1950): 580–594.

68. Khodjasteh, K., and D. A. Lidar. "Fault-Tolerant Quantum Dynamical Decoupling." *Physical Review Letters* 95, no. 18 (2005): 180501.

69. Pham, L. M., N. Bar-Gill, C. Belthangady, D. Le Sage, P. Cappellaro, M. D. Lukin, A. Yacoby, and R. L. Walsworth. "Enhanced solid-state multispin

metrology using dynamical decoupling." *Physical Review B* 86, no. 4 (2012): 045214.

70. Schirhagl, Romana, Kevin Chang, Michael Loretz, and Christian L. Degen. "Nitrogen-Vacancy Centers in Diamond: Nanoscale Sensors for Physics and Biology." *Annual Review of Physical Chemistry* 65, no. 1 (2014): 83–105.
71. Carr, H. Y., and E. M. Purcell. "Effects of Diffusion on Free Precession in Nuclear Magnetic Resonance Experiments." *Physical Review* 94, no. 3 (1954): 630–638.
72. Meiboom, S., and D. Gill. "Modified Spin-Echo Method for Measuring Nuclear Relaxation Times." *Review of Scientific Instruments* 29, no. 8 (1958): 688–691.
73. Horsley, Andrew, Patrick Appel, Janik Wolters, Jocelyn Achard, Alexandre Tallaire, Patrick Maletinsky, and Philipp Treutlein. "Microwave Device Characterization Using a Widefield Diamond Microscope." *Physical Review Applied* 10, no. 4 (2018): 044039.
74. Rable, Jeffrey, Jyotirmay Dwivedi, Nitin Samarth, Paul Stevenson, Arun Bansil, and Swastik Kar. "Flux Channeling Induced Nanoconfinement and Enhancement of Microwaves Imaged by Rabi Oscillation Mapping." *Nano Letters* 25, no. 23 (2025): 9470–9476.
75. Thiel, L., D. Rohner, M. Ganzhorn, P. Appel, E. Neu, B. Müller, R. Kleiner, D. Koelle, and P. Maletinsky. "Quantitative nanoscale vortex imaging using a cryogenic quantum magnetometer." *Nature Nanotechnology* 11, no. 8 (2016): 677–681.
76. Gross, I., W. Akhtar, V. Garcia, L. J. Martínez, S. Chouaieb, K. Garcia, C. Carrétéro, A. Barthélémy, P. Appel, P. Maletinsky, J.-V. Kim, J. Y. Chauleau, N. Jaouen, M. Viret, M. Bibes, S. Fusil, and V. Jacques. "Real-space imaging of non-collinear antiferromagnetic order with a single-spin magnetometer." *Nature* 549, no. 7671 (2017): 252–256.
77. Zhou, Tony X., Rainer J. Stöhr, and Amir Yacoby. "Scanning diamond NV center probes compatible with conventional AFM technology." *Applied Physics Letters* 111, no. 16 (2017): 163106.
78. Dovzhenko, Y., F. Casola, S. Schlotter, T. X. Zhou, F. Büttner, R. L. Walsworth, G. S. D. Beach, and A. Yacoby. "Magnetostatic twists in room-temperature skyrmions explored by nitrogen-vacancy center spin texture reconstruction." *Nature Communications* 9, no. 1 (2018): 2712.
79. Erickson, Adam, Syed Qamar Abbas Shah, Ather Mahmood, Ilja Fescenko, Rupak Timalsina, Christian Binek, and Abdelghani Laraoui. "Nanoscale imaging of antiferromagnetic domains in epitaxial films of Cr_2O_3 *via*

scanning diamond magnetic probe microscopy." *RSC Advances* 13, no. 1 (2023): 178–185.

80. Rana, K. Gaurav, A. Finco, F. Fabre, S. Chouaieb, A. Haykal, L. D. Buda-Prejbeanu, O. Fruchart, S. Le Denmat, P. David, M. Belmeguenai, T. Denneulin, R. E. Dunin-Borkowski, G. Gaudin, V. Jacques, and O. Boulle. "Room-Temperature Skyrmions at Zero Field in Exchange-Biased Ultrathin Films." *Physical Review Applied* 13, no. 4 (2020): 044079.

81. Finco, Aurore, Angela Haykal, Rana Tanos, Florentin Fabre, Saddem Chouaieb, Waseem Akhtar, Isabelle Robert-Philip, William Legrand, Fernando Ajejas, Karim Bouzehouane, Nicolas Reyren, Thibaut Devolder, Jean-Paul Adam, Joo-Von Kim, Vincent Cros, and Vincent Jacques. "Imaging non-collinear antiferromagnetic textures via single spin relaxometry." *Nature Communications* 12, no. 1 (2021): 767.

82. Celano, Umberto, Hai Zhong, Florin Ciubotaru, Laurentiu Stoleriu, Alexander Stark, Peter Rickhaus, Felipe Fávaro De Oliveira, Mathieu Munsch, Paola Favia, Maxim Korytov, Patricia Van Marcke, Patrick Maletinsky, Christoph Adelmann, and Paul Van Der Heide. "Probing Magnetic Defects in Ultra-Scaled Nanowires with Optically Detected Spin Resonance in Nitrogen-Vacancy Center in Diamond." *Nano Letters* 21, no. 24 (2021): 10409–10415.

83. Rohner, D., J. Happacher, P. Reiser, M. A. Tschudin, A. Tallaire, J. Achard, B. J. Shields, and P. Maletinsky. "(111)-oriented, single crystal diamond tips for nanoscale scanning probe imaging of out-of-plane magnetic fields." *Applied Physics Letters* 115, no. 19 (2019): 192401.

84. Simon, Brecht G., Samer Kurdi, Helena La, Iacopo Bertelli, Joris J. Carmiggelt, Maximilian Ruf, Nick De Jong, Hans Van Den Berg, Allard J. Katan, and Toeno Van Der Sar. "Directional Excitation of a High-Density Magnon Gas Using Coherently Driven Spin Waves." *Nano Letters* 21, no. 19 (2021): 8213–8219.

85. Casola, Francesco, Toeno Van Der Sar, and Amir Yacoby. "Probing condensed matter physics with magnetometry based on nitrogen-vacancy centres in diamond." *Nature Reviews Materials* 3, no. 1 (2018): 17088.

86. Roberts, Henry, Hamza Abudayyeh, Xiaoqin Li, and Xiuling Li. "Quantum Sensing with Spin Defects Beyond Diamond." *ACS Nano* 19, no. 25 (2025): 22528–22575.

87. Widmann, Matthias, Sang-Yun Lee, Torsten Rendler, Nguyen Tien Son, Helmut Fedder, Seoyoung Paik, Li-Ping Yang, Nan Zhao, Sen Yang, Ian Booker, Andrej Denisenko, Mohammad Jamali, S. Ali Momenzadeh, Ilja Gerhardt, Takeshi Ohshima, Adam Gali, Erik Janzén, and Jörg Wrachtrup.

"Coherent control of single spins in silicon carbide at room temperature." *Nature Materials* 14, no. 2 (2015): 164–168.

88. Bishop, S. G., J. P. Hadden, R. Hekmati, J. K. Cannon, W. W. Langbein, and A. J. Bennett. "Enhanced light collection from a gallium nitride color center using a near index-matched solid immersion lens." *Applied Physics Letters* 120, no. 11 (2022): 114001.
89. Zhang, Wenlong, Jian Zhang, Junfeng Wang, Fupan Feng, Shengran Lin, Liren Lou, Wei Zhu, and Guanzhong Wang. "Depth-dependent decoherence caused by surface and external spins for NV centers in diamond." *Physical Review B* 96, no. 23 (2017): 235443.
90. Gottscholl, Andreas, Matthias Diez, Victor Soltamov, Christian Kasper, Andreas Sperlich, Mehran Kianinia, Carlo Bradac, Igor Aharonovich, and Vladimir Dyakonov. "Room temperature coherent control of spin defects in hexagonal boron nitride." *Science Advances* 7, no. 14 (2021): eabf3630.
91. Vaidya, Sumukh, Xingyu Gao, Saakshi Dikshit, Igor Aharonovich, and Tongcang Li. "Quantum sensing and imaging with spin defects in hexagonal boron nitride." *Advances in Physics: X* 8, no. 1 (2023): 2206049.
92. Gottscholl, Andreas, Matthias Diez, Victor Soltamov, Christian Kasper, Dominik Krauße, Andreas Sperlich, Mehran Kianinia, Carlo Bradac, Igor Aharonovich, and Vladimir Dyakonov. "Spin defects in hBN as promising temperature, pressure and magnetic field quantum sensors." *Nature Communications* 12, no. 1 (2021): 4480.
93. Chen, Kevin J., Oliver Haberlen, Alex Lidow, Chun Lin Tsai, Tetsuzo Ueda, Yasuhiro Uemoto, and Yifeng Wu. "GaN-on-Si Power Technology: Devices and Applications." *IEEE Transactions on Electron Devices* 64, no. 3 (2017): 779–795.
94. Berhane, Amanuel M., Kwang-Yong Jeong, Zoltán Bodrog, Saskia Fiedler, Tim Schröder, Noelia Vico Triviño, Tomás Palacios, Adam Gali, Milos Toth, Dirk Englund, and Igor Aharonovich. *Bright Room-Temperature Single Photon Emission from Defects in Gallium Nitride*. arXiv, 2016.
95. Geng, Yifei, Jialun Luo, Len Van Deurzen, Huili Xing, Debdeep Jena, Gregory David Fuchs, and Farhan Rana. "Dephasing by optical phonons in GaN defect single-photon emitters." *Scientific Reports* 13, no. 1 (2023): 8678.
96. Sörman, E., N. T. Son, W. M. Chen, O. Kordina, C. Hallin, and E. Janzén. "Silicon vacancy related defect in 4H and 6H SiC." *Physical Review B* 61, no. 4 (2000): 2613–2620.

97. Davidsson, Joel, Viktor Ivády, Rickard Armiento, Takeshi Ohshima, N. T. Son, Adam Gali, and Igor A. Abrikosov. "Identification of divacancy and silicon vacancy qubits in 6H-SiC." *Applied Physics Letters* 114, no. 11 (2019): 112107.

98. Nagy, Roland, Matthias Niethammer, Matthias Widmann, Yu-Chen Chen, Péter Udvarhelyi, Cristian Bonato, Jawad Ul Hassan, Robin Karhu, Ivan G. Ivanov, Nguyen Tien Son, Jeronimo R. Maze, Takeshi Ohshima, Öney O. Soykal, Ádám Gali, Sang-Yun Lee, Florian Kaiser, and Jörg Wrachtrup. "High-fidelity spin and optical control of single silicon-vacancy centres in silicon carbide." *Nature Communications* 10, no. 1 (2019): 1954.

99. Falk, Abram L., Bob B. Buckley, Greg Calusine, William F. Koehl, Viatcheslav V. Dobrovitski, Alberto Politi, Christian A. Zorman, Philip X.-L. Feng, and David D. Awschalom. "Polytype control of spin qubits in silicon carbide." *Nature Communications* 4, no. 1 (2013): 1819.

100. Yan, Fei-Fei, Ai-Lun Yi, Jun-Feng Wang, Qiang Li, Pei Yu, Jia-Xiang Zhang, Adam Gali, Ya Wang, Jin-Shi Xu, Xin Ou, Chuan-Feng Li, and Guang-Can Guo. "Room-temperature coherent control of implanted defect spins in silicon carbide." *npj Quantum Information* 6, no. 1 (2020): 38.

101. Koehl, William F., Bob B. Buckley, F. Joseph Heremans, Greg Calusine, and David D. Awschalom. "Room temperature coherent control of defect spin qubits in silicon carbide." *Nature* 479, no. 7371 (2011): 84–87.

102. Son, N. T., D. Shafizadeh, T. Ohshima, and I. G. Ivanov. "Modified divacancies in 4H-SiC." *Journal of Applied Physics* 132, no. 2 (2022): 025703.

103. Jurgen Von Bardeleben, Hans, Jean-Louis Cantin, Uwe Gerstmann, Wolf Gero Schmidt, and Timur Biktagirov. "Spin Polarization, Electron–Phonon Coupling, and Zero-Phonon Line of the NV Center in 3 *C* -SiC." *Nano Letters* 21, no. 19 (2021): 8119–8125.

104. Zhu, Yizhi, Berk Kovos, Mykyta Onizhuk, David Awschalom, and Giulia Galli. "Theoretical and experimental study of the nitrogen-vacancy center in 4H-SiC." *Physical Review Materials* 5, no. 7 (2021): 074602.

105. Khazen, Kh., H. J. Von Bardeleben, S. A. Zargaleh, J. L. Cantin, Mu Zhao, Weibo Gao, T. Biktagirov, and U. Gerstmann. "High-resolution resonant excitation of NV centers in 6 H – SiC : A matrix for quantum technology applications." *Physical Review B* 100, no. 20 (2019): 205202.

106. Von Bardeleben, H. J., J. L. Cantin, A. Csóré, A. Gali, E. Rauls, and U. Gerstmann. "NV centers in 3 C , 4 H , and 6 H silicon carbide: A variable platform for solid-state qubits and nanosensors." *Physical Review B* 94, no. 12 (2016): 121202.

107. Qian, Chenjiang, Viviana Villafañe, Martin Schalk, G. V. Astakhov, Ulrich Kentsch, Manfred Helm, Pedro Soubelet, Nathan P. Wilson, Roberto Rizzato, Stephan Mohr, Alexander W. Holleitner, Dominik B. Bucher, Andreas V. Stier, and Jonathan J. Finley. "Unveiling the Zero-Phonon Line of the Boron Vacancy Center by Cavity-Enhanced Emission." *Nano Letters* 22, no. 13 (2022): 5137–5142.

108. Healey, A. J., S. C. Scholten, T. Yang, J. A. Scott, G. J. Abrahams, I. O. Robertson, X. F. Hou, Y. F. Guo, S. Rahman, Y. Lu, M. Kianinia, I. Aharonovich, and J.-P. Tetienne. "Quantum microscopy with van der Waals heterostructures." *Nature Physics* 19, no. 1 (2023): 87–91.

109. Stern, Hannah L., Qiushi Gu, John Jarman, Simone Eizagirre Barker, Noah Mendelson, Dipankar Chugh, Sam Schott, Hoe H. Tan, Henning Sirringhaus, Igor Aharonovich, and Mete Atatüre. "Room-temperature optically detected magnetic resonance of single defects in hexagonal boron nitride." *Nature Communications* 13, no. 1 (2022): 618.

110. Mendelson, Noah, Dipankar Chugh, Jeffrey R. Reimers, Tin S. Cheng, Andreas Gottscholl, Hu Long, Christopher J. Mellor, Alex Zettl, Vladimir Dyakonov, Peter H. Beton, Sergei V. Novikov, Chennupati Jagadish, Hark Hoe Tan, Michael J. Ford, Milos Toth, Carlo Bradac, and Igor Aharonovich. "Identifying carbon as the source of visible single-photon emission from hexagonal boron nitride." *Nature Materials* 20, no. 3 (2021): 321–328.

111. Guo, Nai-Jie, Song Li, Wei Liu, Yuan-Ze Yang, Xiao-Dong Zeng, Shang Yu, Yu Meng, Zhi-Peng Li, Zhao-An Wang, Lin-Ke Xie, Rong-Chun Ge, Jun-Feng Wang, Qiang Li, Jin-Shi Xu, Yi-Tao Wang, Jian-Shun Tang, Adam Gali, Chuan-Feng Li, and Guang-Can Guo. "Coherent control of an ultrabright single spin in hexagonal boron nitride at room temperature." *Nature Communications* 14, no. 1 (2023): 2893.

112. Luo, Jialun, Yifei Geng, Farhan Rana, and Gregory D. Fuchs. "Room temperature optically detected magnetic resonance of single spins in GaN." *Nature Materials* 23, no. 4 (2024): 512–518.

113. Rondin, L., J.-P. Tetienne, T. Hingant, J.-F. Roch, P. Maletinsky, and V. Jacques. "Magnetometry with nitrogen-vacancy defects in diamond." *Reports on Progress in Physics* 77, no. 5 (2014): 056503.

114. Radtke, M, E Bernardi, A Slablab, R Nelz, and E Neu. "Nanoscale sensing based on nitrogen vacancy centers in single crystal diamond and nanodiamonds: achievements and challenges." *Nano Futures* 3, no. 4 (2019): 042004.

115. Barry, John F., Matthew H. Steinecker, Scott T. Alsid, Jonah Majumder, Linh M. Pham, Michael F. O'Keefe, and Danielle A. Braje. *Sensitive AC*

and DC Magnetometry with Nitrogen-Vacancy Center Ensembles in Diamond. arXiv, 2023.

116. Jia, Xuguang, Yue Qin, Zhengjie Luo, Shining Zhu, Xin Li, and Hao Guo. "Near-Field Microwave Imaging Method of Monopole Antennas Based on Nitrogen-Vacancy Centers in Diamond." *Micromachines* 15, no. 6 (2024): 679.

117. Zhang, Ning, Chen Zhang, Lixia Xu, Ming Ding, Wei Quan, Zheng Tang, and Heng Yuan. "Microwave Magnetic Field Coupling with Nitrogen-Vacancy Center Ensembles in Diamond with High Homogeneity." *Applied Magnetic Resonance* 47, no. 6 (2016): 589–599.

118. Sasaki, Kento, Yasuaki Monnai, Soya Saijo, Ryushiro Fujita, Hideyuki Watanabe, Junko Ishi-Hayase, Kohei M. Itoh, and Eisuke Abe. "Broadband, large-area microwave antenna for optically detected magnetic resonance of nitrogen-vacancy centers in diamond." *Review of Scientific Instruments* 87, no. 5 (2016): 053904.

119. Bayat, Khadijeh, Jennifer Choy, Mahdi Farrokh Baroughi, Srujan Meesala, and Marko Loncar. "Efficient, Uniform, and Large Area Microwave Magnetic Coupling to NV Centers in Diamond Using Double Split-Ring Resonators." *Nano Letters* 14, no. 3 (2014): 1208–1213.

120. Dong, M. M., Z. Z. Hu, Y. Liu, B. Yang, Y. J. Wang, and G. X. Du. "A fiber based diamond RF B-field sensor and characterization of a small helical antenna." *Applied Physics Letters* 113, no. 13 (2018): 131105.

121. Soshenko, V. V., O. R. Rubinas, V. V. Vorobyov, S. V. Bolshedvorskii, P. V. Kapitanova, V. N. Sorokin, and A. V. Akimov. "Microwave Antenna for Exciting Optically Detected Magnetic Resonance in Diamond NV Centers." *Bulletin of the Lebedev Physics Institute* 45, no. 8 (2018): 237–240.

122. Briegel, Karl D., Nick R. Von Grafenstein, Julia C. Draeger, Peter Blümler, Robin D. Allert, and Dominik B. Bucher. "Optical widefield nuclear magnetic resonance microscopy." *Nature Communications* 16, no. 1 (2025): 1281.

123. Rizzato, Roberto, Nick R. Von Grafenstein, and Dominik B. Bucher. "Quantum sensors in diamonds for magnetic resonance spectroscopy: Current applications and future prospects." *Applied Physics Letters* 123, no. 26 (2023): 260502.

124. Liu, Kristina S., Alex Henning, Markus W. Heindl, Robin D. Allert, Johannes D. Bartl, Ian D. Sharp, Roberto Rizzato, and Dominik B. Bucher. "Surface NMR using quantum sensors in diamond." *Proceedings of the National Academy of Sciences* 119, no. 5 (2022): e2111607119.

125. Jia, Wenfei, Zhifu Shi, Xi Qin, Xing Rong, and Jiangfeng Du. "Ultra-broadband coplanar waveguide for optically detected magnetic resonance of nitrogen-vacancy centers in diamond." *Review of Scientific Instruments* 89, no. 6 (2018): 064705.

126. Mrózek, M., J. Mlynarczyk, D. S. Rudnicki, and W. Gawlik. "Circularly polarized microwaves for magnetic resonance study in the GHz range: Application to nitrogen-vacancy in diamonds." *Applied Physics Letters* 107, no. 1 (2015): 013505.

127. Glenn, D. R., R. R. Fu, P. Kehayias, D. Le Sage, E. A. Lima, B. P. Weiss, and R. L. Walsworth. "Micrometer-scale magnetic imaging of geological samples using a quantum diamond microscope." *Geochemistry, Geophysics, Geosystems* 18, no. 8 (2017): 3254–3267.

128. Alegre, Thiago P. Mayer, Charles Santori, Gilberto Medeiros-Ribeiro, and Raymond G. Beausoleil. "Polarization-selective excitation of nitrogen vacancy centers in diamond." *Physical Review B* 76, no. 16 (2007): 165205.

129. Yaroshenko, Vitaly, Vladimir Soshenko, Vadim Vorobyov, Stepan Bolshedvorskii, Elizaveta Nenasheva, Igor Kotel'nikov, Alexey Akimov, and Polina Kapitanova. "Circularly polarized microwave antenna for nitrogen vacancy centers in diamond." *Review of Scientific Instruments* 91, no. 3 (2020): 035003.

130. Eisenach, E. R., J. F. Barry, L. M. Pham, R. G. Rojas, D. R. Englund, and D. A. Braje. "Broadband loop gap resonator for nitrogen vacancy centers in diamond." *Review of Scientific Instruments* 89, no. 9 (2018): 094705.

131. Opaluch, Oliver Roman, Nimba Oshnik, Richard Nelz, and Elke Neu. "Optimized Planar Microwave Antenna for Nitrogen Vacancy Center Based Sensing Applications." *Nanomaterials* 11, no. 8 (2021): 2108.

132. Li, Xiaofei, Qi Nie, Penghao Xu, Yaowu Lei, Haoliang Hu, Ziyang Zhu, and Yanling Sun. "The Design and Optimization of Radiating Antennas for the Diamond Nitrogen-Vacancy Color Center." In *2024 3rd International Conference on Computing, Communication, Perception and Quantum Technology (CCPQT)*: 344–349. IEEE, 2024.

133. Guo, Yuhang, Jiaxin Zhao, Changfeng Weng, Shengran Lin, Yuanjie Yang, Wei Zhu, Liren Lou, and Guanzhong Wang. "Robust diamond-embedded microwave antenna for optimizing quantum sensing using nitrogen-vacancy center ensembles." *Applied Physics Letters* 123, no. 26 (2023): 264003.

https://doi.org/10.1142/9789819830763_bmatter

Index

www.ingramcontent.com/pod-product-compliance
Ingram Content Group UK Ltd.
Pitfield, Milton Keynes, MK11 3LW, UK
UKHW061932100926
8556IPUK00007B/892

9 789819 83075